HELIOPHYSICS

Space Storms and Radiation: Causes and Effects

Edited by Carolus J. Schrijver and George L. Siscoe

Heliophysics is a fast-developing scientific discipline that integrates studies of the Sun's variability, the surrounding heliosphere, and the environment and climate of planets. Over the past few centuries, our understanding of how the Sun drives space weather and climate on the Earth and other planets has advanced at an ever increasing rate. The Sun is a magnetically variable star and, for planets with intrinsic magnetic fields, planets with atmospheres, or planets like Earth with both, there are profound consequences.

This volume, the second in a series of three heliophysics texts, integrates the many aspects of space storms and the energetic radiation associated with them – from their causes on the Sun to their effects in planetary environments. It reviews the physical processes in solar flares and coronal mass ejections, interplanetary shocks, and particle acceleration and transport, and considers many of the space weather responses in geospace. Historical space weather observations, *in-situ* particle measurement techniques, radiative emissions from energetic particles, and impacts of space weather on people and technology in space are also reviewed. In addition to its utility as a textbook, it also constitutes a foundational reference for researchers in the fields of heliophysics, astrophysics, plasma physics, space physics, solar physics, aeronomy, space weather, planetary science, and climate science. Additional online resources, including lecture presentations and other teaching materials, can be accessed at www.cambridge.org/9780521760515.

CAROLUS J. SCHRIJVER is an astrophysicist studying the causes and effects of magnetic activity of the Sun and of stars like the Sun, and the coupling of the Sun's magnetic field into the surrounding heliosphere. He obtained his doctorate in physics and astronomy at the University of Utrecht in the Netherlands in 1986, and has since worked for the University of Colorado, the US National Solar Observatory, the European Space Agency, and the Royal Academy of Sciences of the Netherlands. Dr Schrijver is currently principal physicist at Lockheed Martin's Advanced Technology Center, where his work focuses primarily on the magnetic field in the solar atmosphere. He is an editor or editorial board member of several journals including *Solar Physics*, *Astronomical Notices*, and *Living Reviews in Solar Physics*, and has co-edited four other books.

GEORGE L. SISCOE received his Ph.D. in physics from the Massachusetts Institute of Technology (MIT) in 1964. He has since held positions at the California Institute of Technology, MIT, and the University of California, Los Angeles – where he was Professor and Chair of the Department of Atmospheric Sciences. He is currently a Research Professor in the Astronomy Department at Boston University. Professor Siscoe has been a member and chair of numerous international committees and panels and is on the editorial board of the *Journal of Atmospheric and Solar Terrestrial Physics*. He is a Fellow of the American Geophysical Union and the second Van Allen Lecturer of the AGU, 1991. He has authored or co-authored over 300 publications that cover most areas of heliophysics.

HELIOPHYSICS

Space Storms and Radiation: Causes and Effects

Edited by

CAROLUS J. SCHRIJVER
Lockheed Martin Advanced Technology Center

GEORGE L. SISCOE
Boston University

CAMBRIDGE UNIVERSITY PRESS
Cambridge, New York, Melbourne, Madrid, Cape Town,
Singapore, São Paulo, Delhi, Tokyo, Mexico City

Cambridge University Press
The Edinburgh Building, Cambridge CB2 8RU, UK

Published in the United States of America by
Cambridge University Press, New York

www.cambridge.org
Information on this title: www.cambridge.org/9780521760515

First published 2010

A catalogue record for this publication is available from the British Library

Library of Congress cataloguing in publication data
Library of Congress Cataloging in Publication Data
Heliophysics : space storms and radiation : causes and effects / edited
by Carolus J. Schrijver, George L. Siscoe.
p. cm.
ISBN 978-0-521-76051-5 (hardback)
1. Solar energetic particles. 2. Heliosphere (Astrophysics) 3. Solar wind.
I. Schrijver, Carolus J. II. Siscoe, George L. III. Title.
QB526.S65H45 2010
523.7′2–dc22
2009050325

ISBN 978-0-521-76051-5 Hardback

Additional resources for this publication at www.cambridge.org/9780521760515

Contents

The plates are to be found between pages 148 and 149.

Preface

Over the past few centuries, our awareness of the couplings between the Sun's variability and the Earth's environment, and perhaps even its climate, has been advancing at an ever increasing rate. The Sun is a magnetically variable star and for planets with intrinsic magnetic fields, planets with atmospheres, or planets like Earth with both, there are profound consequences and impacts. Today, the successful increase in knowledge of the workings of the Sun's magnetic activity, the recognition of the many physical processes that couple the realm of the Sun to our galaxy, and the insights into the interaction of the solar wind and radiation with the Earth's magnetic field, atmosphere and climate system have tended to differentiate and isolate the solar heliospheric and geo-space sub-disciplines of the physics of the local cosmos. In 2001, the NASA Living With a Star (LWS) program was initiated to reverse that trend.

The recognition that there are many connections within the Sun–Earth systems approach has led to the development of an integrated strategic mission plan and a comprehensive research program encompassing all branches of solar, heliospheric, and space physics and aeronomy. In doing so, we have developed an interdisciplinary community to address this systems-science. This has raised awareness and appreciation of the research priorities and challenges among the LWS scientists and has led to observational and modeling capabilities that span traditional discipline boundaries. The successful initial integration of the LWS sub-disciplines, under the newly coined term "heliophysics", needed to be expanded into the early education of scientists. This series of books is intended to do just that: aiming at the advanced undergraduate and starting graduate-level students, we attempt to teach heliophysics as a single intellectual discipline. Heliophysics is important both as a discipline that will deepen our understanding of how the Sun drives space weather and climate at Earth and other planets, and also as a discipline that studies universal astrophysical processes with unrivaled resolution and insight possibilities. The goal of this series is to provide seed materials for the development of new researchers and new scientific discovery.

Richard Fisher, Director of NASA's Heliophysics Division
Madhulika Guhathakurta, NASA/LWS program scientist

Editors' notes

This volume is the second of a three-part series of texts (and an on-line problem set) in which experts discuss many of the topics within the vast field of heliophysics. The texts reference the other volumes by number:

I Plasma Physics of the Local Cosmos
II Space Storms and Radiation: Causes and Effects
III Evolving Solar Activity and the Climates of Space and Earth

The project is guided by the philosophy that the many science areas that together make up heliophysics are founded on common principles and universal processes, which offer complementing perspectives on the physics of our local cosmos. In these three volumes, experts point out and discuss commonalities and complementary perspectives between traditionally separate disciplines within heliophysics.

Many of the chapters in the volumes of this series have a pronounced focus on one or several of the traditional sub-disciplines within heliophysics, but we have tried to give each chapter a trans-disciplinary character that bridges gaps between these sub-disciplines. In some chapters stellar and planetary environments are compared, and in others the Sun is compared with its sister stars or planets are compared with one another; in yet other chapters general abstractions, such as magnetic field topology or magnetohydrodynamic principles, that are applicable to several areas.

The vastness of the heliophysics discipline precludes completeness. We hope that our selection of topics helps to inform and educate students and researchers alike, thus stimulating mutual understanding and appreciation of the physics of the universe around us.

The chapters in this volume were authored by the teachers of the heliophysics summer school following the outlines provided by the editors. In the process of integrating these contributions into this volume, the editors have modified or added segments of text, included cross references, pointed out related segments of text, introduced several figures and moved some others from one chapter to another, and attempted to create a uniform use of terms and symbols, while allowing some differences to exist to remain compatible with the discipline's literature usage. The editors bear the responsibility for any errors that have been introduced in that editing process.

Additional resources

The texts were developed during summer schools for heliophysics, held over three successive years, at the facilities of the University Corporation for Atmospheric Research in Boulder, Colorado, funded by the NASA Living With a Star program. Additional information, including text updates, lecture materials, (color) figures and movies, and teaching materials developed for the school can be found at www.vsp.ucar.edu/Heliophysics. Definitions of many solar–terrestrial terms can be found via the index; a comprehensive list can be found on the web at www.swpc.noaa.gov/info/glossary.html.

Heliophysics

helio-, prefix, on the Sun and environs; from the Greek helios.
physics, n., the science of matter and energy and their interactions.

Heliophysics is the

- *comprehensive new term for the science of the Sun–solar system connection.*
- *exploration, discovery, and understanding of our space environment.*
- *system science that unites all of the linked phenomena in the region of the cosmos influenced by a star like our Sun.*

Heliophysics concentrates on the Sun and its effects on Earth, the other planets of the solar system, and the changing conditions in space. Heliophysics studies the magnetosphere, ionosphere, thermosphere, mesosphere, and upper atmosphere of the Earth and other planets. Heliophysics combines the science of the Sun, corona, heliosphere, and geospace. Heliophysics encompasses cosmic rays and particle acceleration, space weather and radiation, dust and magnetic reconnection, solar activity and stellar cycles, aeronomy and space plasmas, magnetic fields and global change, and the interactions of the solar system with our galaxy.

From NASA's *Heliophysics. The New Science of the Sun–Solar System Connection: Recommended Roadmap for Science and Technology 2005–2035.*

1

Perspective on heliophysics

GEORGE L. SISCOE AND CAROLUS J. SCHRIJVER

1.1 Universal processes: "laws" of space weather

Heliophysics is concerned with laws that give rise to structures and processes that occur in magnetized plasmas and in neutral environments in the local cosmos, both temporal (weather-like) and persistent (climate-like). These laws systematize the results of half a century of exploring space that followed centuries of ground-based observations. During this time spacecraft have imaged the Sun over many wavelengths and resolutions. They have visited every planet, all major satellites and many minor ones, and a selection of comets and asteroids. Beyond this they have traversed the expanse of the heliosphere itself. Out of the vast store of data so accumulated, the laws and principles of heliophysics are emerging to describe structures that are natural to magnetized plasmas and neutrals in cosmic settings and to specify principles that make the heliosphere a realm of numerous, original dynamical modes.

By "the laws of heliophysics" we are not here referring to a subset of the laws of physics that apply to all things everywhere. A discipline that needs to refer back to the fundamental laws of physics to explain its phenomena would be totally derivative, having no synthesizing laws of its own, no regularities peculiar to it, no inherent principles with explanatory power sufficient to link its own distinctive phenomena; in short, no paradigms. To help fix this idea, we list here a few familiar examples from other fields of discipline-specific general laws or principles: chemistry – the periodic table, valence, Le Chatelier–Braun principle; biology – evolution, double helix; geology – "deep time", plate tectonics; astronomy – Kepler's laws, Hertzsprung–Russell diagram, expanding universe; meteorology – Hadley cell, baroclinic instability.

In the case of heliophysics, probably most of its laws have yet to be discovered, since the project of finding them is young. Moreover, heliophysics is a unique hybrid between meteorology and astrophysics with substantial components

Heliophysics: Space Storms and Radiation: Causes and Effects, eds. Carolus J. Schrijver and George L. Siscoe. Published by Cambridge University Press.

of physics and chemistry. Thus, many of the laws of heliophysics that we can list at this time might be subjects for research in meteorology (e.g. the field of aeronomy), astrophysics (e.g. shock waves and cosmic rays), physics (e.g. magnetic reconnection and particle energization), or chemistry (e.g. reaction rates in planetary ionospheres and thermospheres). Other laws are still hiding their full relevance, even as they hint at their existence through the (self-)similarity of processes and scale-free power-law spectra of a wide range of phenomena, in energies of solar flares, coronal mass ejections, and energetic particles, and from geomagnetic storm occurrences to solar-wind turbulence spectra.

Our three volumes on heliophysics, of which this is the second, are intended to lay out the structures and phenomena with which heliophysics is concerned that might be organized under general laws and principles and to indicate how far the field has progressed toward uncovering them. In particular, the present volume is concerned with energy-conversion phenomena with emphasis on the explosive kind that produce solar eruptions and the storms of energetic particles, which on occasion render space a hostile environment for technological, space-faring humanity. These phenomena are, of course, of special interest to space weather.

1.2 Pressure, gravity, and electromagnetism

This volume's emphasis on time-dependent phenomena (commonly captured under the term space weather) follows an emphasis in Volume I on structures and processes that persist over time and that at any time can be seen at one or more places in the heliosphere (examples of which include the solar wind and planetary magnetospheres and ionospheres). In Volume I the emphasis was on the rich variety of such structures and processes, as illustrated in its first chapter by following a wandering proton on an odyssey that began beneath the solar surface and ended in the interstellar medium. In its struggle upward to the photosphere and chromosphere it shuffled through numerous solar structures and processes. Upon reaching the corona it passed through sites of dissipation where, being energized enough to escape the Sun's gravitational hold, it joined the solar wind, only to be caught a few days later by the Earth's magnetic field. Entrained as a member of the magnetosphere's high-pressure plasma, it experienced a tour of ionospherically ruled inner chambers before escaping again to continue its journey to freedom. Eventually it exited through the termination shock and ultimately returned to the interstellar medium from which it had been captured by the proto-heliosphere 4.5 billion years earlier (which is part of the story told in Volume III).

In Volume I we noted that although narrating the odyssey of a proton illustrates well the variety of heliospheric structures and processes, a corresponding narrative exists for the magnetic field – a narrative that features the magnetic field's role in generating space weather. Indeed the magnetic field is the *sine qua non* – "that without which nothing" – of space weather. Without the magnetic field, neither solar activity nor magnetic storms – the solar and terrestrial sources of space weather – would exist. Two properties of the magnetic field initiate its career as a generator of space weather: (1) it has no conserved sources, and (2) it is buoyant. The first of these properties means that the magnetic field must be continually generated. Although in principle fossil magnetic fields could have remained from the creation of the solar system, this appears not to be the case (see Vol. III). Witness the 22-year magnetic cycle of the Sun and the reversals of the Earth's magnetic field. On shorter time scales, the magnetic topography of the solar surface changes so rapidly that it must be monitored constantly as input for space weather forecasts.

The telling comparison is with the gravitational field, **g**, which unlike the magnetic field, **B**, has a conserved source. The conserved source of the gravitational field is mass, as can be seen in the field equations that apply to the gravitational field:

$$\nabla \cdot \mathbf{g} = -4\pi G\rho, \tag{1.1}$$

$$\nabla \times \mathbf{g} = \mathbf{0}, \tag{1.2}$$

where G is the gravitational constant and ρ is the mass density. Thus, gravity is determined by the amount of mass present and its distribution. Since mass is conserved and the gravitational force causes matter to collapse into systems in which the gravitational force is almost perfectly balanced by thermal or inertial forces, gravitationally organized matter tends to be stable over eons (thermally driven instabilities in gravitationally bound gases form an important exception to this generalization, to which we return below). In contrast, the pertinent field equations for the magnetic field are

$$\nabla \cdot \mathbf{B} = 0, \tag{1.3}$$

$$\nabla \times \mathbf{B} = \mu_0 \mathbf{J}. \tag{1.4}$$

The source term for the magnetic field in these equations is electrical current, **J**, which, unlike mass, is not a conserved quantity. Thus we see that **B** is a product of dynamo or other magnetohydrodynamic (MHD) processes that generate current in real time. The crucial distinction is that unlike the gravitational field, which is in effect a byproduct of a conserved, definite quantity of mass and so is inherently persistent, the magnetic field is generated by a variety of plasma motions in the Sun, in the solar wind, and in planetary magnetospheres on time scales shorter than what would be needed to reach an equilibrated state. Hence, the local cosmos is constantly adjusting and attempting to relax, but it never gets to such a quasi-stationary state. The consequence of this is what we call weather, including the focus of this volume: space weather.

The non-steadiness of space weather arises from the dynamical properties of the magnetic force and its interaction with the other force fields – gravity, pressure, and inertia – as given by the MHD momentum equation (see Vol. I, Chapter 3):

$$\rho \frac{\mathrm{d}\mathbf{v}}{\mathrm{d}t} + \nabla p = \rho \mathbf{g} + \mathbf{J} \times \mathbf{B}. \tag{1.5}$$

For a time-independent equilibrium to be possible, or at least one without motion, solutions without the $\mathrm{d}\mathbf{v}/\mathrm{d}t$ term must exist. But there can be no such equilibrium involving the magnetic field because of the different makeup of the stress tensors of the three forces. On this point, Eugene Parker states in his book *Cosmical Magnetic Fields* (1979): "When there is a magnetic field present in a compressible fluid, there can be no equilibrium unless the gravitational force is parallel to the magnetic force" (p. 298), which of course is not possible everywhere within a naturally occurring, autonomous, gravitationally bound magnetized plasma. The role that the gravitational force plays in establishing an equilibrium can be stated in terms of the stress tensors of the three forces: thermal pressure, gravity, and magnetism. Whereas the pressure terms in the stress tensors of the thermal pressure force and magnetic field force are positive definite – they act to force the plasma to expand – the corresponding term for the gravitational force is negative definite – it acts to force the plasma to contract. Therefore, unless restrained by gravity, hot plasmas and magnetic fields would expand indefinitely (note that because of the specific form of the gravitational field equations it is not the pressure term in the gravitational stress tensor that causes matter to contract, it is the tension term – gravity really does pull). That it is the nature of hot plasmas and magnetic fields to expand and of gravitational fields to contract can also be seen from the virial theorem for an isolated system (e.g. Rossi and Olbert, 1970, p. 305):

$$\frac{1}{2} \frac{\mathrm{d}^2 I}{\mathrm{d}t^2} = 2T + M - G, \tag{1.6}$$

where I is the trace of the moment-of-inertia tensor of the system, T is the total kinetic energy (bulk plus thermal), M is the total magnetic energy, and G is the gravitational energy (taken to be positive definite like the other energies). Thus, equilibrium is possible only if the gravitational energy term balances the kinetic and magnetic energy terms, which act to make the system expand, thereby increasing the system's moment of inertia.

Taken together, as in the Sun, the three forces – thermal pressure, gravity, and magnetism – produce a situation in which gravity and pressure are in quasi-equilibrium, but the magnetic field with its positive pressure expands and, being massless, becomes buoyant in the pressure gradient set up by the pressure–gravity equilibrium. Thus begins the odyssey of a magnetic flux tube newly generated in the subsurface solar dynamo. But this story is complex and is better told in chapters dedicated to it (Vol. I, Chapters 4 and 8; and Vol. III).

For the present purpose, the point is that in the Sun the magnetic field necessarily introduces motion, and this motion is subject to instabilities that result in magnetic structures with a wide range of scale sizes. The situation is further complicated by thermally driven motion fields, such as convection cells and differential rotation, that redistribute and concentrate magnetic flux. What, on occasion, raises this interesting but esoteric behavior to a level of importance to space weather is that magnetic structures sometimes reach a dimension so large that the amount of ambient energy that can be tapped explosively by an instability (the nature of which is the subject of ongoing research, see this volume, Chapter 6) is huge enough to disrupt the space between Sun and Earth and beyond.

1.3 Structure and dynamics of the local cosmos

The magnetic field's inherent tendency to expand does not by itself account for its space weather effectiveness. If expansion were enough then the expanding corona unassisted by the magnetic field would manifest storms, which it does not do. An additional property of the magnetic field that adds to its space weather effectiveness is tension. Tension is not a property of thermal pressure but, as noted above, it is a property of gravity. Tension allows gravity and the magnetic field to organize matter into coherent volumes. In the case of gravity, one such volume is called the Sun. As for the magnetic field, tension gives spatial coherence to magnetic flux tubes through transmission of Alfvén waves (a statement that assumes the validity of the MHD condition, i.e. the "freezing" of magnetic flux to the plasma). There is an important difference regarding the types of volumes that the gravitational and magnetic tension forces organize. The gravitational field has no shielding currents ($\nabla \times \mathbf{g} = \mathbf{0}$), and since its source is mass density ($\nabla \cdot \mathbf{g} = -4\pi G\rho$), it has no discontinuities because that would require an infinite mass density. Hence, the gravitational field is relatively homogeneous; it varies smoothly and continuously in space. On the other hand, the magnetic field has shielding currents ($\nabla \times \mathbf{B} = \mu_0 \mathbf{J}$) which spontaneously form discontinuities, called current sheets (Parker, 1994). Therefore the flux tubes into which the magnetic field organizes plasma by tension can have relatively well-defined outer boundaries.

In fact one may picture the heliosphere as being filled rim to rim with more-or-less discrete magnetic flux tubes (Vol. I, Chapters 4 and 6). On the Sun these take the form of filaments, fibrils, and sunspots, to name a few. In the heliosphere flux tubes range in size from the dissipation scale of solar wind turbulence

(Vol. I, Chapter 7) up to global-scale spiral flux tubes that emerge from coronal holes. Planetary magnetospheres are composites of magnetic flux tubes with dimensions from flux-transfer events – small tubes that adhere like parasites to a magnetosphere's "skin" – up to the open-field-line lobes that constitute the magnetotail. The heliosphere's menagerie of flux tubes is tightly packed since a magnetic flux tube expands until stopped by another flux tube expanding in the other direction. For example, magnetospheres are magnetic flux tubes anchored in gravitating planets, which expand until stopped by the momentum-bearing, plasma-filled flux tubes of the solar wind. The dynamics of expanding magnetic flux tubes results in a network of flux tubes separated by current sheets and filling the heliosphere. Current sheets so formed play an important role in space weather by being sites of magnetic dissipation known generally as magnetic reconnection (Vol. I, Chapter 5). Manifestations of magnetic reconnection at current sheets – in some cases quite dramatic manifestations – have been observed on the Sun, in the solar wind, and at various places around magnetospheres.

To recount briefly, from the structure of the stress tensors of the three forces that act on magnetized plasmas in cosmic settings one can deduce these general properties:

(i) On time scales relevant to space weather, gravitational fields are smoothly varying fixtures of space that do not change in time. In contrast, the magnetic field forms a discontinuous, space-filling network of flux tubes that are for the most part in a continual process of non-steady creation and dissipation.
(ii) Plasmas and magnetic fields will expand indefinitely unless held down by a gravitational field or, at a local level, unless restrained by opposing expansions.
(iii) Gravity and pressure create stable, static structures (thermally driven convection and circulation excepted and discussed separately) whereas magnetic fields do not form stable, static structures with gravity or pressure, but in a fluid medium like the Sun form buoyant flux tubes that shred on rising and (passing here from deduction to observation) reform on the surface into filaments, tubes, and loops that cover a great range of sizes.

The manifold sizes, forms, and temporal modes of magnetically derived structures (described here in Chapter 5) are not deducible merely from the structure of the stress tensors of the participating forces. To understand these – indeed, not just the extremes but all space weather phenomena – is part of the business of heliophysical research as covered in these volumes.

Creation, buoyant rise, surface transport, flux-tube formation, stretching, expansion, and dissipation-assisted expulsion name events in the lives of magnetic field structures that at any time and at all times traffic in their thousands from the Sun to the heliopause and create the time-dependent conditions that constitute the solar manifestations of space weather. These magnetic structures are, of course, plasma filled. They form a continuous network of moving magnetic plasma carriers that stretches about 100 AU from the Sun to the border of the heliosphere. The geometry and topology of the network of magnetic plasma carriers reflects the distribution of closed magnetic structures on the Sun and thus changes with the phase of the solar cycle. At solar minimum it usually forms a singly connected wavy sheet, called the streamer belt, forming a low-latitude, Sun-circling band. By the time of solar maximum it has evolved into a multiply connected, pole-to-pole honeycomb-like network. If one could see it, the network would be streaked with flux tubes of variable density and spotted with magnetized plasma blobs of many sizes, mostly advecting outward with the solar wind. Within the heliosphere, but outside the network, we have the domain of the fast, more-or-less unstructured solar wind that emanates from coronal holes and fills most of the three-dimensional heliosphere. Typically one to a few times per day, this picture of orderly outward transport through 3D volumes of fast solar wind interlaced with 2D-like sheets of magnetically structured slow solar wind is disrupted by explosive ejections of quasi-spherical magnetic clouds that expand rapidly and shoot outward. These are coronal mass ejections, CMEs, the bringers of space storms.

CMEs are macroscale magnetic flux tubes that can suddenly, coherently form low in the corona and accelerate to speeds sometimes in excess of 2000 km/s before leaving the Sun (Chapter 5). As presently understood, their high speed is a consequence of unbalanced magnetic forces operating on magnetically organized volumes of plasma (Chapter 6). By contrast – to emphasize again the magnetic field's role as the prime generator of space weather events – the pressure gradient force (aided by waves) is able to propel the solar wind to peak speeds of only about 800 km/s, as measured by the Ulysses spacecraft over the solar poles. The storm that follows a CME's arrival at Earth – a magnetic storm as it has been called since before CMEs were discovered – disturbs the magnetic field everywhere in the magnetosphere and at ground level, sometimes enough to disrupt power and communications systems (Chapter 2). The radiation belts in the magnetosphere are pumped up and become more than usually hazardous to satellites and astronauts (Chapters 13 and 14).

As mentioned, CMEs often move faster than the prevailing solar wind and thus plow through it, sweeping it up and compressing it to form shock waves. CME shock waves can claim importance not just because they signal the arrival of the storm (and so predicting their time of arrival is a high-priority activity of space weather forecasters), but also because they themselves are the generators of one of the most serious hazards of space: solar cosmic rays, or as they are more conventionally called, solar energetic particles (Chapter 8). In the rarefied solar wind, shock waves, viewed at the microscale, are not the result of particles impacting particles with consequent rapid thermalization, as happens in dense media, but rather of waves impacting particles that in turn stimulate the impacting waves through a positive feedback process. Here a subset of the particles can experience multiple, energy-increasing wave encounters before leaving the energy-exchanging shock layer and end up with energies high enough to be of concern to humans who would otherwise have no interest in CMEs. These subjects fall under the headings of particle acceleration in shocks, which Chapter 8 treats in substantial detail, and of the radiation effects on biological materials and spacecraft hardware, which are reviewed in Chapters 13 and 14.

CME shock waves occupy a special place in heliophysical studies because of their sometimes harmful effect on human enterprises outside of heliophysics. But within heliophysics they represent just one example, not especially exceptional, of a large population of shock waves. Shock waves form not only in front of fast CMEs but wherever the solar wind impacts relatively stationary objects, such as planets and the interstellar medium, and where fast solar wind streams are brought into contact with slow solar wind streams by the spiral geometry that the Sun's rotation imparts to all long-lived solar wind structures, so-called corotating interaction regions (CIRs). The frequent occurrences of CMEs, CIRs, and the multitude of planets result in a collection of shock waves that are accessible by spacecraft for detailed study, which is an advantage that puts heliophysics at the forefront in the study of collisionless shocks and particle acceleration at shocks. In this regard Eugene Parker in his *Cosmical Magnetic Fields* has stressed the value of the broader role that heliophysical research plays: "It cannot be emphasized too strongly that understanding of the magnetic activity in the astronomical universe can be achieved only by coordinated study of the various forms of activity that are accessible to quantitative observation in the solar system." Accordingly, Chapter 7 treats shock waves in their heliophysical varieties and, as already noted, Chapter 8 treats particle energization at shocks generally. The heliosphere serves as a laboratory not only to study energization of particles at shocks, but also to look into their transport within the heliosphere once energized, as Chapter 9 describes. Here the coverage includes galactic cosmic rays as well as locally produced energetic particles.

1.4 Energetic particles

On the topic of energetic particles we may segue from the Sun and the solar wind as places where the elements of space weather are generated to magnetospheres as places that lie at the receiving end of all this generation. But not passively – magnetospheres also generate space hazards, especially energetic particles, which, magnetically trapped, fill reservoirs known as radiation belts because they circle the planet equatorially. Earth's radiation belts (also known as the Van Allen belts) are best known, but all magnetized planets have them. They reach their acme of damage potential at Jupiter. The origins and properties of radiation belts are related in Chapter 11, and this concludes the book's description of hazardous space weather elements as such. But recall that all space weather elements entail the conversion of energy between the kinetic and the magnetic forms – kinetic energy of subsurface flows in the dynamo region of the Sun creates magnetic energy, some of which eventually converts explosively into the kinetic energy of CMEs, and some of which drives hydromagnetic shock waves (creating magnetic energy), the dissipation mechanism for which entails the energization of particles (creating kinetic energy). Once CMEs reach Earth and create magnetic storms, the swapping of energy back and forth, starting with the kinetic energy of the CME and ending in part in the radiation belts, becomes even more involved. This story merits its own treatment in Chapter 10, which considers energy conversion at planetary magnetospheres in a fully general way.

The properties of space weather elements that render them hazardous aid in their detection and measurement. Energetic particles can be detected and measured directly *in situ* with instruments carried on spacecraft. *In-situ* measurements have sampled the heliosphere's energetic particles from its inner region around Mercury to its border with the interstellar medium. As Chapter 3 describes, such measurements have determined how electrons and ions from protons to multiply ionized iron are distributed over multiple decades of energy, revealing long tails extending to high energies (the space weather hazardous range) ending finally in cutoffs. These data have led to the discovery that the high-energy tails have a universal slope, instancing Parker's pronouncement on the value of quantitative heliophysical studies. What cannot be measured directly *in situ*, as near the Sun or in the heart of Jupiter's radiation belts, can often be inferred through X-rays and synchrotron radiation and other emissions that energetic electrons cause. Chapter 4 reviews these techniques and shows how the radiative signatures of energetic particles have proven invaluable in probing unreachable environments and, with the advantage of global monitoring of the space environment, of continuously documenting the occurrence of explosive energy conversion events.

1.5 Weather and climate in space

We have named many of the main players in magnetically induced space weather: networks of advecting blobs and ropes of magnetized plasma, CMEs, magnetic storms, shock waves, solar energetic particles, cosmic rays, and radiation belts. As a prelude to this cataloging of space weather elements we emphasized the difference in the behavior of gravitationally organized matter and magnetically organized matter. We return here to this difference as it shows up in time-dependent (weather-like) phenomena; that is, we compare gravitationally organized weather with the magnetically organized type in terms of energy conversions.

Gravitationally organized weather is the kind of weather that occurs in the atmospheres of the Sun and the planets. Weather arises in response to a need to move energy from a source to a sink. On the Sun the source is thermonuclear reactions in the core and the sink is electromagnetic radiation from the photosphere. A small fraction of this energy flow is diverted into the generation of magnetic fields, which is the source of energy for space weather, as already narrated. To represent weather on planets, we will look at the Earth. Here the source of energy is incoming solar radiation, mostly in the visible band of wavelengths, and the sink is outgoing terrestrial radiation, mostly in the infrared (a small amount of energy enters the atmosphere from below, but it is negligible as a source of weather; this is not the case for the Sun, of course, or for the giant planets, which have a significant internal energy source that we are not considering).

The flow of energy from source to sink at both the Sun and the Earth is conveyed in part by the atmosphere carrying the energy from a hot region to a relatively cold region. However, in the case of the Sun, energy is transported mainly by means of radiative diffusion from the core to about 70% of the way to the photosphere. In the outer 30% of the Sun's atmosphere, radiative energy transport gives way to convective transport, the motion field of which takes the form of convection cells, which can be seen as a granular pattern in the photosphere. The photospheric granular pattern is more-or-less homogeneous since there is little variation in the rate of energy outflow over the solar surface to give a variation in the sizes and shapes of the granules; in magnetically active regions, granulation is deformed with little impact on the brightness, but in sunspots it is strongly suppressed, resulting in a pronounced drop in brightness.

A pattern of homogeneous convection cells is not a description that applies to the situation at Earth, where the rates of incoming and outgoing radiative energy vary significantly from equator to pole. Here, atmospheric transport acts to reduce the equator-to-pole temperature difference that would result from local radiative equilibrium. If the Earth did not rotate and solar radiation were nonetheless distributed uniformly in longitude (a highly artificial situation to make a point), there would be one global convection cell rising at the equator and sinking at the pole. But Earth's rotation does not allow an energy conveyor belt to stretch from equator to pole in one loop. Instead it takes three loops, like a gear chain with three gears. The gears are called cells, the Hadley cell in the tropics, the Ferrell cell at mid-latitude, and the polar cell on top. The Hadley cell drives them all. Each hemisphere (north and south) has such a three-celled conveyor belt carrying energy from the equator to the pole. But there is a thermodynamic anomaly in this arrangement. As a convection cell, the mid-latitude Ferrell cell circulates backwards, in that it rises at its cold, high-latitude end and sinks at its hot, low-latitude end. It must do this to mesh with the near-equatorial Hadley cell, which is driving the whole three-celled belt. So instead of carrying energy poleward across the middle latitudes by means of a reversed, vertical convection cell, the atmosphere adopts another option; it does it mainly by means of a horizontal serpentine flow that picks up heat at the low-latitude extreme of its meanders and drops it off at the high-latitude extreme. Now here is the interesting part. The Hadley cell and the polar cell are stable because they circulate in a thermodynamically proper sense. So they do not generate weather, but they generate climate: the equatorial rain belt, trade winds, subtropical deserts, and the polar highs. But the meandering mid-latitude transport system is unstable; it makes weather: eastward migrating high- and low-pressure systems with associated warm and cold fronts. The instability behind this behavior is called the baroclinic instability. The baroclinic instability operates at the discontinuity (called the polar front) between hot air coming poleward across mid-latitudes and cold air coming equatorward in the polar cell. It is the mechanism that allows the cold air and the warm air that have built up at the front between the Ferrell cell and the polar cell to finally exchange places, as the transport of heat from the equator to the pole demands. It is the mechanism that releases the gravitational potential energy inherent in a pool of dense cold air abutting a pool of light warm air.

There are interesting analogies between terrestrial weather and space weather to assist communication between the disciplines. Space weather has, of course, for a long time borrowed heavily from meteorological nomenclature: solar wind, magnetic storms, magnetic clouds, particle precipitation, and others. The analogies we have in mind are not just phenomenological but dynamical. The Hadley cell as the engine that drives the general circulation of the atmosphere is the prototype for the Dungey cell as the main driver for the general circulation of the magnetosphere (Vol. I, Chapter 10). The Hadley cell is fueled by solar radiation, the Dungey cell by the solar wind. At Jupiter we have the Vasyliūnas cell fueled by the planet's rotation. Terrestrial weather is driven by a heat engine operating between the hot equator and the cold poles, with Earth's rotation acting to complicate the process and creating terrestrial weather as a byproduct.

The ionosphere and thermosphere constitute the interface between the domains of gravitationally organized terrestrial weather and magnetically organized space weather. Its gravity waves, tidal waves, and ionospheric stratification are terrestrial features, whereas its space weather features include the imprint of the Dungey cell on thermospheric circulation and traveling atmospheric disturbances (TADs) launched by massive inputs of energy and momentum from the solar wind during magnetic storms (Chapter 12). The ionosphere and thermosphere are where the two weather types come together and affect each other: Dungey circulation in the thermosphere is strongly modified by Earth's rotation and TADs can run from pole to pole and dominate thermospheric dynamics during magnetic storms (see Vol. III, Chapter 15). In terms of energy and momentum, during magnetic storms this border region receives about 90% of the energy and momentum extracted from the solar wind (Chapter 10). In return, through an outflow of ions from the ionosphere driven by an interaction with the magnetosphere during magnetic storms, the ionosphere can supply more than half of the magnetosphere's resident charged particles. There is a still-unknown feedback process regulating the rate of ionospheric outflow. As those who are responsible for understanding the ionosphere and thermosphere and those who specialize in magnetospheric phenomena pursue the mutual influence of the two modes of dynamics interacting in the ionosphere–thermosphere nexus, what is emerging is the realization that the atmosphere–thermosphere–ionosphere–magnetosphere–solar wind must be treated as a strongly coupled system. It seems not unlikely that the lesson learned here in this most accessible space weather system will, upon further investigation, be seen to apply to other such systems, such as those associated with the production of CMEs and the production and transport of energetic particles.

1.6 Universal processes in the local cosmos and instrumentation

As you read the volumes in this series on heliophysics, you will encounter several different perspectives on the theme of "universal processes". The most common perspective is one in which some approximation or conceptualization of the real world is applied to the various environments in the local cosmos: these include MHD, turbulence, reconnection, current sheets, flux tubes and ropes, bow shocks, or various types of waves and instabilities. The essence of such a perspective is that whereas we understand, of course, that the basic laws of physics apply everywhere, we need to transform the real world into an approximation that allows us to think about it in terms of a limited number of simple concepts for which, ideally, we have an essentially intuitive understanding. In other words, this understanding comes from one area in which we have direct experience, but which is applied to other situations that we are trying to understand.

One such concept that appears throughout these volumes is that of "reconnection". This term, widely used, turns out to be very poorly defined. It can be used to refer to the changing connectivity in a vacuum potential field as much as to the decoupling of particle motions from the background magnetic field by any number of concepts, ranging from inertia to wave–particle interactions, or from resistivity to infinitesimal current sheets. It is thus as much a culturally accepted term for something that we really do not understand, as a descriptor of a well-understood consequence: we can say that reconnection occurs whenever the approximation of frozen-in flux fails (Vol. I, Chapters 3–5).

Other "universal processes" are better defined, and thus more directly applicable across discipline boundaries: shocks, turbulence, and associated particle acceleration (Chapters 5, 7, 8, and 9), instabilities (Chapters 5, 6, 11, and 10), and wave–particle interactions (Chapter 11 and Chapter 9 in Vol. I).

Thus our quest for "universal processes" is about finding the common ground between one phenomenon and another to deepen our insight, and to enable the application of a vast area of knowledge from one discipline of physics to another. In this volume, this has a particularly wide reach. Let us take the particular example of energetic particles that interact with matter through which they (attempt to) propagate. These particles lose energy by Coulomb collisions or by direct collisions (depending on their energy and charge state), and as they do so, they cause ionization, dissociation, photon or particle emission, and ultimately the distribution of their energy over many other particles as the energy thermalizes in a multitude of interactions. These interactions occur throughout the local cosmos, be it in the formation of flare ribbons associated with a solar eruption (Chapters 5 and 6) or in the glow of the aurora on Earth (Chapter 12) or any of the other planets with a substantial magnetic field (Chapter 10).

The empirical scientist or instrumentalist will point out that these processes also occur in particle detectors (Chapter 3), and the space weather forecaster and spaceflight engineer will note that they also form the foundation for understanding the impact of space weather on satellites (Chapter 14) and on humans in space (Chapter 13). The concept of "stopping power" by ionization collisions thus appears when we discuss instrument design (Section 3.4), deep-electric spacecraft charging (Section 14.3), and astronaut protection (Chapter 13). But it also appears when we discuss ionization and charge-transfer collisions in the solar and planetary atmospheres that lead to emission of neutral particles (see Fig. 3.23), charged particles (Chapters 4 and 5), or photons (Chapters 4, 5, and 12). One could even view these atmospheres as parts of a detector system in which energetic particles interact with the surrounding medium, which leads to emission of photons or particles that, in turn, can be detected by another part of the detection system, namely the hardware built into optical telescopes and particle detectors, respectively.

It takes a few grams of material to stop most of the energetic particles associated with a solar flare or the Earth's radiation belts (1 keV/nucleon to 1 GeV/nucleon), with little dependence on the properties of the material that is impacted (see Section 3.4). That is therefore about the weight per unit area of aluminum shielding used for astronauts (see Section 14.4). It should come as no surprise that this is also the column depth between the high corona and the chromospheric flare ribbons: the mass column depth for coronal energetic particles totals about 0.05 g/cm^2 when the particles reach the bottom of the chromosphere (at about 500 km above the photosphere) and about 5 g/cm^2 at the photospheric level (e.g. Avrett, 1981). For comparison, note that the column mass of the Earth's atmosphere is slightly larger than 1 kg/cm^2, so that the atmosphere provides an extremely effective shield against almost all of the energetic particles from the Sun, heliosphere, and beyond.

When it comes to learning about our local cosmos, environments such as atmospheres, satellite electronics, and biological tissue are all part of the tool set available to measure properties of the populations of photons and particles, in addition to the specifically designed optical systems or the electric or magnetic deflector systems, with CCDs or charge amplifiers in their focal planes.

In these heliophysics volumes, we do not venture into details of the impacts of solar activity and space weather on human health. If you are interested in these aspects, then PubMed Central,† the searchable archive of the biomedical and life-sciences literature maintained by the US National Institutes of Health, provides an interesting entry point. There, you can find studies that relate (X)(E)UV and energetic-particle radiation to, e.g., fertility, longevity, benign and malignant neoplasms, auto-immune disorders, and even mental health, and – for example – their dependence on geographic latitude as affected by seasonally varying effective atmospheric thickness, the geomagnetic field, and the altitude (particularly important for air crews).

† PubMed Central: www.pubmedcentral.nih.gov

2

Introduction to space storms and radiation

STEN ODENWALD

2.1 Introduction

The opening chapter of Volume I, *Heliophysics: Plasma Physics of the Local Cosmos*, gave an overview of heliophysics that ranged from the deep interior of the Sun to the most distant reaches of the heliopause beyond the orbit of Pluto. The bottom line is that we are talking about a system, knit together by particles and fields, that displays complex behavior at scales from less than seconds to more than centuries, and meters to terameters. The heliosphere may thus appear to be an extremely enriched physical system that contains more than enough phenomenology to keep us focused on an ever-increasing supply of intriguing questions. That is why we need to find patterns in the form of universal processes.

Pure research leads to an increase in our understanding of heliophysics for its own sake. At the same time, this understanding improves our predictive abilities, which help us mitigate financial, technological, and societal impacts. Conversely, as we strive to improve our technological operations in the space weather environment, these help to advance our theoretical understanding of radiation effects and other essential physical phenomena because they drive the modeling process to be more accurate and relevant to engineering issues (see Chapter 13). Heliophysics research is one of the few examples in astronomy where such a direct mutually reinforcing and stimulating relationship is found.

In this chapter, I explore how the human experience of heliophysics has provided certain kinds of interesting boundary conditions to the theoretical modeling of heliophysical phenomenology. I would like you to think of this as a process of going back over the important human impacts of heliophysics research to see whether there are any stones unturned that would be of considerable interest to look beneath in our thriving twenty-first century technological culture.

Heliophysics: Space Storms and Radiation: Causes and Effects, eds. Carolus J. Schrijver and George L. Siscoe. Published by Cambridge University Press.

2.2 Uncovering the Sun–Earth connection

Prior to 1700 CE, explanations for aurorae, sunspots, and Earth's magnetic field more often than not included mythological elements (Fig. 2.1). For aurorae, the advent of magnetic observations in the 1700s led to other more "scientific" possibilities. Anders Celsius (1701–1744) and his assistant Olof Hiorter (1696–1750) had made observations of magnetic fluctuations from Uppsala, Sweden, in 1741, discovering that they occurred at the same local times as aurorae were sighted. The variability of the magnetic field was of great interest to Alexander Von Humboldt who – around 1806 – spent many tedious hours recording its minute changes at his villa in Berlin. These investigations led to the establishment of a dozen magnetic observatories, and millions of observations, from which these magnetic storm changes were found to be global in scale, and nearly simultaneous everywhere (Lovering, 1857). The magnetic effects of aurorae were soon well appreciated in the first-half of the 1800s, and often aurorae would be forecast through their magnetic traces alone even during daylight hours, and at distances quite far from the auroral zone (e.g. Paris).

By 1837, Dennison Olmstead had provided a convincing logical argument that the cause for the aurorae (e.g. the currents that give rise to them) must exist outside Earth due to the global scope of the auroral–magnetic phenomenon and its speed of propagation (Olmstead, 1837). The cyclical rise and fall in sunspot number over time was uncovered in 1843 by Samuel Heinrich Schwabe (1789–1875). Edward Sabine (1788–1883) went on to show that there was a detailed correlation between the sunspot cycle and the frequency of auroral displays (Sabine, 1852). This was a

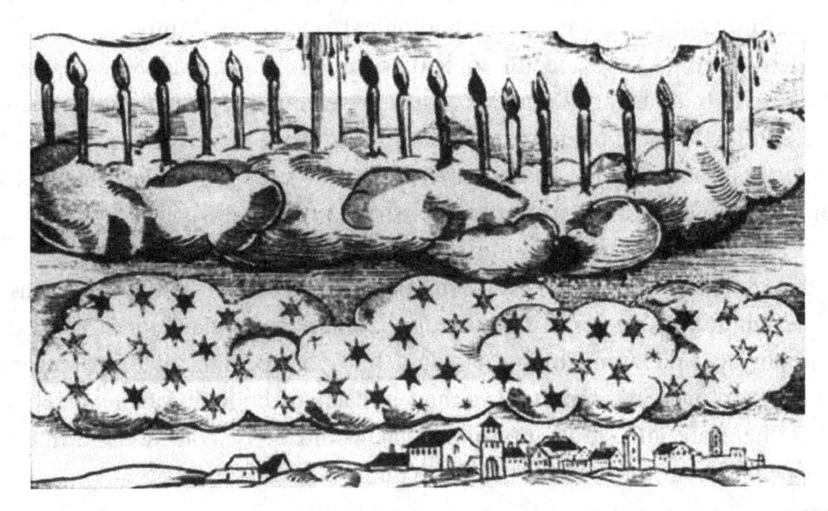

Fig. 2.1. Early drawing of the aurora, depicted as candles in the sky; *c.* 1570 (Original print in Crawford Library, Royal Observatory, Edinburgh.)

spectacular finding because it forced the search for auroral causes away from Earth, and into space. In fact, the Sun, and specifically its spots, had to be implicated in some way as the ultimate cause of the auroral phenomenon.

One of the key events that advanced our thinking about space weather was the double Great Aurora of August 28 to September 2, 1859. One model for the 1859 geomagnetic storm (e.g. Green and Boardsen, 2006) is that a pair of coronal mass ejections (CMEs) was ejected from the Sun on or about August 27 and September 1. The first CME impacted Earth one day later on August 28. A second, faster, CME erupted from the Carrington–Hodgson flare (see Fig. 2.2) observed on September 1; the first solar flare ever to have been recorded by humans. In 17 hours, it collided with the back of the first CME on September 2, which by that time had evacuated a cavity in the interplanetary medium. The sighting of aurorae and magnetic deflections at low geomagnetic latitudes (20°) suggests that magnetotail reconnection propagated from $L = 10\,R_E$ to $L = 1\,R_E$† within a few hours, and may have emptied the outer Van Allen belts (Chapter 11) completely. The CME impacts, meanwhile, caused the magnetopause to be compressed to $R = 3\,R_E$ (19 000 km). While this electromagnetic mayhem was playing itself out, humans below thought their cities were on fire, and lost telegraph service for many days (Green and Boardsen, 2006).

2.3 Human impacts of space weather

While scientists had been busy thrashing out the details about how aurorae, magnetic storms, and solar activity were interrelated, other issues emerged that provided great impetus for a speedy understanding of this problem (see Odenwald, 2000). Beyond the sterile theoretical issues of currents and fields, it is the human-scale drama that tends to galvanize and focus our interests.

> Last evening, while Charles F. Krebs stood outdoors admiring the aurora borealis, the money-drawer was taken from his saloon, and all the cash it contained, to the amount of between $3 and $4, stolen.
>
> *(Chicago Daily Tribune, May 30, 1877, p. 10)*

2.3.1 Magnetic compasses

Magnetic compasses were at the high-tech frontier of the eighteenth and nineteenth centuries, but there are numerous examples of severe magnetic variability

† L is a parameter describing a magnetic surface around the Earth in which L, in R_E, is the distance from Earth's center to the field lines at the magnetic equator for the instantaneous – or, often, unperturbed – geomagnetic field; R_E is a widely used geophysical unit equal to the radius of Earth: 6378 km.

that briefly made navigation using them unworkable. For example, according to Lovering (1857) there were reports of compass disturbances up to 8° reported at Fort Reliance between 1833 and 1835. At Toronto a 2° deviation was seen in May, 1840, and the same disturbance was also recorded at magnetometer stations in Philadelphia and Cambridge, England. On November 18, 1841, the magnetometer in Philadelphia measured a 3° amplitude change over the course of 5 hours. In the face of these changes, daytime navigation was a tricky prospect, since the only guarantee that you were on a stable course would be by repeated compass measurement over the span of several hours. No single measurement could be reliable unless it was a-priori assumed that no magnetic storms were occurring.

During the Great Auroral Display of September 2, 1859, the disturbances of the magnetic needle were very remarkable. At Toronto, in Canada, the declination of the needle changed nearly four degrees in half an hour.

(Harper's New Monthly Magazine, June 1869, Vol. XXXIX, p. 12)

Brussels, Sept. 23 (AP), Budget Minister Joseph Merlot today said "abnormal weather conditions and the aurora borealis" might have put the instruments out of order on the Sabena airlines plane that crashed near Gander, Newfoundland killing 26 persons.

(Los Angeles Times, September 24, 1946, p. 4)

The issue of magnetic navigation led to an increased understanding of the Earth's dipolar field, and from these studies, knowledge blossomed about its long-term variations spanning centuries and geological eras. This led, eventually, to the recognition of the dynamic nature of the field, its drift in space, and, by the twentieth century, its reversals in time and the development of the dynamo theory by Larmor in 1919 and Parker in the 1950s (see Vol. I, Chapter 3; and Vol. III).

The most dynamic interactions between the solar wind and the geomagnetic field occur during times when a CME arrives at Earth. The arriving pressure pulse compresses the geomagnetic field, causing a sudden storm commencement (SSC) that appears on magnetometers as a short-lived increase (a few tens of nanoteslas) in the ground-level field. If the CME field is northward-directed, the SSC will be strong, but may not be followed by an active magnetic storm. If it is southward-directed, the effects will be striking.

Immediately following the SSC, the topology of the geomagnetic field subsequently evolves through the Akasofu–Chapman sequence (Akasofu and Chapman, 1972) during which time magnetotail narrowing and reconnection appear at about 20 R_E. A "Birkeland current" flows from the reconnection region (or the current disruption region near 6 R_E) into the auroral oval and ionospheric E region. The conductivity of the E region increases so that the Birkeland field-aligned current (FAC) completes its circuit by crossing the field lines to form the auroral electrojet.

The current density of the auroral electrojet is time varying, and increases with Kp index, which is a measure of the globally averaged degree of magnetospheric disturbances recorded by mid-latitude magnetometer stations. It can involve approximately 1 million amperes per magnetic substorm. With a typical ionospheric resistance of about 0.1 ohms, the total power dissipated by the current will be about I^2 R = 100 GW per hemisphere (see Chapter 12). This is minute compared to the typical solar irradiance, so aurorae are hard to see in the daytime, except with narrow-band filters tuned to bright auroral lines.

Up to two centuries ago, the only human-observed evidence for these machinations was that, at some point in the ionospheric circuit, the FAC electrons are further accelerated to enough energy (~3 keV) to excite oxygen and nitrogen atoms to produce the familiar auroral curtains, draperies, and other forms. However, an entirely new phenomenon began to be noticed by the middle of the 1800s.

2.3.2 Telegraphy

Earth currents were first proposed by Sir Humphry Davy (1778–1829) in 1821, and later by Michael Faraday in 1831 (Burbank, 1905). It was expected that Earth's magnetic field was created by currents flowing inside the Earth, and in most cases, just below its surface. Telegraphs operated through a single wire strung between poles, with a battery at one of the two stations. The "return current" was provided by connecting the battery/transmitter at one station and the "sounder" at the other to the local ground. The theory was that the battery would drive a closed circuit, in which an "Earth current" would complete the circuit. Little did telegraph engineers realize that such a circuit would also be very efficient in detecting geomagnetically induced currents (GICs) caused by the electrojet.

The advent of the electric telegraph *c.* 1830, and its commercialization *c.* 1838, was followed by the recognition that it could be affected by magnetic storms once the telegraph network had reached a large-enough geographic scale. Carlo Matteucci (1811–1868), the Director of Telegraphs in Pisa observed the electric telegraph connecting Pisa and Florence behave in an unexpected manner during a brilliant aurora on November 17, 1848. The electromagnets remained powered even without the battery attached, and ceased once the aurora dimmed. This is the first documented technological impact of a space weather event. The strongest geomagnetic storms can generate Earth currents, which can induce electric fields from 1 to 10 volts/km. This leads to, potentially, thousands of volts on ocean or ground telegraph and telephone cables.

Related to the very large voltages and currents reported during some exceptional storms have been reports of humans actually being shocked and injured by currents flowing in telegraph wires.

"At its climax [October 31, 1903] there were 675 volts of electricity – enough to kill a man – in the wires without the batteries attached"

(New York Times, November 1, 1903, p.1.)

In the instance of the September 25, 1909, event, a telegraph operator in Luleå, Sweden, actually experienced a severe shock that paralyzed her hand (Stenquist, 1914). A similar injury befell Frederick Royce during the September 2, 1859, storm:

During the auroral display, I was calling Richmond, and had one hand on the iron plate. Happening to lean towards the sounder, which is against the wall, my forehead grazed a ground wire. Immediately I received a very severe electric shock, which stunned me for an instant. An old man who was sitting facing me, and but a few feet distant, said that he saw a spark of fire jump from my forehead to the sounder.

(American Journal of Science, Article XIL, Item 6)

Telegraph systems "over charged" by Earth currents were frequently seen to produce sparks, so it is not surprising to hear of the occasional fire. Frequent mentions of this appear during the 1859, 1882, and 1921 storms; indeed, during the 1921 storm a telegraph office in Karlstad, Sweden was actually burned to the ground (Miami Herald, May 17, 1921, p. 2). Elsewhere during the 1921 storm, voltages exceeding 1000 volts were reported, and electric field strengths in the range of 20 volts/km inferred (Kappenman, 2004). These GICs continued to be a problem in more recent times:

From Newfoundland came reports that magnetism from the aurora has caused the voltage in electric circuits to vary in a range of 320 volts. Utility companies in many parts of the United States reported similar disruptions.

(New York Times, February 11, 1958, p.62.)

Through the numerous magnetic storms that disrupted telegraph services, at times for days, normal commerce was suspended at some economic cost. A snapshot of the US post-office telegraph network *c.* 1858 (Prescott, 1860) would show 50 000 miles installed (128 000 in Europe), serving 1400 stations and employing 10 000 operators. The 5 million US messages relayed each year generated $2 million in revenue. Although a 1-day storm might only cost $5000 in lost revenue to the telegraph company, such was the nature of the commerce that relied on reliable telegraph communication that the collateral impacts were quite large, especially when stock markets were involved (e.g. *Chicago Tribune*, November 18, 1882, "The Electric Storm Caused a General Dearth of News and Orders"). If any damage occurred, replacement or repair could be expensive; this was a particular issue for the tens of thousands of miles of underwater cable that had been installed by 1900:

Three of the eight transatlantic cables owned by Western Union were affected by Earth currents accompanying the aurora. Two of these were in full operation again, but the third, although not entirely out of commission, was not back to normal. The cost of repairing even a small fault in a cable in deep water would reach $200,000.

(New York Times, May 18, 1921, p. 12)

It did not take long before the call went out for scientists to figure out what was going on, and to provide some forecasts of future interruptions. In 1879, William Ellis of the Royal Greenwich Observatory informed the telegraphic community that sunspots are correlated with periods of strong auroral activity, and that the next sunspot cycle was coming to a maximum in 1882. He noted that, in the most recent years, there was little magnetic activity, and that telegraphic technology had taken a turn towards even more sensitive apparatus:

I would therefore ask whether any of the new apparatus possesses such peculiarity in their principle or construction as would render it more liable than were the older forms to be temporarily deranged or interfered with by Earth currents?

(Ellis, 1879)

Amazingly, no one seemed to care about Ellis' prognostication, at least not so that you would notice from the topics of the letters published in the Journal in the months to follow, although his idea and data were later discussed by Professor W. Adams in a lecture at the Royal Institution (Adams, 1881). The idea that you could forecast when these storms would occur seemed not only an activity of wishful thinking, but had no solid basis in fact that could be universally accepted. In 1881, it was still a matter of some controversy that a sunspot cycle and a magnetic storm cycle were causally related to each other. This was largely because there was no theoretical framework in place that convincingly coupled solar conditions to terrestrial ones. By 1882, Lord Kelvin had already proposed that no magnetic action by sunspots could propagate to Earth with anything like the strength of terrestrial magnetic storms, so the connection between sunspots and aurorae simply would have to be presumed to be unreal and an illusion. What made the argument even more compelling was that, although many more magnetic storms rattled their way through the telegraph networks over the decades, not a single one had a solar flare like Carrington's as a "kick off" event. So far as the occasional telegraph outages of the 1800s were concerned, there was still no deep understanding of why solar events should lead to the kinds of disruptions that bedeviled telegraph systems. Only a few of the "dominoes" that needed to fall in the chain could be directly observed:

Enterprising telegraph operators discovered by the 1880's that if you used a pair of lines connecting two stations to complete the circuit, rather than an Earth-ground, the telegraph signals were hardly affected during the most severe magnetic storms. It was clearly Earth itself that was the source of the problem.

(Philadelphia Enquirer, November 18, 1882, "Intense Magnetic Storm")

Meanwhile, the advance of technology did not grind to a stop merely because some occasional problems could not be deeply understood in an about-to-be antiquated technology. No sooner had telephone systems come into play in the late 1800s than these systems also suffered from unwanted GICs. Eventually the solution became the replacement of single-wire networks with dual-wire telegraph and telephone networks using better-quality wire insulation. In essence, the technology evolved and was replaced before a deep understanding of magnetic storm causation became available. There was, however, yet another technology in the making that began showing signs of GIC sensitivity.

2.3.3 Electrical power grids

On October 17, 1879, Thomas Edison created the incandescent bulb, and ran it for 40 hours. By 1882, he had inaugurated the Edison Electric Illuminating Company with the opening of the Pearl Street Station in lower-Manhattan. Within 14 months, it supplied power for 500 paying customers with 12 000 lights. Edison's "DC" grid did not use transformers. This meant that power could not be transmitted for long distances without suffering significant ohmic losses. Meanwhile, George Westinghouse bought a series of patents from Nikola Tesla to develop an AC system for power transmission. A transformer would step-up the voltage at the power plant and, at the higher voltages, less power would be dissipated in the wires operating with lower currents. At the customer's end, a step-down transformer would bring the voltage down to useable levels for applications. In 1893, Westinghouse's AC system was selected to transport power from the hydroelectric facility at Niagara Falls to Buffalo, and the AC system was nationally adopted.

Since then, North America has been thoroughly covered by a patchwork of independent, local power grids that over time have merged to form the five regional "Interconnections" now in existence. One of the largest of these is the Eastern Interconnection, which serves all states east of the Rockies excluding Texas, and also shares power with eastern Canada. Two other Interconnections, Western and Texas, are isolated from the Eastern Interconnection, so that the three behave independently. There are 10 000 generating plants producing 1000 GW. High-voltage lines (765, 500, 345 kV) are used for transmission, and low-voltage (12.4, 13.8 kV) for distribution to neighborhood power poles, where the voltage is stepped-down to 220/110 volts by a power-pole transformer. Three-phase (three hot wires plus ground) transmission moves twice as much power as single-phase (hot, neutral, and ground). In both cases, the transformers are physically connected to the Earth "ground". But, as for telegraph systems, this grounding arrangement leads to infiltration points for geomagnetically induced currents. This problem has been identified by Kappenman *et al.* (1981) and Pirjola *et al.* (2004), who have modeled this effect in increasing detail.

The electrojet current in the ionosphere generates its own magnetic field according to Faraday's law. If we approximate the electrojet as a localized current flow in space, we may use the "wire" approximation to integrate the Biot–Savart law to obtain the usual field strength for an infinitely long wire, $B = \mu I/(2\pi r)$. An electrojet current of 1 million amperes at an altitude of 100 kilometers flowing eastwards produces a north-directed magnetic field with a strength of 2000 nanotesla directly under the electrojet. Typical latitudes for the electrojet in the Northern Hemisphere are near 60°N, so at lower latitudes typical of the United States (43°N) one might expect to realize variations in the north–south component of **B** of about 200 nT relative to Earth's mean field of 60 000 nT.

Because of the time dependence of the electrojet current, the ground-level magnetic field will not be steady-state. We can determine that variations in the magnetic field strength generate ground potentials from the time rate of change of the B_x and B_y components according to Maxwell's equation, $d\mathbf{B}/dt = -\nabla \times \mathbf{E}$. This means that changes in the north–south component of the ground-level magnetic field lead to gradients in the east–west (and vertical) components of the ground-level induced electric field. The response of the Earth to this electric field is to drive "Earth currents" in the conducting upper crust to depths of up to 500 km. The specific details depend on the electrical conductivity of the soil and rock strata and the frequency distribution of the time-varying magnetic field. Computed examples for various storm episodes described by Kappenman (2004) suggest that a dB/dt of 2900 nT/minute resulted in an induced electric field of 20.0 V/km during the May 1921 geomagnetic storm. As a comparison, the March 13, 1989, storm that caused an electrical blackout in Quebec, produced $dB/dt = 800$ nT/minute and an estimated induced electric field of 7.0 V/km. Quantitatively, we might estimate that the 1921 storm was three-fold stronger than the 1989 Quebec Blackout storm, and a proportionately large power outage would have resulted had it occurred in modern times.

Geomagnetically induced currents continue to be a growing concern for continent-spanning, electric power grids. Large power transformers employed to boost the voltages for long-distance transmission are, as were old-style telegraph systems, grounded to the local Earth. These grounding methods allow pathways for pseudo-DC Earth currents to enter the transformers and disrupt their optimal performance at the nominal 50 or 60 Hertz to which they are designed (Kappenman, 2004). Although GICs in the range 1–10 amperes are not uncommon under quiet magnetospheric conditions, the exact levels depend on the kilovolt rating of the transformer. Levels as high as 200–400 amperes can be reached in 765 kV systems. This leads to temperature spikes exceeding 200 °C and the vaporization of transformer coolant fluids, leading to transformer core damage. Outright power grid outages or blackouts are, fortunately, very rare. The earliest event attributable to GICs occurred in Geneva, Switzerland, on October 31, 1903:

In Geneva, all the electrical street cars were brought to a sudden standstill, and the unexpected cessation of the electric current caused consternation at the generating works, where all efforts to discover the cause were fruitless.

(New York Times, November 2, 1903, p. 7)

Impacts on November 2, rail services invariably involve problems with electric signaling equipment, which can apparently be susceptible to GICs, as newspaper accounts seem to attest in 1921 and 1938:

The sunspot which caused the brilliant Aurora on Saturday night and the worst electrical disturbances in memory on the telegraph systems was credited with an unprecedented thing at 7:04 o'clock yesterday morning, when the entire signal and switching system of New York Central railroad below 125th Street was put out of operation, followed by a fire in the control tower at Fifty-seventh Street and Park Avenue. While all outgoing and incoming trains were stopped, the Fire Department extinguished the fire in the tower, but not until residents of many Park Avenue apartment houses were coughing and choking from the suffocating vapors which spread for blocks.

(New York Times, May 16, 1921, p. 2)

The phenomenon was also the cause of delay to express trains on the L.N.F.R. Manchester-Sheffeld line. At 7:48 PM, the signalling apparatus in both the parallel Woodhead Tunnels was found to be out of order. The working of the trains through the tunnels was stopped. An official said that the failure was apparently due to the electrical disturbances caused by the Aurora Borealis.

(The Times, January 27, 1938, p. 2)

The most famous outage occurred in Quebec on March 13, 1989, affecting over 3 million people:

The General Motors car-assembly plant in Boisbraid lost production of $6.4 million worth of automobiles. The Montreal Stock Exchange, located in Place Victoria, was forced to operate on emergency power. Most trades had to be completed manually. Sidbec-Dosco, Inc., a Quebec-owned steel company estimated yesterday's production loss at between $500,000 and $1.5 million, "All the steel that was already on the line in the hot rolling mills is scrap." Cascades, Inc., a pulp and paper company based in Kingsey Falls, said the power shutdown would cost his company between $200,000 and $300,000, the amount doesn't include salaries.

(Montreal Gazette, March 14, 1989, p. A3)

Interestingly, the Quebec Blackout, which is legendary among the space weather community and textbooks, was not covered by any major newspaper in the United States in the days following the event.

Even without transformer damage, added GICs cause saturation of the output waveform during one-half of the nominal 50–60 Hz power cycle. Consequently, magnetostriction causes mechanical expansion and contraction of the core laminae, which can easily be heard as a chattering/clattering cacophony superimposed on the normal transformer "hum" at 60 Hz. This contributes to power being drawn from the primary 60 Hz grid operating mode, and pushed into higher harmonic frequencies for which the transformer and grid are not optimized. To regulate this, reactive power is drawn from the network in an attempt to stabilize the rapidly falling voltages. This leads to increasing power regulation problems across a network.

Reactive power is the portion of the transmitted electricity that establishes and sustains the electric and magnetic fields of alternating-current equipment at the load. Reactive power must be supplied to most types of magnetic equipment, such as motors and transformers. It also must supply the reactive losses on transmission facilities. Reactive power is provided by generators, synchronous condensers, or electrostatic equipment such as capacitors, and it directly influences electric system voltage. It is usually expressed in megavars (MVAR).

Typical grids operate at about 100 MVAR, but during the Quebec Blackout MVAR stresses reached 8000 MVAR, and a superstorm event would conceivably exceed 100 000 MVAR (Kappenman, 2004). Transformers can be designed to be less friendly to GICs by adding resistive shunts to their ground lines, but these are expensive and many thousands would be needed across the North American power grid to provide mitigation for the very rare, once-a-decade storms that might be a problem. Another solution is to increase the mass of the transformer core by about 10 times, suppressing magnetostriction. However, at 200 tons, the largest 765 kV transformers, which generate the largest share of the regulation and MVAR problems, are already at the mass limit for transportation from their manufacturing site (e.g. Japan, Austria) to their operational site, so this is not an option.

2.4 Impacts of solar flares

Geomagnetic storms and the GICs they invariably spawn have been a highly visible part of the human impact equation for over 100 years, beginning with telegraph outages through to the modern era of electrical power grid blackouts. These events are often triggered by irregularities in the solar wind, high-speed coronal-hole solar wind streams and the attendant shock fronts, and of course coronal mass ejections. Meanwhile, a second category of space weather storms has been well known for nearly as long, and has as its root cause solar phenomena of a different type.

The first recorded solar flare was spotted by Richard Carrington and Rodger Hodgson on September 1, 1859, at 11:18 am (see Fig. 2.2). It was a powerful, white-light flare associated with a very large sunspot group near the solar meridian.

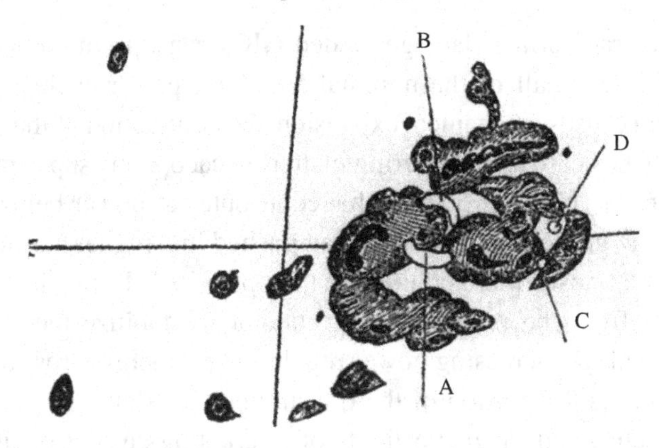

Fig. 2.2. Carrington's sketch at 11:18 GMT on September 1, 1859, of the sunspot and the lettered (white) flaring regions. (From Carrington, 1859.)

Without telescopic aid, and the patience to monitor the Sun for sunspot surveys, the flare would have been missed during its brief 5-minute luminescence at visible wavelengths. Instead, the CME and geomagnetic storm events that followed galvanized scientific interest in this phenomenon.

During the next 30–40 years, however, no further flares were ever sighted, even under nearly identical circumstances. It was only after George Ellery Hale invented the spectroheliograph, in 1892, that the far weaker Hα flares were spotted and studied in detail. The current understanding of the physics of solar flares is reviewed in Chapters 5 and 6 of this volume.

Although the origins of solar flares have been extensively discussed, what is relevant to the human impact equation is how the fluences of X-ray and particle energy arriving 8–30 minutes later can upset magnetospheric and ionospheric systems. The Carrington–Hodgson flare, other than its dazzling white light emission, presented no other impact that could be measured here on Earth, with the important exception of a magnetic disturbance captured at the Kew Magnetic Observatory at the same instant as the flare. The extra ionization in the ionospheric *D* and *E* regions (see Section 12.3.3) allows electric currents to move more easily, which then cause magnetic changes detectable at ground level. Once the flare subsided, the extra ionization vanished within a few hours and the normal geomagnetic field readings returned. These events became known as sudden ionospheric disturbances (SIDs) when detected via radio methods, and as magnetic crochets when detected on a magnetometer trace.

Radio transmissions via short-wave are severely disrupted by changes in the ionospheric *D* layer during solar flares, causing short-wave fade-outs. This mechanism was proposed by John Dellenger (1886–1962) in 1939. Meanwhile, during geomagnetic storms, particle precipitation enhances electron density in the

E and F regions over large geographic areas, and ionospheric currents cause plasma irregularities, which lead to radio wave scattering. These problems, due to separate space weather effects, became increasingly more consequential after *c.* 1930 when higher-frequency broadcasting technology became more commonplace, and increasingly more troublesome for global military communication:

Owing to unfavorable static conditions in the North Atlantic, which have handicapped wireless communication between this country and Germany, the German Government for some time has found it practically impossible to send messages here without having them pass first into the hand of the British censors in London. Germany may thus remain isolated from the rest of the world for several weeks. It has been estimated that the static disturbances now occurring often increase the wireless distance between Nauen and Sayville by the equivalent of 2,000 miles.

(New York Times, May 25, 1915, p. 3)

Sunspots delayed accounts of the Allied landing today (September 3) in Italy. Wireless technicians attributed to the spots the faulty transmission from the Mediterranean area to the United States. Dispatches piled up beside the operators as they tried various wave lengths in an effort to get through.

(New York Times, September 4, 1943, p. 2)

The United Press quoted University of Chicago scientists as calling the cosmic ray shower the greatest ever recorded. The Admiralty speculated today that cosmic disturbances caused a full-scale naval alarm for a British submarine feared missing. The submarine Acheron due to report her position at 10:05 A.M. (5:05 A.M. Eastern standard time) while on an Arctic trial, failed to make radio contact. Four hours later Acheron was heard from and the search was abandoned.

(New York Times, February 25, 1957, p. L21)

Radio Free Europe said yesterday that its engineers found no indication the Kremlin had resumed jamming it or its sister station, Radio Liberty, to block reports on demonstrations in the Soviet Union. Radio Free Europe spokesman Bob Redlich said that an effect similar to jamming could have been caused by recent increases in solar activity, which can hamper radio reception.

(Baltimore Sun, March 15, 1989, p. 4A)

Another impact on radio communications, though not directly related to solar flare X-ray emission, is the polar cap absorption (PCA) event. Polar cap absorption is caused by high-energy protons with energies exceeding 10 MeV. The high-energy protons cause ionization of the D layer so that the layer vigorously absorbs high-frequency (HF) and very-high-frequency (VHF) radio waves. Signals ranging from approximately 3 MHz through 40 MHz are attenuated by the absorption process. During these events, radio blackouts of HF and VHF radio waves (see Fig. 4.1) over the polar areas may result. Although there is no physical danger to pilots or passengers flying through the PCA environment along polar travel routes from North America to Asia (shown in Fig. 2.3), PCA interference with HF radio

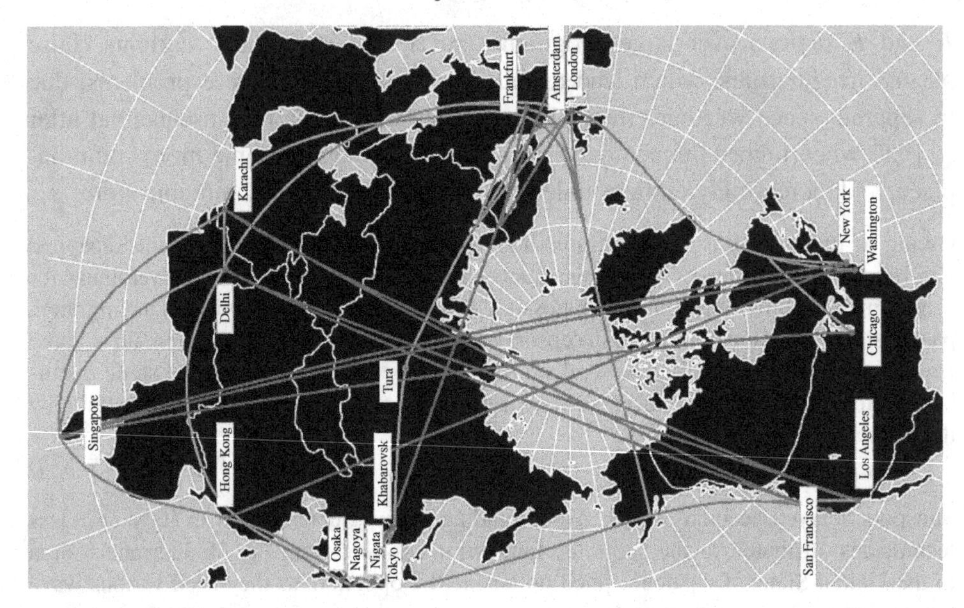

Fig. 2.3. Polar airline routes used by United Airlines *c*. 2006 carrying 1500 flights per year. (Courtesy Hank Krakowski.)

communications between aircraft and ground controllers is considered an unacceptable flight hazard. Some airlines, such as United Airlines, divert flights to lower altitudes and latitudes to escape these blackout conditions, at an unavoidable cost to flight durations and fuel economy. By 2018, United Airlines polar traffic is anticipated to involve some 2 million passengers per year and over 6000 flights. Given the realities of oil pricing and the escalating cost of Jet-A fuel, PCA events and the required diversions to lower altitudes (less fuel efficiency) will be a major financial cost to bear by most airlines (e.g. Krakowski, 2008). Under some scenarios, some fully booked flights will have to be canceled rather than rerouted or delayed to avoid flying at zero or negative profit due to rising fuel prices.

2.5 The satellite era

Solar flares, and the enhanced solar X-ray and extreme ultraviolet radiation during sunspot maximum, are capable of causing other problems. Because of the energy transported by flares to Earth, they are a significant source of heating for the upper atmosphere. Solar flares are classified as A, B, C, M, or X according to the peak flux of 100 to 800 picometer X-rays near Earth, as measured on the GOES spacecraft (see Table 5.1). Two of the largest GOES flares were the X20 events ($2\,\mathrm{mW/m^2}$) recorded on August 16, 1989, and April 2, 2001. However, these events were outshone by a flare on November 4, 2003, that was the most powerful X-ray flare ever recorded. This flare was originally classified as X28 ($2.8\,\mathrm{mW/m^2}$). However, the

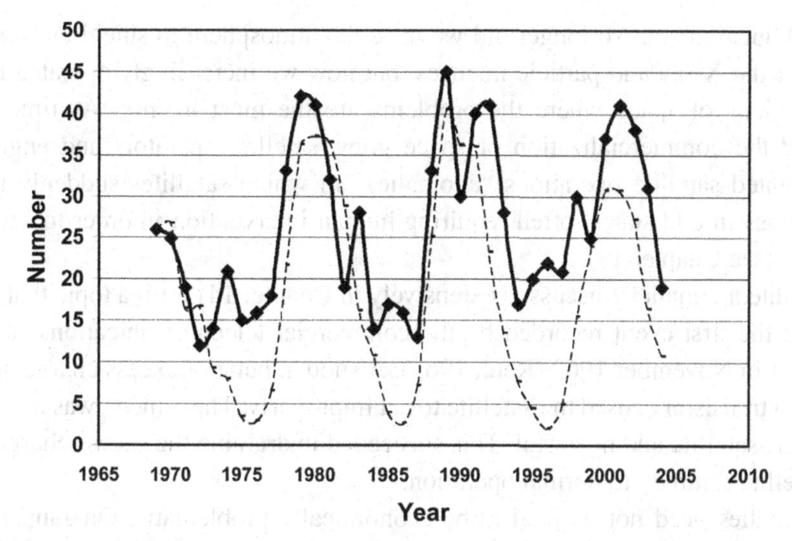

Fig. 2.4. The number of de-orbited satellites in low-Earth orbit compared to the sunspot cycle. (Odenwald *et al.*, 2005.)

GOES detectors were saturated at the peak of the flare, and it is now thought (e.g. Thomson *et al.*, 2005) that the flare was between X40 (4.0 mW/m^2) and X45 (4.5 mW/m^2), based on the influence of the event on the Earth's upper atmosphere.

As a consequence of solar X-ray and flare heating, the upper atmosphere expands. This effect is most noticeable when averaged over the solar cycle. The heating of the thermosphere from 700 °C at sunspot minimum to 1500 °C at sunspot maximum has the effect of increasing the scale height of the atmosphere. The atmosphere literally expands by hundreds of kilometers, causing the density at a fixed distance to increase nearly 50-fold. Figure 2.4 shows the impact that changes in atmospheric solar heating have had on the frequency of 1621 satellite re-entries between 1967 and 2005, based on the low-Earth orbit (LEO) satellite data from Space Track (2005).† A clear correlation is evident in which the largest numbers of re-entries occur during the peak years of sunspot cycles in 1968, 1979, 1989, and 2000.

By the end of the 1950s, it was very clear that solar radio interference was not about to go away through any reasonable means of mitigation humans could deploy. Only by anticipating when the next solar "flare" was to erupt, by monitoring sunspot activity, could the technology stay ahead of the impact at short wavelengths. By the 1950s, short-wave radio communication had become the backbone of international communication services, and hours-long blackouts became more intolerable as time went on. Then, with the launch of Explorer I in 1958, and Van Allen's detection of the "radioactivity of space", did we enter yet another

† www.space-track.org/perl/login.pl

technological arena. No longer did we have the atmosphere to shield us from the worst of the X-ray and particle fluences, but now we increasingly operated in the very regions of space where the problems are the most intense. As time went on, and the commercialization of space grew, satellite operators and engineers documented satellite operations "anomalies" in which satellites suddenly found themselves in odd states, often requiring human intervention in order to save the satellite (see Chapter 13).

Satellite anomalies (discussed extensively in Chapter 14) form a topic that dates back to the first event recorded by the commercial telecommunications satellite Telstar-1 in November 1962 (Reid, 1963). A sudden burst of excess charge on one gate of a transistor caused the satellite to act improperly. The remedy was to power-down the satellite and re-start it. This succeeded in draining the excess charge, and the satellite returned to normal operation.

Anomalies need not be fatal to be economically problematic. On January 20, 1994, the Anik E1 and E2 satellites were severely damaged by electrostatic discharges (ESDs). Although the satellites were not fatally damaged, they required up to $70 million in repair costs and lost revenue, and accrued $30 million for additional operating costs over their remaining life spans (Bedingfield *et al.*, 1996). The Anik satellite problems were apparently the result of a single ESD affecting each satellite (Stassinopoulos *et al.*, 1996), suggesting that large numbers of anomalies are not required to "take out" a satellite. If anomalies are frequent enough, however, the odds of a satellite failure must also increase, as will the workload on satellite operations. According to Futron Corporation (2003), satellite operators ordinarily spend up to 40 percent of their time on anomaly-related activities. Ferris (2001) has estimated the cost of dealing with satellite anomalies as $4300/event leading to overall operations impacts approaching $1 million per satellite per year under apparently routine space weather conditions. Anecdotal reports suggest that, during major solar storms, far higher operator activity can occur on specific satellites. For example, the GOES-7 satellite experienced 36 anomalies on October 20, 1989, during a single, severe solar storm event (Wilkinson, 1994).

Satellite anomaly statistics are not routinely available in the open literature, and there are many good (though not necessarily logical) reasons for this confidentiality. Some satellite owners may regard anomaly data as a sign of failure to maintain satellite operations at the highest state of efficiency. The data could affect a company's competitive edge, serving as an open admission that their resources are not 100% reliable to the end-user or investor. Anomaly data may reveal issues of military vulnerability when traced back to specific space assets. There is also the added economic cost for operators to keep accurate logs of these events, not required by federal regulations. Nevertheless, anomaly logging is undoubtedly a widespread activity among all satellite operators because of its value in mitigating future problems as part of an "institutional memory".

Any direct causal connection between space weather and satellite operations is usually hidden by the fact that no two satellites are identical in design or shielding, and space weather conditions vary enormously in time and space, especially when integrated over the lifetime of the satellite. This leads to perplexing cases in which neighboring satellites report very different anomalies during the same storm-time event. For example, the 3-year-old Telstar 401 located at a longitude of 97° W failed during the September 13, 1997, storm event. However, the nearby 1-year-old satellite Echostar-2 located at 119° W experienced no publicly reported problems. Both satellites were of a similar Lockheed-Martin, AS-7000 bus type.

Among the many citations of peak anomaly rates for satellites in geosynchronous Earth orbit (GEO), TDRS-1 was reported to have a rate of "several hundred per day" during the September/October 1989 solar particle event (SPE; see Rodgers *et al.*, 2000). This especially active satellite is well known to the space weather community, and typically had a baseline anomaly rate of 1–2 anomalies/day. A similar spike in anomalies was recorded by seven commercial GEO communications satellites, requiring 177 manual adjustments during the major geomagnetic storm on March 13–14, 1989 (Wilkinson, 1994) for an average rate of 13 anomalies per satellite per day. The only published long-lasting outcome of this high anomaly rate (consisting mostly of ESDs) for this extreme space weather event, and for these particular satellites, was that the solar panel output on GOES-5 was permanently reduced by 0.5 amps by energetic proton "scouring".

2.5.1 Electrostatic discharges

Cho and Nozaki (2005) investigated the frequency of ESDs on the solar panels of five LANL satellites between 1993 and 2003. During this period, LANL 1989-046 experienced 6038 ESDs/year. Although the cumulative lifetime ESD rates on solar panels can exceed 6000 events/kW over 15 years, the chances of a catastrophic satellite failure involving substantial loss of satellite power steadily increases each year. For example, in 1973, the DSCS-9431 satellite failed as a result of an ESD event. More recently, the Tempo-2 (1998) and ADEOS-2 (2003) satellites were also similarly lost. Koons *et al.* (2000) and Dorman (2005) have shown that ESDs appear to be ultimately responsible for half of all mission failures.

Wahlund *et al.* (1999) and Fennell *et al.* (2000) have studied ESD events on the Freja (MEO) and SCATHA (LEO) satellites and have found that the number of ESDs increases with increasing Kp. These results are consistent with earlier GOES-4 and 5 satellite studies by Farthing *et al.* (1982). In addition to Kp, Fennell *et al.* (2000) and Wrenn *et al.* (2002) identified a correlation between 300 keV electron fluxes and the probability of internal ESDs from the SCATHA satellite. The probability increases dramatically for electron fluxes in excess of 100 000 pfu (particle flux units or particles/cm^2 s). A similar result was found a number of years earlier by Vampola (1987). At daily total fluences of 10^{12} electrons/cm^2 the probability of an ESD occurring on a satellite exponentially reaches 100% (e.g. Baker, 2000). Vampola *et al.* (1992) analyzed CRRES data and identified an onset threshold of 10^6 electrons/cm^2 for deep dielectric discharges. According to a study by Baker (2001) the Anik W1 satellite power system failure occurred when the GOES-8 spacecraft measured 2 MeV electron fluences of 30 billion/cm^2.

Studies have begun to identify the sources of these bursts of super-MeV electrons that lead to ESDs (mainly in the range from $L = 3$ to 5 R_E) with a resonant process occurring between the turbulence spectrum of the impacting solar wind and CME, and the natural gyrofrequencies of particles within the magnetosphere (Barbara Giles (personal communication); Polar/GSFC). At the South Pole, observations of dawn chorus (a magnetospheric electromagnetic phenomenon at radio wavelengths) by Horne *et al.* (2005: British Antarctic Survey) suggest that the EM/plasma waves produced by dawn chorus may be the mechanism that accelerates electrons to MeV energies and makes them suitable agents for generating ESDs when they encounter satellites. At these energies, these high-energy "killer" electrons are capable of penetrating deeply into satellites and contributing to internal charge buildups that eventually discharge and lead to the ESD events themselves.

2.5.2 Energetic particles and solar proton events

Energetic protons (solar proton events: SPEs) are also a cause of satellite anomalies, in particular those identified as single-event upsets or SEUs. According to Brekke (2004), the solid-state recorder of the Solar and Heliospheric Observatory (SOHO) also records SEU events as memory bit-flips, which are corrected by error detection and correction (EDAC) algorithms, and the SEU counter is periodically monitored and reset. During the 1996–2003 period, a clear indication of cosmic-ray correlation was found with an amplitude of 1 SEU/minute near solar minimum and 0.5 SEU/minute near solar maximum. SPEs also produce a clear and consistent signal in the SEU frequencies, typically increasing the SEU rates up to 60 SEU/min for the strongest events (e.g. the Bastille Day Storm of July 14, 2000). For 1996–2003, three events caused SOHO to enter spacecraft safe mode, causing major disruptions of science operations. There were five events when battery discharge regulators went off-line, and seven events in which science instrument boxes were switched off. SEUs also affect the attitude control and pointing system, which employs a star tracker. There were 54 occurrences during 1996–9 when the satellite had to be manually repositioned, with a loss of science data, to recover a nominal pointing state.

Since the initial prediction by Wallmark and Marcus (1962) of software upsets caused by cosmic rays, and the subsequent discovery of "soft errors" on the Intelsat IV satellite by Binder *et al.* (1975) a large body of research has grown up over the years attempting to predict SEU rates for various space conditions (e.g. Adams, Jr. *et al.*, 1981), such as CREME96 (Tylka *et al.*, 1997). These models have their limitations, however. Hoyos *et al.* (2004) compared CREME96 calculations for SEU rates on the SOHO satellite's 2 GBy solid-state recorder during the Bastille Day Storm SPE event. The actual rate was in the range 30 – 60 SEU/minute while the predicted rates ranged from 10 to 3000 SEU/minute depending on the specific assumptions made about the energy spectrum of the SPE itself. The largest modeling uncertainties involve the satellite shielding, assumed device sensitive volume, and the critical charge threshold above which an SEU would be triggered by the accumulated SPE charge buildup.

Space weather conditions can generate thousands of ESDs and SEUs in a satellite each year. The vast majority of these events go unnoticed by satellite operators and lead only to annoying data glitches (harmlessly removed by software) or momentary spikes in a particular satellite housekeeping parameter that do not exceed pre-set operational limits and are ignored as well. Occasionally, an ESD or SEU can lead to an actual operational anomaly in the satellite (phantom commands, etc.) requiring operator intervention.

The problem with both ESD and SEU related anomalies is that the event rates for both categories are extremely large and often exceed 1000 events per satellite per year. These events are caused by an even larger flux of particles passing through the satellite volume and involve up to 10^{15} particles per satellite per year. This larger flux (mostly galactic cosmic rays) can plausibly be modeled once specific environmental and satellite parameters are specified. However, as we see in Fig. 2.5, even the most common and non-critical satellite anomalies only represent 1–2 events per satellite per year. At this rate, the mean time to failure is about 200 years per satellite for an anomaly that proves fatal. These fatality rates from individual particle events represent a reduction by a factor $1{:}10^{15}$ of the original annual fluences of space weather conditions. This means that predicting normal anomaly rates, let alone the still-rarer, fatal events, is based on small-number statistics, which are perhaps impossible to model and estimate convincingly. To do so would require model precisions ($\sim 1{:}10^{15}$) that have heretofore only been approached by such calculational archetypes as relativistic quantum electrodynamics ($1{:}10^{10}$). Conversion of particle fluences/fluxes to satellite anomalies expected for a specific satellite requires a predictive model with accuracies that are essentially unattainable. Instead, the best that space

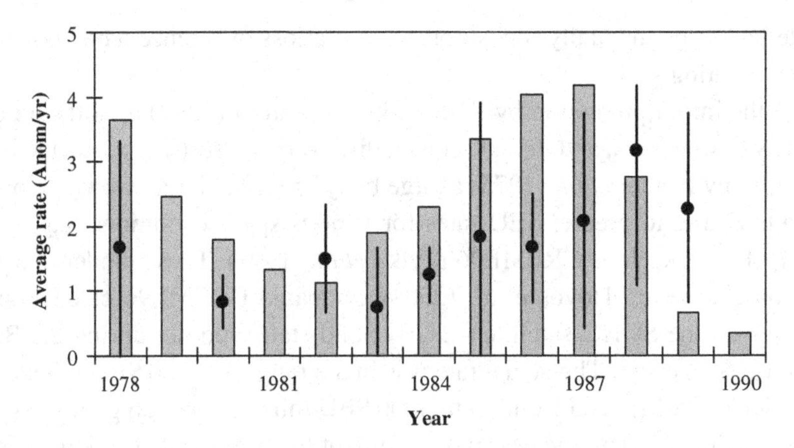

Fig. 2.5. Satellite anomaly rates for satellites in geosynchronous Earth orbit listed in the NGDC anomaly archive. The reference histogram is the annual cosmic ray flux at climax, re-scaled to show phase. (From Odenwald, 2009.)

weather modeling can hope to provide the satellite community is a probabilistic indicator of which days or hours during a space weather event are likely to be the most troublesome for the average satellite, so that satellite operators can be on the alert for an increase in anomalies. Satellite owners, on the other hand, should be knowledgeable about the susceptibility of their particular satellite bus type to a given storm intensity and energy spectrum, which are parameters that may be a reasonable goal for space weather modeling efforts to forecast.

Meanwhile, it is still possible to successfully model the annual power degradation of a satellite due to the collective action of billions of cosmic rays scouring its solar panels. Figure 2.6 shows the power loss (solid line) experienced by the Solar and Heliospheric Observatory (SOHO) since its launch in December 1996 (Brekke, 2004). It shows a steady decline in solar panel output by about 2% per year. Typically, solar panels are oversized at the beginning-of-life to allow for this inevitable degradation, so that satellite systems will be able to survive to the planned satellite end-of-life with adequate power. A list of SPEs between 1976 and 1990 was obtained from the National Geophysical Data Center (NGDC) satellite study (Wilkinson, 1994) and compared with the SOHO power degradation events. Very approximately from the SOHO data, above a threshold flux of 10 000 particles/cm^2 s (i.e. 10 000 pfu), the percentage power loss due to individual SPE events is about 1% per 10 000 pfu, so that a 20 000 pfu event will cause a 2% power reduction, a 30 000 pfu event will cause a 3% power reduction, etc. (see Section 14.4 for further discussion).

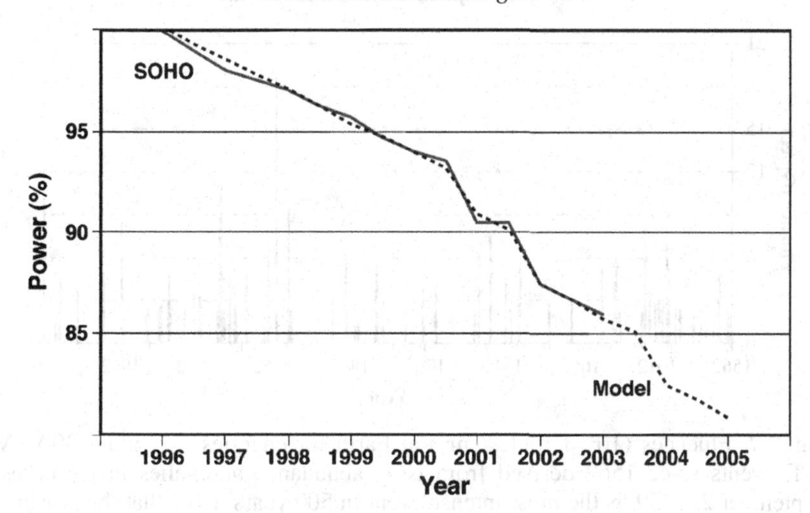

Fig. 2.6. Solar panel power reduction for SOHO from 1996 to 2003. The model (dashed line) includes a 2% GCR decline per year and the effects of known SPE events. The large dip near the center of the curve was the Bastille Day event in July, 2000. A second drop occurred during the intense SPEs on November 4 and 23, 2001. (From Odenwald *et al.*, 2005.)

2.6 How bad can it get?

Satellite designers use sophisticated tools to assess radiation hazards under worst case conditions (e.g. the August 1972 and March 1991 events). However, studies of extreme space weather conditions suggest that the period since 1960 may not be typical.

According to McCracken *et al.* (2001) and Townsend *et al.* (2006), the 1859 storm was the most extreme event observed in the last 500 years, as shown in Fig. 2.7. Smart *et al.* (2006) have identified the Carrington SPE through atmospheric nitrite abundance anomalies in ice core samples. It appears to have been caused by a shock passage past Earth due to solar activity near the Sun's central meridian. No ^{10}Be was detected (see Vol. III) so the spectrum was soft for protons exceeding 30 MeV, for which the fluence was estimated to be about 1.9×10^{10} cm^{-2}. This fluence is about four times greater than the canonical worst-case August 4, 1972, event, which was the strongest solar event during the satellite era. Since 1561, there have been 19 SPEs more intense than the August 1972 SPE (approximately one every 30 years). Nevertheless, a once-a-century or once-a-millennium superstorm would be a disaster to our satellite resources, and its probability of occurrence is far higher than other risks, such as the next San Francisco earthquake, which are taken more seriously.

Odenwald and Green (2007) developed a series of simple Monte Carlo models to assess the economic impacts on this resource caused by various scenarios of

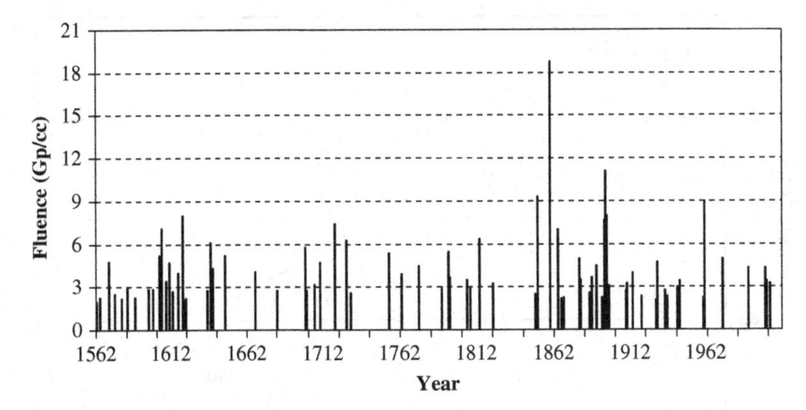

Fig. 2.7. Fluences (10^9 particles/cm^3) at Earth for energies exceeding 30 MeV SPE events since 1562 derived from NOy abundance anomalies in ice cores. September 2, 1859, is the most intense event in 500 years. Note that the November 15, 1960, event (fifth from the right) predates the commercial satellite era that started around 1980. (Data from McCracken *et al.*, 2001.)

superstorm events possible during the sunspot cycle between 2008 and 2018. From a detailed model for transponder capacity and leasing for the entire GEO satellite population available each year from 2008 to 2018, they investigated the total revenue loss over the entire solar cycle as a function of superstorm onset year and intensity.

Figure 2.8 is the result of this calculation. Each of the 1000 models that were run is represented by a single point characterized by the total revenue generated by the satellite population (vertical axis), and the intensity of the storm (horizontal axis). The onset year of the storm at each intensity produces the vertical dispersion of the points at each intensity level, with early onsets in 2008 defining the lower boundary of the dispersion, and later onsets near 2018 defining the upper envelope of the dispersion. This is consistent with earlier superstorm onsets significantly reducing by up to 10 years the total revenue that can be generated from satellites during the 2008–18 period, and later onsets only reducing the revenue from the satellite population by a few years.

The modeling suggested that, by 2018, models that did not include a superstorm event achieved a cumulative transponder revenue of approximately $230 billion (in 2006 dollars). This is seen by examing the far-left edge of Fig. 2.8, where the simulations are based on very weak storms (e.g. near 50 000 pfu), and for which the average models yield $230 billion with a range from $220 billion to $250 billion.

As the strength of the superstorm increases, the far-right edge of Fig. 2.8 shows that the cumulative revenue by 2018 declines to an average level of $205 billion

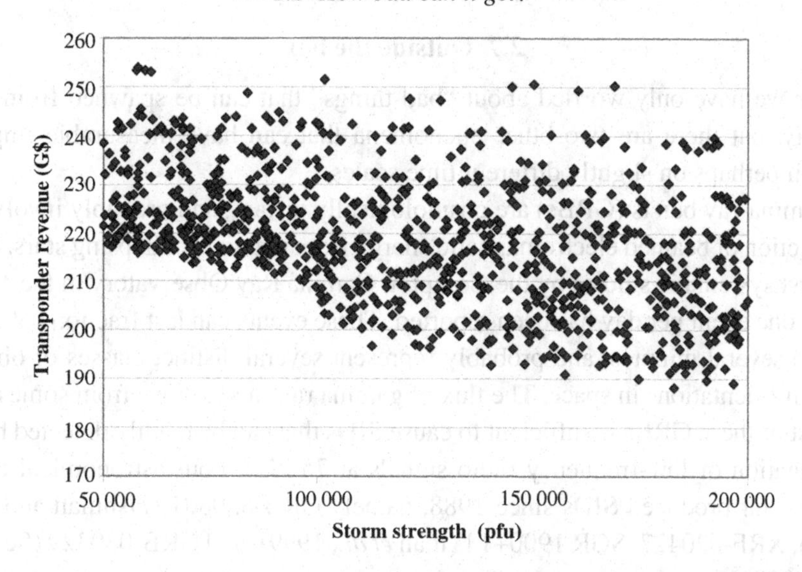

Fig. 2.8. Monte Carlo model results for cumulative transponder revenue versus storm strength in units of particle flux units. The 1859 superstorm event would appear at about 150 000 pfu. The largest SPE events during cycles 22 and 23 equalled 45 000 pfu. The vertical dispersion is due to the variation of the onset year from 2008 to 2018. (From Odenwald and Green, 2007.)

with a range from $230 to $190 billion depending on the onset year. This represents an average decline by about $25 billion over the range of the calculation. Under the worst-case conditions of the Monte Carlo modeling, where the storm arrives early in the solar cycle, the revenues generated would be found along the lower envelope of the data points, and also indicate a loss of $30 billion. As the difference between the upper and lower envelopes of the models indicates (the difference between the high-revenue models near 50 000 pfu and low-revenue models near 200 000 pfu), considerably larger profit losses approaching $60 billion may occur.

If the events of the 1859 Carrington–Hodgson storm serve as a guide, the scope of a contemporary superstorm will most certainly be an awesome event, but one that the vast majority of our satellite resources may reasonably be expected to survive. Perhaps the best indication we have to suggest that our satellites are robust is that we have experienced about a dozen major space weather events since 1980, with no widespread loss of satellite resources. Satellites appear to have evolved into highly resistant systems that seem able to survive the significant storms of the last 20 years. Nevertheless, the occasional once-a-century "superstorm" remains a cause for concern much as a recurrence of the Great San Francisco Earthquake might be in other financial sectors.

2.7 Outside the box

So far we have only worried about "bad things" that can be spawned from solar activity, but there are two other phenomena that can have measurable impacts, though perhaps on slightly different time scales.

Gamma-ray bursts (GRBs) are cosmologically distant, and probably involve the production of beamed electromagnetic energy from massive, collapsing stars. Since the first systematic studies by the Compton Gamma Ray Observatory in the 1990s, about one event per day has been reported. These events can last fractions of a second to several minutes, and probably represent several distinct classes of objects, or their orientations in space. The flux of gamma rays and X-rays from some of the largest of these GRBs is sufficient to cause SIDs that can be readily detected by the propagation of low-frequency radio signals at 75 kHz. Four astronomical events have so far produced SIDs since 1988, namely GRB-830801 (Fishman and Inan, 1988), XRF-020427, SGR 1900+14 (Inan *et al.*, 1999), and GRB-030329 (Schnoor *et al.*, 2003).

On December 27, 2004, a new object called a magnetar, identified as SGR 1806-20, erupted within the Milky Way about 50 000 light years from Earth and produced a flare detected by both ground- and space-based observatories (Palmer *et al.*, 2005; see Fig. 5.14). The radiation blitzed at least 15 spacecraft, knocking their instruments off-scale whether or not they were pointing in the magnetar's direction. One Russian satellite, Coronas-F, detected gamma rays that had scattered off the Moon. The flare also increased the ionization of the daytime ionosphere for five minutes, which was noticed via its disrupting effect on long-wavelength radio communications (Mandea and Balasis, 2006). Nevertheless, these cosmic events are extremely rare and are inconsequential due to their short time scales.

2.8 Space weather awareness

The last 200 years of reporting human and space weather impacts has passed through many stages and fads as new technological problems revealed themselves, and old ideas passed out of scientific fashion. A historical study by Odenwald (2007) shows that earlier accounts in the newspapers were more inclined to report problems because the impacts directly affected how news stories were circulated (telegraph, wireless, teletype). Figure 2.9 shows that, during the 1950s post-war period, there was a broad array of communications media available to transmit and receive news stories, so that any given space weather event caused little interruption in the flow of information. Therefore, the obvious impacts were more subtle and difficult to apprehend. When this is coupled with the lack of timely information on satellite, power grid, or radio anomalies from institutions locked in intense

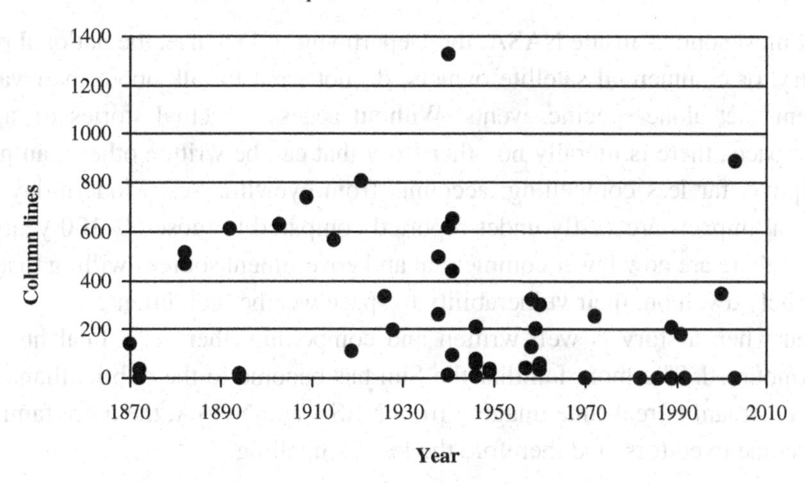

Fig. 2.9. The total number of space weather-related column lines per year published by the *New York Times, Chicago Tribune, Boston Globe, Los Angeles Times* and the *Washington Post* for space storms occurring each year with geomagnetic AA index exceeding 150, showing a sharp decline in coverage after *c.* 1950. (From Odenwald, 2007.)

competitive struggles and attempting to demonstrate high reliability, the present dearth in impact reporting is understandable.

Yet, considering that there are far more technological connections to space weather conditions today than there were 50 years ago, it is puzzling that the "Golden Years" of space weather reportage have indeed passed, and the low-level reporting of today is almost universally considered normal.

An important factor that may explain the lost ground in space weather news is that most reporters do not take the time to dig for the facts themselves. So, if no press release is available on a particular science story, a reporter has to have a strong motivation to write the story from scratch. The Halloween Storm of October 29, 2003, was supported by press releases from NASA, and pro-active work by the NASA Public Affairs Office, and NOAA's Space Weather Prediction Center, through their e-mail distribution network that reaches over 1000 news reporters. Yet despite this effort to get the story out to the news media, the actual number of column-lines that resulted (~888 lines) is only comparable to similar severe storms reported between 1870 and 1950. Moreover, unlike the earlier newspaper stories, the modern-day stories did not report on specific impacts, but focused primarily on the more scientific elements of the phenomenon, reflecting the content provided by the NASA and NOAA/SEC press releases.

A reporter also needs to be savvy enough to call science contacts who can provide new information, in a timely manner, for a novel story that has to be written under a two-hour deadline. This runs into the predictable problem in space weather, in that most sources inside NASA, the Department of Defense, the national power industry, or commercial satellite owners, do not want to talk about their various problems, let alone specific events. Without access to actual stories of significant impacts, there is literally no other story that can be written other than purely descriptive, far less compelling, accounts from eyewitnesses. Modern-day technological impacts are vastly under-reported compared to those 50–100 years ago because there are now fewer commercial and government sources willing to admit, or publicly dwell on, their vulnerability to space weather conditions.

Even when a story is well written and compelling, there is a final hurdle to be surmounted. The more familiar the Sun has become to the public, thanks to a constant stream of real-time imagery from NASA and NOAA, the more familiar it has become to editors, and therefore the less compelling.

2.9 Space weather forecasting

How much advanced warning can we get for a superstorm? Following the lead pioneered by Carrington and Hodgson, it is reasonable to assume that ground-based solar patrols at observatories around the world will catch the brilliant white-light flare. Prior to this stage, the larger-than-typical sunspot that spawned the superstorm event will have been monitored continuously. Much as for the October 2003 "Halloween Storm" we can expect considerable activity from this region in the lead-up to a major outburst. These individual events will trigger X-ray flares and subsequent short-wave outages and ionospheric disturbances.

NASA satellites have been able to see CMEs at the time of launch, and can estimate their arrival times with reasonable accuracy. A superstorm CME will no doubt take 17 hours or less to reach Earth. However, there will be no way to determine the magnetic field orientation of the interplanetary CME (ICME) en route since none of the satellites (assuming ACE is unavailable) will be equipped with magnetometers to study the ICME *in situ*. This means we will have little advance warning of the CME's geo-effectiveness.

For satellites, however, it will not be the ICME that will provide the greatest immediate problem. X-ray flares and SPEs are the most destructive phenomena from the standpoint of satellite operations and power. For large events, CMEs are launched at the same time as the flares occur, so it is important to predict the flare onset event before a CME is even detected.

X-ray and solar magnetic field imagers have provided 24-hour advanced notice on large flares. SPEs, often associated with strong X-ray flares, reach Earth within hours after the X-ray flare is detected. For hard SPE spectra, however, the particles are highly relativistic and arrive at virtually the same time as the X-ray burst, so the onset of the X-ray flare is already too late for SPE mitigation to be effective.

The most famous SPE occurred on November 4, 2003. Forecasts posted by NOAA's Space Weather Prediction Center show that, in the five days prior to the event, the 24-hour X-class flare probabilities were 50%, 40%, 35%, 75%, 75%. On the day of the flare, the probability remained at 75%, and in the three days after, dropped to 10%, 1%, 1%. Although the X-ray class had been anticipated, neither the exact luminosity nor the day of the event had been anticipated for Active Region 10486. Including false-positives, the best estimate of 24-hour forecast reliability is about 80% for X-class flares. This implies a one in five chance that a major X-class superflare is unanticipated.

3

In-situ detection of energetic particles

GEORGE GLOECKLER

3.1 Introduction

Space physics started over 50 years ago with the launches on October 4 and November 3, 1957, of Sputnik I and II by the Soviet Union, and Explorer 1 and 3 by the United States on January 31 and March 26, 1958. Explorers 1 and 3 carried James Van Allen's Geiger counters. He had hoped to measure the low-energy portion of the differential intensity of cosmic rays (particles with energies of hundreds of MeV of non-terrestrial origin), which could not be observed from the ground or with balloons because of atmospheric absorption. Yet the few minutes of data, received whenever the satellite was within range of the tracking station, were puzzling. At low geocentric distances of the 2500 km apogee orbit of Explorer 1 particle intensities were as expected. However, at higher altitudes the intensity dropped to zero. Explorer 3 carried a tape recorder and solved the puzzle. Again, the particle intensity or counting rate was normal at low altitudes, but then it increased rapidly until the maximum transmittable level of 128 counts/s was reached. A constant rate of 128 counts/s was observed for some time but then suddenly dropped to zero, recovering to 128 counts/s later and finally returning to normal at low altitudes. The actual counting rate was increasing rapidly far beyond the 128 counts/s limit, reaching such high rates that the Geiger counter "froze", that is discharged so frequently that it could not properly recover between counts, yielding pulses too small to be detected by the circuitry used. Earth's radiation belt was discovered. The discovery of the radiation belts was confirmed with measurements by more sophisticated instruments on Sputnik III, launched May 12, 1958.

Another, somewhat more recent example was the discovery in the early 1970s of cosmic rays whose composition was vastly different from that of cosmic rays of galactic or even solar origin. The discovery of these so-called anomalous cosmic rays (ACRs) was made with instruments that made measurements at energies below

Heliophysics: Space Storms and Radiation: Causes and Effects, eds. Carolus J. Schrijver and George L. Siscoe. Published by Cambridge University Press.

about 10 to 20 MeV/nucleon. The ACRs were distinguished by a composition that was highly enriched in O, N, Ne, and He, and were eventually explained as, and much later shown to be, accelerated pickup ions created by ionization of interstellar neutrals in the heliosphere, which themselves were discovered by new instruments that made measurements in an even lower energy range and in a new region of space.

Van Allen's discovery of the radiation belts and the discovery of the ACRs provide several important lessons. First, if one explores new territory, be it a new region of space (as was the case with Explorer 1 and 3), a new energy range, or a new type of measurement, one is bound to discover the unexpected. This has happened over and over during the last 50 years and is bound to continue. Second, new observations and discoveries (as was the case with the Earth's radiation belts) drive our understanding of the physics of the system (in this case magnetospheric physics) through theory and modeling, which then make predictions that can be tested by further observations. Successful theories and models that properly explain observations have led to our current understanding of heliophysics as discussed in these volumes. As new observations are made, theories and models must be modified, or replaced by new ones that can account for all of the observations. Finally, there is no ideal instrument or detection device. All have their limitations. If you get strange or completely unexpected results, first suspect your instrument and dig hard to find the cause for the peculiar results. Only after you have convinced yourself that it was not an instrumental effect should you dare to believe your observations, which could well be some new discovery.

The fact that energetic particles are observed in all regions of space explored so far (see Chapters 7, 9, and 11 for details) implies that acceleration mechanisms that energize these particles abound. What these mechanisms are and where the acceleration takes place are key questions that are being pursued. Particles are accelerated in various regions of the heliosphere, heliosheath, and the galaxy beyond and some are transported from the acceleration site to the observer. During this propagation the energy spectra are modified from the original spectra at the acceleration site.

Figure 3.1 shows the differential intensities (see Eq. (3.3) for its definition) of protons and oxygen as a function of energy/nucleon, measured in the heliosphere at 1 AU. Before the space age there were no particle measurements below ~300 MeV per nucleon. All of the spectra shown below these energies were observed using satellite instruments and are classified as either quasi-steady state (quiet, shown in black) or transient (shown in grey). The energy range and intensity (dynamic) range over which these spectra need to be observed are huge, over 6 and 19 decades, respectively, and at least three different techniques are required to obtain these measurements, as will be discussed. In addition to the

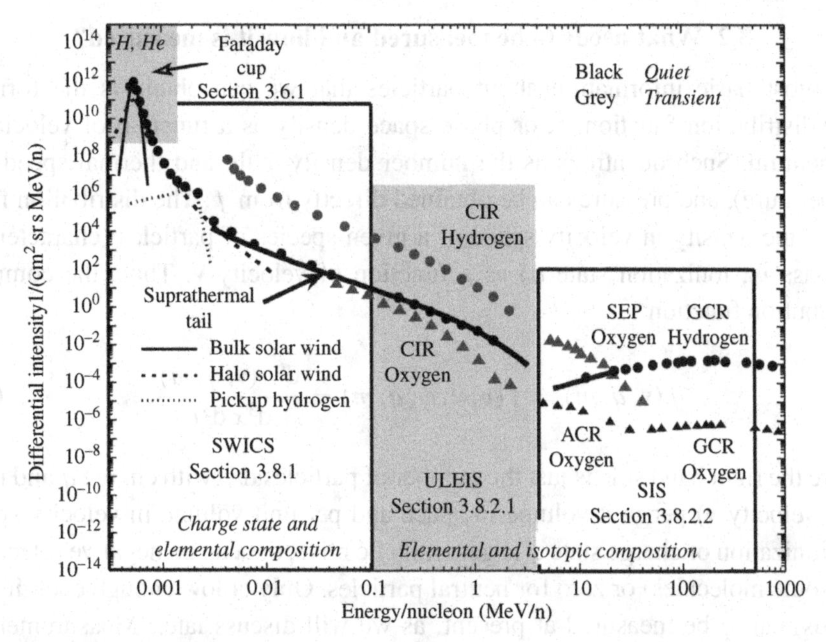

Fig. 3.1. Differential intensities of protons (circles) and oxygen ions (triangles) for various particle populations (see Chapter 9 for a description of the populations) observed in the heliosphere. The four boxes represent the energy and dynamic ranges of four instruments (described in Section 3.8) that measure the energy spectra and composition of ions and nuclei (see Fig. 9.1). Symbols: black, quasi-steady state; grey, transient states.

differential intensity spectra, the elemental, isotopic, and charge state composition of the particles provides important clues concerning their origin, as well as physical and chemical characteristics of the acceleration site region. For example, the anomalous composition of the ACRs, similar to that of interstellar pickup ions, pointed to their origin as ionized and then accelerated interstellar gas. But only now, with Voyager observations in the heliosheath, are we beginning to understand where and how ACRs are accelerated. Similarly, measurements of the charge state of the solar wind provide information about coronal temperatures.

In this chapter, I describe how plasma, suprathermal, and energetic (~300 eV per charge to 300 MeV/nucleon) particles are detected and their properties measured, beginning with a description of what needs to be measured and how it is measured. Next I discuss briefly how particles interact with matter. Following this, simple detectors and devices that are commonly used in present space instruments are described, and finally I discuss four instruments that together cover the entire energy (over 6 orders of magnitude) and dynamic range (over 19 orders of magnitude) shown in Fig. 3.1, ending with a description of techniques to detect and measure spectra and composition of energetic neutral atoms (ENAs).

3.2 What needs to be measured and how it is measured?

The most basic information about particles that we can obtain is the form of their distribution function f, or phase space density as a function of velocity or momentum. Such quantities as the number density, bulk and thermal speeds (or temperature), and pressure can be obtained directly from f. The distribution function is the density in velocity space of a given species of particles (characterized by mass m, ionization state q) as a function of velocity $\mathbf{v}$. Thus, the complete distribution function

$$f(\mathbf{v}, q, m) = f(v, \theta, \phi, q, m) = \frac{\mathrm{d}^6 n(\mathbf{v}, q, m)}{\mathrm{d}^3 x \, \mathrm{d}^3 v}, \tag{3.1}$$

where the right hand side is just the number of particles dn (with charge q and mass m at velocity $\mathbf{v}$) per unit volume in space and per unit volume in velocity space. The ionization or charge state q is generally positive, but can be negative (electrons and some molecules) or zero for neutral particles. Only at low enough energies (or speeds) can q be measured at present, as we will discuss later. Measurement of mass m or, more often, the nuclear charge Z, of nuclei, establishes the identity of the particle (e.g. proton, carbon, iron, H_2O molecule, etc.). A precise measurement of mass is necessary to identify molecules (e.g. $m = 18$ amu – atomic mass units – for water molecules) or isotopes (e.g. ^{13}C) of chemical elements.

To illustrate how the distribution function may be measured, consider an ideal detector looking in a direction (θ, ϕ) measuring the distribution function of particles of mass m and charge q (e.g. protons with $m = q = 1$). The area of the detector perpendicular to its look direction is dA, and it has a narrow field of view of dΩ steradian centered on its look direction (see Fig. 3.2). Our ideal detector counts (detects) each particle with 100% efficiency and measures its kinetic energy, $E = mv^2/2$. From the definition of the distribution function

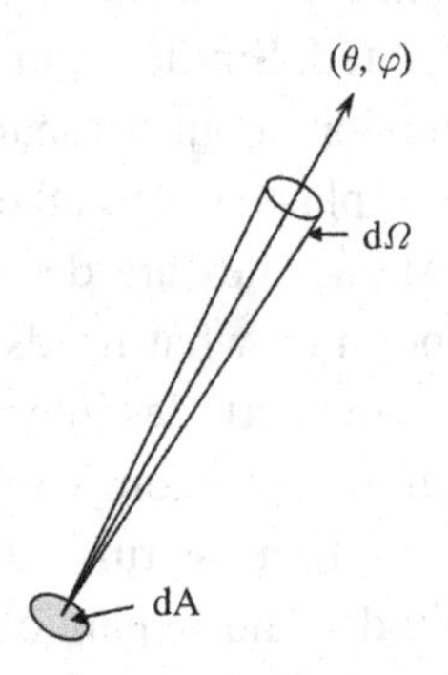

Fig. 3.2. Ideal small detector of area dA with a narrow conical field of view dΩ centered on its look direction.

$$f(v, \theta, \phi, q, m) = [\mathrm{d}n(v, q, m)/\mathrm{d}t]/[\mathrm{d}A\,\mathrm{d}\Omega v^3\,\mathrm{d}v]$$
$$= [m^2/2] \cdot [\mathrm{d}n(E, q, m)/\mathrm{d}t]/[\mathrm{d}A\,\mathrm{d}\Omega E\,\mathrm{d}E], \qquad (3.2)$$

where $\mathrm{d}n(E, q, m)/\mathrm{d}t$ is the counting rate of particles with mass m and charge q in the energy range between E and $E + \mathrm{d}E$, and $\mathrm{d}n(v, q, m)/\mathrm{d}A \cdot \mathrm{d}r = [\mathrm{d}n(v, q, m)/\mathrm{d}t]/[\mathrm{d}A \cdot \mathrm{d}r/\mathrm{d}t] = [\mathrm{d}n(v, q, m)/\mathrm{d}t]/[v\,\mathrm{d}A]$, where $\mathrm{d}r$ is along the look direction of the detector. Equation (3.2) can also be expressed in terms of the differential intensity $\mathrm{d}j/\mathrm{d}E$, which is commonly used to describe the energy spectra of particles with energies above several hundred keV/nucleon and is defined to be

$$\mathrm{d}j/\mathrm{d}E = [\mathrm{d}n(E, q, m)/\mathrm{d}t]/[\mathrm{d}A\,\mathrm{d}\Omega\,\mathrm{d}E], \qquad (3.3)$$

becoming

$$f(v, \theta, \phi, q, m) = M^2(m_\mathrm{p}c^2)^2[\mathrm{d}j/\mathrm{d}E(E, \theta, \phi, q, m)]/[2c^4E], \qquad (3.4)$$

where M is the atomic mass number and $m_\mathrm{p}c^2$ is the proton rest energy. The common units for the differential intensity are [(number of particles)/(s cm^2 sr MeV/nucleon)], or [(number of particles)/(s cm^2 sr keV/nucleon)]. If the units for the distribution function are [(number of particles)(s^3/km^6)] then, with $\mathcal{E}$ in keV/nucleon,

$$f(v, \theta, \phi, q, m) = 0.545\,[\mathrm{d}j/\mathrm{d}\mathcal{E}]/\mathcal{E}. \qquad (3.5)$$

Now we can calculate $f(v, \theta, \phi, q, m)$ from the counting rate $\mathrm{d}n(E, q, m)/\mathrm{d}t$ of (m, q) particles with energies between E and $E + \mathrm{d}E$ in a detector of area $\mathrm{d}A$ looking in direction (θ, ϕ) with a narrow field of view (FOV) $\mathrm{d}\Omega$. The ion speed v (in units of km/s) is computed from its measured energy E (in units of keV/nucleon) using

$$v \approx 438\sqrt{\mathcal{E}}. \qquad (3.6)$$

3.3 Geometrical factor of detectors

The quantity $\mathrm{d}A\,\mathrm{d}\Omega$ is the geometrical factor (GF) of the infinitesimally small detector shown in Fig. 3.2. The common units are cm^2 steradian (cm^2 sr). For a detector system of finite area and FOV the GF is calculated by integrating over all possible look directions that intersect the detector area as we shall illustrate with a few simple examples. For more complicated detector geometries, Monte Carlo methods or forward models are used. The forward model is especially useful when coordinate frame transformations (e.g. from the solar wind to the spacecraft frame) are required.

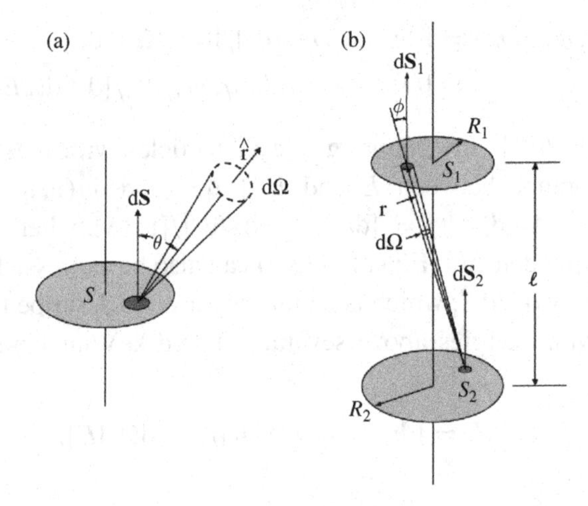

Fig. 3.3. Relationship (a) between **dS**, dΩ, **r**, and Θ for a single planar detector and (b) between $\mathbf{dS}_1$, $\mathbf{dS}_2$, dΩ, **r**, and ϕ for a two-detector telescope.

The simplest finite detector is shown in Fig. 3.3a. It is a planar detector of area S with a conical FOV along its symmetry axis, with cone angle Θ. The geometrical factor

$$\mathrm{GF} = \int_S \int_\Omega (\hat{\mathbf{r}} \cdot \mathbf{dS}) \mathrm{d}\Omega = \pi S[1 - \cos^2(\Theta)]. \tag{3.7}$$

A particle telescope, consisting of two planar detectors of areas S_1 and S_2, separated by l, is shown in Fig. 3.3b. Its geometrical factor is

$$\mathrm{GF}_2 = \int_{S_2} \int_\Omega (\hat{\mathbf{r}} \cdot \mathbf{dS}_2) \mathrm{d}\Omega. \tag{3.8}$$

The domain in Ω is limited by the top detector. Integrating Eq. (3.8), the geometrical factor of a telescope consisting of two circular planar detectors is

$$\mathrm{GF}_2 = (\pi^2/2) \cdot \left[R_1^2 + R_2^2 + l^2 - \sqrt{((R_1^2 + R_2^2 + l^2)^2 - 4R_1^2 R_2^2)} \right], \tag{3.9}$$

where R_1 and R_2 are the radii of detectors 1 and 2 respectively. In the limiting case where $R_1 \ll l$ and $R_2 \ll l$, Eq. (3.9) reduces to

$$\mathrm{GF}_2 \approx \pi^2 R_1^2 R_2^2 / l^2 = S_1 S_2 / l^2. \tag{3.10}$$

3.4 Energy loss of energetic particles by ionization

When an energetic particle (or photon) passes through a slab of material it gives up some of its energy to eject electrons and ions from the surfaces of the slab and to ionize or excite some of the atoms or molecules of the material, or create

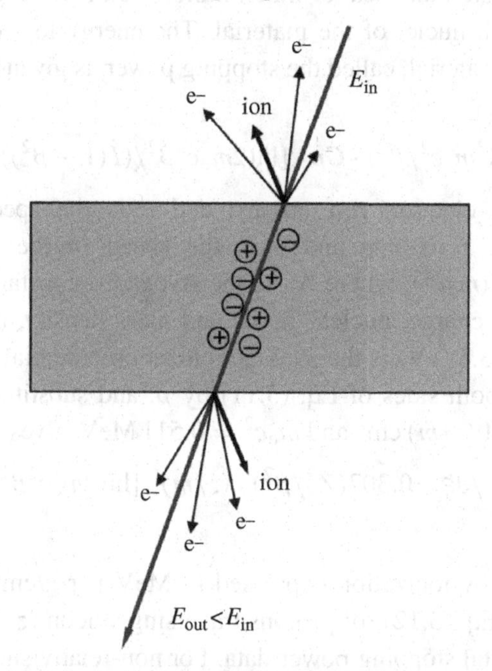

Fig. 3.4. Interaction of an energetic particle with a slab of material showing the ejection of electrons and ions from the two surfaces and ionization of matter inside the slab. The average charge state of the particle leaving the slab will depend primarily on its energy and not on its initial degree of ionization. For example, an energetic neutral atom will most likely be charged when exiting the slab.

charge carriers. This is illustrated in Fig. 3.4. We make use of these interactions of energetic particles with matter in order to detect the particles and measure the energy they deposit in the material. For example, electrons (called secondary electrons) ejected from solid surfaces are used to detect particles and thin foils are used to ionize neutral atoms.

Particles are detected in gas counters (such as the Geiger counters used by Van Allen) by recording current pulses generated by electrons and ions that are created by ionization of some of the gas (see Section 3.5.1). The energy a particle deposits in the material may be measured by determining the total charge liberated in the material by ionization. This process of losing energy is called energy loss by ionization.

Heavy charged particles of mass M (atomic mass units) and nuclear charge Z (electron charge units) interact with the material they are traversing by distant collisions with electrons in the material. In each interaction the heavy particle loses a small amount of its energy, but deviates little from its original straight path (small angle scattering). Energetic electrons, however, lose a much larger fraction of their energy in each collision and scatter much more, even reversing their direction in direct collisions with nuclei of the material. The energy loss $\mathrm{d}E$ in traversing a thickness $\mathrm{d}x$ of the material, called the stopping power, is given by (see, e.g. Rossi, 1961, for derivation)

$$-\mathrm{d}E/\mathrm{d}x = (4Z^2 m_e c^2/\beta^2) \cdot C\rho \cdot [\ln(2m_e c^2 \beta^2/(\bar{I}(1-\beta^2))) - \beta^2], \qquad (3.11)$$

where $m_e c^2$ is the electron rest energy, and c is the speed of light. Z is the nuclear charge, β is v/c and v is the speed of the energetic particle. $C = \pi N_0 e^4 (z/m)/(m_e^2 c^4)$, where N_0 is the Avogadro constant, z, m, and ρ are the average nuclear charge, nuclear mass, and mass density, respectively, of the material, and $\bar{I}$ ($\sim$13.5z eV) is the average ionization potential of electrons in the material. Dividing both sides of Eq. (3.11) by ρ, and substituting the numerical values for $C = 0.150(z/m)\,\mathrm{cm}^2$ and $m_e c^2 = 0.511$ MeV, gives

$$-\mathrm{d}E/(\rho\,\mathrm{d}x) \equiv -\mathrm{d}E/\mathrm{d}\xi = 0.307(Z^2/\beta^2) \cdot (z/m) \cdot [\ln\big(2m_e c^2\beta^2/(\bar{I}(1-\beta^2))\big) - \beta^2] \qquad (3.12)$$

for the energy loss by ionization expressed in MeV per g/cm^2 units. Figure 3.5 compares a plot of Eq. (3.12) for protons traversing silicon ($z = 14$) with a curve based on experimental stopping power data. For non-relativistic particles ($v \ll c$) the stopping power for protons ($Z = 1$) reduces to

$$-\mathrm{d}E/\mathrm{d}\xi = 0.153(m_p c^2/E) \cdot (z/m) \cdot [11.93 - \ln(z) - \ln(m_p c^2/E)]. \qquad (3.13)$$

The dominant energy dependence is contained in the first term of Eqs. (3.12) and (3.13). The dependence on the material (z) is in the $\ln(z)$ term and is weak since, to a good approximation, $(z/m) \approx 0.5$. For the same particle speed (or energy per

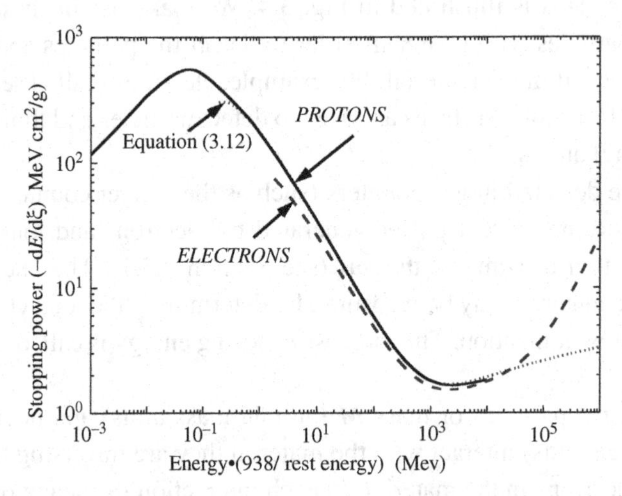

Fig. 3.5. Comparison of the stopping power for protons traversing silicon computed using Eq. (3.12) with that based on experimental stopping power data. (Berger *et al.*, http://physics.nist.gov/PhysRefData/Star/Text/contents.html.)

nucleon E/A) the energy loss per g/cm^2 will be less in heavy (high z) material such as copper, than in light materials such as silicon.

An important feature of Eqs. (3.12) and (3.13) is that above a few hundred keV/nucleon, the ionization loss for non-relativistic heavy ($Z > 1$) particles at some given speed can be obtained by multiplying the stopping power of protons at the same speed by Z^2. At energies below a few hundred keV/nucleon laboratory data are used to determine the stopping power for different particles in various materials. Figure 3.6 shows the energy dependence of the stopping power for the indicated elements in carbon.

From the stopping power equation one can compute the range of a particle with energy/nucleon E_0/A. The range R is defined to be the distance the particle travels in an absorber before stopping and losing all of its energy.

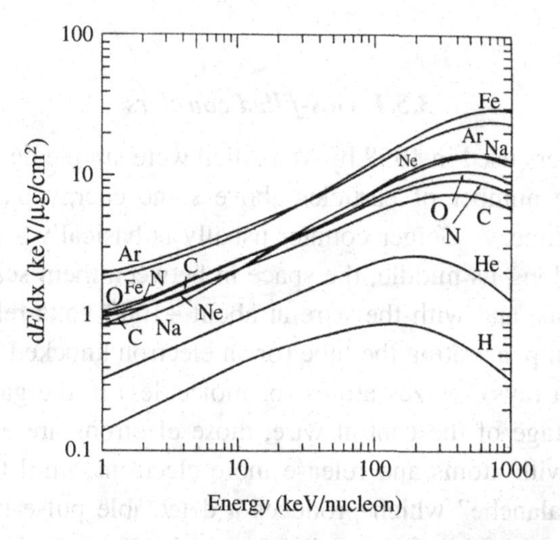

Fig. 3.6. Energy loss in thin carbon foils for energetic particles of various elements. (From Gloeckler *et al.*, 1980, private communication.)

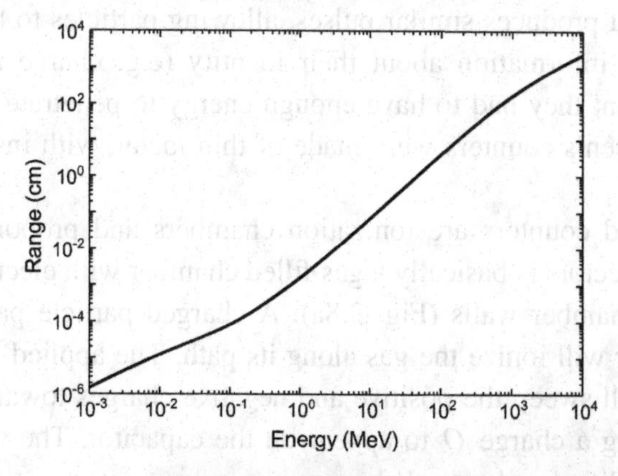

Fig. 3.7. Range of protons in silicon, based on experimental data compiled by Berger *et al.* (See http://physics.nist.gov/PhysRefData/Star/Text/contents.html.)

$$R = -\int_{E_0/A}^{0} (\mathrm{d}E/\mathrm{d}\xi)^{-1}\mathrm{d}E. \tag{3.14}$$

Figure 3.7 shows the dependence of R on energy for protons in silicon.

3.5 Simple particle detectors

Simple particle detectors can be used to simply count particles in some energy range or, in some cases, to also measure some of their characteristics, such as energy. However, a combination of simple detectors and other devices such as energy analyzers can be configured to not only count particles efficiently but also to measure many of their properties such as ionization state and direction of travel.

3.5.1 Gas-filled counters

The Geiger counters used in 1958 by Van Allen were simple detectors. They basically counted the number of particles above some energy passing through the counter per unit time. A Geiger counter usually is basically a metal tube with a thin metal wire along its middle, the space in between them sealed off and filled with a suitable gas, and with the wire at about +1000 volts relative to the tube. An ion or electron penetrating the tube (or an electron knocked out of the wall by X-rays or gamma rays) ionizes atoms (or molecules) in the gas. Because of the high positive voltage of the central wire, those electrons are accelerated toward it. They collide with atoms and release more electrons, until the process snowballs into an "avalanche" which produces a detectable pulse of current. With a suitable filling gas, the flow of electricity stops by itself, otherwise the electrical circuitry is designed to stop it. The Geiger tube is a "counter" because every particle passing through it produces similar pulses, allowing particles to be electronically counted with no information about their identity (e.g. charge and/or mass) or energy (except that they had to have enough energy to penetrate the walls of the counter). Van Allen's counters were made of thin metal, with insulating plugs at the ends.

Other gas-filled counters are ionization chambers and proportional counters. Each of these detectors is basically a gas-filled chamber with electrodes well insulated from the chamber walls (Fig. 3.8a). A charged particle passing through a gas-filled counter will ionize the gas along its path. The applied voltage between the electrodes will sweep the positive and negative charges toward the respective electrodes causing a charge Q to appear on the capacitor. The resulting voltage pulse is then amplified and recorded.

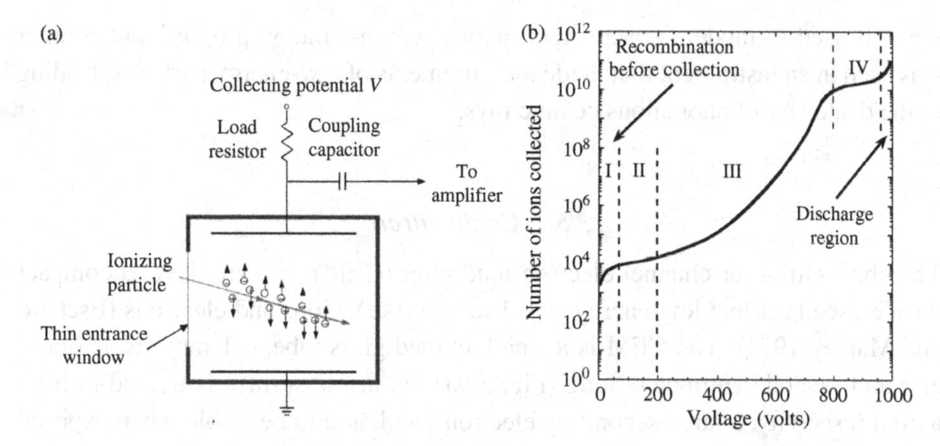

Fig. 3.8. (a) Schematic diagram of a gaseous ionization chamber. (b) Collected charge or pulse height as a function of applied voltage, illustrating the regions of operation of a gas chamber. Regions II, III, and IV correspond respectively to the ionization chamber, the proportional counter, and the Geiger counter mode of operation. (From Gloeckler, 1970.)

The amount of charge that is collected, and thus the amplitude of the pulse, will depend on the applied voltage V as is illustrated in Fig. 3.8b. For low V (region I) the electric field is so low that recombination of charges takes place long before they can drift apart and be collected, and no pulses are produced. As V is increased to a level where loss of ions due to recombination becomes small (region II), the charge collected $Q = Ne$, where N is the number electron–ion pairs produced by the incident energetic particle. Region II is called the ionization chamber region. At higher voltages (region III) the charge collected is increased by a factor of M through gas multiplication. The electric field is now strong enough that electrons released in the primary ionization will be sufficiently accelerated to produce additional ionization and thus add to the total charge. The multiplication factor M is independent of the initial ionization at the onset of region III and the output pulses will be proportional to the amount of primary ionization. This is the proportional counter range. The proportionality (constant M) breaks down at the high voltage range of region III where the pulses tend to be nearly independent of the initial ionization. This range is called the region of limited proportionality. In region IV, the Geiger region, the collected charge is completely independent of the initial ionization, governed only by the characteristics (e.g. filling gas) of the chamber and electronics. Proportional counters not only measure the flux (the number of particles per second per unit area) of particles that pass through the device, but also the energy deposited by these particles. Furthermore, their relatively large sensitive area and thin entrance window allows measurements of low fluxes and low energies. Even though gas-filled counters were invented more than 75 years ago, they are still used in modern space experiments. A thin-window proportional counter was used in an instrument that made measurements of low-energy particles, leading to the discovery of anomalous cosmic rays.

3.5.2 Channeltron

The channeltron, or channel electron multiplier (CEM), is a simple and compact device used to detect low-energy (~0.1 to ~100 keV) ions and electrons (Eschard and Manley 1971). The CEM is a small, curved glass tube, ~1 mm inside diameter and several centimeters long (Fig. 3.9a). Its inside surface is treated to have a high resistivity, a large secondary electron yield, and to be stable when exposed to air. When a several kilovolts potential is imposed from one end to the other, a single electron produced at the low-potential end will be accelerated down the tube and, at every collision with the tube wall, will produce several secondary electrons that continue that process. Straight CEMs are unstable at gains of more than 10^4 because of ion feedback caused by cascading electrons that ionize some of the residual gas inside the device toward the high-potential end of the devices. Positive ions are then accelerated toward the low-potential input, where they could initiate a new cascade producing ion feedback, making the straight tube CEM unstable. In curved (or "c" shaped, spiral, helical) CEMs, any ion that is created will strike the tube wall before gaining sufficient energy to re-initiate an electron cascade.

CEMs require a 2 to 4 kV bias voltage to achieve gains of 10^6 to $>10^8$. Higher bias voltages are not used, to minimize background counts and to maintain CEM lifetime. For a fixed voltage, the gain depends on length to diameter ratio, which sets the number of secondary electron multiplications. The gain and detection efficiency are weakly dependent on the incident particle mass and energy above some threshold energy. Incident electrons require several hundred eV and ions require several keV to obtain good detection efficiency. Uniform gain is observed for count

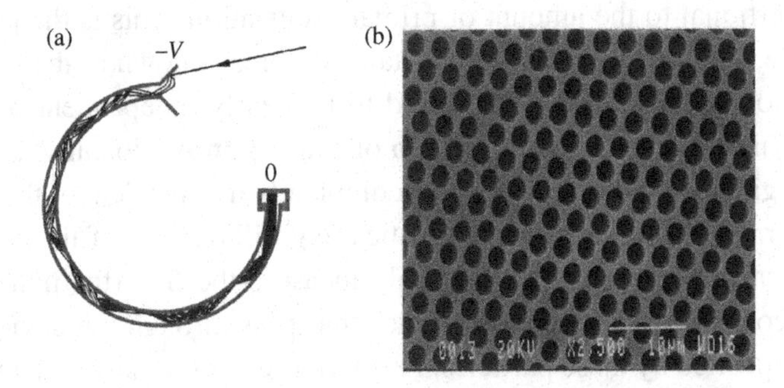

Fig. 3.9. (a) Cross section of a channeltron. (b) Photograph of a section of the surface of a microchannel plate. (Courtesy of Burle Industries, Inc.)

rates whose pulse current is over 10% of the nominal CEM bias current. Operating pressures below 10^{-5} mbar are required. Background rates decrease significantly as pressures drop below 10^{-6} mbar.

CEMs are generally operated in pulse saturated counting mode with gains of $\sim 10^7$ to 10^8. Detection thresholds can then be set to some small fraction of the nominal gain to eliminate dark current counts. CEMs can also be operated in analog mode, where variations in the CEM current are used to measure the particle flux rather than counting individual events. A review of CEMs can be found in Kurz (1979).

3.5.3 Microchannel plate

Microchannel plate (MCP) detectors began replacing CEMs for low-energy ion and electron detection in most plasma instruments beginning in the mid 1980s. MCPs are compact front-end particle or photon detectors with a high signal to noise ratio allowing individual event counting and relatively low background rates ($< 1\ \mathrm{cm}^{-2}\ \mathrm{s}^{-1}$). MCPs require operating pressures of less than 10^{-5} mbar. A most important feature of MCPs that is now commonly used is that they can also be used to obtain a spatial distribution of ions.

As in CEMs, electron multiplication, produced by voltage bias across a resistive glass tube, generates an electron cascade through secondary electron production (Wisa, 1979). MCPs consist of an array of microscopic glass tubes (typically 12 to 25 μm spacing), hexagonally packed (Fig. 3.9b) and sliced as thin wafers (0.5 to 1.0 mm thick) with typical microchannel length to diameter (l/d) ratios between 40:1 and 80:1. The wafers are treated by high-temperature (250–450 °C) reduction in a hydrogen atmosphere to produce a resistive coating along the microchannels, and the top and bottom surfaces are metallized (Lampton, 1981).

MCP wafers are sliced at a small bias angle (typically 8–12°) relative to the microchannel axis. They are stacked in pairs (chevron configuration) or in triplets (Z-stack), with adjacent wafers having opposite bias angles (Fig. 3.10) to prevent ion feedback. Typical bias voltages are ~1 kV per plate and typical gains are ~1000 per plate.

Chevron configurations (Fig. 3.10a) produce charge pulses of $\sim 10^6$ electrons, which are readily detected with charge-sensitive preamplifiers. The voltage required for these gains of 10^6 depends upon the l/d ratio and the micropore diameter. The l/d ratio generally sets the number of electron multiplications for a fixed bias voltage. However, at high gains the micropores will saturate and the saturated gain will depend on pore diameter and the number of micropores that fire. A chevron pair of 1 mm plates with $l/d = 80/1$ will typically require several hundred volts more bias than a pair of 1 mm plates with $l/d = 40/1$. In a Z-stack

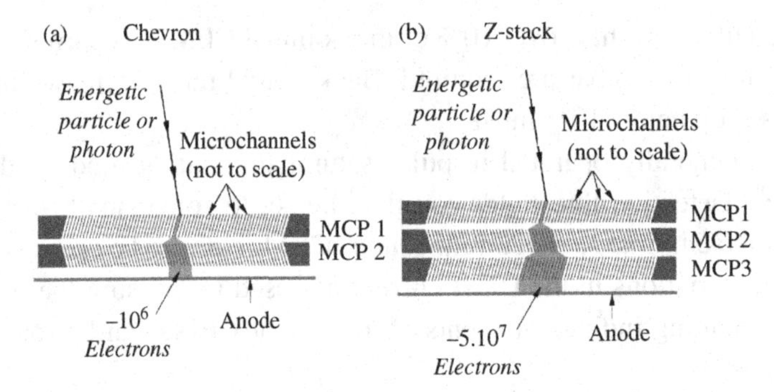

Fig. 3.10. Schematic drawing of MCPs in a chevron (a) and in a Z-stack arrangement (b). A single energetic particle enters one microchannel of MCP 1 and initiates a secondary electron avalanche in that channel. The cloud of electrons leaving that MCP then spreads to several microchannels of the next plate starting avalanches in each of these channels. About 10^6 electrons are collected by the anode behind the chevron configuration and $\sim 5 \times 10^7$ in the Z-stack arrangement.

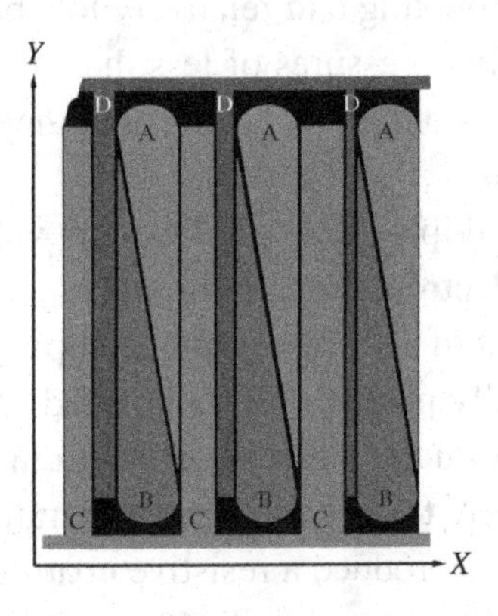

Fig. 3.11. A portion of a four-electrode wedge and strip anode. Black regions are insulators, the rest conductors. Each set of wedges and strips, labeled A, B, C, and D, is tied together to four separate conductors on the back of the anode. The X and Y positions are given by the ratio of signals: $X = C/(C + D)$ and $Y = A/(A + B)$.

configuration (Fig. 3.10b), more microchannels in the back plate fire, resulting in a much higher gain of $\sim 5 \times 10^7$, but lower imaging resolution.

Charge pulses from MCPs can be registered in several ways. The highest counting rates with coarse position resolution are obtained using discrete anodes with separate preamplifiers. Trading off high counting rates for better position resolution, resistive anodes (Lampton and Carlson, 1979; Fraser and Mathieson, 1981), delay line anodes (Lampton *et al.*, 1987; Siegmund *et al.*, 1994), or wedge and strip anodes (Martin *et al.*, 1981) are used. Wedge and strip anodes (Fig. 3.11), in particular, offer extremely fine position sensing, approaching that of the microchannels. These position sensing or imaging systems typically allow count rates of 10^5 to 10^6 counts per second, depending upon the spatial resolution desired. To obtain counting rates as high as 10^8 counts per second requires use of discrete anodes, each with its own preamplifier.

3.5.4 Semiconductor detectors

Semiconductor or solid-state detectors (SSDs) are basically semiconductor ionization counters. An electric field is set up within a semiconductor crystal by a voltage applied across opposite faces of the crystal. Radiation (particles or energetic photons) penetrating the crystal produces secondary electrons, which, in turn, produce further ionization in the crystal until no electron has enough energy left to ionize the atoms in the crystal any further. The electric field inside the crystal sweeps out the liberated charges, their number being proportional to the energy lost by the primary radiation.

There are several advantages that solid-state detectors have over gas-filled counters. First, because of the higher density of the detection material, SSDs are far more effective in stopping particles, and hence absorbing their energy, than are gas-filled counters. Second, solid-state detectors can be made thin and essentially windowless. This is an important advantage in their application as $\mathrm{d}E/\mathrm{d}x$ detectors that will be discussed later. Finally, because it takes on average less energy for the production of an ion pair, the measurement of the incident particle energy in solid-state detectors is more precise than in gas-filled or scintillation counters described in Section 3.5.5.

Energetic particles will not only ionize some of the detector material, which results in the production of ion–hole pairs, but will also lose some of their energy by interacting with the nuclei of the material. This fraction of energy loss is not measured, resulting in what is called the total energy defect. The energy defect becomes significant for incident energies below $\sim$1 MeV/nucleon and is most pronounced for heavy particles as shown in Fig. 3.12.

Semiconductor material must have certain special properties before it can be used in the fabrication of detectors. The resistivity of the material must be high enough to support the required electric field without creating an excessive leakage current. The charge carriers must have a high mobility in the crystal in order to be collected in a reasonably short time. The trapping rate for the carriers must be low, in order to maintain a high efficiency for charge collection and to reduce space-charge buildup within the crystal. The average energy required to produce an ion pair must be low, in order to give better energy resolution. Finally, the crystal must be uniform and its properties stable over prolonged time periods.

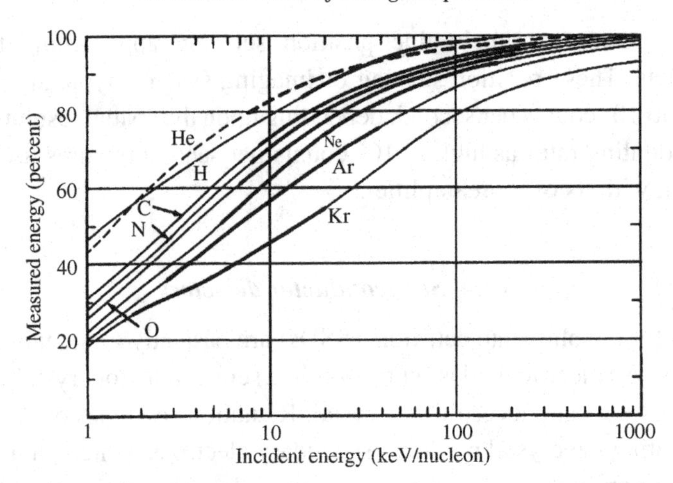

Fig. 3.12. Total energy defect in solid-state detectors with 19 μg/cm^2 gold front surface for ions H through Kr between 1 and 1000 keV/nucleon. Bold curves are least squares fits of measurements. (From Ipavich *et al.*, 1978.)

Silicon and germanium semiconductors have properties that make them useful in the fabrication of radiation detectors. For room-temperature operation only silicon can be used, therefore we shall primarily discuss detectors produced from this material. To understand how a semiconductor detector works, it is necessary to examine first some basic properties of the material. In the band theory of solids, a semiconductor is represented by a completely filled valence band and an empty conduction band, separated by a forbidden energy gap. Inorganic scintillation crystals (described below) are described in a similar fashion, except that their energy gap is larger than in semiconductors. At absolute-zero temperature the semiconductor has infinite resistivity since the conduction band is empty. As the temperature is increased, a number of electrons jump the energy gap. The vacancies in the conduction band behave very much as positively charged carriers whose mass and mobility depend on the properties of the crystal. One refers to these positive carriers as holes. In a perfectly pure or intrinsic semiconductor the number n of electrons per cm^3 in the conduction band is equal to the number p of holes in the valence band and is given by

$$n = p \approx 10^{19} \exp(-E_g/2kT), \tag{3.15}$$

where E_g is the energy gap in eV, k is the Boltzmann constant, and T is the absolute temperature. At room temperature, $kT = 0.026\,\text{eV}$, $E_g \approx 1\,\text{eV}$ in silicon, and $n \approx 2 \times 10^{10}\,\text{cm}^{-3}$ (compare this with $\sim 2 \times 10^{22}$ for a metal). It can be shown that the theoretical resistivity of silicon at room temperature is around 10^5 ohm cm. This resistivity is not enough to support the required electric fields without admitting large currents through the crystal.

There is another way in which electron–hole pairs may be produced in a semiconductor. A charged particle passing through the crystal loses energy by ionization, and in the collision process lifts electrons from the valence band or deeper-lying electronic bands to the conduction band or higher-lying unoccupied bands. The highly excited states quickly decay ($\sim 10^{-12}$ s) until the electrons are near the bottom of the conduction band and the holes are near the top of the valence band. Decay of these highly excited states produces additional electron–hole pairs. This process is shown schematically in Fig. 3.13. On average, for every 3.6 eV a particle (or photon) loses in the crystal, one electron–hole pair is produced. Assuming that the resistivity of the material is high enough (for example, T is low enough) to support an applied electric field E, the electrons and holes will drift toward the respective electrodes with drift velocities, $v_n = \mu_n E$ and $v_p = \mu_p E$, respectively. For silicon the electron and hole mobilities, μ_n and μ_p, are $\sim$1500 and $\sim$500 cm^2/V s, respectively. Under ideal conditions the total charge collected will be Ne, where N is the number of electron–hole pairs released. There are, however, several processes that tend to remove a fraction of the carriers and reduce the total charge collected. One way to lose carriers is by recombination of electron–hole pairs. Another way is for carriers to be trapped long enough to prevent their collection at the electrodes.

Incomplete charge collection in a detector is undesirable for several reasons. First, pulse size may become a function of applied voltage. Second, detector energy resolution is degraded. Third, if charge carriers are not quickly removed, internal electric fields are set up opposing the applied field. This further reduces charge-collection efficiency.

It is not possible at this time to grow pure or intrinsic silicon; there are always small amounts of impurity atoms that produce energy levels in the forbidden

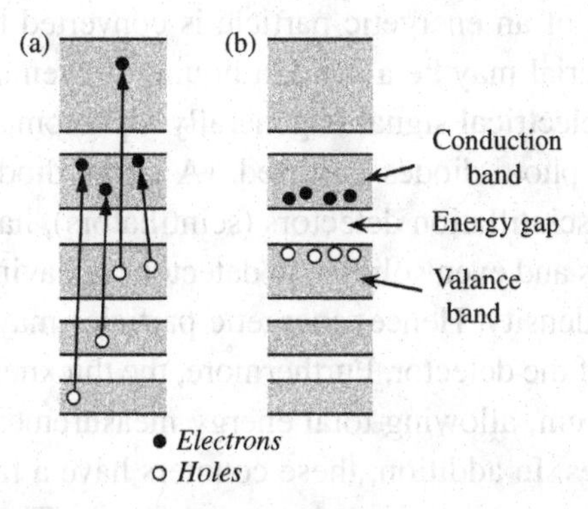

Fig. 3.13. Production of electrons and holes in a semiconductor resulting from passage of an energetic particle. (a) Initial conditions with highly excited states; (b) residual excitation after about 10^{-12} s.

gap and contribute additional charge carriers. Impurity atoms can supply either additional electrons (donor impurities) or additional holes (acceptor impurities); both types of impurity may be present in the same crystal at the same time. Donors have energy levels in the forbidden zone near the conduction band; acceptors near the valence band. Because the energy gap between impurity levels and the conduction (valence) band is small, the number of free carriers is large (see Eq. 3.15). These extra carriers add significantly to the conductivity of the material.

Semiconductor material is classified as either n-type or p-type depending on whether it contains more donor or more acceptor impurities. If the number of donors N_D is equal to the number of acceptors N_A the crystal behaves like an intrinsic semiconductor. This property is used to increase the resistivity of a crystal by doping or compensating it with the appropriate impurity to make $N_D \approx N_A$. Unfortunately, compensation also increases the number of traps. This, as we saw earlier, is undesirable since the presence of traps tends to reduce collection efficiency and carrier lifetime and enhance the buildup of space charge.

The most successful semiconductor detectors are those employing a p-n junction under reverse bias (see Gloeckler, 1970). In this manner a charge-deficient region, called the depletion region, is produced in the vicinity of the junction interface. Semiconductor detectors that are most commonly used today are ion-implant detectors (Mattsson and Holmén, 1971) and lithium-drifted detectors (Gibbons and Blamires, 1965). Further descriptions of solid-state radiation detectors may be found in Kleinknecht (1998), Lutz (1999), Knoll (2000), and Spieler (2005).

3.5.5 Scintillation detectors

In the scintillation detector, the light emitted by atoms excited directly or indirectly by the passage of an energetic particle is converted to an electrical signal. The scintillation material may be a solid, a liquid, or even a gas. The device converting light into an electrical signal is generally a photomultiplier tube, although in some applications photo-diodes are used. (A photo-diode is essentially a thin window SSD). Solid scintillation detectors (scintillators), have the obvious advantage over gas counters and even solid-state detectors of having a detecting medium with a high electron density. Hence, energetic particles may be stopped in a relatively small volume of the detector. Furthermore, the thickness of solid scintillators can be as large as 10 cm, allowing total energy measurements of several hundred MeV/nucleon particles. In addition, these counters have a faster response and can operate at much higher counting rates than gas counters. Their main disadvantages are due to problems of light collection and conversion to electrical signals, which require high voltage supplies, phototube, etc. The use of photo-diodes tends to

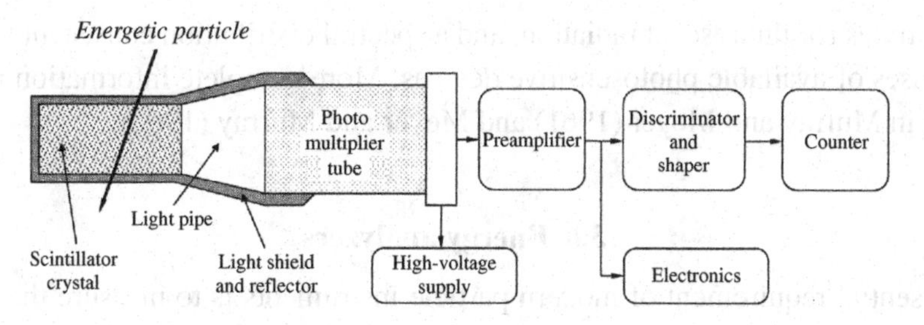

Fig. 3.14. Diagram of a scintillation detector with typical electronic circuitry.

reduce some of these problems. A typical scintillation counter assembly is shown in Fig. 3.14.

The initial step leading to the formation of an output pulse is the interaction of a charged particle with the atoms of the scintillator, causing the particle to give up part or all of its energy. Only a small fraction η of the lost energy, ΔE, is converted into light, which is characterized by some spectral distribution with E_{ph}, the average energy of the emitted photons. The number of photons N_0 produced is $N_o = \eta \Delta E / E_{\mathrm{ph}}$. The number of photons that reach the photocathode of the tube is less than N_0 because of losses in reflections from the walls and in passage through the crystal and optical coupling, or light pipe. Thus, $N = \omega N_0$ where ω is the optical efficiency. If a photomultiplier tube is used, the ejection of an electron from the photocathode by a photon occurs with a probability θ, the photoelectrons reach the first dynode with an efficiency $\mu < 1$ and, by secondary emission, the number of electrons is increased K times. Combining all these processes leads to a charge Q produced at the photomultiplier tube output given by

$$Q = \Delta E(\eta\omega\theta\mu K / E_{\mathrm{ph}})e, \tag{3.16}$$

where e is the charge of an electron. Clearly the constant of proportionality between Q and ΔE depends critically on the type of scintillator, light coupling, photomultiplier tube, etc., that are used. In general, it is almost impossible to calculate this constant. For some of the more efficient inorganic scintillator assemblies it takes about 50 to 300 eV to produce one photoelectron, compared to 3.6 eV in silicon SSDs.

Two types of scintillators are used: organic and inorganic. The organic scintillator material consists of aromatic hydrocarbons whose molecules contain benzene-ring structures along with various non-aromatic substitutions. Examples of organic scintillators are anthracene and stilbene crystals. Inorganic scintillators are crystals of inorganic salts, primarily alkali halides containing small amounts of impurities as activators for light emission. The desirable properties of a good scintillator are high conversion efficiency η, high transparency to its fluorescent radiation, short decay times for fluorescent radiation, and a spectral distribution consistent with the responses of available photosensitive devices. More complete information may be found in Murray and Meyer (1961) and Meyer and Murray (1962).

3.6 Energy analyzers

An essential requirement of modern particle instruments is to measure the energy of particles. As discussed previously, proportional counters, semiconductor detectors, and scintillation counters are able to do this provided that the particles have enough energy to penetrate beyond the "windows" of these devices and deposit enough of their energy to exceed the threshold energy of the detector. The average deposited energy is then measured with some uncertainty ΔE. For silicon surface barrier detectors the minimum energy required is about 30–40 keV. Recently it has become feasible to use small, passively cooled, thin-window silicon semiconductor detectors, coupled to state-of-the-art front-end electronics, to detect electrons down to ~2 keV, and protons (and neutral hydrogen atoms) down to ~5 keV (Wang *et al.*, 2008).

To measure or, more accurately stated, select the energy of low-energy charged particles, typically below ~100 keV, various types of electrostatic and magnetic energy analyzers are used in space instruments. Magnetic analyzers are bulky and heavy compared to electrostatic analyzers, and they require magnetic shielding to prevent interference with magnetic field measurements on the same spacecraft. These disadvantages have limited their use and I will not describe them here. Electrostatic analyzers often use high voltages that require special care to prevent discharges.

There are three classes of electrostatic analyzers (ESAs): (1) retarding potential analyzers, best suited for particles below ~1 keV/*e* (keV per unit charge), (2) spherical and cylindrical section analyzers, used for particles with energies between ~0.1 and ~20 keV/*e* and (3) small-angle deflection analyzers, that can filter charged particles with energies as high as several MeV/*e*. There is a huge variety in the configuration of these analyzers, but they all operate by allowing only particles in a selected energy/charge ($\mathcal{E}$) window ($\mathcal{E}_1 < \mathcal{E} < \mathcal{E}_2$) to pass through the system. Particles outside this window are rejected.

3.6.1 Faraday cup and retarding potential analyzer

Some of the earliest space instruments made use of the retarding potential analyzer (RPA) as an energy filter for the Faraday cup, a device that measures current. A modern version of the Faraday cup and RPA (Faraday cup sensor), currently flying on the Wind spacecraft, is used to measure distribution functions and basic flow parameters of solar wind protons and alpha particles. A Faraday cup sensor is flying on the Voyager spacecraft and is now measuring solar wind properties in the heliosheath.

Figure 3.15 shows a simplified cross-sectional view of the Faraday cup detector, illustrating its principle of operation. The RPA is the front section of the instrument (left of the central aperture stop); the Faraday cup is the structure to the right of the central aperture stop. The RPA consists of three highly transparent metal grids, central aperture stop, and support structure. The aperture stop defines the field of view of the instrument.

The energy of incoming ions is selected by applying a time-varying positive potential to the central metal grid. The time-varying potential is generated by a modulator, which produces a dc-biased, 200 Hz square wave. The resulting waveform can then vary from V to $V + \Delta V$. Both V and ΔV can be changed, up to 8 and 1 kV respectively. The result of applying the square wave potential to the central grid is to reject ions with "perpendicular" energies, $mv_\perp^2 < 2qV_1$, accept particles with $mv_\perp^2 > 2qV_2 = 2q(V_1 + \Delta V)$, and accept or reject ions with energies in between, depending on whether the modulating voltage is in its high or low step.

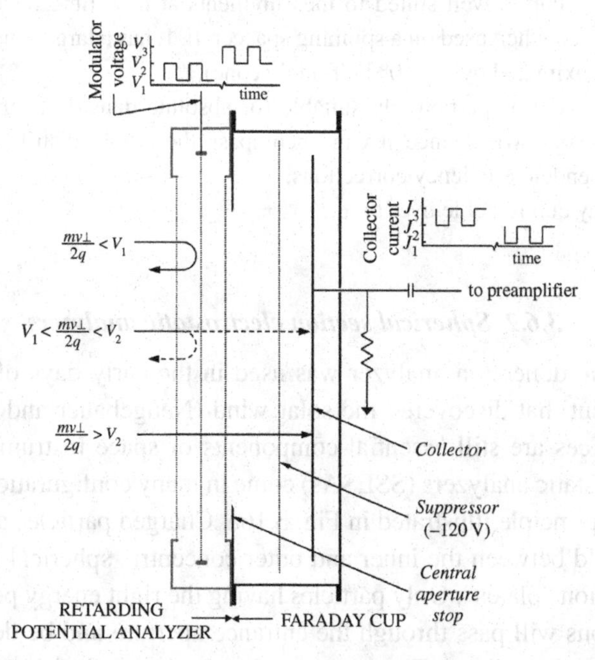

Fig. 3.15. Simplified cross-sectional view of the Faraday cup sensor on the Wind spacecraft. The retarding potential analyzer section is on the left. The Faraday cup section is shown on the right. (From Ogilvie *et al.*, 1995, with kind permission of Springer Science and Business Media.)

The RPA is a simple device with a large acceptance area and geometrical factor. Its only disadvantage is that its upper energy range is limited to ~6 to 8 keV/*e*.

The Faraday cup is a simple detector that is made of a collector, a highly transparent metal suppressor grid, biased at -120 volts, and support structure. The suppressor grid repels secondary electrons ejected from the collector plate by ions impinging upon it. Ions transmitted by the RPA deposit their charge on the collector, creating a current. The current (in the range from 10^{-13} to 10^{-8} A) from the collector is synchronously detected and integrated on a capacitor for a fixed time interval. The resulting voltage is converted to a digital signal using a logarithmic analog to digital (A/D) converter. The current also has a square wave pattern (see top right of Fig. 3.15) as the RPA transmits alternately more or less particles during the low or high step of the modulating voltage.

The Faraday cup sensor system has several advantageous properties:

(i) Since V is variable, the energy/charge bandwidth of the detector is variable whereas it is fixed by geometry in the case of electrostatic analyzers.
(ii) The flow direction can be determined to better than one degree by using segmented collectors.
(iii) The Faraday cup is well suited to measurements at high time resolution (typically seconds), even when used on a spinning spacecraft. It has a large sensitive acceptance angle (approximated by a ~60° half-angle cone).
(iv) The Faraday cup is particularly suitable for absolute density determinations in the supersonic solar wind since it can encompass the whole distribution and has no energy-dependent efficiency corrections.
(v) The Faraday cup is stable over time.

3.6.2 Spherical section electrostatic analyzers

An electrostatic deflection analyzer was used in the early days of the space age in an instrument that discovered the solar wind (Neugebauer and Snyder, 1962), and these devices are still essential components of space instruments. Spherical section electrostatic analyzers (SSESAs) come in many configurations but all work on the simple principle illustrated in Fig. 3.16a. Charged particles are deflected by the electric field between the inner and outer concentric spherical (or cylindrical) section deflection "plates". Only particles having the right energy per charge $\mathcal{E}$ and arrival directions will pass through the entrance aperture and be detected without first hitting one of the plates. The mean energy per charge $\mathcal{E}$ of the particle arriving at the detector is

$$\mathcal{E} \approx 0.5(V_{\text{out}} - V_{\text{in}})/\ln(R_{\text{out}}/R_{\text{in}}) \approx 0.5\Delta V/(\Delta R/\langle R\rangle), \qquad (3.17)$$

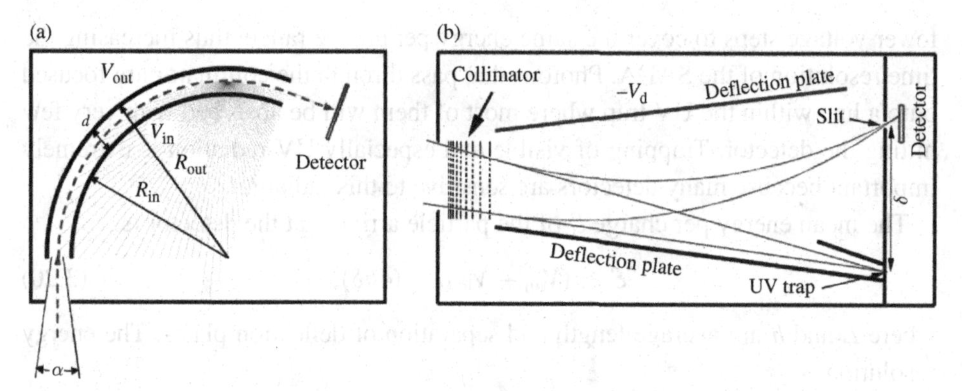

Fig. 3.16. Schematic representations of (a) a typical spherical or cylindrical section electrostatic deflection analyzer illustrating its principle of operation, and (b) the cross section of a typical small-angle deflection analyzer.

where $\Delta R = R_{out} - R_{in}$ and $\langle R \rangle = (R_{out} + R_{in})/2$. The energy per charge resolution is

$$\Delta\mathcal{E}/\mathcal{E} \approx \Delta R/\langle R \rangle, \tag{3.18}$$

and the acceptance angle α is approximately $\Delta R/\langle R \rangle$.

The ratio of the energy (in eV) of a particle that passes through the analyzer to the potential difference (in volts) between its deflection plates is the analyzer constant K,

$$K = \mathcal{E}/\Delta V \approx 0.5\langle R \rangle/\Delta R. \tag{3.19}$$

For example, an analyzer constant of 20 can be achieved with $\langle R \rangle = 20$ cm and a gap between the deflection plates $\Delta R = 0.5$ cm.

3.6.3 Small-angle deflection analyzer (SADA)

Electrostatic analyzers that can deflect particles up to several MeV/charge have a configuration as shown in Fig. 3.16b. Charged particles pass through a multi-slit collimator (which defines their incoming directions) and are then deflected by the electric field between the upper and lower deflection plates. A multi-slit collimator combined with large-area deflection plates (which may be slightly curved) gives the small-angle deflection analyzer (SADA) a reasonably large geometrical factor. The gap between the plates can also be large enough to support fairly high voltages (tens of kV), resulting in a large analyzer constant. Only particles having the right energy per charge $\mathcal{E}$ will pass through the narrow slit and be recorded by some detector behind the slit. Different mean energy per charge values are selected by varying or stepping the deflection voltage. Alternatively, a position-sensitive detector could be used to measure the amount of deflection and thus $\mathcal{E}$. This scheme would require fewer voltage steps to cover the same energy per charge range, thus increasing the time resolution of the SADA. Photons that pass through the collimator are focused onto a line within the UV trap where most of them will be absorbed with very few hitting the detector. Trapping of visible and especially UV radiation is extremely important because many detectors are sensitive to this radiation.

The mean energy per charge $\mathcal{E}$ of the particle arriving at the detector is

$$\mathcal{E} \approx (V_{\rm up} - V_{\rm lo})L^2/(4h\delta), \tag{3.20}$$

where L and h are average length and separation of deflection plates. The energy resolution is

$$\Delta\mathcal{E}/\mathcal{E} \approx \Delta\delta/\delta, \tag{3.21}$$

where $\Delta\delta$ is the slit width. The analyzer constant is

$$K = L^2/(4h\delta), \tag{3.22}$$

and the acceptance angle is

$$\alpha \approx h/(2L). \tag{3.23}$$

3.7 Time-of-flight telescopes

The speed of a low-energy particle may be determined by measuring the time it takes to travel the distance from a start detector to a stop detector. Figure 3.17 shows the schematic configuration of a typical time-of-flight (TOF) telescope consisting of a very thin foil (usually grid-supported carbon foil) and a detector separated by L. An ion enters the TOF telescope from the left, passes through the thin foil (that may change the direction of the particle by small-angle scattering) and then travels a distance d before entering the detector. Secondary electrons are ejected from the foil as well as the detector front surface (see Section 3.4 and

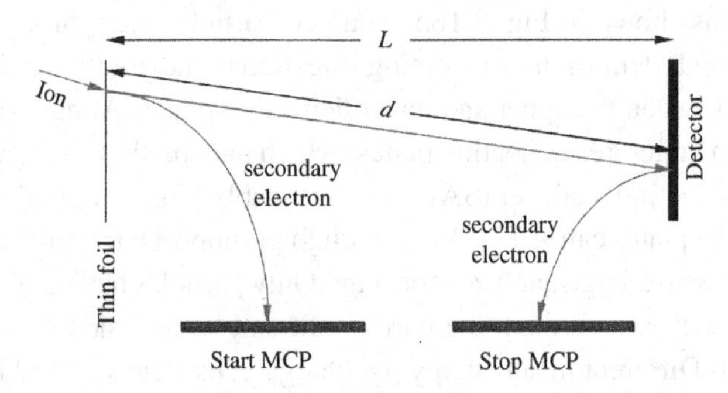

Fig. 3.17. Diagram of the cross section of a typical time-of-flight (TOF) telescope illustrating its principle of operation.

Fig. 3.4). These secondary electrons are accelerated (to ∼1 keV) and deflected onto the start and stop microchannel plates (MCPs), respectively. From the measured time difference τ between the signals of the start and stop MCPs, combined with knowledge of the distance d traveled ($d \approx L$) the speed of the particle ($v = d/\tau$) is determined.

3.8 Space instruments measuring composition

The velocity distributions (or differential energy spectra) as well as the elemental and isotopic compositions of plasmas and energetic particles may be determined by combining the simple detectors and energy analyzers discussed previously in clever ways. I will briefly describe the principle of operation of several ion composition instruments now operating in space that together span the entire energy range shown in Fig. 3.1. These instruments made use of the most advanced technology available at the time they were developed, especially in electronics, on-board data processing systems and, as required, low-mass materials and construction techniques.

3.8.1 Plasma composition spectrometer

3.8.1.1 The Solar Wind Ion Composition Spectrometer (SWICS)

While many of the physical properties (bulk speed, thermal speed, and density) and the variability of the solar wind have been measured since the beginning of the space age, its full chemical properties (ionization states, elemental and isotopic composition) could not be measured before 1990 with conventional solar wind instruments (such as the Faraday cup detector or the electrostatic analyzer–channeltron instrument). To overcome these limitations and to observe for the first time interstellar pickup ions and the unexplored suprathermal (∼5 to ∼100 keV) portion of the energy spectra, the Solar Wind Ion Composition Spectrometer (SWICS) investigation was proposed for the Ulysses mission in 1977. SWICS instruments (Gloeckler *et al.*, 1992) are flying on Ulysses and ACE (Gloeckler *et al.*, 1998), and instruments of almost identical or somewhat modified design are on the Wind spacecraft (Gloeckler *et al.*, 1995), and have flown on the Active Magnetospheric Tracer Explorer (AMPTE) mission. The PLASTIC instrument on the two STEREO spacecraft is also based on the SWICS design, although the configuration of the subsystems is different from SWICS. These instruments measure the distribution functions and charge state, elemental, and isotopic (^{3}He) composition of plasmas in the energy range of ∼100 eV/e to ∼100 keV/e at all conceivable ambient conditions, and have contributed much to our current knowledge of the

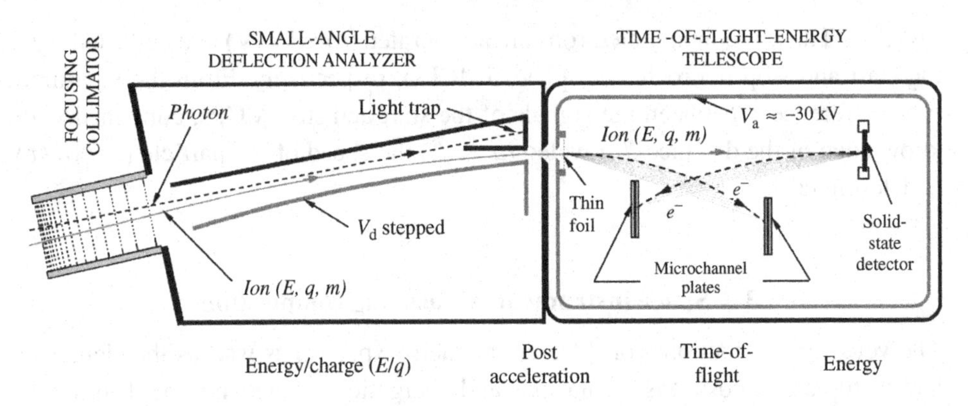

Fig. 3.18. Diagram of the measurement technique used in SWICS.

solar wind, magnetospheric ion populations (such as ring currents), pickup ions, and interstellar physical processes and chemical characteristics.

The SWICS sensor is based on the technique of particle identification using a combination of electrostatic deflection, post-acceleration, and a time-of-flight (TOF) and energy measurement (Gloeckler and Hsieh, 1979). Figure 3.18 shows schematically the operating principle of the sensor and the functions of the five basic sensor elements employed:

- Ions of kinetic energy E, mass m, and charge (ionization state) q enter the sensor through a large-area, multi-slit collimator of the small-angle deflection analyzer (see Fig. 3.16) that selects proper entrance trajectories for the particles and serves as an energy-per-charge (E/q) filter, allowing only ions within a given energy-per-charge interval (determined by a stepped deflection voltage) to enter the TOF vs. energy system.
- Ions are post-accelerated by a $\sim$30 kV potential drop just before entering the TOF vs. energy system. The energy they gain is sufficient to be measured adequately by the solid-state detectors, which typically have a $\sim$30 keV energy threshold. An energy measurement is essential for determining the elemental composition of an ion population, and ions with energies below $\sim$30 keV must be accelerated if their mass is to be identified.
- In the time-of-flight (TOF) system (Section 3.7 and Fig. 3.17) the velocity of each ion is determined by measuring the travel time τ of the particle between the start and stop detectors separated by typically 10 cm.
- Particle identification is completed by measuring the residual energy of the ions in a conventional thin-window solid-state detector.

From simultaneous measurements of the time-of-flight τ and residual energy E_{meas} and a knowledge of the deflection system voltage, and hence the E/q, and of the post-acceleration voltage V_a, the mass (m), charge state (q) and incident energy (E) of each ion is determined as follows:

$$
\begin{aligned}
m &= 2(\tau/d)^2(E_{\mathrm{meas}}/\alpha),\\
m/q &= 2(\tau/d)^2(V_a + E'/q) \approx 2(\tau/d)^2 V_a), \qquad (3.24)\\
q &= (E_{\mathrm{meas}}/\alpha)/(V_a + E'/q) \approx (E_{\mathrm{meas}}/\alpha)/V_a,\\
E &= q(E/q),
\end{aligned}
$$

where d is the flight path. E'/q takes account of the small energy loss of ions in the thin foil of the start-time detector (see Fig. 3.6 and Ipavich *et al.*, 1982) and $\alpha(< 1)$ is the nuclear defect in solid-state detectors (see Fig. 3.12 and Ipavich *et al.*, 1978). The approximate expressions for q and m/q hold for typical solar wind ions.

The photograph of the SWICS instrument (Fig. 3.19a) shows the configuration of the main sensor and the small-angle deflection analyzer with the collimator opening covered by a dust/acoustic protective cover that swings open after launch. The gold-plated, cylindrically shaped container houses the -30 kV post-acceleration supply. The sensor is mounted on the Sun-facing side of the spinning Ulysses (ACE) spacecraft platform, in the same orientation as shown in the photograph. The total mass of the instrument is 3.992 kg and the average power it consumes is 2.45 W. SWICS measures the mass (from 1 to ~50 amu), the mass/charge (from 1 to ~30 amu/e) of ions in the energy range of ~0.6–100 keV/e. The fractional energy resolution is $\Delta E/E \approx 0.06$. The mass and mass/charge resolutions of SWICS for typical ions are summarized in Table 13.1. The directional geometrical factor is 2 mm^2. Since shortly after launch on October 6, 1990, SWICS operated flawlessly (and since May 2008 at its full post-acceleration voltage of 30 kV) until the mission was terminated in mid 2009.

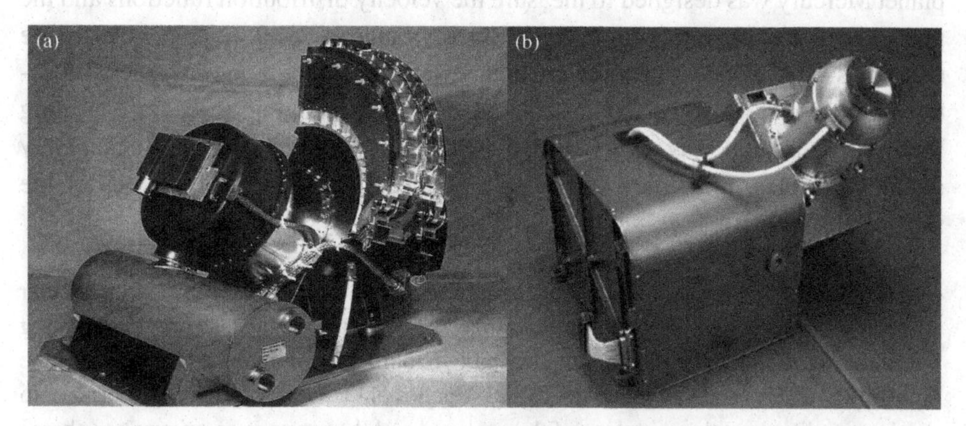

Fig. 3.19. Photographs of (a) the Solar Wind Ion Composition Spectrometer (SWICS) on Ulysses and ACE (from Gloeckler *et al.*, 1992), and (b) of the Fast Imaging Plasma Spectrometer (FIPS) on MESSENGER (from Andrews *et al.*, 2007).

Table 3.1. *Resolution characteristics of SWICS*[a]

Element	Mass	Charge	Energy (keV)[b]	Time-of-flight (ns)	$\Delta(m/q)/(m/q)$ (FWHM)	$\Delta m/m$ (FWHM)
H	1	1	19	48.8	0.054	0.742
He	4	2	38	66.7	0.042	0.397
C	12	6	103	66.3	0.039	0.223
N	14	7	117	66.3	0.039	0.224
O	16	6	91	75.7	0.038	0.265
Ne	20	8	116	73.2	0.034	0.305
Si	28	9	122	80.6	0.033	0.302
S	32	10	133	81.6	0.033	0.305
Fe	56	11	111	98.9	0.030	0.353

[a] For 440 km/s solar wind speed and 23 kV post-acceleration
[b] Measured by solid-state detector

The double and especially the triple coincidence techniques (TOF and TOF plus energy) used in SWICS significantly reduce background noise compared to earlier solar wind instruments that simply count particles. This led to the discovery of suprathermal power law tails with a common spectral index of -5 (e.g. Gloeckler *et al.*, 2008) and of interstellar pickup ions, and enabled measurements of the very rare interstellar ^{3}He/^{4}He ratio (Gloeckler and Geiss, 1996).

3.8.1.2 The Fast Imaging Plasma Spectrometer

The Fast Imaging Plasma Spectrometer (FIPS) on the MESSENGER spacecraft to planet Mercury was designed to measure the velocity distribution functions and the mass/charge composition of ions in the energy range of $\sim$0.05 to $\sim$20 keV/charge (Andrews *et al.*, 2007). The main challenge was to meet these measurement objectives with a less than 1.5 kg instrument on a non-spinning spacecraft orbiting Mercury. This goal was achieved using the sensor design shown schematically in Fig. 3.20. The principle of operation of the sensor and the functions of the subsystems used are:

- Ions of kinetic energy E, mass m, and charge state q enter the sensor through a circular opening of the cylindrically symmetric imaging deflection analyzer (IDA), consisting of three hemispherical deflection plates and two multi-slit collimators, that (a) maps the polar (25° to 75°) and azimuthal (0° to 360°) angles of the incoming ion trajectories onto polar coordinates at the exit plane of the analyzer, and (b) serves as an energy-per-charge (E/q) filter, allowing only ions within a given energy-per-charge interval (determined by a stepped deflection voltage) to enter the TOF system. The unique design of IDA provides a wide and instantaneous field-of-view as well as strong UV suppression, achieved

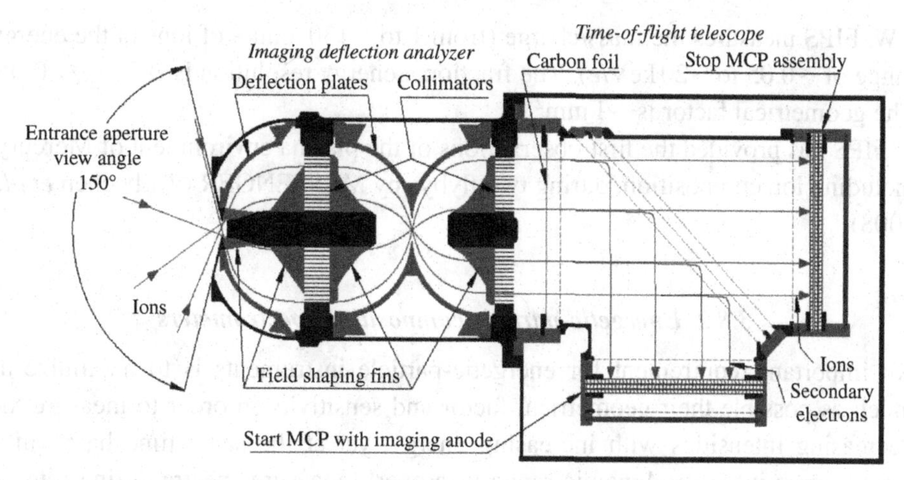

Fig. 3.20. Schematic cross section of the Fast Imaging Plasma Spectrometer (FIPS). The distance between the carbon foil and the Stop MCP is 7 cm. (From Andrews *et al.*, 2007.)

by using two collimators, three deflection sections, plus black coatings and serration of the deflection plates.

- Ions are post-accelerated by a ~15 kV potential drop just before entering the TOF telescope. The energy they gain is sufficient for even the lowest energy incoming ions to traverse the thin carbon foil of the TOF telescope and reach the stop MCP.
- The TOF system (Section 3.7 and Fig. 3.17) determines the velocity of each ion by measuring the travel time τ of the particle between the start and stop detectors separated by a distance of 7 cm. Unlike in the SWICS TOF system, the FIPS TOF telescope uses an electrostatic mirror (see Fig. 3.20) to reflect secondary electrons from the start foil by 90° onto a position-sensitive wedge and strip MCP detector (see Section 3.5.3 and Fig. 3.11), thus recording (imaging) the arrival directions of the incoming ions.

From simultaneous measurements of the time-of-flight τ and knowledge of both the deflection system voltage, and hence the E/q, and of the post-acceleration voltage V_a, the mass/charge (m/q) of each ion is determined using

$$m/q = 2(\tau/d)^2(V_a + E'/q), \tag{3.25}$$

where d is the 7 cm flight path. $E'/q = E/q - \Delta(E)$, where $\Delta(E)$ is the small energy loss of ions in the thin foil of the start-time detector (see Fig. 3.6 and Ipavich *et al.*, 1982).

Figure 3.19b shows a photograph of the FIPS sensor attached to the box containing the sensor electronics and high-voltage power supplies. The total mass of the instrument shown in Fig. 3.19 is 1.41 kg and the average power it consumes is

2 W. FIPS measures the mass/charge (from 1 to ~130 amu/*e* of ions in the energy range of ~0.05 to ~20 keV/*e*). The fractional energy resolution is $\Delta E/E \approx 0.05$. The geometrical factor is ~1 mm^2 sr.

FIPS has provided the first observations of the plasma environment of Mercury, including ion composition, during two flybys by MESSENGER (Zurbuchen *et al.*, 2008).

3.8.2 Energetic-particle composition spectrometers

An important requirement for energetic-particle instruments is to maximize as much as possible their geometrical factor and sensitivity in order to measure the decreasing intensities with increasing energy, yet at the same time have sufficiently large intensity dynamic range to properly measure spectra during intense solar particle events. Instruments observing ultra-low energetic particles (~0.1 to ~10 MeV/nucleon) use some combination of TOF vs. energy E, or $\mathrm{d}E/\mathrm{d}x$ vs. E to measure the particle mass and energy. To determine the charge states of ultra-low-energy ions electrostatic deflection analysis must also be used. Position sensitive detectors are essential in many of these instruments. Below is a very brief description of the concepts used in two types of energetic-particle spectrometers on ACE. Other energetic-particle composition instruments are described in the ACE book (Russell *et al.*, 1998).

3.8.2.1 The Ultra-Low-Energy Isotope Spectrometer

The Ultra-Low-Energy Isotope Spectrometer (ULEIS) measures the differential energy spectra and isotopic composition of low-energy particles from protons to Fe with energies of ~0.1 to ~2 MeV/nucleon (Mason *et al.*, 1998) The large geometrical factor of ~1 cm^2 sr and its excellent mass resolution of $\Delta m/m$ of ~0.01 are achieved by the sheer size (~70 cm long) and consequently large mass (~19 kg) of the telescope.

The ULEIS telescope (Fig. 3.21) consists of a sunshade, a UV-opaque entrance foil, a TOF telescope and an array of three large- and four small-area solid-state detectors to measure the particle energy E. The TOF telescope uses three (two "start" and one "stop") identically constructed timing detectors (similar to the FIPS start detector), consisting of a thin, grid-supported foil, a 45° electrostatic harp-mirror, chevron-stacked MCPs and a wedge-and-strip position-sensitive anode. The telescope is mounted at 60° from the spin axis (pointing within a few to ~10° of the Sun direction) of ACE, thus preventing direct sunlight from illuminating the entrance foil, which also serves as the start foil of the first "start" detector. From simultaneous measurements of the two times-of-flight, τ_1 and τ_2, and energy E one can determine the particle mass m:

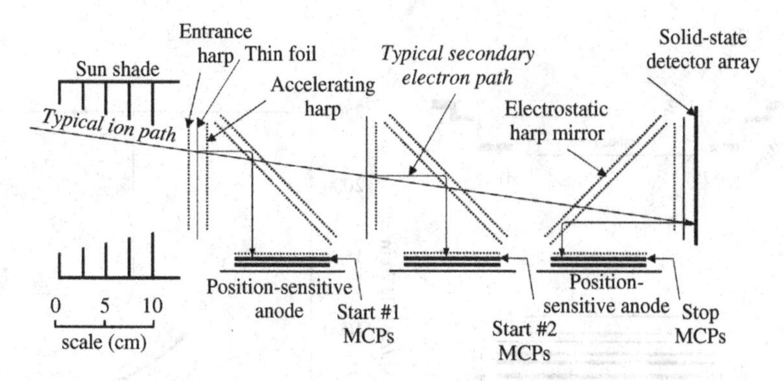

Fig. 3.21. Schematic cross section of the Ultra-Low-Energy Isotope Spectrometer. (From Mason *et al.*, 1998.)

$$m \approx 2E(\tau_1/d_1)^2 \approx 2E(\tau_2/d_2)^2, \tag{3.26}$$

where τ_1 and τ_2, and d_1 and d_2 are the times-of-flight and flight paths between the first "start" and "stop" and the second "start" and "stop" foils respectively. The redundant determination of mass reduces background.

The ULEIS instrument provides the most precise and detailed measurements of the spectra and composition of ultra-low-energy particles in SEP and CIR events (Mason *et al.*, 2008).

3.8.2.2 *The Solar Isotope Spectrometer*

The Solar Isotope Spectrometer (SIS) was designed to measure the differential energy spectra and isotopic composition of energetic nuclei from H to Zn ($Z = 1$ to 30) over the energy range from ~10 to ~100 MeV/nucleon (Stone *et al.*, 1998). SIS has a large geometry factor (~40 cm^2 sr) and excellent mass resolution ($\Delta m/m$ of ~0.01 for O and ~0.006 for Fe) that, as in ULEIS, are the result of the large size (30 by 42 by 28 cm) and mass (22 kg) of the instrument.

The SIS sensor uses stacks of large-area semiconductor detectors of various thicknesses arranged as shown in Fig. 3.22a. The top two detectors, M1 and M2, are thin position-sensitive detectors that measure the energy losses and determine the trajectory of a particle of atomic number Z and mass m with sufficient energy E to penetrate into the stack of the thicker T detectors. The T stack is thick enough to stop ~100 to 200 MeV/nucleon particles. Entrance foils provide UV suppression.

The nuclear charge Z, mass m, and energy E of particles are determined using the dE/dx versus E technique first introduced in the 1960s. From Eq. (3.11) or (3.12)

$$\mathrm{d}E/\mathrm{d}x \approx \Delta E/\Delta x \propto Z^2 m/E. \tag{3.27}$$

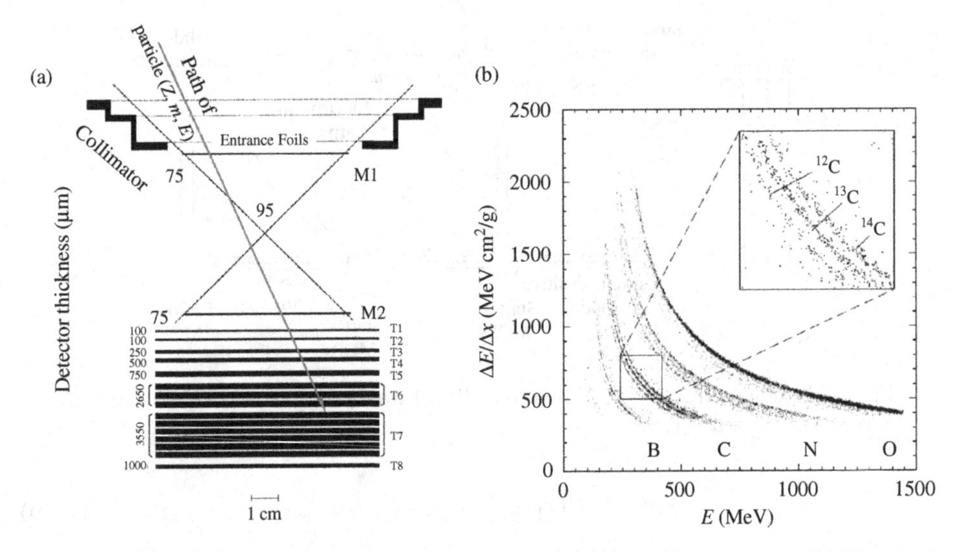

Fig. 3.22. (a) Schematic cross section of the Solar Isotope Spectrometer. (b) Illustration of the $\mathrm{d}E/\mathrm{d}x$ by E analysis technique using data acquired during an accelerator calibration of SIS. (From Stone *et al.*, 1998.)

A plot of $\Delta E/\Delta x$ (approximating $\mathrm{d}E/\mathrm{d}x$) as a function of the total particle energy E for a particular nuclide will be a hyperbola. The hyperbola for different particles will be separated by Z^2m, as illustrated in Fig. 3.22b.

In the SIS sensor ΔE is measured by both the M1 and M2 detectors; $\Delta x = l/\cos(\Theta)$, where Θ, the angle from perpendicular direction with respect to the detectors, is determined from the trajectory information obtained from the position sensitive M1 and M2 detectors; and the total energy is found by adding the energy losses in all detectors penetrated by the particle. The expected curves for each isotope can be approximately calculated using tabulated forms of Eq. (3.12) (see also Figs. 3.5 and 3.6), but are best determined by calibration using particle accelerators, or even from in-flight calibrations using SEP particles.

3.8.3 Energetic neutral atom composition spectrometers

Energetic neutral atoms (ENAs) are created by charge exchange of energetic ions with ambient neutral gas. ENAs have the energy of the original ion and travel in ballistic trajectories, unaffected by magnetic or electric fields. This property makes it possible to use ENAs to derive properties of energetic ions, such as their energy spectra and composition, in distant regions, determine the dependence of these properties on look direction, and image the spatial distributions of energetic ions in these regions. Examples of ENA images of Saturn's ring current (Krimigis *et al.*, 2007) are shown in Fig. 3.23.

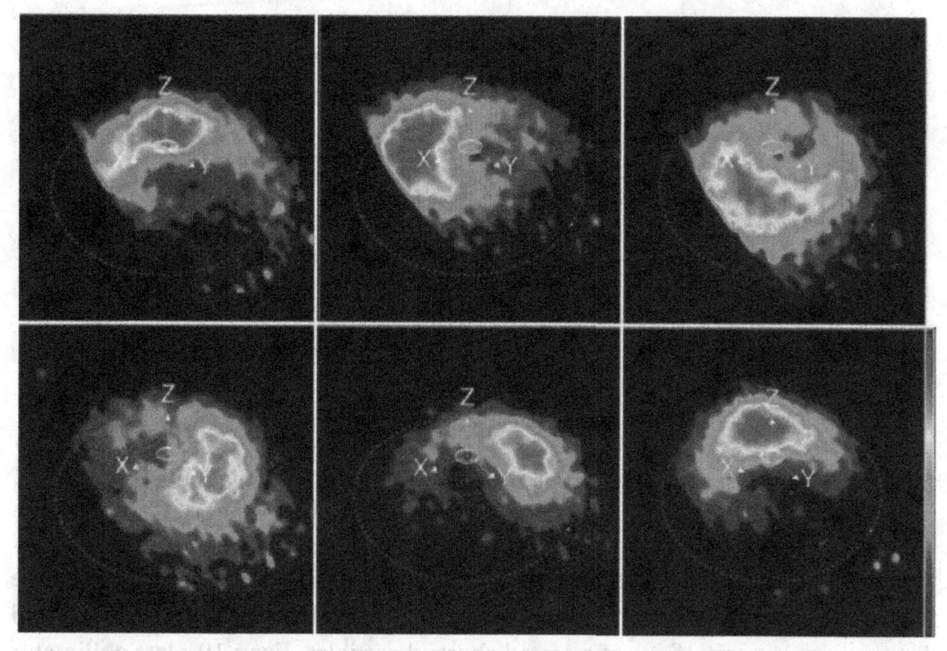

Fig. 3.23. Neutral-hydrogen (20–50 keV) images of Saturn's ring current taken with the INCA sensor on Cassini at 2.13-hour intervals showing counterclockwise rotation of the plasma. (Courtesy S. M. Krimigis. See also Plate 1 in the color-plate section.)

To measure the energy spectra and composition of ENAs, especially in the presence of much higher fluxes of ambient energetic ions and electrons, one uses instruments that are similar or even identical to those that measure spectra and composition of energetic ions of comparable energies. However, the entrance system of these instruments must prevent ions and electrons, which generally have much higher fluxes than the ENAs one wishes to measure, from entering the composition spectrometer. Two different technical approaches will be described below to measure characteristics of ENAs of greater than ~50 keV and less than ~50 keV.

3.8.3.1 *High-energy ENA composition spectrometer*

The schematic cross section of an instrument measuring the energy spectra and composition of ENAs in the energy range from ~10 eV to ~100 keV/nucleon is shown in Fig. 3.24. The spectrometer consists of two subsystems, the charged particle rejection collimator and the time-of-flight versus energy telescope, which is nearly identical to the ULEIS instrument (Section 3.8.2.1 and Fig. 3.21). The size of the instrument could be large, as shown, or small, depending on available resources. Clearly, a smaller instrument will have a smaller geometry factor than a larger instrument. The particle rejection collimator has 19 planar, serrated deflection plates with alternating plates (shown in grey) biased at a fixed positive DC

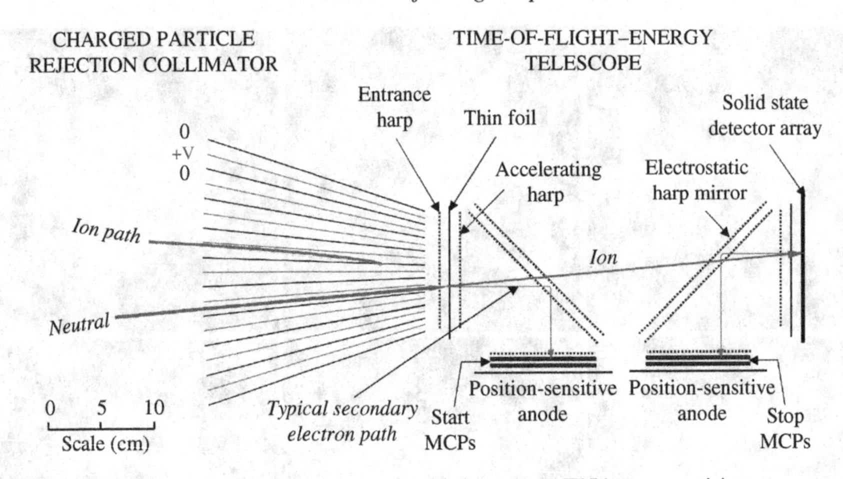

Fig. 3.24. Schematic cross section of a high-energy ENA composition spectrometer.

voltage. Charged particles will be deflected and will hit one of the plates where they will be absorbed with minimum scattering. Equation (3.20) allows us to find the maximum energy, $\mathcal{E}_{max}$, of rejected charged particles. For a 19-plate collimator of the size shown (~20 cm long) and $V = 10\,\text{kV}$, ions below $2\,\text{MeV}/e$ will be excluded. Neutral particles (and photons) will be transmitted by the collimator and will reach the TOF telescope.

ENAs having energies above ~50 keV/nucleon will easily traverse the UV-opaque entrance foil and measurements of their energy, mass, and arrival directions within the ~20° by 20° field of view of the instrument is then done exactly as for charged particles, described in Section 3.8.2.1.

3.8.3.2 Low-energy ENA composition spectrometer

The schematic cross section of an instrument measuring the energy spectra and composition of ENAs in the energy range below ~50 keV/nucleon is shown in Fig. 3.25. The spectrometer consists of three subsystems, the charged particle rejecter (CPR), the small-angle deflection analyzer (SADA) with an exit collimator, and the time-of-flight versus energy telescope, which is nearly identical to the ULEIS instrument (Section 3.8.2.1 and Fig. 3.21). The CPR and SADA subsystems are at a high positive potential (~30 kV) to accelerate ions leaving the collimator of the SADA. The size of the instrument could be large or small, depending on available resources, with the smaller instrument having a smaller geometry factor than the larger instrument. The entrance system consisting of three highly transparent conductive grids biased (from left to right) at 0, ~–6 and ~30 kV, respectively, will repell less than ~6 keV electrons and less than ~30 keV/charge ions. Higher-energy electrons and ions will be deflected toward and absorbed by the traps or

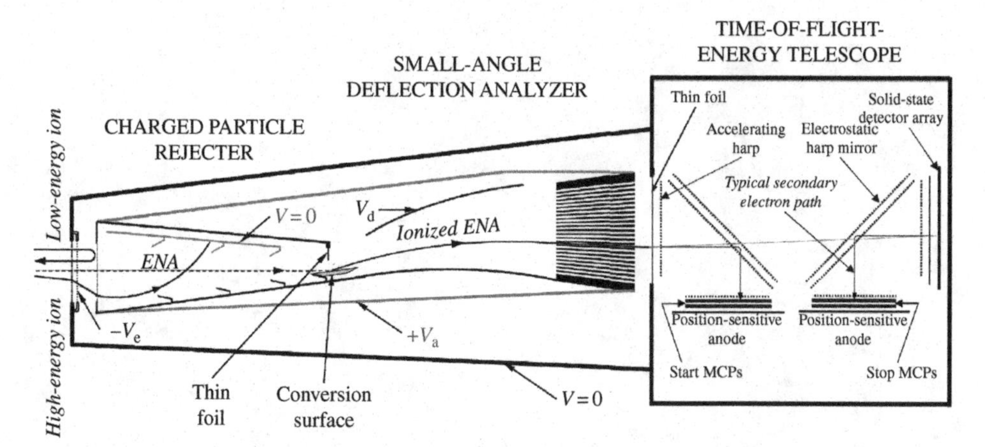

Fig. 3.25. Schematic cross section a low-energy ENA composition spectrometer. The charged particle rejecter and the small-angle analyzer are at high positive potential.

serrated surfaces of one of the deflection plates (at 30 kV potential difference) with minimum scattering.

The CPR will, of course, not deflect neutrals and a means must be found to ionize them. One way to accomplish this is to have them pass through the thin foil located at the top exit end of the CPR. However, for the low-energy ENAs that cannot penetrate the thinnest foils available, this technique will not work. Instead, these low-energy ENAs are reflected at near grazing angles from a specially prepared conversion surface (e.g. highly polished tungsten), at the bottom exit of the CPR as indicated in Fig. 3.25, where a reasonably large fraction of ENAs becomes ionized. The energy per charge of the ionized atoms entering the electrostatic analyzer can now be selected and, in combination with post-acceleration and a TOF versus E analysis (see Section 3.8.1.1 and Eq. 3.24), the incoming energy and mass of the ENAs can now be measured.

Many different configurations of the basic measurement techniques described above have been implemented in a variety of ENA instruments that have been flown or are flying today (see reviews by Gruntman, 1997; Wurz, 2000). The first measurement of the density, velocity, and temperature of interstellar neutral He was made with the GAS instrument (Witte *et al.*, 1992) on the Ulysses spacecraft. Several instruments on the Earth- orbiting IMAGE satellite (http://pluto.space.swri.edu/IMAGE/) measured ENAs from the Earth's ring current, and the MIMI instrument (Krimigis *et al.*, 2004) on Cassini provides spectacular images of Saturn's ring current (see Fig. 3.23). The payload of the IBEX spacecraft† is devoted entirely to measurements of ENAs from the heliosheath.

† IBEX: www.ibex.swri.edu/mission/payload.shtml

[illegible] RN[illegible] [illegible] potential.

separate the [illegible] of the deflection plates (at 30 kV potential) [illegible]

The CPR will of course not deflect neutrals, and a means must be found to [illegible] them. One way [illegible] to have them pass through the [illegible] of the [illegible] the CPR. However, for the low-energy ENAs [illegible] available [illegible] ENAs are reflected at near grazing angles from [illegible] conversion surface [illegible] at the bottom [illegible] the [illegible] indicated in Fig. 3.5 [illegible] a reasonably large fraction of ENAs becomes ionized. The energy [illegible] of the ionized atoms [illegible] the electrostatic [illegible] TOF [illegible] analysis (see [illegible] Fig. 3.2) [illegible] energy and mass of the ENAs can now be measured.

Many different [illegible] have been flown [illegible] (see [illegible]). The first measurement of the [illegible] interstellar neutral [illegible] was made with the GAS instrument [illegible] Ulysses spacecraft [illegible] instrument [illegible] Earth [illegible] the [illegible] current and the [illegible] instrument [illegible] ring current [illegible] The payload [illegible] ENAs [illegible]

4

Radiative signatures of energetic particles

TIM BASTIAN

Sunspots, the white light manifestation of solar active regions, were first observed through telescopes in 1610 by Galileo and Harriot, but it was not until nearly 250 years later that the flare phenomenon was first observed by Carrington (1859), who noted a white light "conflagration" in a sunspot group (see Fig. 2.2) and speculated (correctly) that it originated above the sunspot group. It was not for almost another century – during the post-war years of the late 1940s – that solar observations strayed outside the confines of the visible spectrum into invisible wavelengths both longward (radio) and shortward (X-rays) of visible light. The launch of the Sputnik 1 satellite in late 1957 ushered in the space age which, for the first time, made vast portions of the electromagnetic spectrum accessible for sustained study. Over the last half-century, extraordinary progress has been made in developing successive generations of both ground- and space-based instrumentation to characterize and to understand radiative signatures of both quiescent and energetic processes on the Sun and in the heliosphere, although gaps remain. In this chapter we discuss radiation from energetic particles, with an emphasis on the radio, hard X-ray (HXR), and γ-ray wavelength bands. For it is in these bands that radiative signatures of the most energetic particles, those accelerated by violent processes on the Sun and in the heliosphere, are detected. In other words, the focus is on the extreme frontiers of the electromagnetic spectrum, on the extreme departures from equilibrium conditions, and on the extreme energies involved.

4.1 Overview of the electromagnetic spectrum

Electromagnetic radiation is carried by photons, particles with zero rest mass and charge that mediate electromagnetic interactions between charged particles. Photons propagate at a constant speed c in a vacuum. They are characterized by both

Heliophysics: Space Storms and Radiation: Causes and Effects, eds. Carolus J. Schrijver and George L. Siscoe.
Published by Cambridge University Press.

particle- and wave-like properties: a photon characterized by a wavelength λ carries an energy

$$E = hc/\lambda = h\nu, \tag{4.1}$$

where h is the Planck constant and ν is the cyclic frequency, with $\lambda\nu = c$. The photon momentum vector is given by $\mathbf{p} = \mathrm{h}\mathbf{k}/2\pi$, the magnitude of which is $p = h/\lambda = h\nu/c$. Photons are also characterized by spin angular momentum, which can occupy one of two states – parallel or anti-parallel to the momentum vector $\mathbf{k}$ – corresponding to two circular polarization states.

Photons are produced by a number of physical processes, including the acceleration of charged particles; bound–bound and free–bound transitions in molecules, atoms, and nuclei; the annihilation of matter against anti-matter (e.g. electrons and positrons); or by the decay of other subatomic particles (e.g. π^0 mesons). Prior to the era of quantum physics, electromagnetic radiation was well characterized by its wave-like properties in most known circumstances. In the modern era, a classical wave treatment can often still be used as a convenient, accurate, and intuitive description of the electromagnetic radiation while a quantum mechanical treatment of matter is required (semi-classical treatment). Other wavelength regimes and mechanisms may require a full quantum description of the interaction of both radiation and matter.

The visible portion of the electromagnetic spectrum, to which the human eye is sensitive, spans a mere octave in wavelength (380–750 nm or 3800–7500 Å), neatly spanning the Sun's spectral maximum near 500 nm (Section 4.2.2). Yet electromagnetic radiation in general, and electromagnetic radiation from the Sun in particular, extends to wavelengths many orders of magnitude greater than and less than the visible band. Figure 4.1 illustrates the portion of the electromagnetic spectrum that is currently accessible to measurement by modern instrumentation in terms of wavelength, frequency, and energy units, spanning 18 orders of magnitude. The total electromagnetic energy emitted by the Sun amounts to 3.8×10^{26} W (or 3.8×10^{33} erg s^{-1}). At Earth, the mean solar irradiance at the top of the atmosphere is given by the "solar constant": 1.366 kW m^{-2}. The energy contained in the energetic particles and the radiation they produce can be impressive; but it is tiny compared with the Sun's overall luminosity. Yet the particles and radiation produced by violent events on the Sun (flares, coronal mass ejections – CMEs) and in the heliosphere (interplanetary CMEs and shocks) nevertheless have a profound effect on the solar atmosphere, the interplanetary medium, the near-Earth environment, and the environment near other planets.

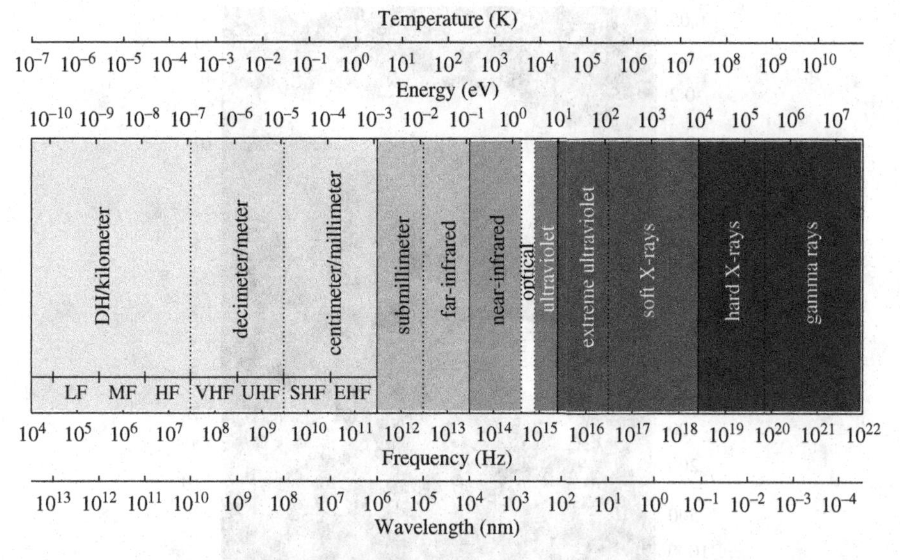

Fig. 4.1. Overview of the electromagnetic spectrum with energy in electronvolts and the equivalent temperature in kelvin (top axes), frequency in hertz, and wavelength in nanometers (bottom axes). Note that the AM band lies in the low- and medium-frequency (LF-MF) range and the FM band in the very-high-frequency (VHF) range.

4.1.1 Radio emission

Radio wavelengths are defined here to include all wavelengths, $\lambda > 1$ mm or frequencies $\nu < 300$ GHz. For the purposes of discussion the radio spectrum is, in turn, subdivided into millimeter–centimeter, decimeter–meter, and decameter–kilometer wavelengths.† The longest wavelengths, corresponding to radio frequencies below 30 MHz are, in turn, often divided into the dekameter–hectometer wavelength band (DH-λ; 300 kHz to 30 MHz) and kilometric emission (km-λ; below 300 kHz). These emissions are for the most part only observable from satellites, well above the Earth's ionosphere and its associated frequency cutoff ($\nu_0 \sim 5$–15 MHz). Radio emission at wavelengths shortward of ~ 20 m (15 MHz) are accessible to study from the ground. Loosely speaking, the higher the radio frequency, the lower in the solar atmosphere it originates. Therefore, emission at DH- to km-λ originates in the interplanetary medium (IPM), that at dm- to m-λ originates in the corona (Fig. 4.2), and that at mm- to cm-λ originates in the

† Unlike other wavelength regimes, radio technologies have a strong heritage in commercial and military applications: telecommunications, radar, and other uses. The associated terminology sometimes carries over into the radio astronomical realm, with reference to particular wavelength bands, for example: the VLF, VHF, UHF bands; the AM or FM bands; the L, S, C, X, and Ku bands, etc.

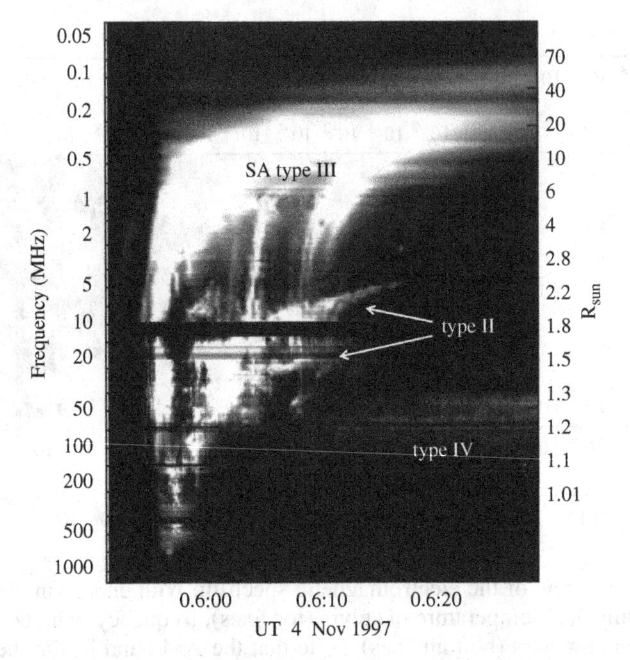

Fig. 4.2. An example of a dynamic radio spectrum formed from the Culgoora spectrographs on the ground and the Wind/WAVES spectrographs from space. The spectrum shows radio emission from a radial range spanning the low corona to $\sim 70\,R_\odot$. The spectrum shows the presence of interplanetary type III radio bursts (see Table 4.1 for descriptive definitions of radio burst types I–IV), coronal type II radio bursts, and so-called "shock-associated" (SA) type III radio bursts. (After Dulk *et al.*, 2000.)

chromosphere and low corona. A variety of emission mechanisms are involved in radio emission from the Sun, but the dominant mechanisms associated with energetic processes are coherent plasma radiation at dm-λ and longer, and incoherent gyrosynchrotron radiation at cm-λ and shorter.

Dynamic spectroscopy, recording the radio spectrum as a function of time, is performed by a large number of ground-based spectrometers, including those at Green Bank in the United States; Hiraiso, Japan; Culgoora, Australia; Izmiran, Russia; Tremsdorf, Germany; and Bleien, Switzerland. Space-based radio spectrometers include those on board the Interplanetary Sun-Earth Explorer (ISEE) satellites, the Wind spacecraft, Ulysses, and STEREO. Ground-based radio telescopes used for solar imaging over the past several decades include the Culgoora Radioheliograph, the Westerbork Synthesis Radio Telescope, the Clark Lake Radio Observatory, and the Very Large Array (VLA). The Nobeyama Radioheliograph (NoRH) and the Nançay Radioheliograph (NRH) began playing a central role in radio imaging in the 1990s. Imaging spectroscopy was pioneered by the Owens Valley Solar Array

Table 4.1. *Classification of radio bursts (see examples in Fig. 4.2)*

Type	Description
I	A noise storm composed of many short, circularly polarized, narrow-band bursts in the decimeter-meter range (300–50 MHz).
II	Narrow-band emission that begins in the meter range (300 MHz) and sweeps slowly (tens of minutes) toward dekameter wavelengths (10 MHz), often with harmonics at twice the frequency.
III	Narrow-band bursts that sweep rapidly (seconds) from decimeter to dekameter wavelengths (500–0.5 MHz).
IV	A smooth continuum of broad-band bursts primarily in the meter range (300–30 MHz), with moving and stationary (in frequency) subtypes.
V	Broad-band continuum radiation following a type-III burst.

Adapted from www.swpc.noaa.gov/info/glossary.html, and Zirin (1988).

(OVSA), a technique that will be fully exploited by the Frequency Agile Solar Radiotelescope (FASR).

4.1.2 Far-IR to submm-λ emission

Continuum radiation from the quiet Sun at submillimeter to far-infrared wavelengths is dominated by thermal bremsstrahlung emission (Section 4.3) from free electrons interacting with protons and neutral hydrogen. Interest in both solar IR emission (taken here to be ~1–100 μm) and submm (100 μm to 1 mm, or 300 GHz to 3 THz) emission was strong in the 1960s and 1970s and motivated many photometric measurements of the quiet Sun brightness across the IR and submm-λ bands (see review by De Jager, 1975) as a means of probing the low chromosphere. The usefulness of continuum IR and submm-λ to study energetic processes was also anticipated (Hudson and Ohki, 1972), yet little progress has been made in the intervening decades and observations of flare-associated far-IR and submm-λ emissions are extremely sparse. This has been due, in part, to the fact that the required technologies and techniques have been slow to mature. Solar observations in these bands have been limited to single dish antennas (e.g. the James Clerk Maxwell Telescope, the Caltech Submillimeter Observatory (see Fig. 4.3), and the Solar Submillimeter Telescope). With the construction of the Atacama Large Millimeter Array (ALMA), the submm-λ window will be opened for systematic study for the first time although access to wavelengths in the far-IR remains problematic.

4.1.3 Near-infrared, optical, and ultraviolet emission

Observations of the Sun in the optical band have been a mainstay of solar physics since the time of Galileo. More recently, key ideas concerning the flare

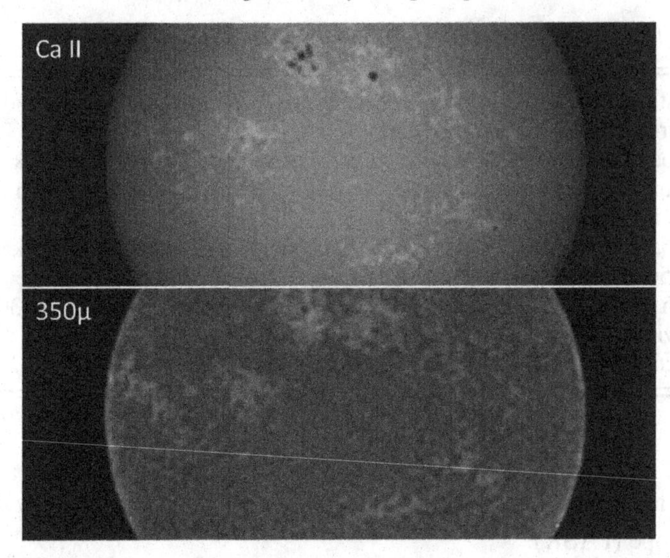

Fig. 4.3. *Top*: chromospheric emission in the Ca II K line at 393.2 nm. *Bottom*: the corresponding image at a wavelength of 350 μm (850 GHz). This image was obtained by Lindsey *et al.* (1995) using the JCMT on February 9, 1991. (Reproduced by permission of the AAS.)

phenomenon find their basis in optical line and continuum observations (see the example of a flare observed in the strong Hα line shown in Fig. 4.4), enlarged in more recent decades to include ultraviolet (120–380 nm or 1200–3800 Å) and near-IR ($\sim$1–10 μm) observations. Similarly, white light coronagraph observations of Thomson-scattered light form the basis for much of what we know about the CME phenomenon. Analyses of optical and UV spectral lines and continuum (as well as IR continuum) have played critical roles in deducing the structure of the quiescent photosphere and chromosphere (e.g. the celebrated semi-empirical models of Vernazza, Avrett, and Loeser, 1981, and subsequent models), and the flaring chromosphere via semi-empirical (e.g. Ricchiazzi and Canfield, 1983) and self-consistent synthetic models (e.g. Hawley and Fisher, 1994). Significant inroads are being made into the near-IR, which holds promise for extending solar magnetographic measurements into the corona (Lin *et al.* 2004). Despite their importance in understanding the flare phenomenon, emphasis here is placed on direct radiative signatures of energetic particles whereas optical and UV emissions originate in thermal or quasi-thermal plasmas with temperatures $T \sim 10^4$ K. Even so, it must be recognized that the interpretation of radiation from energetic particles (particularly hard X-ray and γ-ray radiation) relies on a detailed understanding of the nature of the relatively cold target material with which energetic particles interact – its temperature and density structure, ionization state, and abundances – which must be deduced from observations in the IR/O/UV bands. Significant advances

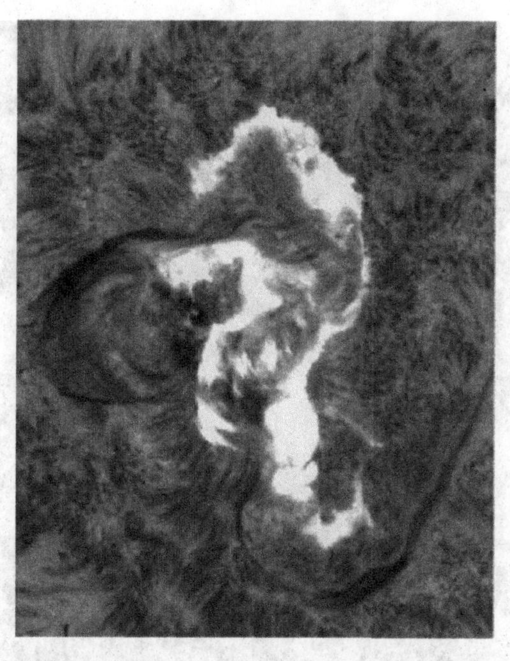

Fig. 4.4. A two-ribbon solar flare in the chromospheric Hα line at 656.28 nm.

are anticipated with the construction of the Advanced Technology Solar Telescope (ATST) in the coming decade.

4.1.4 EUV/soft-X ray emission

Copious emission at extreme UV (~10–120 nm, or 100–1200 Å) and soft X-ray (~0.1–10 nm, 1–100 Å, or 0.1–10 keV) wavelengths is produced by thermal/quasi-thermal plasma during flares. The EUV/soft X-ray flare spectrum is characterized by a wealth of emission lines superposed on a continuum that increases in importance with decreasing wavelength and with increasing plasma temperature. Spectral lines dominate the spectrum up to a plasma temperature $\sim 10^7$ K whereas continuum emission dominates for higher temperatures. Continuum emission is produced by free–free and recombination radiation, as well as two-photon decay. Spectral lines are produced as a result of collisional excitation of ions and their subsequent spontaneous decay. Line emission offers rich diagnostic possibilities for establishing the thermodynamic state of the hot plasma, elemental abundances, and its ionization state, although a detailed quantitative understanding of the relevant atomic physics (e.g. collision and absorption cross sections, oscillator strengths, excitation rates, spontaneous emission rates) is required. Modeling tools such as the CHIANTI software suite are invaluable in this regard (Dere *et al.*, 1997b; Landi *et al.*, 2006).

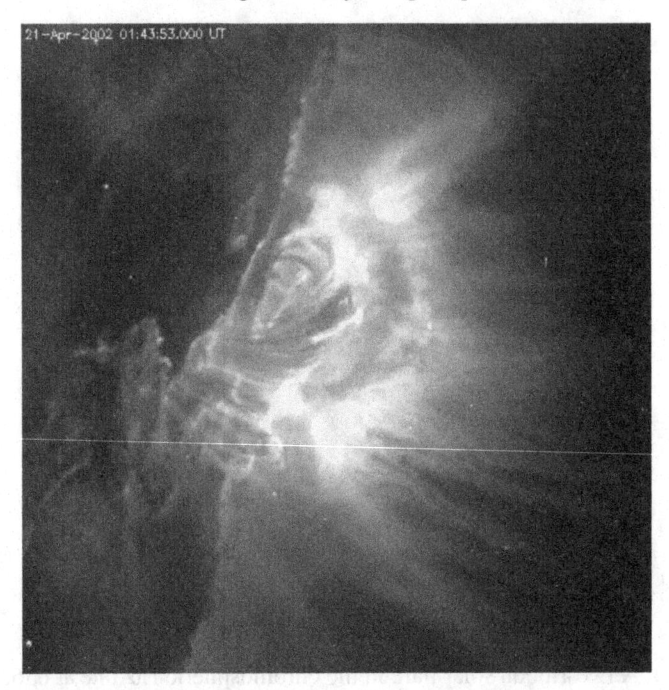

Fig. 4.5. A TRACE observation of the 171Å, or 17.1 nm, EUV emission from the X1.5 solar flare (see Table 5.1 for the flare magnitude scale) on 21 April 2002. (From Gallagher *et al.*, 2002.)

Solar physics has enjoyed remarkable opportunities in exploring this wavelength regime over the past three decades. Instruments include the Solar Maximum Mission UV Spectrometer and Polarimeter and soft X-ray Polychromator; the Yohkoh Soft X-ray Telescope (SXT); the Solar and Heliophysics Observatory (SOHO) Coronal Diagnostic Spectrometer, EUV Imaging Telescope, UV Coronal Spectrometer, and Solar UV Measurements of Emitted Radiation; the Transition Region and Coronal Explorer (TRACE; see example image in Fig. 4.5), and the Hinode EUV Imaging Spectrometer. The Solar Dynamics Observatory (SDO) continues to build on this foundation.

4.1.5 Hard X-ray/γ-ray emission

Similarly to high-frequency radio emission, continuum hard X-ray (~10–300 keV) and γ-ray (>300 keV) emissions originate from the most energetic electrons accelerated in a solar flare, although the physical mechanisms differ (Section 4.3). In addition, in the case of γ-rays, a host of additional energetic particles produce radiative signatures – positrons, pions, protons, neutrons, and heavy ions – thereby enabling additional diagnostic possibilities. These are discussed in greater detail in

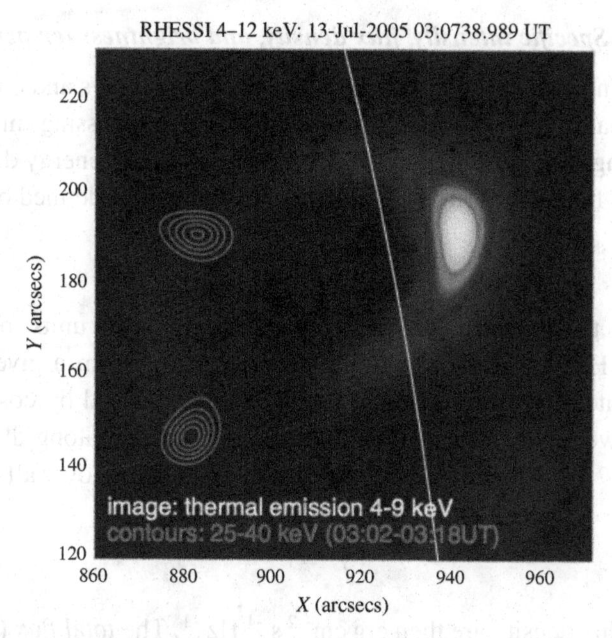

Fig. 4.6. An example of hard X-ray emissions from a solar flare from 4 to 9 keV (background image) and 25 to 40 keV (grey contours) by the RHESSI satellite near the solar limb (white curve). (From Bastian *et al.*, 2010.)

Section 4.3.2. The first hard X-ray observations (>20 keV) date back to Peterson and Winckler (1959), whereas the first γ-ray line observations were those of Chupp *et al.* (1973). Extensive work in hard X-ray and γ-ray spectroscopy has since been conducted. The first spatially resolved hard X-ray observations were made with a balloon-borne modulation collimator in the range 30–60 keV (Takakura *et al.*, 1971). Dedicated hard X-ray imaging experiments were subsequently carried by successive missions: SMM Hard X-ray Imaging Spectrometer and Hinotori Solar X-ray Telescope in the 1980s, followed by the highly successful Yohkoh Hard X-ray Telescope (HXT) in the 1990s. Prominent among γ-ray experiments were the SMM Gamma-ray Spectrometer and the Compton Gamma-Ray Observatory (CGRO). A key mission to both hard X-ray and γ-ray studies is the Ramaty High Energy Solar Spectroscopic Imager (RHESSI), launched in 2002. An example of hard X-ray radiation from a solar flare in HXR is shown in Fig. 4.6.

4.2 Preliminaries

Before discussing specific radio, hard X-ray, and γ-ray emission mechanisms and the circumstances under which they occur, it is worth digressing in order to introduce some essential concepts and terminology that are used in later sections.

4.2.1 Specific intensity, flux density, and brightness temperature

For the moment, think of radiation as propagating along rays and consider a small area dA normal to a given ray $\mathbf{r}$. Now consider all rays passing through dA with directions lying within a solid angle $d\Omega$ centered on $\mathbf{r}$. The energy dE crossing dA in a time dt, a frequency range $d\nu$, in the solid angle $d\Omega$ is defined by

$$dE = \mathcal{I}_\nu \, dA \, dt \, d\Omega \, d\nu, \tag{4.2}$$

where $\mathcal{I}_\nu$ is the *specific intensity* or *brightness*. The units of $\mathcal{I}_\nu$ are erg cm^{-2} s^{-1} sr^{-1} Hz^{-1}. The flux of radiation through dA from a given direction is the specific intensity times the solid angle $d\Omega$, but reduced by $\cos\theta$, where θ is the angle between the normal of dA and the ray directed along $d\Omega$. So we have $dF_\nu = \mathcal{I}_\nu \cos\theta \, d\Omega$. The *flux density* is given by integrating over all directions:

$$F_\nu = \int \mathcal{I}_\nu \cos\theta \, d\Omega. \tag{4.3}$$

The units of flux density are then erg cm^{-2} s^{-1} Hz^{-1}. The *total flux* (erg cm^{-2} s^{-1}), obtained by integrating over frequency (or a restricted frequency or energy range), is also sometimes of interest as is the *fluence* (erg cm^{-2} or counts cm^{-2}), which is the total flux integrated over a given time interval.

Now consider a distant source of radiation observed by a telescope. If the source is resolved, we have a measure of the radiation flux density per resolution element. The resolution element, determined by the properties of the instrument, subtends a solid angle Ω_0. Then, because $F_\nu = \mathcal{I}_\nu \Omega_0$, $\mathcal{I}_\nu = F_\nu / \Omega_0$. It represents an imperfect measure of the specific intensity of the source – imperfect because it is limited by the finite angular resolution Ω_0 of the instrument. Other imperfections arise from the practical limitations imposed by finite integration times, finite wavelength or energy bandwidths, the presence of noise, and of measurement errors.

In practice, a variety of units are used to represent measurements of specific intensity, flux density, and related quantities. For example, radio flux density is expressed in units of jansky, where 1 Jy = 10^{-26} W m^{-2} Hz^{-1} = 10^{-23} erg cm^{-2} s^{-1} Hz^{-1}. In the case of solar radio observations, the flux density is typically expressed in *solar flux units*, where 1 SFU = 10^4 Jy. The resolution element of radio imaging data is referred to as the "beam". Hence, radio brightness or specific intensity is often expressed in units of Jy beam^{-1}. Observations in other wavelength regimes often find it convenient to express flux densities in terms of wavelength (e.g. erg cm^{-2} s^{-1} Å^{-1}) or energy (e.g. erg cm^{-2} s^{-1} keV^{-1}). Observations at higher energies (EUV, soft X-ray, hard X-ray, and γ-ray), where measurements are based on photon counts, sometimes report flux densities as counts cm^{-2} s^{-1} Å^{-1} or photons cm^{-2} s^{-1} per energy unit (e.g. keV, MeV, or GeV).

4.2.2 Thermodynamic equilibrium and departures therefrom

A system of matter and radiation is in *thermodynamic equilibrium* when it is characterized everywhere by a single temperature T. In the case of a gas with a total particle number density N that is isotropic in particle velocities, the differential energy distribution of the particles is described by the *Maxwell–Boltzmann* distribution,

$$n(E)\mathrm{d}E = \left(\frac{2}{\pi}\right)^{1/2} \frac{N}{k_\mathrm{B}T}\left(\frac{E}{k_\mathrm{B}T}\right)^{1/2} \exp\left(\frac{E}{k_\mathrm{B}T}\right)\mathrm{d}E, \tag{4.4}$$

also referred to as a *Maxwellian* distribution. Here, E is the particle energy and k_B is the Boltzmann constant.

The specific intensity of the radiation from a system in thermal equilibrium is referred to as *blackbody radiation*; it is fully specified under conditions of thermal equilibrium by a universal function of T and ν; i.e. $I_\nu = B_\nu(T)$. The function $B_\nu(T)$ is called the *Planck function*. A derivation of the Planck function, based on thermodynamic and quantum mechanical arguments, can be found in Rybicki and Lightman (1979) and many other textbooks. It is given by

$$B_\nu(T) = \frac{2h\nu^3/c^2}{\mathrm{e}^{h\nu/k_\mathrm{B}T} - 1}, \tag{4.5}$$

where h is the Planck constant. It is worth noting that while thermodynamic equilibrium does not rigorously hold true in the Sun's atmosphere, the Sun's gross spectral distribution is well described by a blackbody with a temperature of 5780 K which, via *Wien's displacement law*, has its maximum at 500 nm.

At radio wavelengths the frequency of the radiation is such that $h\nu \ll kT$ so that $\mathrm{e}^{h\nu/kT} - 1 \approx h\nu/kT$. Therefore, to a high degree of accuracy we have the simple expression referred to as the *Rayleigh–Jeans law*:

$$B_\nu(T) \approx \frac{2\nu^2}{c^2}kT. \tag{4.6}$$

Given the simplicity of the expression for the specific intensity in the Rayleigh–Jeans regime, it is useful to characterize the specific intensity at a particular frequency by the temperature of the blackbody having the same brightness at that frequency. We refer to this temperature as the *brightness temperature* T_B; it is defined through the expression

$$\mathcal{I}_\nu = \mathcal{B}_\nu(T_B) = \frac{2\nu^2}{c^2}kT_B. \tag{4.7}$$

In addition to having the advantage of being related to the physical properties of the source of radiation, it also has the advantage of simple units (kelvin, as opposed to erg cm^{-2} s^{-1} sr^{-1} Hz^{-1}).

In the case of coronal plasmas the particle distribution function is locally Maxwellian even though it is largely decoupled from the radiation field, which is dominated by photospheric emission. The particle distribution is determined by collisions, as are the excitation and ionization states. Radiation from a plasma with a Maxwellian distribution is generally referred to as *thermal* radiation and, if the plasma is optically thick, the radiation is well approximated by the Planckian. In contrast, *non-thermal* radiation refers to radiation that arises from any non-equilibrium, non-thermal (i.e. non-Maxwellian and/or anisotropic) distribution of particles. As an important example, a high-energy non-thermal "tail" of the electron distribution function produces non-thermal radio and hard X-ray/γ-ray emission. The distribution is often well described by a power-law distribution. For a power-law distribution with a low-energy cutoff E_c,

$$n(E)\,\mathrm{d}E = CNE^{-\delta}\,\mathrm{d}E, \tag{4.8}$$

where in this case N is the total number density of particles with energies $E > E_c$ and $C = (\delta - 1)E_c^{\delta-1}$.

4.2.3 Radiative transfer

Rays traveling in free space have a constant specific intensity: $\mathrm{d}I_\nu/\mathrm{d}s = 0$. Only emission, absorption, and/or scattering along the ray path can change the specific intensity. Scattering is discussed briefly in Section 4.3.1 (radio) and Section 4.3.2 (HXR). The *emission coefficient*, j_ν, is defined as the energy emitted per unit time per unit solid angle per unit volume per unit frequency: $\mathrm{d}E = j_\nu\,\mathrm{d}V\,\mathrm{d}\Omega\,\mathrm{d}t\,\mathrm{d}\nu$. It therefore has units of erg cm^{-3} s^{-1} sr^{-1} Hz^{-1}. Comparing with the definition of $\mathcal{I}_\nu$, it's seen that for rays traveling a distance $\mathrm{d}s$, a beam of cross section $\mathrm{d}A$ travels through a volume $\mathrm{d}V = \mathrm{d}A\,\mathrm{d}s$ and the incremental intensity added to the beam is $\mathrm{d}I_\nu = j_\nu\,\mathrm{d}s$. The beam loses energy by absorption as it travels a distance $\mathrm{d}s$. The *absorption coefficient*, α_ν, is defined by $\mathrm{d}\mathcal{I}_\nu = -\alpha_\nu \mathcal{I}_\nu\,\mathrm{d}s$ where the convention is $\alpha_\nu > 0$ for energy removed from the beam. To see this phenomenologically, consider a random distribution of absorbing particles with a number density n in some volume. Suppose the effective absorbing area (cross section) of each particle is σ_ν cm^2. For a beam traversing an area $\mathrm{d}A$ the total absorbing area A_{abs} over a path length $\mathrm{d}s$ is $n\sigma_\nu\,\mathrm{d}A\,ds$ and the incremental energy removed from the beam by the absorbing particles is

$$-\mathrm{d}\mathcal{I}_\nu\,\mathrm{d}A\,\mathrm{d}\Omega\,\mathrm{d}t\,\mathrm{d}\nu = \mathcal{I}_\nu A_{\mathrm{abs}}\,\mathrm{d}\Omega\,\mathrm{d}t\,\mathrm{d}\nu = \mathcal{I}\nu(n\sigma_\nu\,\mathrm{d}A\,\mathrm{d}s)\,\mathrm{d}\Omega\,\mathrm{d}t\,\mathrm{d}\nu. \tag{4.9}$$

We then have $\mathrm{d}\mathcal{I}_\nu = -n\sigma_\nu\mathcal{I}_\nu\,\mathrm{d}s$ and we identify $\alpha_\nu = n\sigma_\nu$ in the present case. The units of the absorption coefficient are cm^{-1}. An equation describing the change in specific intensity along a ray can now be written as

$$\frac{d\mathcal{I}_\nu}{ds} = -\alpha_\nu \mathcal{I}_\nu + j_\nu. \tag{4.10}$$

This is the *radiative transfer equation*. It describes the macroscopic behavior of radiation in an emitting and absorbing medium, hiding all of the microscopic physics in α_ν and j_ν. It is useful to recast the equation in a more intuitive form. Defining the *optical depth* τ_ν through $d\tau_\nu = \alpha_\nu\, ds$ and the *source function* as $\mathcal{S}_\nu = j_\nu/\alpha_\nu$, the transfer equation can be rewritten as

$$\frac{d\mathcal{I}_\nu}{d\tau_\nu} = -\mathcal{I}_\nu + \mathcal{S}_\nu. \tag{4.11}$$

For the simple case of an isolated source with a constant source function the solution is

$$\mathcal{I}_\nu(\tau_\nu) = \mathcal{I}_\nu(0)e^{-\tau_\nu} + \mathcal{S}_\nu(1 - e^{-\tau_\nu}). \tag{4.12}$$

It is seen that as τ_ν becomes large, $\mathcal{I}_\nu$ approaches $\mathcal{S}_\nu$. We refer to media where $\tau_\nu \gg 1$ as being *optically thick* and those where $\tau_\nu \ll 1$ as being *optically thin*. Note that in thermodynamic equilibrium the source function $\mathcal{S}_\nu = \mathcal{B}_\nu$, resulting in *Kirchoff's Law*: $j_\nu = \alpha_\nu \mathcal{B}_\nu(T)$.

Given the relationship between $\mathcal{I}_\nu$ and T_B it is also useful to identify an *effective temperature* $T_{\rm eff}$ with $\mathcal{S}_\nu$, defined through $\mathcal{S}_\nu = kT_{\rm eff}\nu^2/c^2$. The radiative transfer equation can then be written as

$$\frac{dT_B}{d\tau_\nu} = -T_B + T_{\rm eff}, \tag{4.13}$$

which, when $T_{\rm eff}$ is constant, has the solution $T_B = T_{\rm eff}(1 - e^{-\tau_\nu})$ in the absence of background emission. When the source is optically thick $T_B = T_{\rm eff}$. Note that when the emitting particles are Maxwellian, $T_{\rm eff} = T$. If the emitting particles are non-Maxwellian, $T_{\rm eff}$ represents the mean energy of the emitting particles. When the source is optically thin the exponential can be expanded and we have $T_B \approx \tau_\nu T_{\rm eff}$.

4.2.4 Polarization

The polarization properties of radiation carry the imprint of the emission mechanism that produced it and its subsequent interaction with particles and fields as it propagates to a distant observer. Consider a monochromatic plane wave propagating along direction $\mathbf{k}$ and choose a coordinate system such that $\mathbf{k}$ is parallel to $\hat{\mathbf{z}}$. In the classical picture, the electric field simply oscillates in a fixed direction perpendicular to $\mathbf{k}$, which we take to be $\hat{\mathbf{y}}$. More generally, radiation may be described as a superposition of two such waves that are orthogonal; i.e. the electric field of one is aligned with $\hat{\mathbf{y}}$ and the other is aligned with $\hat{\mathbf{x}}$. In the general case, then, the electric field can be expressed as

$$\mathbf{E} = (E_x\hat{\mathbf{x}} + E_y\hat{\mathbf{y}})e^{-i2\pi\nu t}. \tag{4.14}$$

Both E_x and E_y are complex quantities to express the fact that each has a *phase*. That is,

$$E_x = \mathcal{E}_x e^{i\phi_x}, \qquad E_y = \mathcal{E}_y e^{i\phi_y}. \tag{4.15}$$

In the general case, the tip of the electric field vector traces out an ellipse in the x–y plane. The polarization state is fully characterized by the *Stokes parameters*:

$$\mathcal{I} = \mathcal{E}_x^2 + \mathcal{E}_y^2, \tag{4.16}$$

$$\mathcal{Q} = \mathcal{E}_x^2 - \mathcal{E}_y^2, \tag{4.17}$$

$$\mathcal{U} = 2\mathcal{E}_x\mathcal{E}_y \cos(\phi_x - \phi_y), \tag{4.18}$$

$$\mathcal{V} = 2\mathcal{E}_x\mathcal{E}_y \sin(\phi_x - \phi_y), \tag{4.19}$$

where $\mathcal{I}$ represents the total intensity. Consider two special cases: if the phase difference between $\mathcal{E}_x$ and $\mathcal{E}_y$ is such that $\phi_x - \phi_y = 0$ or π, then $\mathcal{V} = 0$ and the ellipse collapses into a line and the radiation is said to be *linearly* polarized. The ratio of $\mathcal{U}$ to $\mathcal{Q}$ describes the orientation of the electric field in x–y plane. The *degree of linear polarization* is $\rho_l = \sqrt{\mathcal{Q}^2 + \mathcal{U}^2}/\mathcal{I}$. If $\mathcal{E}_x = \mathcal{E}_y$ and $\phi_x - \phi_y = \pi/2$ then $\mathcal{U} = \mathcal{Q} = 0$ and the ellipse becomes a circle and the electric field vector rotates counterclockwise when $\mathbf{k}$ points toward the observer. In this case the wave is said to be *right circularly polarized* (RCP). If $\phi_x - \phi_y = -\pi/2$, then the electric field vector rotates clockwise when $\mathbf{k}$ points toward the observer and the wave is said to be *left circularly polarized* (LCP). The *degree of circular polarization* is $\rho_c = \mathcal{V}/\mathcal{I}$.

Polarization at radio wavelengths warrants further elaboration. The propagation of radio waves in a magnetoactive plasma is complex but the magnetoionic approximation, wherein the plasma is taken to be cold and the ion motion is ignored, often suffices. The dispersion relation yields four electromagnetic modes, two of which can escape the plasma: the extraordinary mode (x-mode) and the ordinary mode (o-mode). Under most circumstances the so-called quasi-circular approximation applies (e.g. Melrose 1980). The two modes are approximately circularly polarized and propagate independently of each other. Under certain circumstances the quasi-circular approximation breaks down and *mode coupling* may occur (e.g. Cohen 1960; Section 4.3.1). Under some circumstances, radio waves may be produced that are intrinsically linearly polarized. A linearly polarized signal propagating in a magnetoactive plasma experiences *Faraday rotation*, where the plane of polarization rotates as the wave propagates. The angle through which it rotates is $\phi = RM\,\lambda^2$, where RM is the *rotation measure*, given by

$$RM = \frac{e^3}{2\pi m_e^2 c^4} \int n_e B_{\parallel} \, ds. \tag{4.20}$$

Faraday rotation measurements, sensitive to the electron number density and the (longitudinal) magnetic field along the line of sight are an important diagnostic. Unfortunately, Faraday rotation is extremely large in the solar corona, so much so that over typical observing bandwidths the electric field vector executes many turns, thereby washing out any linearly polarized signal. Solar radio emission is not expected to be linearly polarized, in general, but it is often circularly polarized. Stokes $\mathcal{I}$ and $\mathcal{V}$ are therefore the relevant observables.

In contrast to solar radio emission, non-thermal hard X-ray radiation is expected to be significantly linearly polarized under some circumstances (i.e. $\mathcal{Q}, \mathcal{U} \neq 0$), as discussed in Section 4.3.2. It is not expected to be circularly polarized ($\mathcal{V} = 0$), however.

4.2.5 Incoherent and coherent radiation

Emission from an ensemble of particles is said to be *incoherent* if they each radiate independently of one another. If this is not true, and emission is correlated between particles, the emission is said to be *coherent.* In other words, the emitting particles no longer emit independently and collective effects must be taken into account. Coherent radiation should not be confused with non-thermal radiation. An example of circumstances that can yield coherent emission is when inverted level populations in an atom or molecule occur. In such cases, the absorption coefficient can be negative. Rather than decreasing along a ray path, the intensity increases exponentially (negative absorption). Free electrons can also produce coherent radiation when they bunch in phase and/or have an anisotropic distribution function (e.g. beam, loss cone). Most electromagnetic emissions encountered in astrophysics are the result of incoherent emission mechanisms but there are some outstanding exceptions. Pulsars emit coherent radio emission, for example, as do molecular masers in the outer envelopes of late-type giant stars. On the Sun, coherent mechanisms also play a prominent role at radio frequencies below a few GHz, as described in the next section. Coherent emission mechanisms do not play a significant role outside of the radio regime on the Sun.

4.3 Radiation from energetic particles

Energetic particles represent non-equilibrium distributions of particles that are produced on the Sun and in the heliosphere, notably by flares and the shocks driven by fast CMEs. Our interest in radiation from energetic particles is to use it to infer, as best we can, the nature of the emitting particles and the environment in which they occur. In so doing, we wish to gain insights into the deeper questions of *how* and *why* such distributions of energetic particles form. These questions are by no means parochial; they are relevant to the whole of astrophysics. In this section we discuss the dominant emission mechanisms at radio, hard X-ray, and γ-ray wavelengths. These include non-thermal gyrosynchrotron and coherent plasma radiation at radio wavelengths, non-thermal bremsstrahlung at hard X-ray and γ-ray wavelengths, and a variety of processes involving nuclear interactions and their secondary products at γ-ray wavelengths. In all cases, non-equilibrium, non-thermal distributions of energetic electrons, protons, and ions are involved.

4.3.1 Radio emission

Radio emission is produced by accelerating free electrons. No bound–bound or free–bound transitions occur in molecules, atoms, or nuclei that produce photons in the radio frequency range. Therefore, no spectral lines are available for study at radio frequencies. Radio emission tells us about *energetic electrons* and their environment.

Consider a single charge of mass m and charge q undergoing an acceleration $\mathbf{a}$. The power emitted into a solid angle $\mathrm{d}\Omega$ is

$$\frac{\mathrm{d}P}{\mathrm{d}\Omega} = \frac{q^2}{4\pi c^3}\mathbf{a}^2 \sin^2\theta, \tag{4.21}$$

where θ is the angle relative to the vector along which the particle is accelerated. The power radiated is proportional to (charge $\times$ acceleration)^2 and the *radiation pattern* is dipolar ($\sin^2\theta$ pattern). Furthermore, the emission is peaked in the direction perpendicular to the acceleration vector. Since acceleration $a = F/m$ it is easy to see that, all other things being equal, electrons emit $(m_\mathrm{p}/m_\mathrm{e})^2$ times the power that protons do. Integrating over all solid angles $\mathrm{d}\Omega = \sin\theta\,\mathrm{d}\theta\,\mathrm{d}\phi$ yields the *Larmor formula* for the total radiation power emitted by a single accelerating charge:

$$P = \frac{2q^2\mathbf{a}^2}{3c^3}. \tag{4.22}$$

The relativistic counterparts to Eqs. (4.21) and (4.22) are given by

$$\frac{\mathrm{d}P}{\mathrm{d}\Omega} = \frac{q^2}{4\pi c^3}\frac{(a_\perp^2 + \gamma^2 a_\parallel^2)}{(1-\beta\cos\theta)^4}\sin^2\theta, \tag{4.23}$$

$$P = \frac{2q^2}{3c^3}\gamma^4(a_\perp^2 + \gamma^2 a_\parallel^2), \tag{4.24}$$

where $a_\perp$ and $a_\parallel$ are the acceleration perpendicular and parallel to the velocity vector of the charged particle q, respectively, and $\gamma = 1/\sqrt{1-(v/c)^2}$ is the *Lorentz factor*. The chief effect of relativistic particle speeds is beaming.

(a)

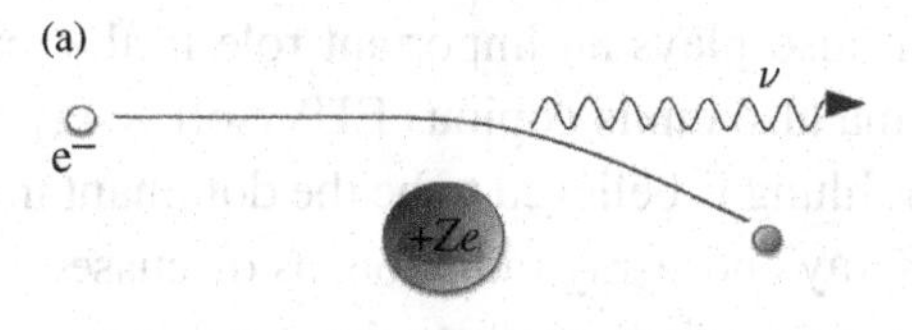

(b)

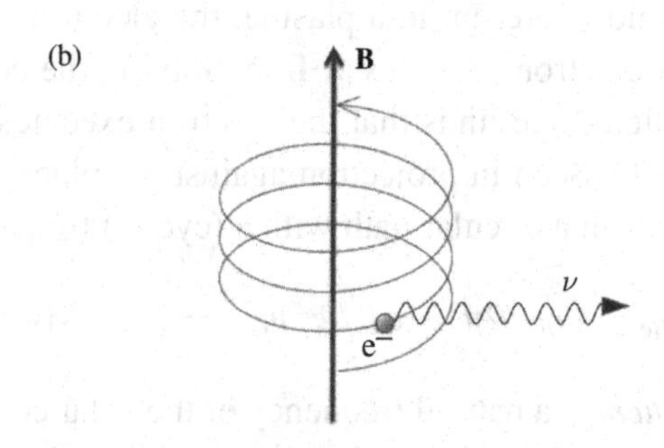

Fig. 4.7. (a) Schematic illustration of bremsstrahlung, or free–free, emission resulting from the collision of an electron with an ion; (b) schematic illustration of gyromagnetic emission resulting from the gyration of an electron in a magnetic field.

One of the most familiar radiation mechanisms is electron *bremsstrahlung*, or *free–free* radiation. An electron passing near a proton experiences an impulsive acceleration as a result of the Coulomb force and consequently emits radiation (see Fig. 4.7a). A convenient and intuitive means of formulating the problem is first to consider the radiation power from a single electron–proton (denoted e^-–p here) collision via Larmor's formula and then to compute the collision rate and sum over all such collisions. The rate ν_C at which an electron with a speed v scatters on protons in a volume with a number density n can be expressed as $\nu_C = n\sigma(v)v$ where $\sigma(v)$ is the speed-dependent (or more generally, energy-dependent) differential cross section (units $\mathrm{cm^2\,keV^{-1}}$). This approach can be carried over to many other particle–particle interactions. In particular, the use of cross sections (generally differential in both energy and solid angle) is a convenient way to parameterize a variety of particle interactions leading to radiation, excitation, and ionization processes, as well as the reverse processes (absorption, de-excitation, and recombination, respectively).

Integration over the contributions of electrons in a thermal or non-thermal energy distribution yields thermal bremsstralung or non-thermal bremsstrahlung, respectively. Non-thermal bremsstrahlung is not a significant factor at radio wavelengths but thermal bremsstrahlung from the quiet solar atmosphere is ubiquitous at all wavelengths.[†] It also plays an important role in the gradual phase of flares when hot, dense plasma also emits copious EUV/soft X-ray radiation. In contrast, non-thermal bremsstrahlung is believed to be the dominant mechanism responsible for continuum hard X-ray and γ-ray radiation, as discussed in Section 4.3.2.

A second radiation mechanism involves the acceleration of free electrons by a *field*. If a magnetic field is present in a plasma, the electrons and ions experience the *Lorentz* force; for electrons, $\mathbf{F} = e\mathbf{v} \times \mathbf{B}/c$. Solving the equation of motion in a uniform magnetic field, one finds that the electron executes a helical trajectory (see Chapters 9 and 11). Seen in projection against the plane perpendicular to the magnetic field it moves in a circular path with a (cyclic) frequency

$$\nu_{Be} = \omega_{Be}/2\pi = eB/2\pi m_e c = 2.8B \text{ MHz}, \tag{4.25}$$

the electron *gyrofrequency*, a natural frequency of the solar corona and heliosphere that falls within the radio frequency range. Similarly, an ion also executes gyromotion in a magnetic field, but in the opposite sense to an electron. Since the ion gyrofrequency, $\nu_{Bi} = ZeB/2\pi m_i c \ll \nu_{Be}$, the ion gyromotion can usually be neglected. Note that the electron mass is modified as γm_e if the electron energy is relativistic. An electron in a circular (or helical) trajectory undergoes continuous acceleration, the acceleration vector perpendicular to the instantaneous velocity, and it therefore radiates (see Fig. 4.7b). Substitution of the acceleration term into the Larmor formula yields the total power radiated by a single electron. Note that as the energy of the electron increases, relativistic beaming increases, leading to a distortion of the dipolar radiation pattern. As a consequence, the electron emits into frequencies that are integer multiples s of ν_{Be}; i.e. harmonics $\nu = s\nu_{Be}$.

The specific form of the emission and absorption coefficients therefore depends on the energy distribution of the electrons in question. When the electrons are non-relativistic, gyromagnetic radiation is called *gyroresonance* or *cyclotron* radiation. Thermal gyroresonance absorption – that is, gyroresonance absorption by a thermal plasma – is sufficient to render the corona optically thick in active regions with strong magnetic fields, above 150 G, at low harmonics of the electron gyrofrequency, i.e. frequencies $\nu = s\nu_{Be}$, $s = 1, 2, 3, \ldots$ Thermal gyroresonance emisson is an important radio emission mechanism in solar active regions and is a unique diagnostic of coronal magnetic fields. Gyromagnetic emission from mildly relativistic electrons ($\sim$100 keV to a few MeV) is referred to as *gyrosynchrotron* radiation and harmonics with $s \sim 10$–100 contribute. The spectral width of the harmonics increases with s, causing their joint contributions to merge into a broadband

[†] While thermal free–free emission is dominated by collisions between electrons and ions (mostly protons) at low frequencies, collisions between free electrons and neutral H can dominate the opacity, so-called H^- opacity, at mm, submm, and IR wavelengths.

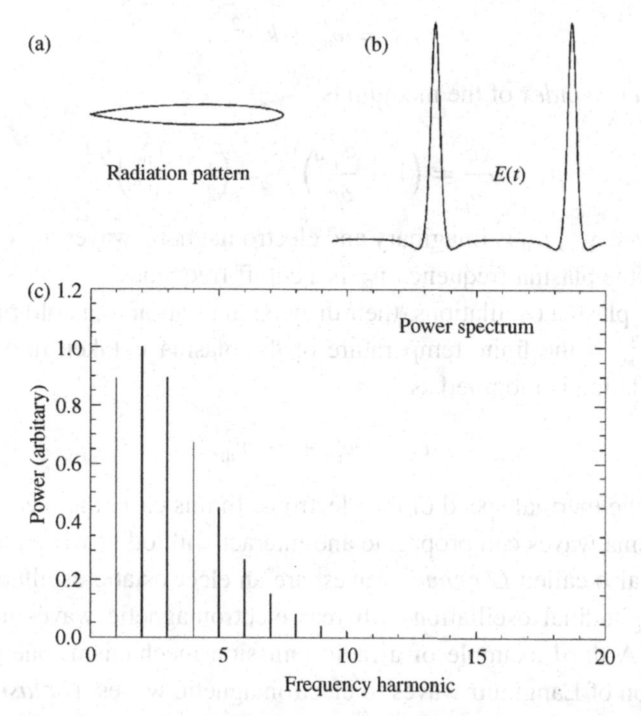

Fig. 4.8. (a) Schematic illustration of the radiation pattern of a mildly relativistic electron ($\gamma \sim 2$) gyrating in a magnetic field. The radiation is strongly beamed along the instantaneous velocity vector; (b) the time variation of the electric field measured by a distant observer; (c) the power spectrum of the free–free emission resulting from the collision of an electron with an ion.

continuum. Non-thermal gyrosynchrotron radiation is the dominant incoherent radiation mechanism during impulsive flares from dm to mm-λ. Gyrosynchrotron emission is discussed in more detail in the next section. Gyromagnetic radiation from fully relativistic electrons ($\gamma \gg 1$) is called *synchrotron* radiation, which plays a prominent role in astrophysics (e.g. radio galaxies, supernova remnants, and the galactic background).

A second natural frequency of the corona and heliosphere, the electron *plasma frequency*, also falls within the radio range. Consider a simple quasi-neutral plasma composed of protons and electrons. If a mean charge separation occurs locally, an electric field results that serves as a restoring force and an electrostatic oscillation results. The plasma oscillation frequency is

$$\nu_{\mathrm{pe}} = \frac{\omega_{\mathrm{pe}}}{2\pi} = \left(\frac{e^2 n_{\mathrm{e}}}{\pi m_{\mathrm{e}}}\right)^{1/2} \approx 9 n_{\mathrm{e}}^{1/2} \text{ kHz.} \tag{4.26}$$

Now the dispersion relation for electromagnetic waves *in vacuo* is just $\omega^2 = k^2c^2$ where $\omega = 2\pi\nu$. In a plasma, the dispersion relation is

$$\omega^2 = \omega_{\rm pe}^2 + k^2c^2. \tag{4.27}$$

and the *refractive index* of the medium is

$$\mu_\nu = \frac{kc}{\omega} = \left(1 - \frac{\omega_{\rm pe}^2}{\omega^2}\right)^{1/2} = \left(1 - \frac{\nu_{\rm pe}^2}{\nu^2}\right)^{1/2}. \tag{4.28}$$

Note that if $\nu < \nu_{\rm pe}$, μ_ν is imaginary and electromagnetic waves are evanescent. In other words, the plasma frequency $\nu_{\rm pe}$ is a cutoff frequency.

Turning to plasma oscillations, their dispersion relation in a cold plasma is simply $\omega^2 = \omega_{\rm pe}^2$. If the finite temperature of the plasma is taken into account, the dispersion relation is modified as

$$\omega^2 = \omega_{\rm pe}^2 + 3k^2v_{\rm th}^2, \tag{4.29}$$

where $v_{\rm th}$ is the thermal speed of the electrons. In this case, the medium is dispersive and plasma waves can propagate and interact with other wave modes. Plasma oscillations, also called *Langmuir* waves, are an electrostatic oscillation. As such, they are longitudinal oscillations whereas electromagnetic waves are transverse oscillations. A third example of a radio emission mechanism, one that involves the conversion of Langmuir waves to electromagnetic waves, is *plasma radiation*. Plasma radiation is a coherent mechanism that plays a dominant role in radio bursts from dm- to DH-λ, and even km-λ. Given the importance of gyrosynchrotron radiation above dm-λ and plasma radiation below cm-λ, we discuss each in somewhat more detail.

4.3.1.1 Non-thermal gyrosynchrotron radiation

The gyrosynchrotron emission and absorption coefficients have been given by Ramaty (1969) and by Benka and Holman (1992); the latter includes the correction pointed out by Trulsen and Fejer (1970). The expressions for the emission and absorption coefficients are extremely cumbersome and represented a computational challenge. This is because the expressions involve lengthy sums of terms involving Bessel functions J_s of order s and their derivatives over many harmonics s. Hence, approximate expressions were derived with varying domains of applicability and accuracy (e.g. Petrosian, 1981; Dulk and Marsh, 1982; Dulk, 1985; Robinson, 1985; Klein, 1987). While these expressions can be calculated quickly and easily, and can be used to constrain various types of models, they are often too limiting in practice. With modern computational power, the full gyrosynchrotron emission and absorption coefficients can be readily computed for simple models. Fits to more elaborate source models may still require the use of approximate expressions, however. In this section some of the essential properties of non-thermal gyrosynchrotron emission are summarized.

To begin, we illustrate the spectral characteristics of gyrosynchrotron emission from an idealized source, a homogeneous slab viewed from above at 1 AU. It has a thickness d and an area A and it is permeated by a uniform magnetic field. The angle between the magnetic field vector and the line of sight is θ. The slab is filled with a background plasma with a number density n_{th} and non-thermal electrons that are described by a simple isotropic power-law distribution with a low-energy cutoff $E_{\mathrm{c}} = 100\,\mathrm{keV}$ and an index δ. The number density of energetic electrons with $E > E_{\mathrm{c}}$ is given by n_{rl}.

Figure 4.9 shows the flux density spectrum from 1 to 30 GHz for a number of cases. The solid line in each case shows a reference spectrum for the set parameters given in the caption. Panel (a) shows the variation of the total intensity spectrum

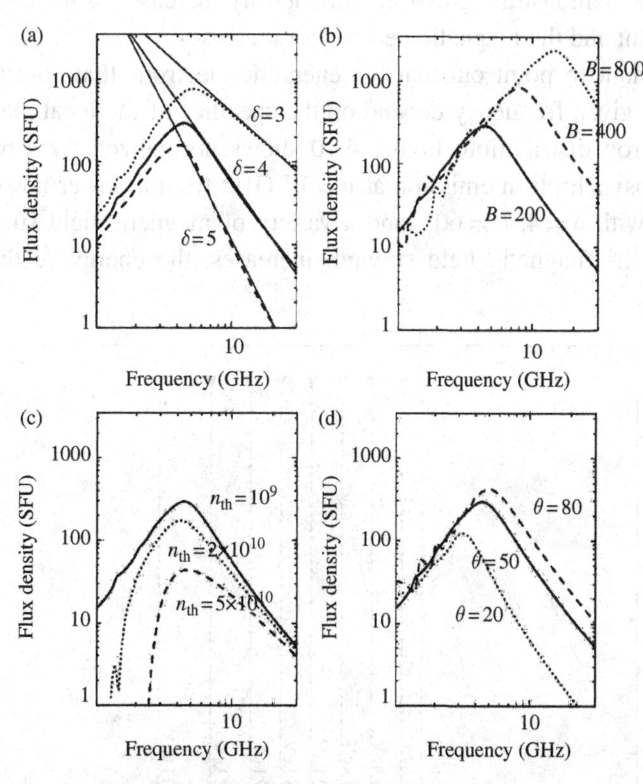

Fig. 4.9. The flux density spectrum of gyrosynchrotron emission from a power-law distribution of electrons in an ambient thermal plasma. The source area and depth are held constant at 3×10^{18} cm^2 and 10^9 cm, respectively. The solid line represents the same reference spectrum in all panels, where the magnetic field is $B = 200$ G, the low-energy cutoff is $E_{\mathrm{c}} = 100\,\mathrm{keV}$, the spectral index is $\delta = 4$, the thermal number density is $n_{\mathrm{th}} = 10^9\ \mathrm{cm}^{-3}$, and the number density of electrons with $E > E_{\mathrm{c}}$ is $n_{\mathrm{rl}} = 10^5\ \mathrm{cm}^{-3}$. Specific parameters are allowed to vary in panels (a)–(d) as shown.

with spectral index δ of the power-law electron energy distribution. The index of the resulting photon spectrum is $\alpha \approx 0.9\delta - 1.22$ (compare with $\alpha = 0.5\delta - 0.5$ for synchrotron radiation). Panel (b) shows the spectral variation of the flux density with the magnetic field strength. The emission increases dramatically with B, as does the frequency of the spectral maximum. Since, at a given frequency, lower harmonics of ν_{Be} are responsible for the emission as B increases, a richer variety of harmonic structure appears in the spectrum with increasing B. Panel (c) shows the variation of the spectrum with ambient plasma density. When the density of the background plasma is increased it is seen that, while the high-frequency emission is largely unaffected, the low-frequency emission is increasingly suppressed. The effect, called *Razin suppression*, becomes important for frequencies $\nu < \nu_{\rm R} \sim \nu_{\rm pe}^2/\nu_{Be} \approx 30 n_{\rm th}/B$. Panel (d) shows the variation of the spectrum with viewing angle, demonstrating that the flux density increases as the angle between the line of sight and the magnetic field vector increases.

It is important to point out that the energetic electrons that contribute to the emission at a given frequency depend on the specifics of the local magnetic field and the electron distribution. Figure 4.10 shows normalized contribution functions for gyrosynchrotron emission at $\nu = 17$ GHz from a power-law distribution of electrons with $\delta = 4$, $\theta = 60°$, and a variety of magnetic field strengths. It is seen that as the magnetic field strength increases, the energy of the electrons

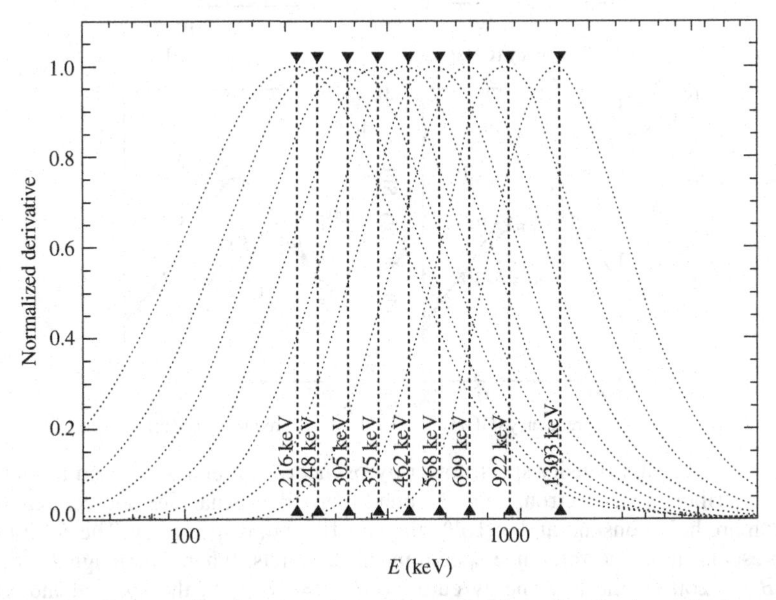

Fig. 4.10. The contribution function of a power-law distribution of electrons to gyrosynchrotron radiation at a fixed frequency of 17 GHz for values of the magnetic field varying from 200 G to 1000 G in steps of 100 G.

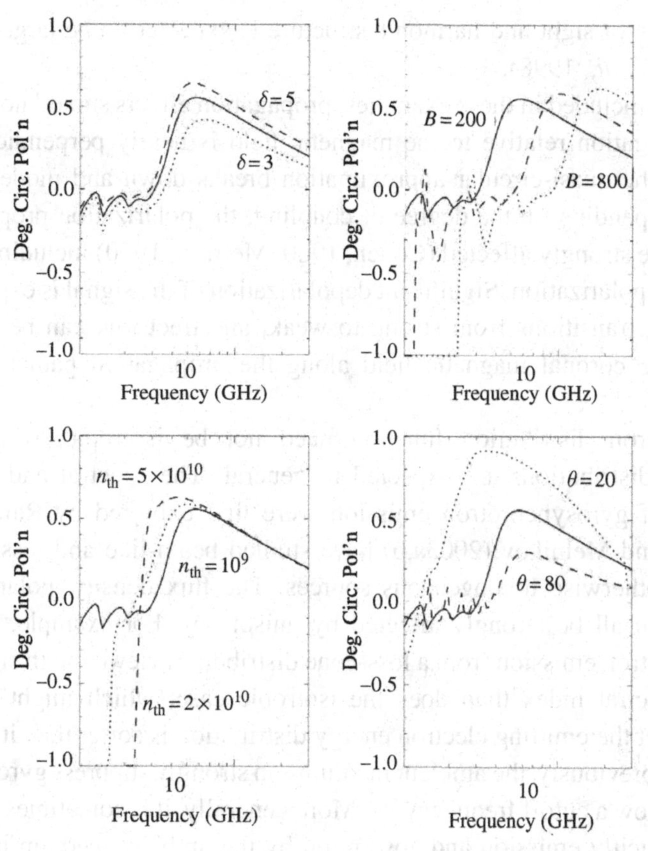

Fig. 4.11. The degree of polarization (Stokes I/V) for the cases shown in Fig. 4.9.

making the largest contribution to the emission decreases. The width of the contribution function in each case is relatively broad, with $\Delta E/E \sim 4$. Non-thermal gyrosynchrotron emission from a magnetically inhomogeneous source – a coronal magnetic loop, for example – involves emission from electrons with a broad range of energies. Spatially resolved observations of the radio spectrum are therefore of critical importance.

Gyrosynchrotron radiation is not expected to be linearly polarized for reasons given in Section 4.2.4. Nevertheless, the degree of circular polarization ρ_c is a powerful constraint on the magnetic field. Figure 4.11 shows ρ_c for each of the cases considered in the previous subsection. It is seen that when the magnetic field is strong and/or the angle between the magnetic field vector and the line of sight θ is small, the source can be significantly circularly polarized in the sense of the x-mode. The optically thick portion of the spectrum can be weakly polarized in the sense of the o-mode for a homogeneous source. A real source is inhomogeneous along the line of sight and harmonic structure is expected to be largely smoothed out (Bastian *et al.*, 1998).

While not included in these examples, propagation effects should not be ignored. When propagation relative to the magnetic field is nearly perpendicular (quasi-transverse) the quasi-circular approximation breaks down and mode coupling is possible. Depending on the degree of coupling, the polarization properties of the source can be strongly affected (Cohen, 1960; Melrose, 1980) including the degree and sense of polarization. Significant depolarization of the signal is expected where the coupling transitions from strong to weak, an effect that can be exploited to constrain the coronal magnetic field along the propagation path (e.g. Ryabov, 2004).

The electron distribution function need not be isotropic, of course, and anisotropic distributions are expected in general. The spectral and polarization properties of gyrosynchrotron emission were first explored by Ramaty (1969). Fleishman and Melnikov (2003a,b) have studied beam-like and loss-cone distributions in otherwise homogeneous sources. The flux density, polarization, and spectrum can all be strongly affected by anisotropy. For example, non-thermal gyrosynchrotron emission from a loss-cone distribution viewed with small θ shows a larger spectral index than does the isotropic case, which might lead one to conclude that the emitting electron energy distribution is softer than it actually is.

As noted previously, the ambient medium can strongly suppress gyrosynchrotron emission below a cutoff frequency ν_R. More generally, it is sometimes necessary to include explicitly emission and absorption by the ambient medium in the source function. For example, Ramaty and Petrosian (1972) approximated the source function as the ratio of the gyrosynchrotron emissivity to the thermal free–free absorption coefficient $\mathcal{S}_\nu \approx j_{gs}/\alpha_{th}$ in order to explain flat-spectrum bursts from dense plasmas; Benka and Holman (1992) developed a "thermal–non-thermal" model of microwave bursts which took account of the hot background plasma so that the source function took the form $\mathcal{S}_\nu = (j_{gs} + j_{th})/(\alpha_{gs} + \alpha_{th})$; Bastian *et al.* (2007) used an analogous source function in forward fitting model spectra to a time series of microwave spectra (see Section 4.4.2). These compound source functions introduce additional richness to the emitted spectrum which, in turn, embodies additional information about the source.

4.3.1.2 Plasma radiation

Plasma radiation is a coherent radiation mechanism wherein Langmuir waves are converted via non-linear wave–wave interactions to transverse electromagnetic waves with a frequency near the electron plasma frequency ν_{pe} or its harmonic at $2\nu_{pe}$. It is thought to be responsible for radio bursts of type II and type III, for example (see below). The production of plasma radiation is complex and, despite several decades of work, the theory must be regarded as "semi-quantitative" although real progress has been made in several areas. A relatively accessible discussion of plasma radiation mechanisms can be found in Melrose (1985). Plasma radiation is fundamentally a two-stage process: first, a spectrum of Langmuir waves must be produced; second, these must be converted to transverse electromagnetic waves.

Langmuir waves can be excited in a plasma by a number of mechanisms, but perhaps the best-studied mechanism is the propagation of a suprathermal electron beam in a background plasma (the corona or the IPM). An electron beam will produce a "bump-on-the-tail" velocity distribution function and the positive gradient will be unstable to the production of a spectrum of Langmuir waves (two-stream instability). Other types of distributions can produce Langmuir waves; for example, loss-cone and gap distributions have also been studied extensively in the literature. If the total energy density of the Langmuir waves is u_{L}, the effective temperature of the Langmuir waves, T_{L}, can be defined through

$$u_{\mathrm{L}} = \int \frac{\mathrm{d}^3 k}{(2\pi)^3} k_{\mathrm{B}} T_{\mathrm{L}}(k). \tag{4.30}$$

The effective temperature T_{t} of the transverse waves can be similarly defined.

More than one conversion mechanism is available to convert Langmuir waves to transverse electromagnetic waves with $\nu \approx \nu_{\mathrm{pe}}$. Induced scattering on thermal ions, which is analogous to Thomson scattering and can lead to exponential growth in brightness, is possible but inefficient for coronal conditions and is therefore not thought to be relevant to the solar case. Direct mode conversion on plasma density inhomogeneities is also possible in principle, but it is not believed to play a significant role in practice. The most efficient and therefore the most likely conversion process in the corona and IPM involves scattering Langmuir wave L on low-frequency ion-sound waves S to produce a transverse electromagnetic wave T: that is, $L + S \rightarrow T$. The derivation of the radiative transfer equation for fundamental plasma radiation is rather involved and requires a grounding in plasma physics. The interested reader is referred to Melrose (1980) and references therein. It can be shown that the T_{t} can grow until it saturates at $T_{\mathrm{t}} \sim T_{\mathrm{L}}$, which can be quite high ($T_{\mathrm{L}} \gtrsim 10^{12}$ K and, in some cases, much higher). The observed brightness temperature of plasma radiation is $T_B \lesssim T_{\mathrm{t}}$ because it typically suffers absorption and/or scattering (Section 4.3.1.3) in the overlying plasma as it propagates from the source. Since the optical depth to thermal free–free absorption increases as ν^2, plasma radiation is largely confined to frequencies below a few GHz (e.g. Benz, 2000).

In contrast to the production of fundamental plasma radiation, where several mechanisms are available in principle, the only viable mechanism for harmonic plasma radiation is through nearly head-on coalescence: $L + L' \rightarrow T$. The direct production of a suitable angular distribution of Langmuir waves can be problematic. For example, the streaming instability produces Langmuir waves that are collimated with the stream velocity. A suitable spectrum of secondary Langmuir waves can build up, however; Langmuir waves can decay to a daughter Langmuir wave and an ion-sound wave ($L \rightarrow L' + S$). Langmuir waves can then scatter on the ion-sound waves to produce fundamental plasma radiation and Langmuir waves can coalesce to produce harmonic radiation.

The natural bandwidth of both fundamental and harmonic plasma radiation from a uniform, unmagnetized plasma is expected to be very small (Melrose, 1980). In practice, the spectral signature of plasma radiation is closely tied to the mechanism responsible for the spectrum of Langmuir waves and to the medium in which they occur (corona or IPM). Two cases are briefly described: (1) electron beams, which are responsible for type III radio bursts; (2) MHD shocks, which drive coronal and interplanetary type II radio bursts.

Suprathermal electron beams, with speeds of $\sim$0.1–0.3c ($\sim$3–20 keV) result from coronal energy release. These are unstable to the production of Langmuir waves. Plasma radiation produced by an electron beam is common in the solar corona and the IPM. It is referred to as a type III radio burst. The radio spectral signature is distinctive. As the electron beam propagates through the corona or the IPM, it traverses a gradient in the electron number density. The frequency of the plasma radiation excited by the electron beam therefore changes with time since the location where Langmuir waves are excited changes with time. The frequency drift rate can be expressed as

$$\dot{\nu} = \frac{\mathrm{d}\nu_{\mathrm{pe}}}{\mathrm{d}t} = \frac{1}{2}\left(\frac{e^2}{\pi m_{\mathrm{e}} n_{\mathrm{e}}}\right)^{1/2} \frac{\mathrm{d}n_{\mathrm{e}}}{\mathrm{d}s}\frac{\mathrm{d}s}{\mathrm{d}t}. \tag{4.31}$$

Taking $v_{\mathrm{b}} = \mathrm{d}s/\mathrm{d}t$ as the speed of the electron beam along its trajectory s and $H_n = |n_{\mathrm{e}}(\mathrm{d}n_{\mathrm{e}}/\mathrm{d}s)^{-1}|$ as the density scale height, the frequency drift rate can be written as $\dot{\nu} = -\nu_{\mathrm{pe}} v_{\mathrm{b}}/2H_n$ if the density decreases along the beam trajectory, which is the case for coronal and interplanetary type III radio bursts. Thus, the spectral signature of a type III radio burst is a fast drift from high to low frequencies as it propagates away from the Sun. Observations indicate instantaneous bandwidths $\Delta\nu/\nu$ of a few tenths or greater. The bandwidth at any given instant is determined by the density inhomogeneity of the source volume encountered by the beam, the velocity dispersion of the beam Δv_{b}, and propagation effects (Robinson and Cairns, 1998).

A second type of disturbance that occurs in the corona and IPM is an MHD shock caused by the explosive energy release of a flare (blast wave), flare ejecta, or a fast CME. Shocks can also yield Langmuir waves that commonly produce fundamental and harmonic plasma radiation. Shock-driven radio bursts are called type II radio bursts. While coronal and interplanetary shocks are super-Alfvénic, they propagate at speeds far lower than those of type III radio bursts. Replacing v_{b} with the shock speed v_s in Eq. (4.31) yields frequency drift rates that are much slower than those associated with type III radio bursts. Typical frequency drift rates of coronal type II radio bursts are of order 1 MHz s^{-1} or less. Type II bandwidths are smaller than those of type III bursts, typically 15–25%, although the emission lanes often show substructure that is of much smaller bandwidth. An example of an event displaying both coronal type II and type III radio bursts is shown in Fig. 4.12.

The presence of a magnetic field leads to circularly polarized plasma emission. The refractive index μ_ν of a magnetized plasma is characterized by cutoffs and resonances. A cutoff corresponds to a zero in the refractive index whereas a resonance corresponds to an infinity. In the absence of a magnetic field, plasma radiation is unpolarized and electromagnetic waves are subject to a single cutoff at $\omega_{\rm pe}$. In the presence of a magnetic field, the o-mode cutoff is $\omega_{\rm o} = \omega_{\rm pe}$ whereas the x-mode cutoff is

$$\omega_{\rm x} = \frac{1}{2}\left(\omega_{B{\rm e}} + \sqrt{\omega_{B{\rm e}}^2 + 4\omega_{\rm pe}^2}\right). \tag{4.32}$$

Consider fundamental plasma radiation. In the presence of a magnetic field the frequency of the Langmuir waves is

$$\omega_{\rm L} \approx \omega_{\rm pe}\left(1 + \frac{3k^2 v_{\rm th}^2}{\omega_{\rm pe}^2} + \frac{\omega_{B{\rm e}}^2}{\omega_{\rm pe}^2}\sin^2\theta\right)^{1/2}, \tag{4.33}$$

where $\omega_{B{\rm e}} = 2\pi\nu_{B{\rm e}}$ is the angular electron gyrofrequency. If $3k^2 v_{\rm th}^2/\omega_{\rm pe}^2 \lesssim \omega_{B{\rm e}}/\omega_{\rm pe}$, a condition that should be easily satisfied, then Langmuir waves are emitted *below* the x-mode cutoff, in which case only the o-mode propagates and the fundamental plasma radiation is expected to be 100% circular polarized in the sense of the o-mode. In reality, while fundamental plasma radiation is observed to be polarized in the sense of the o-mode, as expected, it is rarely observed to be strongly polarized, with the exception of type I noise storms (Kai *et al.*, 1985). Fundamental type III bursts have been observed to be as much as 60% circularly polarized. It is speculated that a depolarization mechanism is typically operative along the propagation path (Melrose, 1985).

The treatment of polarization for harmonic plasma radiation is complex because it balances several small effects. Melrose (1980) summarized the results. Briefly, for harmonic radiation produced by Langmuir waves collimated along the magnetic field and $\omega_{B{\rm e}} \ll \omega_{\rm pe}$ the degree of polarization is $\rho_{\rm c} \approx 0.2\omega_{B{\rm e}}/\omega_{\rm pe}$ in the sense of the o-mode. If the Langmuir waves are more nearly isotropic $\rho_{\rm c} \approx 1.8\omega_{B{\rm e}}|\cos\theta|/\omega_{\rm pe}$ in the sense of the x-mode.

4.3.1.3 Refraction and scattering

Coronal and flare plasmas are sufficiently rarefied that at most frequencies they have little effect on the propagation of radiation. However, at radio frequencies, the plasma medium can have profound effects on the propagation of radiation. Three effects have already been discussed: (1) frequency cutoffs, which prevent the propagation of radiation at angular frequencies below ω_{pe} and ω_x; (2) Razin suppression, which strongly suppresses gyrosynchrotron radiation for frequencies $\nu < \nu_R$; and (3) mode coupling, which can strongly modify the observed polarization of the radiation emitted. Additional propagation effects can affect radio waves as they propagate through the corona and IPM. These result from large-scale density gradients and random density inhomogeneities in the corona and IPM. Variations in the electron number density result in variations in the refractive index μ_ν that cause refraction and scattering of radio waves, particularly at m-λ and longer. These effects can, in turn, affect the apparent source size, position, brightness temperature, polarization, and other measured properties of the radio emission (see e.g. Bastian (2001) for a review).

4.3.2 Hard X-ray and γ-ray radiation

Hard X-ray (~10–300 keV) and γ-ray (>300 keV) photons are produced by the most energetic particles accelerated in solar flares. Electrons can be accelerated to energies above 100 MeV and nuclei can be accelerated to several GeV nucleon^{-1}. The details of the photon spectrum higher than 10 keV emitted by these particles and/or secondary particles produced by their interaction with the ambient medium therefore depend on the energy spectrum of the accelerated electrons and ions, the ratio of electrons to ions accelerated, the abundances of the ions, and the nature of the ambient plasma with which the accelerated particles interact.

4.3.2.1 Non-thermal bremsstrahlung

Bremsstrahlung was discussed briefly in Section 4.3.1. Thermal bremsstrahlung is ubiquitous at radio wavelengths but non-thermal bremsstrahlung does not play a significant role. In contrast, non-thermal electron bremsstrahlung is believed to be the dominant radiation mechanism responsible for the production of continuum hard X-ray and γ-ray emission. We first consider hard X-ray radiation, where photon energies are less than a few hundred keV, produced by electrons that are essentially non-relativistic. Consider an energetic electron with an energy E_0 incident on a uniform, fully ionized hydrogen target that has a proton number density n_p. The rate at which photons are produced with energy ϵ is just $n_p \sigma_\epsilon(E_0) v(E_0)$, where $v(E_0)$ is the electron speed and $\sigma_\epsilon(E_0)$ is the electron bremsstrahlung cross section. Note that in order to produce a photon of energy ϵ the electron must have a kinetic energy $E_0 > \epsilon$. Now consider a distribution of energetic electrons with a differential number density $N(E)\,dE$ incident on the target volume V. Since the emission is optically thin for target number densities $n_p \lesssim 10^{17}\ \mathrm{cm}^{-3}$, the total

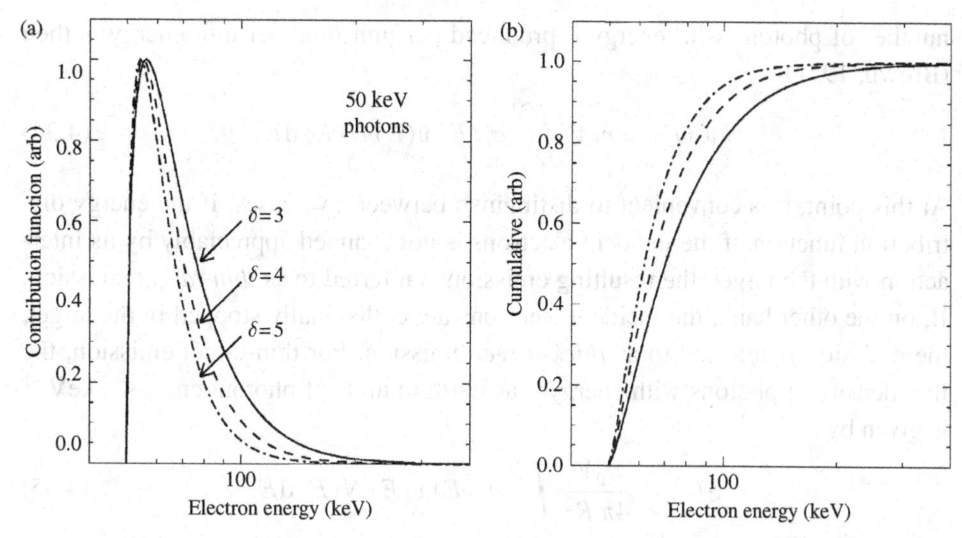

Fig. 4.12. The normalized contribution function of a power-law distribution of electrons to HXR photons with an energy $\epsilon = 50\,\mathrm{keV}$ for several values of the spectral index δ in the thin-target case, computed from the integrand of Eqn. (4.38). The corresponding cumulative distribution functions are shown in panel b. See Fig. 4.13 for the thick-target case.

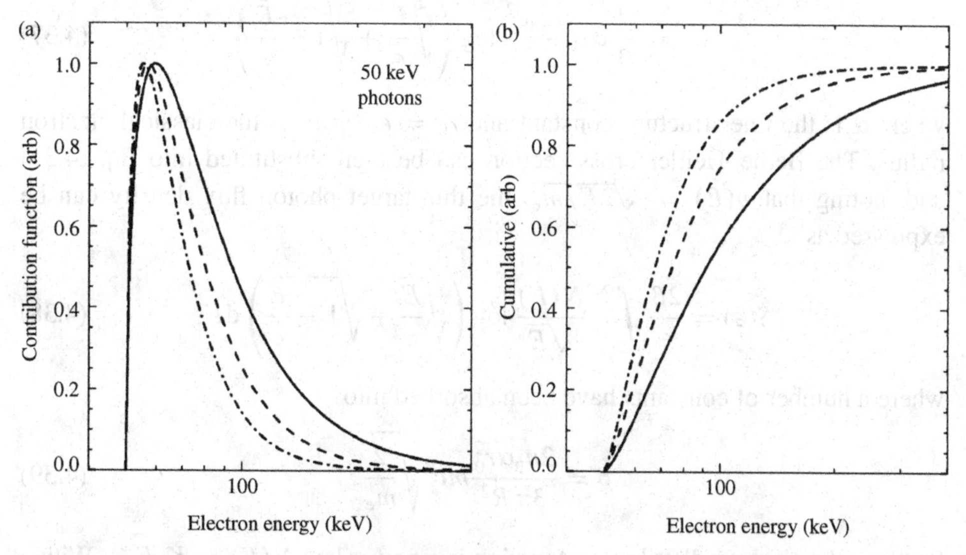

Fig. 4.13. The normalized contribution function of a power-law distribution of electrons to HXR photons with an energy $\epsilon = 50\,\mathrm{keV}$ for several values of the spectral index δ in the thick-target case, computed from the integrand of Eqn. (4.47). The corresponding cumulative distribution functions are shown in panel b. See Fig. 4.12 for the thin-target case.

number of photons with energy ϵ produced per unit time per unit energy is then (Brown, 1971)

$$\nu_\epsilon(E) = n_{\rm p} V \int_\epsilon^\infty \sigma_\epsilon(E)\, v(E)\, N(E)\, {\rm d}E. \tag{4.34}$$

At this point, it is convenient to distinguish between two cases. If the energy distribution function of the incident electrons is not changed appreciably by its interaction with the target, the resulting emission is referred to as *thin-target* emission. If, on the other hand, the incident electrons are collisionally stopped in the target, the emission is referred to as *thick-target* emission. For thin-target emission, the flux density of photons with energy ϵ at Earth in units of photons $\mathrm{cm^{-2}\,s^{-1}\,keV^{-1}}$ is given by

$$S(\epsilon) = \frac{n_{\rm p} V}{4\pi R^2} \int_\epsilon^\infty \sigma_\epsilon(E)\, v(E)\, N(E)\, {\rm d}E. \tag{4.35}$$

For the non-relativistic regime considered here, the electron bremsstrahlung cross section for non-relativistic electrons is given by the Bethe–Heitler cross section (Koch and Motz, 1959):

$$\sigma_\epsilon(E) = \frac{8}{3}\alpha r_0^2 \frac{m_{\rm e}c^2}{\epsilon E} \log \frac{1+\sqrt{1-\epsilon/E}}{1-\sqrt{1-\epsilon/E}} \tag{4.36}$$

$$= \frac{16}{3}\alpha r_0^2 \frac{m_{\rm e}c^2}{\epsilon E} \log \left(\sqrt{\frac{E}{\epsilon}} + \sqrt{1-\frac{E}{\epsilon}} \right), \tag{4.37}$$

where α is the fine structure constant and $r_0 = e^2/m_{\rm e}c^2$ is the classical electron radius. The Bethe–Heitler cross section can be then substituted into Eq. (4.35) and, noting that $v(E) = \sqrt{2E/m_{\rm e}}$, the thin-target photon flux density can be expressed as

$$S(\epsilon) = \frac{2\beta}{\epsilon} \int_\epsilon^\infty \frac{N(E)}{\sqrt{E}} \log \left(\sqrt{\frac{E}{\epsilon}} + \sqrt{1-\frac{E}{\epsilon}} \right) {\rm d}E, \tag{4.38}$$

where a number of constants have been absorbed into

$$\beta = \frac{2 n_{\rm p} \alpha r_0^2}{3\pi R^2} m c^2 \sqrt{\frac{2}{m_{\rm e}}}. \tag{4.39}$$

Suppose the electron distribution function is a power law $N(E) = K_1 E^{-\delta}$. With a change of variables from E to $u = E/\epsilon$ we then have $K_1 E^{-\delta} = K_1 \epsilon^{-\delta} u^{-\delta}$ and the photon flux density at 1 AU is recast as

$$S(\epsilon) = 2\beta K_1 \epsilon^{-(\delta+1/2)} \int_1^\infty u^{-(\delta+1/2)} \log \left(\sqrt{u} + \sqrt{1-u} \right) {\rm d}u. \tag{4.40}$$

It is seen that for thin-target emission,

$$S(\epsilon) \propto \epsilon^{-(\delta+1/2)}. \tag{4.41}$$

The result was obtained by Brown (1971), although the problem was instead formulated in terms of the observed photon flux density spectrum, taken to be a power law $S_{\text{obs}}(\epsilon) = K_2\epsilon^{-\kappa}$ from which $N(E)$ was inferred through the *inversion* of Eq. (4.38) to be $N(E) \propto E^{-(\kappa-1/2)}$. The key result is that for thin-target hard X-ray emission from a power-law electron distribution function with a spectral index δ, the photon spectrum has an index $\delta + 1/2$. The emitted photon spectrum is a power law with a spectral index that is softer than that of the electron distribution by 1/2.

Figure 4.12a shows the normalized contribution function of a power law distribution of electrons to HXR photons with an energy $\epsilon = 50\,\text{keV}$ for several values of the spectral index δ in the thin-target case, computed from the integrand of Eqn. (4.38). Electrons with energies near 50 keV contribute the bulk of the 50 keV photons. The corresponding cumulative distribution functions in Fig. 4.12b show that $\approx 90\%$ of the 50 keV photons are emitted by electrons with energies $E \lesssim 2\epsilon$.

Turning to the case of thick-target emission, the collisional energy loss of the incident electrons must be taken into account. For a fully ionized hydrogen plasma, electron energy loss is dominated by collisions with the ambient electrons. The energy loss rate for an incident electron is given by

$$\frac{\mathrm{d}E}{\mathrm{d}t} = -n_{\text{p}}v(E)\,\sigma_{\text{ee}}(E)\,E, \tag{4.42}$$

where

$$\sigma_{\text{ee}}(E) = \frac{2\pi e^4}{E^2}\Lambda_{\text{ee}}(E) \tag{4.43}$$

is the energy loss cross section, $\Lambda_{\text{ee}}(E) = \log(Eb_0/e^2)$, and b_0 is the maximum impact parameter of the electron collisions. $\Lambda_{\text{ee}}(E)$ can be taken to be approximately constant over the energy range considered here. A single electron injected into the target with an initial energy E_0 will emit photons with energy ϵ until the electron energy falls below ϵ. Therefore, using Eq. (4.42), the rate at which it emits photons of energy ϵ is given by

$$\nu_\epsilon(E) = \int_{t(E=E_0)}^{t(E=\epsilon)} n_{\text{p}}\sigma_\epsilon(E)v\,(E)\,\mathrm{d}t = \frac{1}{C}\int_\epsilon^{E_0} E\sigma_\epsilon(E)\,\mathrm{d}E, \tag{4.44}$$

where $C = 2\pi e^4 \Lambda_{\text{ee}}(E)$. If $F(E_0)$ is the *injected* spectrum of electrons per second, the total photon emission rate is

$$\int_\epsilon^\infty F(E_0)\,\nu_\epsilon(E_0)\,\mathrm{d}E_0 \tag{4.45}$$

and the photon flux density at 1 AU is then

$$S(\epsilon) = \frac{2\beta}{\epsilon}\frac{1}{Cn_{\mathrm{p}}}\sqrt{\frac{m_{\mathrm{e}}}{2}}\int_{\epsilon}^{\infty} F(E_0)\left[\int_{\epsilon}^{E_0} \log\left(\sqrt{\frac{E}{\epsilon}}+\sqrt{1-\frac{E}{\epsilon}}\right)\mathrm{d}E\right]\mathrm{d}E_0. \tag{4.46}$$

Exchanging the order of integration,

$$S(\epsilon) = \frac{2\beta}{\epsilon}\frac{1}{Cn_{\mathrm{p}}}\sqrt{\frac{m_{\mathrm{e}}}{2}}\int_{\epsilon}^{\infty} \phi(E)\log\left(\sqrt{\frac{E}{\epsilon}}+\sqrt{1-\frac{E}{\epsilon}}\right)\mathrm{d}E, \tag{4.47}$$

where

$$\phi(E) = \int_{E}^{\infty} F(E_0)\,\mathrm{d}E_0. \tag{4.48}$$

Using the same change of variables as used in Eq. (4.40) and again assuming a power-law injection function

$$\phi(E) = \frac{K_1}{\delta-1}E^{-(\delta-1)} = \frac{K_1}{\delta-1}\epsilon^{-(\delta-1)}u^{-(\delta-1)} \tag{4.49}$$

and the photon flux density at 1 AU becomes

$$S(\epsilon) = \frac{2\beta K_1}{\epsilon}\frac{1}{Cn_{\mathrm{p}}}\sqrt{\frac{m_{\mathrm{e}}}{2}}\frac{\epsilon^{-(\delta-1)}}{\delta-1}\int_{1}^{\infty} u^{-(\delta-1)}\log\left(\sqrt{u}+\sqrt{1-u}\right)\mathrm{d}u. \tag{4.50}$$

For thick-target emission, therefore, the injection of a power-law distribution of energetic electrons yields a power-law photon spectrum

$$S(\epsilon) \propto \epsilon^{-(\delta-1)} \tag{4.51}$$

or, equivalently, if $S(\epsilon) \propto \epsilon^{-\kappa}$ then Eq. (4.50) can be inverted, yielding $F(E) \propto E^{-(\kappa+1)}$. The emitted photon spectrum is a power law with a spectral index that is harder than that of the electron distribution function by 1.

Figure 4.13a shows the normalized contribution function of a power-law distribution of electrons to HXR photons with an energy $\epsilon = 50$ keV for several values of the spectral index δ in the thick-target case, computed from the integrand of Eqn. (4.47). Electrons with energies from 50 to 60 keV contribute the bulk of the 50 keV photons although the contribution function is broader than it is for the thin-target case (Fig. 4.12a). The corresponding cumulative distribution functions in Fig. 4.13b show that $\approx 90\%$ of the 50 keV photons are emitted by electrons with energies $E \lesssim 2 - 5\epsilon$, depending on the spectral index δ; i.e. electrons with energies several times the photon energy contribute significantly to their emission in the thick-target case.

While the thin- and thick-target models for non-thermal bremsstrahlung emission produce power-law photon spectra from the injection of power-law electron distributions the reality is more complex – and more interesting. Most hard X-ray spectroscopic observations do not lend themselves to analytical inversion and the recovery of simple functional forms. Modern observations have high spectral resolution and require more sophisticated techniques. Powerful numerical data inversion schemes are now employed (see Brown *et al.* (2006) for a comparative assessment) and/or forward-fitting schemes that include multiple spectral components (e.g. Holman *et al.*, 2003). Moreover, additional complexities such as density inhomogeneity and the ionization fraction of the source plasma must be considered, as well as their evolution in time.

As is the case for radio emission, the polarization of hard X-ray radiation can yield additional information about the energetic electrons and their target. Hard X-ray bremsstrahlung radiation is expected to be polarized in the plane of emission, defined by the electron momentum **p** and the photon **k** (Haug, 1972). Brown (1972) investigated the polarization of thick-target bremsstrahlung when electrons were guided vertically by a magnetic field to the chromosphere, finding that while flares at disk center should be unpolarized, the degree of polarization should increase to ~30% for flares near the limb. An analysis by Bai and Ramaty (1978) suggested even higher degrees of polarization for limb flares involving directed beams of electrons. Isotropic electron distributions are not expected to be significantly polarized. Attempts to perform hard X-ray polarimetry over the years have been frustratingly ambiguous, however. Dennis (1988) emphasized the need for an imaging polarimeter and with the advent of RHESSI, the first tentative measurements are being reported, as will be described briefly in Section 4.4.

An important effect that needs to be considered when analyzing hard X-ray observations is Compton scattering. Photons can scatter off free or bound electrons. A familiar example is Thomson scattering, wherein low-energy photons elastically scatter off low-energy electrons. When the photon energy is large, the term *Compton scattering* is used. Another case of astrophysical interest is when the electron energy is large, in which case the term *inverse Compton scattering* is used. While the hard X-ray source region is optically thin to hard X-ray photons, the dense photosphere is not. Downward propagating hard X-ray photons in the energy range ~10–100 keV Compton backscatter from the photosphere (Tomblin, 1972). The problem has been studied in detail by many authors (e.g. Bai and Ramaty, 1978; Alexander and Brown, 2002; Kontar *et al.*, 2006). The cross section for Compton backscatter has a broad maximum at ~30–40 keV and the reflectivity can approach unity for some energies and scattering angles. The area of the photosphere from which hard X-ray photons are backscattered is called the "albedo patch". Clearly, an albedo correction must be made to hard X-ray spectra observed over the 10–100 keV range. Compton scattering depends on the polarization of the incident photons and since the incident hard X-rays may be linearly polarized, the observed polarization would also need to corrected for the polarized albedo flux.

So far, bremsstrahlung radiation from energetic electrons scattering from protons and ions has been discussed (e^-–p bremsstrahlung). Brown and Mallik (2008) pointed out that free–bound particle interactions (recombination) can contribute a significant flux to hard X-ray spectra. It is also important to note that as the photon energy increases to the γ-ray regime ($\gtrsim$300 keV), electron–electron ($e^- - e^-$) bremsstrahlung can be as important as e^-–p bremsstrahlung. In contrast to an electron–proton system, an electron–electron system has no dipole moment and therefore does not radiate at non-relativistic energies, for which the dipole approximation applies. At relativistic energies, however, higher-order terms become important and e^-–e^- bremsstrahlung can be significant (Haug, 1975, 1998). The maximum photon energy emitted by e^-–e^- bremsstrahlung depends on the angle between the incoming fast electron and the outgoing photon. Hence, for beamed electron distributions the e^-–e^- bremsstrahlung photon spectrum depends on viewing angle (see an analysis by Kontar *et al.* 2007). As discussed below, high-energy nuclear reactions (π^0 decay) produce relativistic positrons. Hence, contributions from e^+–e^+ and e^+–e^- bremsstrahlung are also possible (Haug, 1985). Finally, since protons are accelerated to high energies in flares and are incident on ambient electrons, proton–electron (p–e^-) bremsstrahlung (sometimes called *inverse bremsstrahlung*) occurs (Emslie and Brown, 1985; Heristchi, 1986; see Haug, 2003 for cross sections).

4.3.2.2 Gamma-ray emission processes

Gamma-rays are produced by the interaction of energetic protons, α-particles, and heavy nuclei with the ambient chromospheric and photospheric plasma. Consider a particle species j that has been accelerated to a high energy and is incident on an ambient particle species i. The interaction rate can be written in general as

$$\nu_{ij} = n_i \int_0^\infty N_J(E)\, \sigma_{ij}(E)\, v(E)\, \mathrm{d}E. \tag{4.52}$$

As in the case of hard X-ray bremsstrahlung radiation, we can formulate the problem in terms of thin- or thick-target emission. Thin-target processes are relevant to particles that escape into the IPM. Here, we discuss thick-target processes. The yield of *particles* (e.g. neutrons, positrons, pions) from a particular interaction in the thick target case is given by (Ramaty, 1986):

$$Q = \frac{1}{m_\mathrm{p}} \sum_{ij} \frac{n_i}{n_\mathrm{H}} \int_0^\infty \bar{N}_k(E)\, \mathrm{d}E \int_0^\infty \frac{\sigma_{ij}(E')}{(\mathrm{d}E'/\mathrm{d}x)_j} \mathrm{d}E'. \tag{4.53}$$

In our treatment of hard X-ray bremsstrahlung we specified the dominant energy loss term (collisions with electrons) as a function of time. Here, the energy loss term(s) are left unspecified and are expressed as a function of *range*, or depth into the source. Since energy loss by protons and ions is dominated by losses on H and He,

$$\left(\frac{dE}{dx}\right)_j \approx \left(\frac{dE}{dx}\right)_{j,\mathrm{H}} \left[1 + \frac{n_{\mathrm{He}}}{n_{\mathrm{H}}} \frac{m_{\mathrm{He}}}{m_{\mathrm{p}}} \frac{(dE/dx)_{j,\mathrm{He}}}{(dE/dx)_{j,\mathrm{H}}}\right]. \tag{4.54}$$

The terms in the square brackets amount to ~1.13 and are nearly independent of energy. The differential energy loss of particles on protons is

$$\left(\frac{dE}{dx}\right)_{j,\mathrm{H}} \approx 630 \left(\frac{Z_{\mathrm{eff}}^2}{A}\right)_j E^{-0.8}, \tag{4.55}$$

where the atomic mass of particle j is A and its effective charge is

$$Z_{\mathrm{eff}} = Z[1 - \mathrm{e}^{-v(E)/c\alpha Z^{2/3}}]. \tag{4.56}$$

Again, the detailed physics of the interaction between particles i and j is contained in the cross section $\sigma_{ij}(E)$.

A number of interactions between energetic protons and ions with ambient material yield γ-rays. First, high-energy protons and α-particles incident on the dense solar chromosphere produce both neutral and charged π-mesons; e.g.

$$\mathrm{p} + \mathrm{p} \rightarrow \pi^+ + {}^2\mathrm{H}.$$

Neutral pions (π^0) have a rest mass of about 135 MeV and charged pions ($\pi^\pm$) have a rest mass of about 140 MeV. The threshold energy for pion production is therefore roughly 300 MeV nucleon^{-1}. Most neutral pions (~99%) decay directly into two photons,

$$\pi^0 \rightarrow 2\gamma,$$

each with an energy of approximately 67 MeV (half the π^0 rest mass) although the photon energies are strongly Doppler broadened in the observer's frame. The decay of π^- and π^+ mesons is somewhat more circuitous as they first decay into muons,

$$\pi^+ \rightarrow \mu^+ + \nu_\mu,$$
$$\pi^- \rightarrow \mu^- + \bar{\nu}_\mu,$$

and thence to positrons and electrons,

$$\mu^+ \rightarrow \mathrm{e}^+ + \nu_\mathrm{e} + \bar{\nu}_\mu$$
$$\mu^- \rightarrow \mathrm{e}^- + \bar{\nu}_\mathrm{e} + \nu_\mu.$$

The relativistic secondary electrons resulting from π^- decay produce non-thermal $e^- - p$ and $e^- - e^-$ bremsstrahlung radiation via scattering on ambient ions and electrons, as described in the previous subsection.

Second, collisions of high-energy protons and α-particles on the chromosphere also produce radioactive isotopes of C, N, O, and Ne that subsequently β-decay to produce positrons. The threshold energies for production of these isotopes range from several MeV to several tens of MeV. In contrast, the threshold energies for π^+ production via proton collisions with ^{1}H or ^{2}He are 292.3 and 185 MeV, respectively. In both cases, positrons are produced with energies much greater than the thermal energies of particles in the ambient plasma: up to ~1 MeV from isotope decays and tens to hundreds of MeV from pion decay. As they propagate, positrons continuously lose energy through Coulomb collisions with electrons and through ionization/excitation of neutrals until they slow sufficiently to annihilate with an electron, producing two photons, each with an energy equal to the rest mass of an electron: $\epsilon = m_e c^2 = 511$ keV. Alternatively, and more likely, positronium (Ps), a bound electron–positron atom, forms via charge exchange with neutral atoms or through radiative recombination with free electrons, and then annihilates. The details of Ps annihilation depend on the spin configuration of the atom in the ground state, with 1/4 of Ps formed with a total spin 0 (singlet state ^{1}Ps: parapositronium) and 3/4 of Ps formed with a total spin 1 (triplet state ^{3}Ps: orthopositronium), the spin in units of $\hbar$. In practice, the ratio of triplet to singlet Ps is modified by the depletion of ^{3}Ps (quenching) through collisions with neutral H. The annihilation of ^{1}Ps results in two photons, each with an energy of 511 keV in the rest frame. The annihilation of ^{3}Ps yields three photons, each with energy $\epsilon < 511$ keV. Therefore, direct and ^{1}Ps annihilation produce a spectral line at 511 keV whereas ^{3}Ps annihilation produces a continuum below 511 keV. A comprehensive analysis of the physics of positron annihilation in solar flares can be found in the excellent paper by Murphy *et al.* (2005).

Third, excited nuclear states are produced when heavy nuclei of C, N, O, Ne, Mg, Al, Si, S, Ca, and Fe in the ambient chromospheric plasma are bombarded by high energy protons and α-particles. Prompt de-excitation, usually to the ground state, produces narrow emission lines, the most prominent lying between 1 and 8 MeV. The inverse process also occurs: heavy nuclei accelerated in the flare collide with ambient protons and α-particles. Excited states are also produced via spallation reactions where energetic protons or α-particles break apart heavy nuclei into lower mass fragments that emerge from the reaction in excited states. An important example is the ^{12}C line at 4.44 MeV, to which an important contribution is the spallation reaction ^{16}O(p, pα)^{12}C. The widths of lines emitted by ambient heavy nuclei are of order 10–100 keV and are largely determined by the recoil velocity of the nucleus upon emission of the γ-ray. The inverse excitation of nuclei – fast nuclides

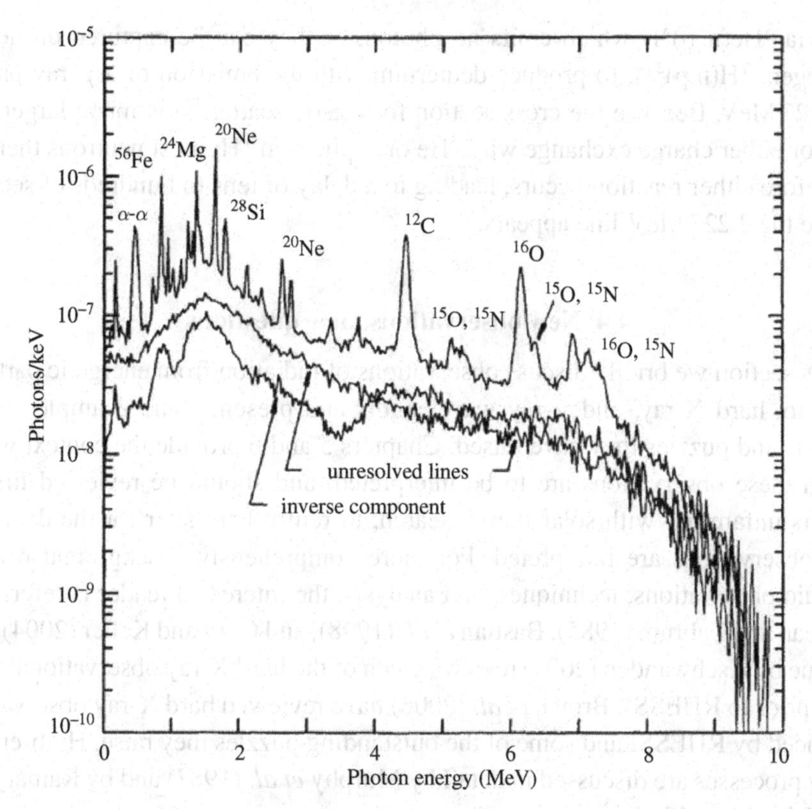

Fig. 4.14. A theoretical spectrum of nuclear de-excitation emission showing the prominent narrow line emission and components corresponding to unresolved lines and broad lines emitted by heavy ions colliding with the target medium. (From Murphy *et al.*, 1990. Reproduced by permission of the AAS.)

excited by collisions with ambient protons and α-particles – produces lines that are strongly Doppler broadened. Their combined contributions yield a continuum that is particularly prominent between 3 and 7 MeV.

Finally, high-energy reactions produce neutrons in a variety of ways. The main neutron production modes are pp, pα, αp, and $\alpha\alpha$, as well as reactions between protons and α-particles on heavy nuclei (e.g. ^{13}C(p, n)^{13}N, ^{13}C(α, n)^{16}O) and the inverse reactions. The energy thresholds for neutron production range from ~1 MeV nucleon^{-1} for reactions involving heavy nuclei to a few hundred MeV nucleon^{-1} for pp reactions. The neutron spectrum produced by these reactions depends on the incident spectrum of ions, their composition, and the ambient thermal plasma. High energy neutrons that escape the Sun can be observed directly in space or by neutron monitors on the ground; those that decay en route can be detected indirectly through their decay protons. There are three possible fates for those neutrons that remain on the Sun: they can decay, charge exchange with

^{3}He via ^{3}He(n, p)^{3}H which emits no photon, or they can be captured on neutral hydrogen, ^{1}H(n,p)^{2}H, to produce deuterium with the emission of a γ-ray photon at 2.223 MeV. Because the cross section for elastic scattering is much larger than that for either charge exchange with ^{3}He or capture on ^{1}H, most neutrons thermalize before either reaction occurs, leading to a delay of tens to hundreds of seconds before the 2.223 MeV line appears.

4.4 New observations, new questions

In this section we briefly discuss observations of radiation from energetic particles at radio, hard X-ray, and γ-ray wavelengths and present some examples of the insights and puzzles they have raised. Chapters 5 and 6 provide the context within which these observations are to be interpreted, and should be reviewed first by readers unfamiliar with solar flare research, to return here later for the details of how observations are interpreted. For more comprehensive background reviews of radio observations, techniques, and analysis, the interested reader is referred to McLean and Labrum (1985), Bastian *et al.* (1998), and Gary and Keller (2004). The volume by Aschwanden (2004) reviews much of the hard X-ray observational work done prior to RHESSI. Brown *et al.* (2006) have reviewed hard X-ray observations produced by RHESSI and some of the outstanding puzzles they raise. High-energy γ-ray processes are discussed in detail by Murphy *et al.* (1987) and by Ramaty and Mandzhavidze (1994).

The ideal of any telescope is to measure the specific intensity of the radiation of interest. Observers want to resolve the emission in time, frequency, and space in each polarization mode with high sensitivity. Due to both practical (access, technology, budget) and inherent (noise) limitations, compromise is both necessary and inevitable and instruments must be designed with narrower goals in mind. Radio, X-ray, and γ-ray instrumentation has gone through many successive generations. While the simplest experiments obtain the time variation of the flux density or a photon count rate at a particular frequency or energy bin (revealing a surprising wealth of information), observations have progressed significantly in the areas of spectroscopy and imaging.

At radio wavelengths, fixed-frequency observations are still obtained by the USAF/RSTN array and by the polarimeters at Nobeyama and elsewhere. Radio spectroscopic observations have mostly emphasized dm- and m-λ from the ground and DH- and km-λ from space. One instrument (OVSA) performs microwave spectroscopy in the range 1–18 GHz. Most radio spectrometers are designed to perform dynamic spectroscopy in order to identify and to study coronal and interplanetary radio bursts. They typically employ single radio antennas on the ground (e.g. the spectrometers operated by IAP/Potsdam, NRAO/Green Bank, STRC/Hiraiso), or dipole antennas on satellites in space (e.g. ISEE-3, Wind/WAVES, STEREO). Imaging observations at radio wavelengths mostly exploit Fourier synthesis imaging and Earth rotation aperture synthesis techniques. Briefly, arrays of antennas are used to measure the Fourier transform of the radio brightness distribution on the sky. For an array of N antennas, there are $N(N-1)/2$ independent pairs of antennas. Each pair measures a Fourier component. Examples of modern imaging radio telescopes include the VLA, as well as a number of solar dedicated telescopes (radioheliographs), notably the NoRH and the NRH. The NoRH images the Sun at 17 and 34 GHz whereas the NRH images the Sun at five discrete frequencies between 150 and 450 MHz.

The development of observations at hard X-ray and γ-ray wavelengths have followed a similar path, albeit from space-based platforms. Spectroscopy of energetic emissions has been emphasized from the beginning with SMM HXRBS, GRANAT, and CGRO playing central roles in the 1980s and 1990s. Although important progress was made on hard X-ray imaging observations with SMM HXIS and Hinotori in the 1980s, it was not until the launch of Yohkoh HXT that hard X-ray imaging came into its own, using 64 collimated grids to perform essentially what ground-based radioheliographs do: Fourier synthesis imaging. Another significant leap forward was made with the launch of RHESSI. With its cooled germanium detectors and enormous energy bandwidth (3 keV to 17 MeV), it can perform high-resolution spectroscopy of hard X-ray and γ-ray emissions. RHESSI also exploits collimated grids to image hard X-ray and, now, γ-ray emissions.

Radio and hard X-ray/γ-ray instrumentation and techniques are therefore converging, with important commonalities emerging in both observational techniques and the interest in the energetic particles responsible for the emissions.

4.4.1 Radio observations of flares and CMEs

Several examples of observations and their interpretation are discussed as a means of illustrating how radio observations are used in practice to study energetic electrons accelerated by flares and by CMEs. We begin with an example of joint ground- and space-based observations of interplanetary radio bursts.

4.4.1.1 Shock-associated type III radio bursts

Shock-associated or shock-accelerated (SA) type III radio bursts were first identified by Cane *et al.* (1981) as possible type-III-like signatures of electron beams accelerated at interplanetary shocks. Type II radio bursts are believed to be the result of plasma radiation associated with coronal and interplanetary shocks and are therefore used as a proxy to identify times when shocks are present. Dulk *et al.* (2000) used a sample of joint ground- and space-based observations of SA type

IIIs to suggest that the electron beams are indeed accelerated by interplanetary shocks. The ground-based observations were obtained by the Culgoora Radioheliograph whereas the low-frequency space-based observations were obtained by the Wind/WAVES experiment. Figure 4.2 shows an example from the study with the type II (showing both fundamental and harmonic plasma radiation) and the SA type IIIs. The SA type IIIs appear to originate at the times and frequencies when the type II is present, a characteristic of each of the events studied. Note that the observed emission spans heights from low in the corona to 70 $R_\odot$.

4.4.1.2 Radio CME

Type II radio bursts have long been associated with CMEs, but direct imaging of incoherent non-thermal radio emission from white light CMEs has not been reported until relatively recently. Using the NRH, Bastian *et al.* (2001) made time-resolved imaging observations of synchrotron emission from a fast CME (April 20, 1998) at 164, 237, 327, and 421 MHz. While the frequency coverage was sparse, crude imaging spectroscopy was possible. Spectral fits were made to the data using a simple gyrosynchrotron model with Razin suppression. The data are consistent with synchrotron emission from MeV electrons entrained in the CME magnetic field. The fits enabled measurements of the CME magnetic field and the ambient plasma density to be made (Fig. 4.9). The magnetic field in the CME was found to vary from 1.5 G at a radius of 1.45 $R_\odot$ to 0.33 G at a radius of 2.8 $R_\odot$. An example of a higher frequency radio CME that occurred on April 15, 2001 has been reported by Maia *et al.* (2007) – shown in Fig. 4.15

4.4.1.3 Electron injection and transport in flares

Radio and hard X-ray observations are highly complementary to the extent that hard X-ray observations are mostly from electrons with energies of $\sim$10–100 keV whereas cm- and mm-λ emission is from electrons with energies of $\sim$100 keV to MeV. Hard X-ray emission is largely non-thermal thick-target bremsstrahlung emission due to electrons colliding with cold, dense, chromospheric material at the magnetic footpoints of flaring loops, whereas radio emission is due to non-thermal gyrosynchrotron emission from electrons that are injected into coronal magnetic loops (see Chapters 5 and 6). A particularly useful model for understanding many of the properties of hard X-ray and radio emissions during flares is the trap-plus-precipitation (TPP) model of Melrose and Brown (1976) and modifications introduced by Aschwanden *et al.* (1998, 1999; see below). However, key issues remain in disentangling electron acceleration, injection, and transport effects.

The Owens Valley Solar Array enables time-resolved spectroscopy of gyrosynchrotron emission from 1 to 18 GHz. Lee and Gary (2000) constructed a model of the time evolution of a distribution of trapped electrons, including the effects of

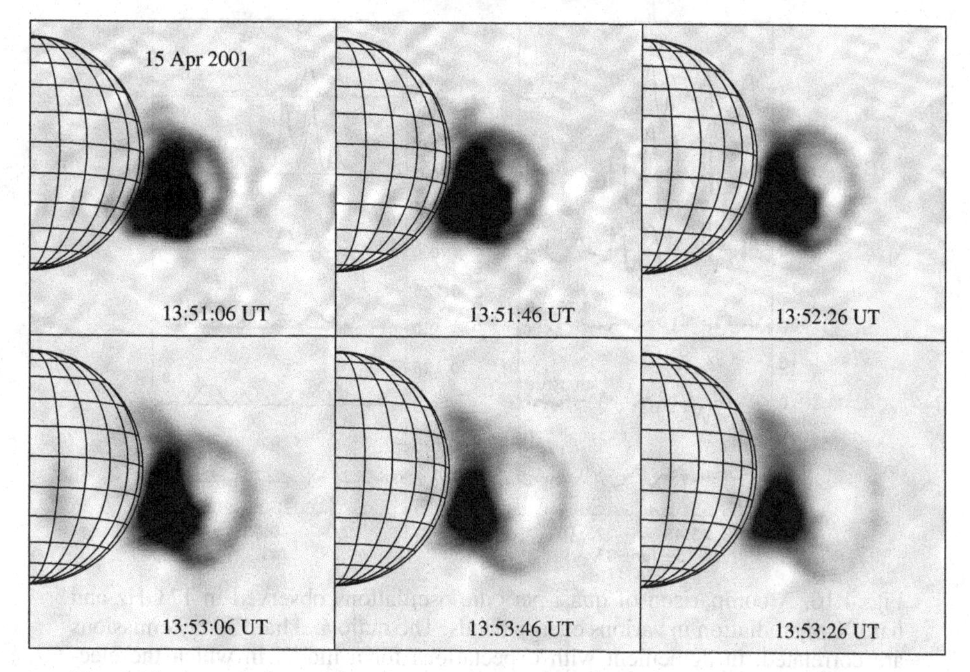

Fig. 4.15. A sequence of images of the fast CME of April 15, 2001 made by the NRH at a frequency of 421 MHz. (After Maia *et al.*, 2007.)

Coulomb collisions, pitch angle diffusion, and magnetic mirroring. They were able to fit the observations to the model using a χ^2-minimization scheme, showing that the initial injection of electrons was beamed perpendicular to the magnetic field and that the ambient density of the trap was low. A study by Melnikov *et al.* (2002) of flares observed by the NoRH showed that gyrosynchrotron-emitting loops display loop-top sources at 17 and 34 GHz, as would be expected for anisotropic electron distributions of the kind deduced by Lee and Gary.

4.4.1.4 X-ray poor flare

Joint radio and hard X-ray observations from OVSA, NoRP, NoRH, and Yohkoh HXT were used to study an X-ray poor flare. The simple source morphology enabled Bastian *et al.* (2007) to build a model with a compound source function to include both the gyrosynchrotron emission from energetic electrons and the absorption properties of the cool, dense, ambient plasma. They fit a sequence of 25 successive radio spectra to the data using a χ^2 minimization scheme. The peculiar spectral evolution of the event could be accounted for in terms of optical depth effects as the ambient plasma was collisionally heated by fast electrons. The time evolution of the emission could not be adequately described in terms of a TPP model, however. The authors suggest that stochastic electron acceleration may play a role in this event.

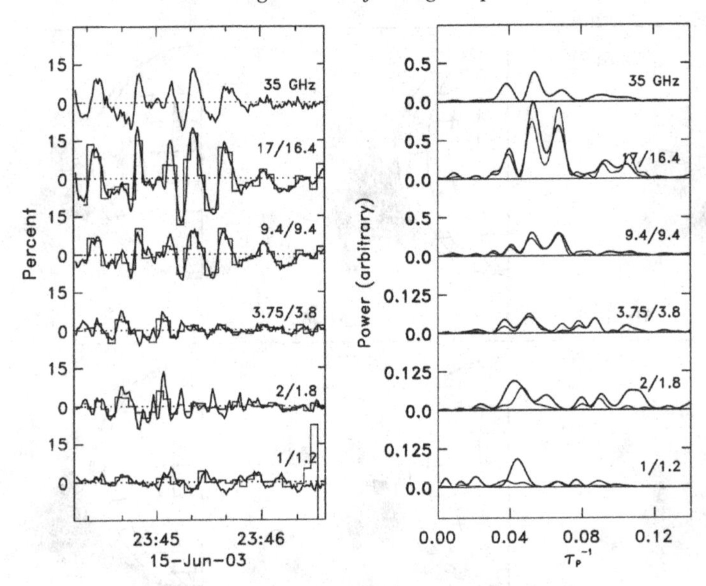

Fig. 4.16. A comparison of quasi-periodic oscillations observed in 17 GHz and hard X-ray radiation in various energy bands. The radio and hard X-ray emissions are correlated, in agreement with expectations for a model in which the electron acceleration and injection is modulated. A model wherein the oscillations result from MHD oscillations in a magnetic loop would yield an anti-correlation between the radio and hard X-ray emissions. (From Fleishman *et al.*, 2008. Reproduced by permission of the AAS.)

4.4.1.5 Quasi-periodic oscillations

Joint radio and hard X-ray observations by OVSA, NoRP, NoRH, and RHESSI by Fleishman *et al.* (2008) considered whether quasi-periodic oscillations of the radio and hard X-ray emission from a powerful solar flare (Fig. 4.16) were due to MHD oscillations of the flaring loop or to quasi-periodic (QPP) modulation of the electron acceleration/injection. Based on a Fourier analysis of the radio and hard X-ray normalized modulations, and correlations between the hard X-ray and radio flux, as well as the modulation and phasing of the radio spectral index and circularly polarized emission, Fleishman *et al.* were able to show that MHD oscillations could be ruled out as the cause of the QPPs. Instead, the QPPs were attributed to a modulation of the acceleration and injection of fast electrons into the emitting source.

4.4.1.6 Mm/submm-λ observations of flares

Progress has been made at mm-λ and even submm-λ. The image of a flaring source at 3 mm was first reported by Silva *et al.* (1996). While no imaging is yet available at submm-λ, the SST has reported several dual-frequency (212 and 405 GHz) emissions from flares. Kaufmann *et al.* (2004) report the puzzling phenomenon of a sub-THz emission component with a spectrum that is *increasing* with frequency, a component that Silva *et al.* (2007) suggest may be due to a synchrotron source that is distinct from the lower-frequency radio emission. Confirmation of this interpretation awaits additional examples of the phenomenon and more extensive observational coverage.

4.4.2 Hard X-ray/γ-ray observations of flares

The RHESSI mission has opened a new era of hard X-ray observations of energetic electrons in flares. It provides exceptional spectral resolution (1–3 keV) over an extremely broad photon energy range, a few keV to 15 MeV. Moreover, RHESSI provides an imaging capability over this entire energy range, enabling hard X-ray/γ-ray imaging observations in both line and continuum at energies >100 keV.

4.4.2.1 Imaging γ-ray line emission

Hurford *et al.* (2003) report the first imaging observations in a γ-ray line. In particular, they report imaging in the strong 2.223 MeV neutron capture line during the flare of July 23, 2002 as well as the non-thermal bremsstrahlung continuum. The surprise is that the neutron line source and the non-thermal bremsstrahlung source are significantly displaced (Fig. 4.17), implying that the responsible electrons and ions are themselves displaced. Emslie *et al.* (2004) suggest that the effect

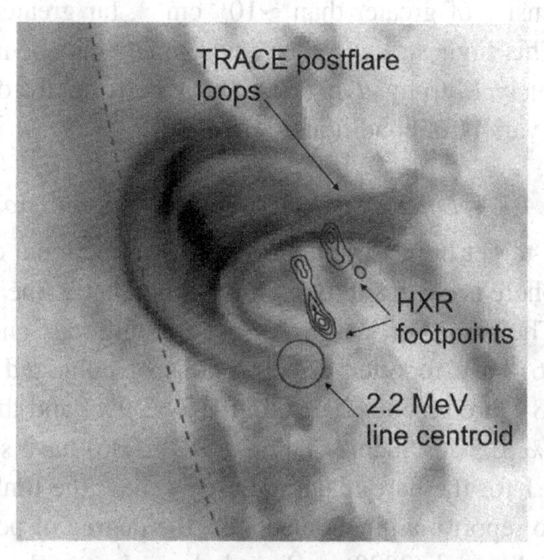

Fig. 4.17. Energetic emissions from the X4.8 flare on July 23, 2002. The 2.223 MeV neutron capture line and the 50–100 keV hard X-ray emission is compared with the EUV loops observed by TRACE (background negative image). (From Brown *et al.*, 2006, based on results from Hurford *et al.*, 2003. Reproduced by permission of the AAS.)

(confirmed in three more flares by Hurford *et al.* 2006) may result from the tendency for stochastic acceleration to favor larger structures for ion acceleration and smaller structures for electron acceleration.

4.4.2.2 Anomalous hard X-ray spectral feature

A second surprise from RHESSI, and one that caused a certain degree of consternation, was the appearance of non-monotonic features – "dips" – in deconvolved mean source electron spectra for certain flares, which had not been encountered previously because of the relatively poor spectral resolution of previous generations of hard X-ray spectrometers (e.g. Piana *et al.*, 2003). If the spectral feature is real, then the non-thermal bremsstrahlung thick-target model is called into question. Kontar *et al.* (2006) have analyzed Compton backscatter in detail and find that the albedo contribution to the hard X-ray flux may resolve the issue. However, more work remains before definitive conclusions can be drawn.

4.4.2.3 Spectrally resolved annihilation line

A third surprising result from RHESSI is the first spectrally resolved observations of the electron–positron annihilation line at 511 keV, reported for three flares by Share *et al.* (2004). These observations raise a number of puzzles, not least of which is the large width of the line, which suggests the source resides in plasma with $T \sim 10^5$ K, yet the lack of a measurable positronium continuum implied an ambient source density of greater than $\sim 10^{14}\,\text{cm}^{-3}$, far greater than expected at this temperature. This suggests the source environment may be more dynamic than appreciated previously. Murphy *et al.* (2005) have performed a detailed analysis of the line for a wide variety of conditions.

4.4.2.4 Linearly polarized hard X-ray emission

Non-thermal thick-target bremsstrahlung from fast directional electrons colliding with the chromosphere is expected to produce significantly linearly polarized hard X-ray emission. The results to date have been ambiguous and they remain so. Boggs *et al.* (2006) have reported 2σ detections of polarized 0.2–1 MeV emission from the X4.8 Flare on July 23, 2002 (21% ± 9%) and the X17 flare on 28 October 2003 (11% ± 5%). Suarez-Garcia *et al.* (2006) have studied six X-class flares (see Table 5.1 for the flare-magnitude scale) near the limb between 0.1 and 0.35 MeV and also report marginal detections, the degree of polarization ranging from 2 to 54%. Emslie *et al.* (2008) made an independent analysis of the X4.8 flare on July 23, 2003, which obtained a polarization of 15% but the polarization vector is non-radial, contrary to expectations.

5

Observations of solar and stellar eruptions, flares, and jets

HUGH HUDSON

5.1 Introduction

A solar flare is narrowly defined as a sudden atmospheric brightening, traditionally in chromospheric Hα emission but more practically now as a coronal soft X-ray source. The physical processes resulting in a flare include restructurings of the magnetic field, non-thermal particle acceleration, and plasma flows. Flares have intimate relationships with other observable phenomena such as filament eruptions, jets, and coronal mass ejections (CMEs). Chapter 6 discusses our current theoretical understanding, and in this chapter we review the observational aspects of these phenomena.

The phenomena associated with the term "solar flare" dominate our thinking about energy conversion from magnetic storage to other forms in the solar corona on time scales below a few minutes.[†] The distinction between a gas dominated by hydrodynamic forces and a magnetized plasma becomes obvious in the solar atmosphere and in the solar wind. At first glance we do not need plasma physics to explain the basic (interior) structure of a star; hydrodyamics, nuclear physics, and the theory of radiative transfer seem to do quite well. Nevertheless, this apparently simple medium drives the currents that result in the violent and beautiful phenomena we see so readily above its surface (see Vol. III). We need plasma physics to describe them.

Understanding the flaring solar atmosphere (photosphere, chromosphere, and corona; see Chapter 8 in Vol. I for descriptions of these regions), since it involves electrodynamics, requires a strong overlap with magnetospheric physics as well as with astronomical techniques useful for studying stellar atmospheres. For some purposes one can accept the standard spherically symmetric, gravitationally stratified approximation to the structure of a stellar atmosphere (e.g. Vernazza *et al.*,

[†] Examples of solar flares, CMEs, and other explosive or eruptive events can be found at www.vsp.ucar.edu/HeliophysicsScience.

Heliophysics: Space Storms and Radiation: Causes and Effects, eds. Carolus J. Schrijver and George L. Siscoe. Published by Cambridge University Press.

1981), but this approach has become obsolete for most problems of current interest. Chapter 8 in Vol. I gives a good grounding in modern approaches to the problems involved in physically characterizing the solar atmosphere; see also the lecture notes by Steiner (2007). The advancement of numerical techniques allows much-improved treatment of three-dimensional structure and time variability, including the study of shock waves. Numerical simulations are now linking the corona to the convection zone self-consistently (e.g. Abbett, 2007).

We begin the chapter with a historical overview, which follows the development of observational capability. Solar flares involve the whole depth of the solar atmosphere, and are associated with heliospheric events extending far past the Earth's orbit. Accordingly, the observing techniques span the entire range of human capability for classical astronomical remote sensing (see Chapter 4), often with optimization for bright objects, plus the whole range of *in situ* techniques (see Chapter 3). Because solar flares are directly observable only by remote-sensing techniques, there are many important things that we simply cannot know empirically. The results of the observations consist of a sometimes patchy coverage of parameter space, leaving room for many new discoveries even in such a well-observed system (see e.g. Harwit, 1981, and Hudson, 1987, for discussions of how to quantify "discovery" in this respect). This chapter discusses basic flare phenomena in Section 5.3, analogous astrophysical processes in Section 5.5, and interpretations of the flare observations in terms of large-scale magnetic reconnection scenarios in Section 5.6 as a separate item of great interest.

Confusion often comes from trying to understand these disparate kinds of observation as a whole (e.g. Hudson and Cliver, 2001). To link the pieces of the puzzle together often involves a sketch or cartoon,[†] and as technology improves it also involves large-scale numerical simulations. The simulations can be used as a kind of forward-fitting tool, with the comparison done in terms of the observations. Often, though, they are more useful simply as numerical experiments that help to guide the framework of the eventual theory.

The energy release in a solar flare is dominated by particle acceleration, both of electrons (Lin and Hudson, 1976) and of ions (Ramaty *et al.*, 1995; Emslie *et al.*, 2005). This means that the most direct observations are in the X-ray and γ-ray domains; note that non-thermal processes also usually dominate the emission signatures in the radio range (10^7–10^{12} Hz; meter–submillimeter wavelengths). Please refer to Chapter 4 of this volume for a fuller discussion of the remote-sensing signatures. We will simply comment here that in general the hard X-ray spectrum ($h\nu \gtrsim 10\,\mathrm{keV}$) is dominated by bremsstrahlung from electrons of this energy or greater, while the soft X-ray spectrum ($h\nu \lesssim 10\,\mathrm{keV}$) also includes the free–bound

[†] See http://solarmuri.ssl.berkeley.edu/~hhudson/cartoons.

and bound–bound transitions of a thermal plasma with assumed Maxwellian distribution functions, and also usually assuming the electron and ion temperatures to be equal, i.e. $T_e = T_i$.

5.2 Overview of flare properties

5.2.1 Chronological/chromatic history

Our observational knowledge of the phenomena of solar activity has grown immeasurably since the first flare observation (Carrington, 1859). The development of observational knowledge has of course followed the growth of technical capability. For example, the Carrington flare occurred prior to Röntgen's discovery of X-rays or Heaviside's recognition of the ionosphere, and so its "geo-effective" significance could not really be assessed (see Chapter 2).

It is instructive to follow the history of this development (Švestka and Cliver, 1992), which is roughly chromatic (in the sense of new wavebands becoming accessible to observation; see Chapter 4 for more details about techniques): the original observations were in white light, done visually through broad-band filters. These observations began with Galileo and extended into the nineteenth century, mainly oriented towards the morphology of sunspots. We now interpret these observations in terms of dynamo theory, a subject discussed in Vol. III. Carrington was measuring sunspots when the 1859 flare intruded itself.

Towards the end of the nineteenth century, spectroscopy and photography improved (e.g. Hale, 1930), and the study of solar activity became much richer through access to the chromospheric lines such as Hα. Indeed, flares had been observed spectroscopically by Young, Lockyer, Secchi and presumably others within a decade or so of Carrington's pioneering observation (Švestka and Cliver, 1992). This made it possible to study prominences at the limb, for example, since the spectroscope could suppress the glare of the photosphere and reveal these structures in the corona directly. During this period, a solar flare was a "chromospheric flare", observed by Hα "flare patrol" telescopes around the world. The importance of a flare could be judged from its Hα area (S, 1, 2, 3, where the "S" stands for "subflare") and brightness (F, N, B for "faint", "normal", and "brilliant").

Finally, a third chromatic epoch began in the mid-twentieth century with the development of radio astronomy (e.g. Hey *et al.*, 1948), and then X-ray (Dellinger, 1935; Friedman *et al.*, 1951) and γ-ray astronomy (Peterson and Winckler, 1959; Chupp *et al.*, 1973). Via these techniques the emphasis in solar-flare research has shifted into the corona, where the magnetic energy release results in "loop prominence systems" (a somewhat archaic term referring to Hα arcade structures),

Table 5.1. *Flare classifications*

GOES class	1–8 Å peak (W/m^2)	Hα class	Hα area (millionths of hemisphere)	CME fraction[a] (percentage)	Events/year (max/min)
A	$>10^{-8}$	—	—	—	—
B	$>10^{-7}$	S	<200	—	—
C	$>10^{-6}$	1	>200	20	>2000/300
M	$>10^{-5}$	2	>500	50	300/20
X	$>10^{-4}$	3	>1200	90	10/one?
—	$>10^{-3}$	4	>1200	100	few?/none?

[a] Yashiro *et al.* (2005) (approximate values)

closely related to the "sporadic coronal condensations" (a definitely archaic term describing these structures seen in optical coronal emission lines, e.g. from Fe^{13+} – spectroscopically referred to as Fe XIV). The modern view of these structures is via the soft X-ray monitoring by the GOES and other "operational" spacecraft. We now routinely classify solar flares by their GOES classes: A, B, C, M, and X in decades, with the X class signifying 1–8 Å energy fluxes greater than 10^{-4} W/m^2, on the order of 0.01% of the solar luminosity. Table 5.1 summarizes these and other properties, with approximate correspondences between the Hα and GOES X-ray systems, and approximate ranges for the number of flares that occur per year at maximum and minimum of the solar cycle.

These stages in the development of observational capability have essentially changed the meaning of the word "flare", for example. Hale used the term "eruption" and recent decades have seen some confusion about nomenclature (Cliver, 1995). We now know that the physics of a flare, or other form of solar activity, requires rapid restructuring of the coronal magnetic field where energy has been built up much more gradually. In summary, the chronological/chromatic history of solar flare research has generally proceeded through visible light (the photosphere), spectroscopy of the chromosphere, and finally X-rays and radio waves (the corona). At present it appears that the most important region physically is the chromosphere (e.g. Hudson, 2007a), because it mediates the dramatic changes of state between the photospheric and coronal plasmas.

5.2.2 Flare phases

The release of energy can either be "impulsive" (Kane and Anderson, 1970), with time scales sometimes shorter than 1 s, or "gradual". The impulsive and gradual

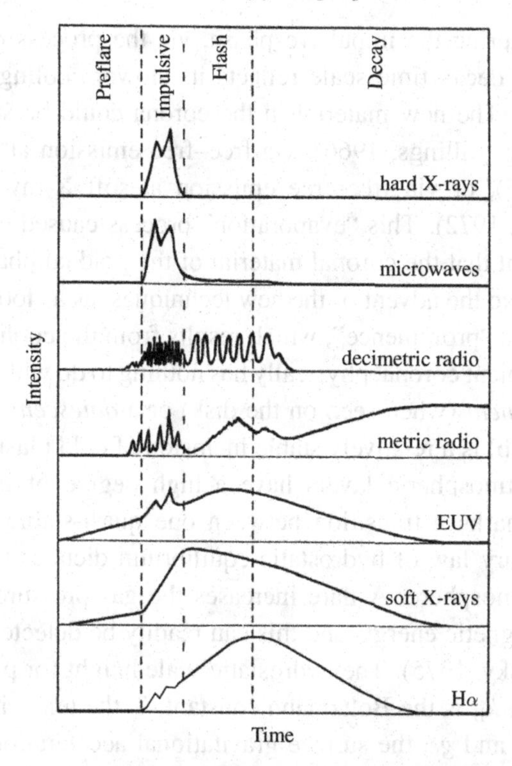

Fig. 5.1. Schematic view of the evolution of flare emissions in different wavelengths, showing the intermingling of impulsive-phase and gradual-phase signatures across the spectrum. (From Benz, 2002.) Note the wide variety of radio signatures.

signatures of a flare extend across the entire electromagnetic spectrum in a complicated way, as illustrated in Fig. 5.1 (see also Fig. 6.3). The terminology may not seem appropriate when one considers a slowly developing flare-like event, such as a quiet-Sun filament eruption (Harvey *et al.*, 1986; Hudson *et al.*, 1995a); in such a case the "impulsive phase" may take tens of minutes to evolve, and the hard X-ray emission may be below the detection level. Thus we don't know how "impulsive" the energy release really is in such an event, but in other respects it has the morphology of an ordinary active-region flare.

We understand the impulsive and gradual phases to show the main energy release and its aftermath (secondary effects), with the proviso that it is really not just that simple. The most prominent "aftermath" is the action of coronal magnetic loops as an energy reservoir, with cooling time scales that can approach hours. This reservoir function is often described as the "Neupert effect" (Neupert, 1968; Dennis and Zarro, 1993): the coronal manifestations of a flare tend to lag behind its chromospheric ones. This results from the finite time scale associated with the coronal density increase during the impulsive phase, via the process of "chromospheric evaporation". The decay time scale reflects its slower cooling and return to the lower atmosphere. The new material in the corona could be seen in the coronal emission lines (e.g. Billings, 1966), via free–free emission at radio wavelengths (e.g. Kundu, 1965), or via free–free emission at soft X-ray wavelengths (e.g. Hudson and Ohki, 1972). This "evaporation" process caused confusion from the outset, to the extent that the coronal material of the gradual phase of a flare could best be seen, prior to the advent of the new techniques, as a "loop prominence system" in Hα. Such a "prominence", which results from the cooling of plasma even hotter than the ambient corona, physically has nothing to do with a true solar prominence: an *Hα filament* (when seen on the disk) or a *quiescent prominence* (when seen above the limb) is a relatively stable inclusion of cold plasma in the corona.

The different atmospheric layers have a high degree of interconnectedness. Because a flare marks a transition between one quasi-stable configuration and another, the ordinary law of hydrostatic equilibrium dictates the run of pressure up through the atmosphere. A flare increases the gas pressure in the corona, at the expense of magnetic energy, and this can readily be detected at all levels (e.g. Machado and Linsky, 1975). The hydrostatic scale height for pressure is given by $2k_{\rm B}T/mg_{\odot}$, where $k_{\rm B}$ is the Boltzmann constant, T the temperature, m the mean molecular weight, and $g_{\odot}$ the surface gravitational acceleration. For a flare temperature of 10^7 K, this scale height is a large fraction of the solar radius, much larger than the flare loop structures. Thus the vertical structure is isobaric in the upper chromospheric and coronal regions, and the chromosphere acts as a reservoir of mass to maintain this isobaric state as the flare loops cool and lose pressure quasi-statically.

5.2.3 Before the flare

The physical condition of the corona prior to a flare must contain the information one needs to predict its occurrence, but it remains to be established which properties are most telling. For example, many flares, as seen in GOES soft X-ray or microwave light curves, have a pre-event increase, mainly seen in the free–free (bremsstrahlung) continuum. This can be unambiguously identified with an increase of the emission measure ($\int n_{\rm e}^2\,{\rm d}V$) of hot plasma in the corona. But is such a precursor physically related to the flare that is going to happen? Is it indirectly related, or is it a coincidence made more likely by frequent flare occurrence in a given active region? These questions are convolved with the appearance of flickering, swelling, rising, and other signs of activity in a filament that is about to erupt (e.g. Crockett *et al.*, 1977; Webb, 1985; Gaizauskas, 1989; Fárník *et al.*, 2003; Chifor *et al.*, 2007).

Sometimes there is virtually no early activity and so it is difficult to accept this as a prerequisite for flare occurrence. The bright flare loops themselves usually appear at new locations as identified by their "line-tied"† footpoint locations (Fárník *et al.*, 1996; Fárník and Savy, 1998; Hudson *et al.*, 2008). In such cases we assume that the magnetic flux tubes anchored at the same footpoints as the flaring loops were empty and dark prior to the flare.

In the lower solar atmosphere, and especially in the magnetograph and chromospheric observations, there are patterns that anticipate flare occurrence (e.g. Rust *et al.*, 1994; Schrijver, 2007). Zirin and Liggett (1987) found an almost one-to-one correspondence between the "δ spot"‡ sunspot configuration and the occurrence of X-class flares. The most important of these is "flux emergence", revealed in Hα as an "arch filament system" or simply as an "emerging flux region" (e.g. Vorpahl, 1973; Nitta and Hudson, 2001). We can interpret this as one of the ways in which the coronal field can be stressed, i.e. to carry field-aligned currents, for the duration of the energy buildup that precedes the flare itself. The time scale for this buildup and release – not yet observed as a true relaxation oscillator – appears to be a few hours.

5.2.4 Flare types

For the most part, solar flares have similar properties, and their extensive parameters tend to scale together in a systematic way. This is one view of the "big flare syndrome" (Kahler, 1982). This suggests that all flares fit one pattern, and that the energy release is just a matter of energy scale. Pallavicini *et al.* (1977) identified two types of solar flare, which we refer to as "confined" and "eruptive" here. No solar property appears to have a bimodal distribution that clearly distinguishes these two categories, and so this classification remains somewhat arbitrary. However, in the domain of solar energetic particles (SEPs) there is a bimodal separation into "impulsive" and "gradual" events (e.g. Reames, 1999). The names given to these categories may not exactly match the observed properties. Extremely impulsive flares may certainly be eruptive as well (e.g. Nitta and Hudson, 2001). The extensive properties of flares (for example, CME kinetic energy and soft X-ray peak brightness; see Section 5.3.5, but there are many other examples) generally correlate over four to five decades with an rms scatter of about a factor of two. This means that the dynamics of the solar atmosphere during a disruption follows some

† The concept of (field) line tying refers to the anchoring of coronal field in the photosphere where it enters a much denser plasma; consequently, the photospheric field does not immediately respond to coronal impulsive changes, and the field must behave as if "tied" to a base; see also Section 6.2.1.

‡ Sunspot groups are classified as α, β, or γ depending on their polarity structure (called the Mt. Wilson magnetic classification); the added qualifier δ characterizes a sunspot with two or more dark umbral cores of different polarities that lie within 2° of each other and are contained within a single encircling penumbra.

regulated development that generally ignores the distinction between confined and eruptive properties. We do not yet have theories or numerical simulations that are sufficiently model-independent to explain this broad regulation of flare properties.

5.2.5 Flare–microflare occurrence patterns

The frequency distribution of flare energies has a featureless power-law distribution $dN/dE \propto E^{-\alpha}$ (Akabane, 1956; Drake, 1971; Crosby *et al.*, 1993). This distribution extends over several decades of energy, from the domain of major flares with energy of order 10^{32} ergs down to the "microflare" domain around 10^{26} ergs. Many extensive parameters associated with solar flares show this kind of power-law distribution, which implies scale invariance. This property probably has an important physical significance, but it is deceptive: *average* properties of such distributions only reflect the sensitivity of the observation, not anything physically significant.

The slope of the flare–microflare power law ($\alpha < 2$) suggests that the microflares do not contribute in a dominant manner to the total energy in flaring; indeed, the flare–microflare occurrence distribution must steepen above some total energy in order not to diverge in total energy (Hudson, 2007b). Figure 5.2 shows a distribution of hard X-ray peak fluxes, taken here to serve as a proxy for total flare energy. Crosby *et al.* find a power-law index of $\alpha = 1.732 \pm 0.008$ for this sample, in

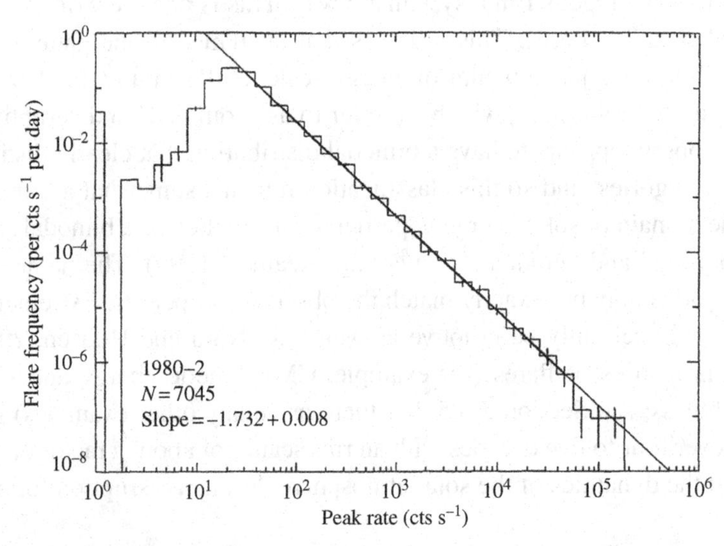

Fig. 5.2. Distribution of peak counting rates of 7045 hard X-ray bursts observed over 1980–2 by the HXRBS instrument on board the Solar Maximum Mission (Crosby *et al.*, 1993). Note the fidelity of the power law, down to a low-rate rolloff due to selection effects; also note the lack of a high-rate rolloff in this range of observations.

good agreement with Akabane's original estimate of $\sim$1.8 using peak microwave fluxes. It appears that the peak flux of the burst, whatever the wavelength, may scale in a similar way with the total event energy. This is consistent with the "big flare syndrome" scaling of extensive parameters noted in Section 5.2.4.

Physically, the microflares look like less-energetic versions of major flares (e.g. Christe *et al.*, 2008; Hannah *et al.*, 2008). However, at least two clear distinctions do appear as one goes along the distribution of flare magnitudes. First, the major flares tend to have a strong association with CMEs. This becomes almost one-to-one for X-class GOES ratings (e.g. Yashiro *et al.*, 2005; see Table 5.1). Second, the minor events tend to have more clearly recognizable soft X-ray jets (e.g. Shimojo *et al.*, 1996; see the illustration in Fig. 5.4). There may be a tendency for arcades to form in more energetic events, as compared with the more common appearance of a single dominant coronal loop in a less-energetic event.

5.3 The basic phenomena of a solar flare

In the photospheric spectrum we see solar flares as brief flashes of white light and UV continuum. At present these sources are often not resolved either in space (Mm scales) or time (few seconds scales) (Hudson *et al.*, 2006). The bright emission regions are embedded in the "ribbon" regions that become more prominent in the chromospheric and EUV coronal lines. In the coronal emissions one sees bright coronal loops developing slowly, with those from the highest temperatures appearing first and then cooling down through generally longer wavelengths, while at the same time shrinking in length (Švestka *et al.*, 1987).

In the following sections we outline the basic phenomena of a flare, including the development of a coronal mass ejection (CME). More energetic flares almost always have this association, whereas weaker flare events usually do not. The exception to this rule is the class of major CME events from quiet-Sun filaments, for example the "polar crown" filaments at latitudes well above those of the sunspot regions. Such events may have spectacular CMEs but only barely detectable large-scale chromospheric/soft X-ray signatures (Harvey *et al.*, 1986). Furthermore the soft X-ray jets discovered with Yohkoh (Shibata *et al.*, 1992; Strong *et al.*, 1992) invariably are associated with *microflares*, discussed separately below. These are less-energetic events. The jets are essentially plasma motions parallel to the magnetic field, whereas the more energetic flares are better associated with CMEs, which have the appearance of loop expansion and hence perpendicular plasma motion. Note that these perpendicular plasma motions usually begin in active regions where the plasma $\beta = 2nk_{\mathrm{B}}T/(B^2/8\pi)$ (ratio of gas pressure to magnetic pressure) is low (see Gary (1989) for a review of coronal β values). Microflares, flares, and CME-related major flares all look similar in many respects, except for scale, but the major CME-related flares tend to have the LDE ("long-decay event" or "long-duration event") characteristic of long-lived arcade sources, as discussed below in Section 5.3.3.

5.3.1 Flare luminosity and mechanical energy

Solar flares are not luminous on the scale of the total solar irradiance ("solar constant"), although they may produce a localized brightening seen against the bright photosphere. The powerful flare of November 4, 2003 was the first that could actually be detected in the total solar irradiance, by the radiometer on board the SORCE spacecraft (Woods *et al.*, 2006). The signal, at roughly 5σ significance, amounted to about 300 ppm of the total signal, or 0.3 millimagnitudes in astronomical terms. There is a solar background noise level for such a measurement due to convection and oscillations; this amounts to some 50–100 ppm spread out over a bandwidth of a few mHz (e.g. Hudson, 1988).

The localized brightening of a flare is much easier to see, of course, via an image even in white light. Carrington described his 1859 discovery as resembling the brilliance of Vega (α Lyrae), for example. Although it has been difficult to obtain comprehensive photometric observations across the entire spectrum of a flare, we now know enough about the energy distribution to know that what Carrington saw was a major fraction of the flare luminosity. Soft X-ray emission, for example, contains only 5–10% as much luminosity. This gradual component, as discussed below, results from a thermal distribution (hot gas) for which the X-ray emission itself is a dominant cooling term. The non-thermal tail of the X-ray spectrum ($h\nu > 10\,\mathrm{keV}$), on the other hand, is due to bremsstrahlung from stopping particles. The bremsstrahlung mechanism is very inefficient, providing a fraction of order 10^{-5} of the energy losses. The rest of the energy winds up in longer-wavelength radiation, notably the visible/UV continuum (Hudson, 1972; Fletcher *et al.*, 2007).

We must also consider the bulk kinetic aspects of flare luminosity, since for major events a CME almost invariably results. CME kinetic energies can rival flare luminosities (e.g. Emslie *et al.*, 2005) in such cases. In rare cases a CME can occur in the absence of a major perturbation of the lower atmosphere. The least ambiguous example of such an occurrence was discussed by Webb *et al.* (1998). The partition of energy in a flare/CME event remains unclear physically and hard to determine observationally.

5.3.2 The impulsive phase (hard X-rays, footpoints)

The impulsive phase of a flare marks the period of intense energy release and strong non-thermal effects, including the launching of the CME. The traditional

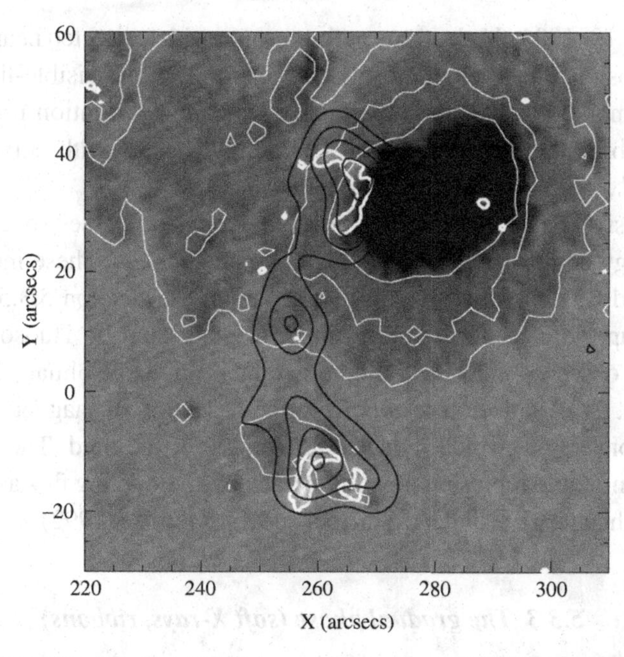

Fig. 5.3. TRACE white-light (dark grey contours) and RHESSI hard X-ray (light grey contours; 25–50 keV) observations of a flare of July 24, 2002 (Fletcher *et al.*, 2007). Note the extremely compact (arcsec), and temporally unresolved ($\sim$10 s), white-light patches in the north and south footpoint regions. The RHESSI source in between the footpoint regions is not associated with the white-light emission. (Reproduced by permission of the AAS.)

observational tools for the impulsive phase are hard X-ray emission and gyrosynchrotron emission at cm to mm radio wavelengths (see Chapter 4). The hard X-rays normally show two dominant footpoints embedded in ribbon regions of opposite magnetic polarity, but we do not presently understand why there are usually just two. The sources are compact and rapidly variable, and we associate them with the UV and white-light continuum emissions that also come from the footpoint regions, as illustrated in Fig. 5.3. Other wavelengths (see Fig. 5.1) show impulsive emission components as well as gradual ones. A clear impulsive-phase signature also appears even in the total irradiance, but rarely exceeds the background variability, because it requires the most energetic of events to outshine the Sun as a whole.

The hard X-ray spectrum above about 10 keV plays a central role in our understanding of the impulsive phase because the collisional energy losses of the bremsstrahlung-emitting electrons rival the total flare energy itself. This relationship can be established directly by inverting the hard X-ray spectrum, under model assumptions. The "collisional thick-target model" (Kane and Donnelly, 1971; Brown, 1971; Hudson, 1972; see also Section 4.3.2.1) envisions a black-box accelerator of 10–100 keV electrons in the corona, with a directed beam penetrating to the chromosphere or even photosphere to excite UV and visible-light emission. This simple model has become less tenable as spatial resolution improves, since the WL/UV brightenings seen by TRACE imply beams with extreme intensity (Hudson *et al.*, 2006; Fletcher *et al.*, 2007).

The impulsive phase also corresponds to global processes, even though the radiated energy comes from exceedingly compact sources. These include coronal dimmings and CMEs, which we discuss separately in Section 5.3.5. In addition, there is the appearance of an "implosion", as suggested by Hudson (2000) and possibly now observed in RHESSI and other data (Sui and Holman, 2003; Veronig *et al.*, 2006a). The implosion results from the reduction of magnetic pressure via the energy conversion, which reduces the volume of the field. The characteristic inward motions could represent flows associated with Poynting flux as the magnetic equilibrium changes (Emslie and Sturrock, 1982; Melrose, 1992).

5.3.3 The gradual phase (soft X-rays, ribbons)

"Gradual phase" refers to the thermal emission from the hot coronal material evaporated during the impulsive phase, plus the strong transition-region and chromospheric emissions driven by the cooling of these coronal loops. The loops connecting the roughly parallel ribbons form a semi-cylindrical *arcade* structure, divided into many unresolved loops. These hot X-ray and EUV loop structures were first seen in early optical observations of coronal forbidden lines. The loops were also termed *sporadic coronal condensations* (e.g. Billings, 1966). The hot regions eventually cool to form the Hα loop prominence system, whence thermal instability leads to the phenomenon of "coronal rain." The cooling also corresponds to shrinkage, as the gas pressure diminishes; shrinkage may also relate to the gradual release of energy as the coronal equilibrium returns to a stable configuration (Švestka *et al.*, 1987; Forbes and Acton, 1996). This is the process termed "dipolarization" in the geomagnetic community and basically resembles the impulsive-phase implosion noted in Section 5.3.2.

5.3.4 Jets (parallel motions)

Soft X-ray jets were discovered with the Yohkoh soft X-ray telescope (Shibata *et al.*, 1992; Strong *et al.*, 1992; Shimojo *et al.*, 1996). They found an immediate interpretation in terms of the emerging-flux reconnection scenario (Heyvaerts *et al.*, 1977). The jet material is hot plasma projected along magnetic flux tubes that may open out into the heliosphere or close on large scales without entering the solar wind. These are plasma flows parallel to the apparent field direction. The jet

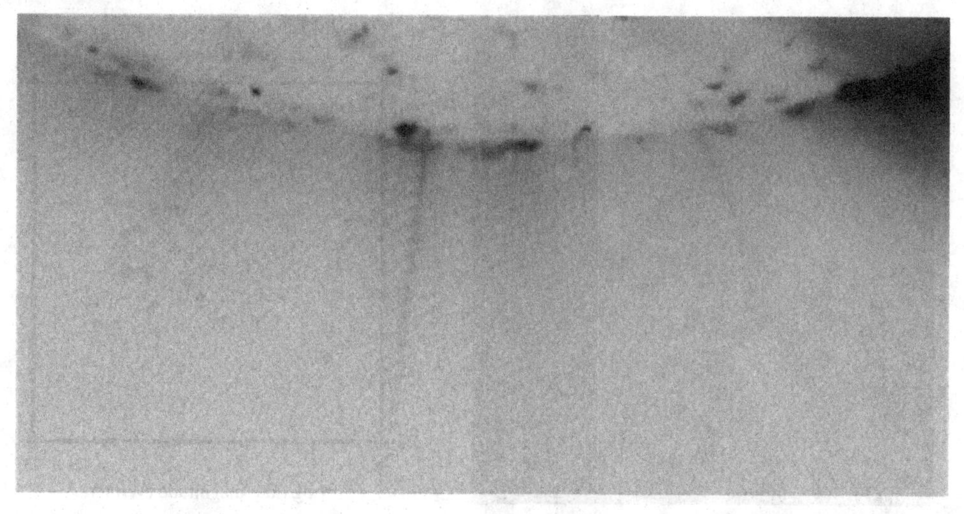

Fig. 5.4. Soft X-ray image of the Sun's south polar region, with an inverted grey scale, showing a highly collimated polar jet structure. (Courtesy P. Grigis.) Note that this is a coronal-hole jet, but that similar features often occur in active regions in association with microflares (see Section 5.2.5).

sources have a strong association with radio type III bursts – known to come from non-thermal electrons streaming outwards along open flux tubes (Aurass *et al.*, 1994; Kundu *et al.*, 1995) – and also with electron events observed in interplanetary space (Lin, 1974; Krucker *et al.*, 2007; see also Nakajima and Yokoyama, 2002). Invariably a compact flare appears near the jet's point of origin near the chromosphere (Shimojo *et al.*, 1996).

The jet-associated microflares have magnetic connectivity that permits access to the heliosphere, and they have other occurrence patterns linking them to emerging (or disappearing) magnetic flux (Shimojo *et al.*, 1998). The jet-associated microflares seem to be compact and less energetic flares, and Hinode observations clearly show them to be part of a continuum of weaker and weaker jet-like events (Shibata *et al.*, 2007) found in the quiet Sun and especially visible in the polar regions (Fig. 5.4). Soft X-ray jet structures are seldom as visible in major flares.

5.3.5 Coronal mass ejections (perpendicular motions)

Major flare events almost invariably involve the "opening" of the magnetic field as a CME (e.g. Hundhausen *et al.*, 1994); see Table 5.1 for the statistics. This involves the unstable expansion of the field (equivalent to a motion perpendicular to the field). Note that at low plasma β, the gas whose emission we observe (the mass of the CME) has little influence on the dynamics. Observationally, in the Thomson-scattering brightness measurements made by a coronagraph, we often

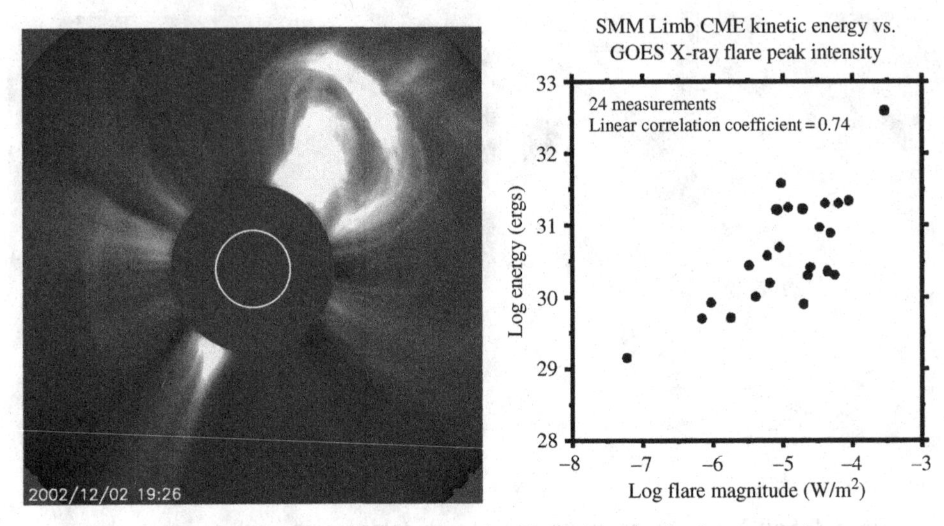

Fig. 5.5. *Left*: Coronagraph observation of a CME that nicely shows the three-part structure: front, cavity, and (the bright core) filament (this is a file image taken from the LASCO database, presented in a reverse grey scale). *Right*: Correlation between inferred CME kinetic energy and peak GOES soft X-ray flux. (From Burkepile *et al.*, 2004.)

see a characteristic three-part structure: front, cavity, and filament (Figs. 5.5, 6.1, 6.2; see also e.g. Hundhausen, 1999). This pattern makes it clear that the CME originated in a filament cavity near the surface of the Sun. A filament cavity (see e.g. Engvold, 1989; but note that there seems to be no recent review of this important subject) consists of long, basically horizontal field, presumably more intense than its overlying "tie-down" field that is more potential (Gibson and Fan, 2006; Martin *et al.*, 2008; Schrijver *et al.*, 2008).

The interpretation of the front structure of a CME is complicated. One expects, from the standard models (see Chapter 6), that this would incorporate coronal material contained in the overlying magnetic flux tubes as they are expelled from the corona and become "open". There should also be a sweeping-up of ambient coronal or solar-wind material, and we would expect the occurrence of a bow wave analogous to that of the Earth in the solar wind flow. The presence of such a bow wave is consistent with observations of type II radio signatures at hectometric–kilometric wavelengths. The emission from these bursts requires the shock condition to have been met (e.g. Kundu, 1965), and their propagation velocities are consistent with the known outward velocities of the CMEs that provide the driver gas for this large-scale shock. To clinch the picture, we also observe the shock when it impacts the magnetosphere with the classic geomagnetic signature of a storm sudden commencement (SSC; see Fig. 10.1) signature (e.g. Chapman and Bartels, 1940).

Much of the mass of a CME comes from below the occulting edge of the LASCO C2 coronagraph. Indeed, a glance at LASCO movies readily available on the Internet† shows mass flow long after the three-part structure has vanished. This late flow certainly originated in the lowest corona or even chromosphere.

Modern images in coronal emissions such as soft X-rays allow a comparison of the coronal state before and after a CME event. Such comparisons revealed "dimmings", readily interpreted as the evacuation of the mass of the corona by the CME eruption (Hudson *et al.*, 1995b; see also Rust, 1979, for the earlier *Skylab* observation). The soft X-ray dimmings presumably correspond to the coronal depletions found via similar before/after comparisons of the visible corona (Hansen *et al.*, 1974).

5.3.6 Global waves (coronal and other)

There are at least five types of large-scale wave structures associated with solar flares and CMEs, perhaps not all distinct: helioseismic, metric type II, Moreton, interplanetary type II, and EIT. The Moreton waves (Athay and Moreton, 1961) can now be detected at several different wavelengths. Originally discovered in Hα (the chromosphere), they are fast (of order 1000 km/s) waves radiating, generally into restricted sectors, from the flare site. The standard hypothesis of Uchida (1974) describes these chromospheric waves as the skirts of global fast-mode shock waves actually propagating in the corona; the wave energy refracts into the chromosphere because of its lower Alfvén speed.

Large-scale coronal shock waves had long been known from meter-wave radio astronomy, where the radio signatures clearly imply that the shock condition has been met (Wild *et al.*, 1963). The type II burst (Fig. 5.6) is relatively rare, and it is observed best at the frequencies below ~200 MHz. As with the "fast-drift" type III bursts, the assumption of emission at the local plasma frequency or its harmonic (see Section 4.3.1) allows for a height estimation by assuming a coronal density model. The derived motions point to an origin in the impulsive phase of the flare, but this requires an extrapolation because of the shock "ignition" requirement (Vršnak and Lulić, 2000). We also know directly of *interplanetary* shock waves driven by CMEs as bow waves, both from longer-wavelength radio astronomy and also from the *in situ* observations (and the SSC response of the Earth's magnetosphere to the impulse).

The EUV observations from SOHO (Extreme-Ultraviolet Imaging Telescope) disclosed a rich assortment of "EIT waves" (Moses *et al.*, 1997; Thompson *et al.*, 1999). The EUV signature is somewhat complicated, and it appears that multiple

† http://sohowww.nascom.nasa.gov/data/data.html

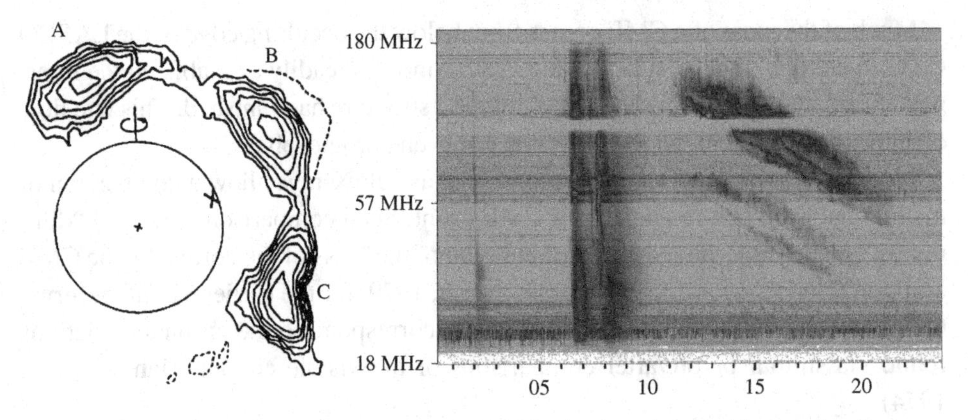

Fig. 5.6. *Left*: Culgoora image of a type II burst associated with a major flare (Palmer and Smerd, 1972). Note how this plasma-frequency radiation appears to wrap around a concentric spherical surface, presumably at the right mean density. *Right*: Radio spectrogram (frequency versus time in minutes) of a different major flare, illustrating type III bursts (fast drift, produced by streams of energetic electrons) and a type II burst (slow drift, fundamental/harmonic structure produced during the propagation of a large-scale shock wave), also from Culgoora.

causes can produce wave-like disturbances (Biesecker *et al.*, 2002), including the classical Moreton wave.

Finally, the helioseismic waves discovered by Kosovichev and Zharkova (1998) seemed rare at first, but now there are several examples. Figure 5.7 shows the original event, that of July 9, 1996. These waves result from energy coupled into the interior by the flare process. The excitation of such a wave is thus closely associated with the dynamics of the deepest atmospheric layers that we can see into. This probably involves the most energetic aspects of a flare. In this context we note the 1.56 μm "opacity minimum" observations of Xu *et al.* (2004) and also the γ-ray observations of Share *et al.* (2004; see also Schrijver *et al.*, 2006, for further discussion).

5.3.7 Magnetic signatures

The observation and interpretation of solar magnetic signatures has improved dramatically in the past decade, with new facilities such as the ground-based SOLIS and the Hinode satellite providing vector Zeeman measurements, for example. Such measurements show clear flare-associated lasting (stepwise) changes (Kosovichev and Zharkova, 1999; Wang *et al.*, 2002; Sudol and Harvey, 2005), which would be expected if the stresses in the coronal field had their origins in motions below the photosphere ("energy buildup"). In addition there are vigorous activities related to interpreting the data in terms of the coronal field, which is almost unobservable (but see Lin *et al.*, 2000; Tomczyk *et al.*, 2008) and in any case is optically thin. The extrapolations have an excellent chance to be extremely

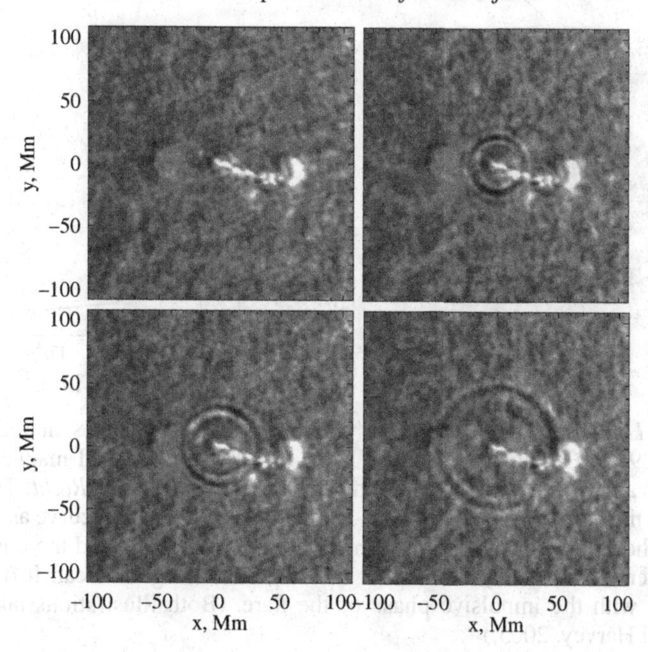

Fig. 5.7. The original helioseismic wave observed from the singular solar-minimum flare of July 7, 1996 (Kosovichev and Zharkova, 1998), from the "last best active region" of that solar cycle (Hudson *et al.*, 1995b). The figure shows the wave via Doppler images, with the wave representation enhanced but based on the observed Fourier components. More recent helioseismic waves are directly visible in the filtered images.

informative in active regions in particular, since the active-region corona has low plasma β values.

Figure 5.8 (left) shows the stepwise magnetic changes derived by Sudol and Harvey (2005) for the X10 flare of October 29, 2003. These are well defined and appear to delineate the general regions of the flare ribbons, and within the time resolution of the data they tend to happen in coincidence with the impulsive phase of the flare. There is thus no reason not to associate these changes with the source of flare energy. Liu *et al.* (2005) report similar changes and show how one could interpret them in terms of simple global changes in the coronal field (e.g. Hudson, 2000).

The implications of these new developments are clear: when we can do the same thing with vector fields, and in addition do the measurement well above the photosphere, we will be able to reconstruct the before/after 3D field structure in an active region and learn quite directly about the exact geometry of the instability. The measurement of the chromospheric field, as opposed to that of the photosphere which is not force-free, is important to minimize the effects of stresses imposed by photospheric flows. Note that future "frequency-agile" imaging spectroscopy in the microwave band offers a precise and complementary way of

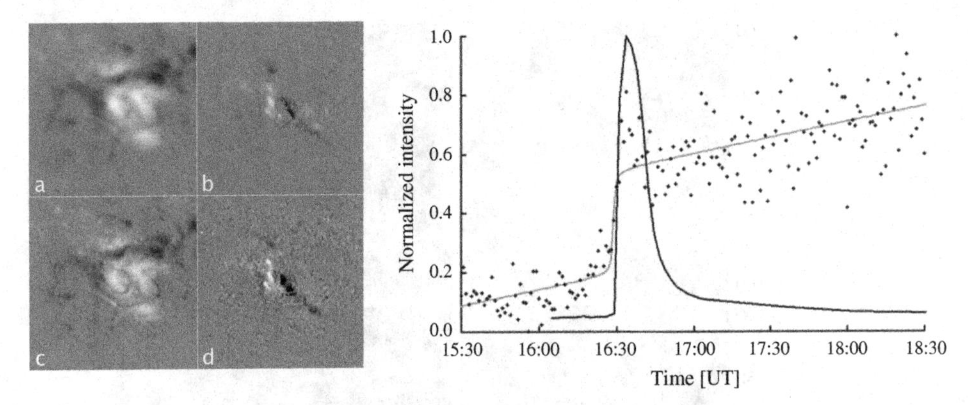

Fig. 5.8. *Left*: Map of the stepwise photospheric field changes in the flare of October 29, 2003. Panels (a) and (c) show the GONG and MDI magnetograms; panels (b) and (d) show their before/after changes, respectively. *Right*: Time variations for the flare of August 25, 2001, showing the GOES light curve as a smooth line and the GONG data as points. The fluctuations are large and there is a background trend, as in many events, but the stepwise change is clear. It (typically) coincides with the impulsive phase of the flare. (Both illustrations taken from Sudol and Harvey, 2005.)

checking the observations and extrapolations (White, 2005), since this wavelength range includes the electron cyclotron (Larmor) frequency of these fields. Until the advent of these new capabilities, it is unlikely that a quantitative understanding of the actual field restructuring will be possible.

5.3.8 Coronal non-thermal events

Prior to Yohkoh (1991–2001) and RHESSI (2002 onward), meter-wave radio astronomy was the main source of knowledge about non-thermal processes in the corona (e.g. shock waves and particle acceleration). The radio observations are very sensitive and result from interesting physical processes (see Kundu, 1965, for much interesting detail, or Bastian *et al.*, 1998, for a more recent review of radio techniques). The type II bursts, for example, involve many small-scale accelerations of ambient electrons to few-keV energies (the "herringbone" structure). However, the meter-wave telescopes have low angular resolution and the emission mechanisms (except for the free–free mechanism) have complicated dependences on the physical parameters of the emitting region and its environment. Thus it would be extremely valuable to detect some of these sources in X-radiation, which is more direct.

Krucker *et al.* (2008) review the current observational status of coronal hard X-ray observations. As more sensitive data become available, it is clear that the corona is a rich source of hard X-ray emission, as expected, but the details are in some cases unexpected. For example, one would confidently expect that the electron streams commonly observed at 1 AU (Lin, 1974) would produce at least thin-target bremsstrahlung near their point of acceleration in the corona (Lin and Hudson, 1971). We still do not have clear observations of this emission (Krucker *et al.*, 2007). On the other hand, coronal hard X-rays associated with CME eruptions appear to be common Krucker *et al.* (2008), and these may be related in some manner to the radio type IV bursts. Type II burst sources can be observed in soft X-rays (Khan and Aurass, 2002; Hudson *et al.*, 2003), but not yet in hard X-rays (the signature of the non-thermal particles) because of lack of sensitivity.

One of the most striking of the new RHESSI coronal hard X-ray sources is shown in Fig. 5.9 (Krucker *et al.*, 2008). The high energy of observation (250–500 keV shown in the figure) means that the source electrons were relativistic. Footpoint sources appeared early in the event, but the coronal source remained

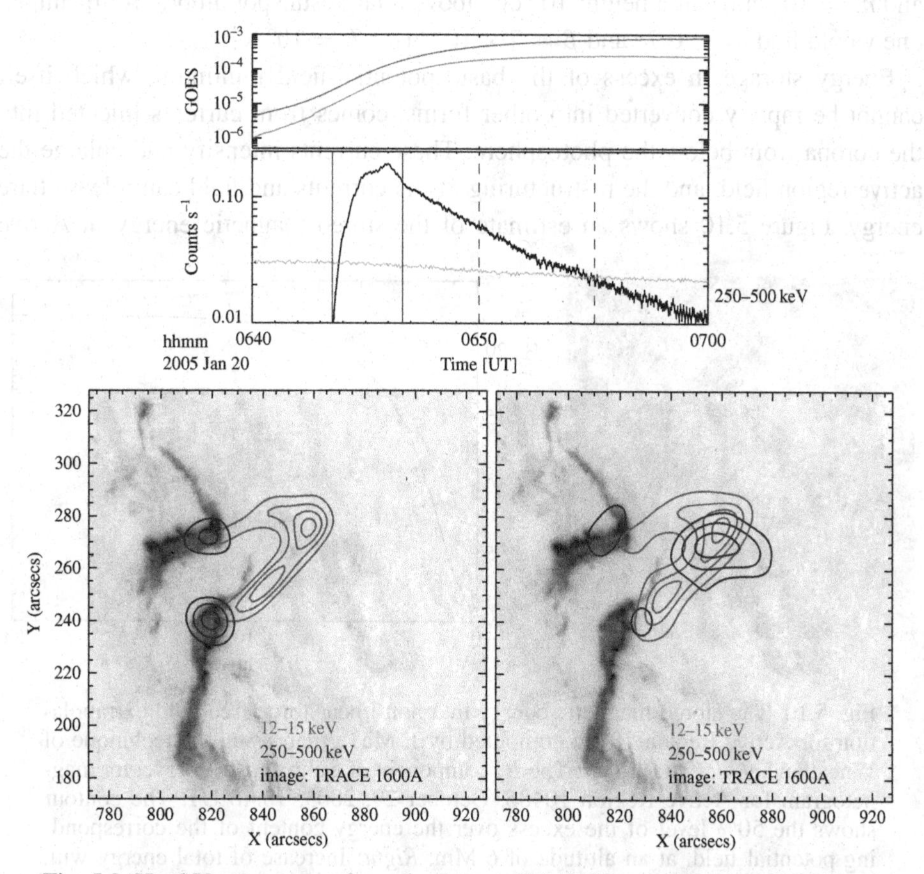

Fig. 5.9. Hard X-ray sources from the January 20, 2005 event. *Upper*: GOES and RHESSI light curves. *Lower:* Early image showing well-developed footpoints at 250–500 keV (dark contours), and a later image showing the persistent coronal hard X-ray coronal source. The light contours show large-scale loop structures with thermal spectra.

bright and decayed in flux with a nearly exponential decay with a time constant of about 5 minutes, similar to that observed in the prototype coronal hard X-ray event of March 31, 1969, described by Frost and Dennis (1971).

5.4 Flare energetics

5.4.1 Magnetic energy storage

An active region with large sunspots creates a localized region of strong magnetism in the corona. The basic potential-field description of the sunspot fields already predicts strong fields at altitudes comparable to the spot diameter, and in fact microwave observations do show such fields (e.g. Brosius and White, 2006). Extreme values of the Alfvén speed and plasma β could result; for $|\mathbf{B}| = 10^3$ G and $n_e = 10^8$ cm^{-3} at a height 10^9 cm above a large sunspot umbra, for example, one would find $v_A = 0.7c$ and $\beta = 7 \times 10^{-6}$ (for $T = 10^6$ K).

Energy storage in excess of the basic potential-field minimum, which itself cannot be rapidly converted into other forms, comes from currents injected into the corona from below the photosphere. These currents intensify and enlarge the active-region field, and the restructuring of the currents and field can release flare energy. Figure 5.10 shows an estimate of the stored magnetic energy in Active

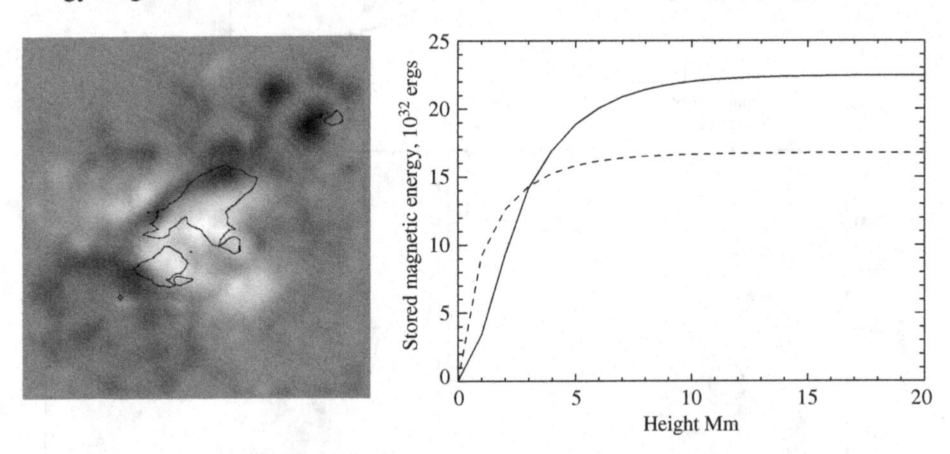

Fig. 5.10. The stored magnetic energy in a non-linear force-free field extrapolation for Active Region 10486 computed by J. McTiernan using the technique of Wheatland *et al.* (2000). *Left*: The B_z component of a chromospheric vector magnetogram for Active Region 10486, October 29, 2003, 18:46 UT. The contour shows the 50% level of the excess over the energy content of the corresponding potential field, at an altitude of 6 Mm. *Right*: Increase of total energy with height in the data cube of the extrapolation (dimension 65^3 arcsec). The dashed and solid lines show the integrated energies for a potential field model and for the non-potential field model, respectively. The 50% level gives a rough idea about the location of stored magnetic energy; it is higher for the non-potential field but still located close to the base of the corona.

Region 10486, which produced the flare detected bolometrically by Woods *et al.* (2004). Note that the excess magnetic energy, using the non-linear force-free model of Wheatland *et al.* (2000), apparently can exceed the potential-field energy even though strongly twisted coronal structures are not often seen.

5.4.2 Partition of energy release

The energy released from its magnetic storage is lost to the corona either as radiation or in the form of mass motions. Note that thermal conduction should generally lead to excess radiation at transition-region or chromospheric temperatures (e.g. Emslie *et al.*, 2005). The initial energy release is dominated by the acceleration of high-energy particles (Lin and Hudson, 1976), which are relatively easy to detect from their hard X-ray, γ-ray, and radio signatures. There is also presumably some direct heating in the sense of Ohmic dissipation or adiabatic compression, but this is harder to recognize observationally. The energy that appears in the corona ultimately increases the temperature, which in turn enables heated chromospheric material to expand and rise into the corona. The radiation signatures at these different stages spread across the electromagnetic spectrum, as sketched out in Fig. 5.1.

The partition of the energy release must also include bulk terms (kinetic energy, gravitational potential energy, and enthalpy). We can readily estimate the kinetic energy of the CME ejecta (e.g. Vourlidas *et al.*, 2000), but the magnetic energy – the dominant term, because electrodynamic forces drive the whole process – is much more difficult. Indeed, a plausible extension of the Aly–Sturrock theorem (Aly, 1991; Sturrock, 1991) suggests that the creation of a CME actually absorbs magnetic energy rather than releasing it, because the open fields it creates are maximally non-potential in nature. So, even the sign of this component of CME energy remains ambiguous. In any case, by order of magnitude, a major flare/CME event may have comparable radiant and bulk kinetic "emissions" (e.g. Emslie *et al.*, 2005; see also Fig. 5.5).

5.4.3 Nanoflares

The nature of coronal heating may involve flare-like processes, even outside the times of actual flares or microflares. Parker (1988) introduced the term "nanoflare", implying that just this kind of non-thermal process might be involved in ordinary coronal heating. Here the "nano" implies an event energy on the order of 10^{-9} of that of a major flare, and the suggestion was that a swarm of such tiny events might not be recognizable from a continuous heating process. In general, the possibility that individual elements of a structure are unresolved by a given observation strongly affects its interpretation (Sturrock *et al.*, 1990; Cargill and Klimchuk, 2004).

Hudson (1991) noted that such an occurrence pattern of tiny events would necessarily differ from the "hard" power law seen for true flares (see Section 5.2.5). A single nanoflare could not be detected directly, but the nanoflaring process could be detected statistically from the fluctuation spectrum. In practice most workers ignore this distinction and just view nanoflares as still smaller microflares that can still have individually recognizable signatures.

The concept of nanoflare heating lies close to the interpretation of a flare as an assembly of semi-independent filamentary substructures. This might be expected from the anisotropy of plasma transport properties in the presence of a magnetic field. The arcade structure of many flares indeed shows their inherently filamentary structure, albeit on observable scales. Aschwanden *et al.* (2001), for example, decomposed a major arcade structure into about 100 individually recognizable strands. This has made "multithread modeling" of flare structures possible (Hori *et al.*, 1998; Warren, 2006), with substantial implications for the physics.

5.5 Flare analogs

In this section we discuss the possible analogies between the forms of solar activity and non-solar phenomena. These often seem striking enough to beg for a common model, but even without success in developing such a model (it would be fair to say that no predictive models for flares now exist) we can certainly use paradigms from one domain as frameworks for understanding another. The two major areas of overlap are the terrestrial aurora and stellar flares, but there are other possible patterns as well.

5.5.1 Other patterns of flare activity

As we have seen, there is a rather well-defined basic observational template for solar flares, both eruptive and confined. The key features include intense non-thermal radiations in an impulsive phase that leads to a gradual phase via the formation of a coronal reservoir (the Neupert effect). The hard X-ray emission characteristically follows the soft–hard–soft pattern of spectral variation in the impulsive phase. The gradual phase has temperatures characteristically one to two orders of magnitude higher than those of the quiet corona. There is a weak correlation between temperature and emission measure $\int n_e^2 \, dV$. The chromospheric signatures (e.g. Hα) are dominated by the formation of ribbon structures that tend to spread apart in the gradual phase, reflecting the arcade structure of the flare loops. These properties, and possibly a few others, describe the solar flare

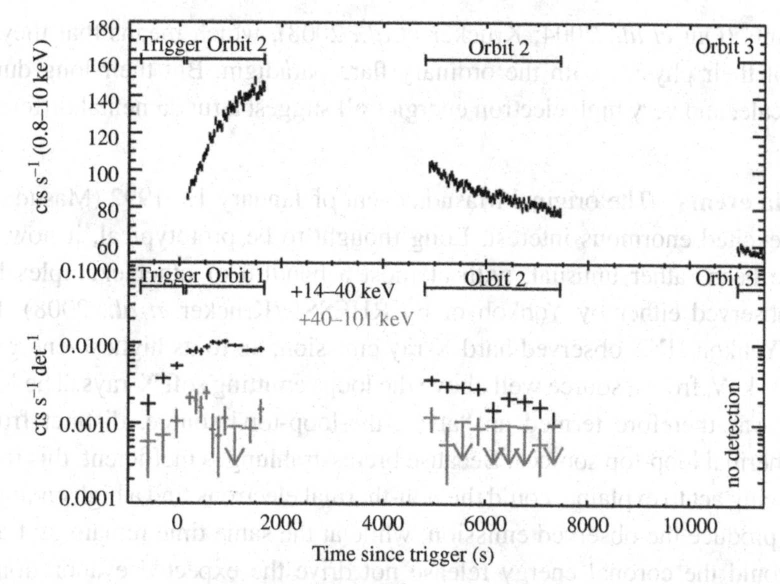

Fig. 5.11. Powerful stellar flare observed December 16, 2005, on the active binary system II Pegasi (Osten *et al.*, 2007). The upper curve shows 0.8–10 keV counting rate from the XRT instrument on board SWIFT (Burrows *et al.*, 2005), and the lower curves show two hard X-ray channels (14–40 keV and 40–101 keV) from the BAT instrument. One can see the clear progression of a Neupert-effect analogy, with the highest-energy channel (lighter shading) showing an impulsive-phase excess in the first ks of the observation.

paradigm. The stellar flare shown in Fig. 5.11 has a clear Neupert-effect time profile. This does not mean very much in terms of the physical distinctions between this event and a solar flare, unfortunately, except to confirm that a coronal energy reservoir can also form in the vicinity of this star (II Pegasi) as well.

Other patterns of solar activity exist, and these may be more relevant to some non-solar conditions than the standard paradigm. These would include the following (Hudson and Micela, 2006):

Extended events In major flares, especially those associated with solar energetic particles, an extended non-thermal phase sometimes develops on time scales of tens of minutes following the impulsive phase. These events have a close relationship with the meter-wave type IV emission, which reveals the presence of relativistic electrons via synchrotron emission (Boischot and Denisse, see Wild *et al.*, 1963). In the hard X-ray band we see a *soft–hard–harder* spectral evolution (Frost and Dennis, 1971; Hudson, 1978; Cliver *et al.*, 1986) rather than the clear soft–hard–soft evolution of the impulsive phase. Kiplinger (1995) found that this hard X-ray spectral pattern tends to accompany solar proton events. The coronal structures associated with such events are now known to have bright footpoints (Qiu *et al.*, 2004; Krucker *et al.*, 2008), which means that they share some of their physics with the ordinary flare paradigm. But their long duration, great scale, and very high electron energies all suggest a fundamental difference in origin.

Masuda events The original Masuda event of January 13, 1992 (Masuda *et al.*, 1994) excited enormous interest. Long thought to be prototypical, it now seems to have been rather unusual, with at most a handful of other examples having been observed either by Yohkoh or by RHESSI (Krucker *et al.*, 2008). In this event, Yohkoh/HXT observed hard X-ray emission, up to its highest–energy band at 53–93 keV, from a source well above the loops emitting soft X-rays. The Masuda source was therefore termed an "above-the-loop-top" source, distinct from the usual thermal loop-top sources. Because bremsstrahlung is inefficient, this required a balancing act to explain – could the non-thermal electrons find a high enough density to produce the observed emission, while at the same time remaining trapped? How could the coronal energy release not drive the expected evaporation? The physics remains unclear because of these discrepancies.

Non-thermal ejecta The meter-wave radio observations provide several examples of distinctly different high-energy processes operating in the solar corona. These include the types I–V bursts (Wild *et al.*, 1963) and now probably some of their counterparts in hard X-rays (Hudson *et al.*, 2001; Krucker *et al.*, 2008). These have great interest at the present time because of their association with CMEs and therefore with disturbances in the Earth's environment.

Coronal thick-target events In the ordinary flare paradigm, the collisional thick-target model places the target (the hard X-ray source) in the chromosphere. Recently events have been found for which the best interpretation is that the fast electrons actually do not propagate as far as the chromosphere, but instead brake collisionally in the corona (Veronig and Brown, 2004). This development was unexpected because of the general success of the standard model (see Chapter 6), and it suggests that the powerful electron acceleration of the impulsive phase can take place in a relatively high-density medium ($n_e > 10^{10}$ cm^{-3}), in order to provide enough coronal column density to bring a $\sim$50 keV electron to rest.

Shock waves This mechanism is of particular interest in astrophysics, where there is hardly a domain on any scale in which shock physics is not invoked. In the case of the solar flare, we are particularly interested in large-scale waves that accompany the basic restructuring of the field needed to release energy. Note that in 2D Petschek reconnection (Section 5.3.1 in Vol. I) it is precisely the large-scale shock waves that convert the magnetic energy; the reconnection point itself is of little consequence for energy release. We do not know yet whether or not this logic carries over to non-steady 3D magnetic reconnection. The large-scale shock waves

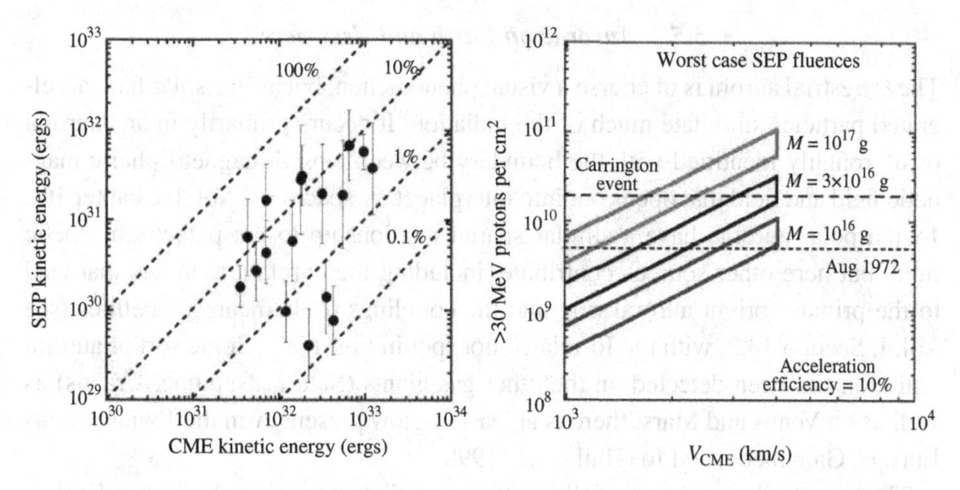

Fig. 5.12. *Left*: Energy converted by interplanetary (CME-driven) shock waves into solar energetic particles (SEPs). The efficiency of conversion can exceed 10%. *Right*: Comparison of particle fluences for model CME masses and speeds, relating the Carrington event to better-observed recent examples (Mewaldt *et al.*, 2007).

in solar flares can readily be detected via their radio emission. We understand the physics of a type II burst well enough to identify it as the emission signature of the product of Langmuir turbulence scattering energy into electromagnetic radiation near the local plasma frequency (Wild *et al.*, 1963) or its harmonic. The shock can occur either near the surface of the Sun, where it may be a blast wave propagating through the ambient, undisturbed corona, or it may be an interplanetary wave driven by the CME. Recently, Mewaldt *et al.* (2007) have obtained the results shown in Fig. 5.12. High-energy particles play a major role in the dissipation of energy at such a collisionless shock because their fraction of the total CME energy may exceed 10%. Note that such a mode of energy dissipation is basically a long-range effect: energy is removed from the shock but not converted to heat locally. This means that an ideal MHD simulation will not correctly localize the eventual sink of the shock energy.

Impulse response White *et al.* (1992) observed a solar radio burst with quite remarkable properties. With high-resolution VLA observations, the event was located in an active region and had an oblong shape about 5000 km in length and 1500 km in width, thus presumably a compact loop. This, plus oddities in the radio spectrum, place it and a few similar events in a separate category. The small scale presumably means that the event took place in the lowest atmosphere, below the chromosphere–corona interface region. "Impulse response" refers to the emission time profile, which had a nearly unresolved rise time and a brief, exponential, and frequency-independent decay (about 20 s) at 15 GHz.

5.5.2 Aurorae on Earth and elsewhere

The terrestrial aurora is of course a visual phenomenon, but as in a solar flare accelerated particles stimulate much of the radiation. It occurs primarily in an "auroral oval" roughly identified with the boundary between closed magnetospheric magnetic field and field that opens out into interplanetary space (see Vol. I, Chapter 10). Jovian polar aurorae have a similar spatial relationship to the planet's magnetic field, but here other sources contribute, including the corotation effects that lead to the primary bright auroral ring and the couplings to the nearest satellites (see Vol. I, Section 13.2, with the Io-related hot spot in Fig. 13.7). Some sort of auroral emission has been detected on the other gas giants (Saturn, Neptune, Uranus) as well as on Venus and Mars; there is at least airglow present even the Jovian moons Europa, Ganymede, and Io (Hall *et al.*, 1998).

The terrestrial aurora, especially its "substorm" development, has several points of similarity to the phenomena of solar flares. This has long been noted to be of interest (e.g. Obayashi, 1975; Bratenahl and Baum, 1976; Schindler, 1976; Akasofu, 1979, 2001). Properties that might be related include the acceleration of non-thermal electrons and the identification of N–S conjugate auroral zones as ribbon-like structures. One can note the gradual buildup of stored energy, and its sudden release, as the system evolves past the point of marginal stability. This theoretical idea plus the attractiveness of magnetic reconnection as an energy source for each process have also encouraged this kind of thinking.

Over the decades these possible analogies have retained their fascination, but putting them to use in learning about flares has proven difficult. Why is this? Presumably the answer is to be found in the very different physical conditions in the corona and in the geomagnetic tail, even though parameters such as the Alfvén speed in the geotail and in the active-region corona may be similar (e.g. Obayashi, 1975; see Chapter 10 for a discussion of the physics of dynamic planetary magnetospheres). We note that the boundaries of the magnetosphere are the ionosphere and the magnetopause. Along the flanks of the magnetosphere there is a solar wind flow that creates a large convective electrical potential. This would not be present in the solar corona. The not-so-analogous boundaries of the solar corona are the photosphere/corona transition zone, mainly the chromosphere, and a rather nebulous and ill-understood process that creates the solar wind. These boundaries have some commonalities (Haerendel, 2006) but some major differences as well. The chromosphere and the ionosphere have different conductivity tensors, and the ionosphere has a non-conducting lower boundary, for example. As for the solar wind flow around the magnetosphere, there is simply no solar analog. This flow is thought to be the source of the substorm energy, and so the "flare buildup" process, which for the Sun lies below the solar surface, seems not to be analogous.

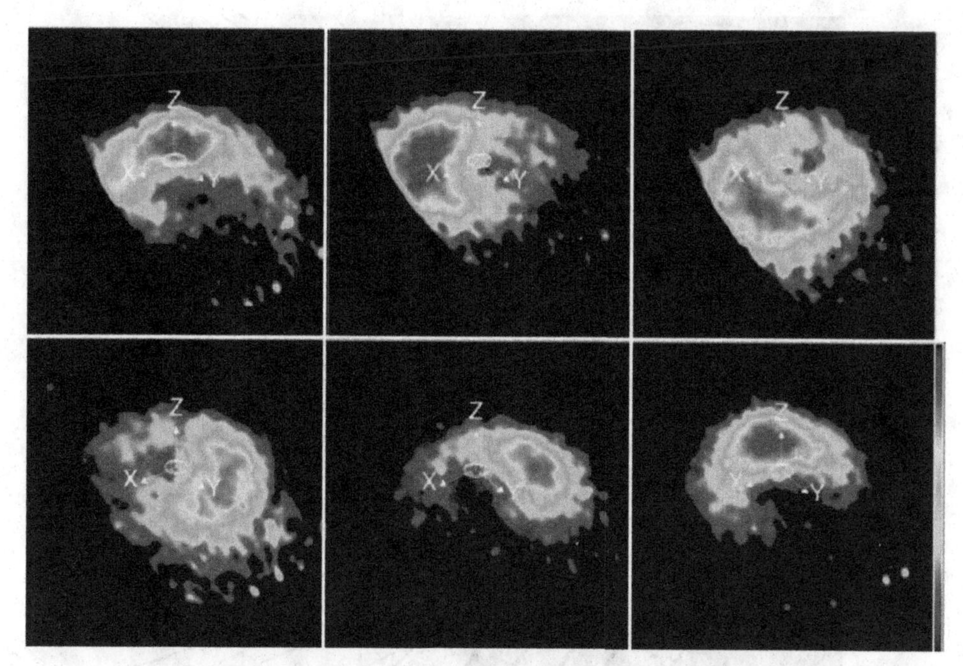

Plate 1 (Fig. 3.23). Neutral-hydrogen (20–50 keV) images of Saturn's ring current taken with the INCA sensor on Cassini at 2.13-hour intervals showing counterclockwise rotation in the plasma. (Courtesy S. M. Krimigis.)

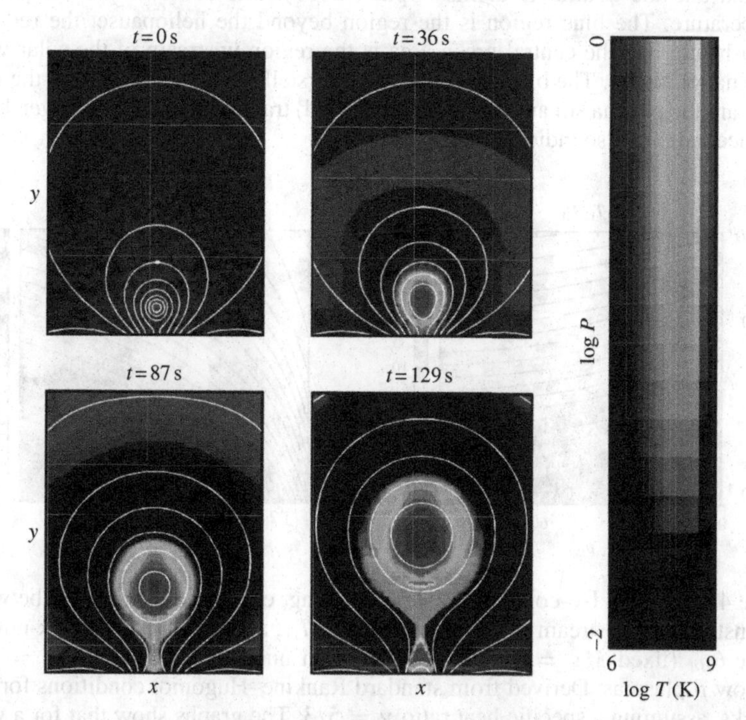

Plate 2 (Fig. 6.12). Numerical simulation of an erupting flux rope. The color hue indicates the temperature, while the color intensity indicates the pressure. The white lines are contours of the flux function.

Plates 1-21 are available for download in colour from www.cambridge.org/9780521760515

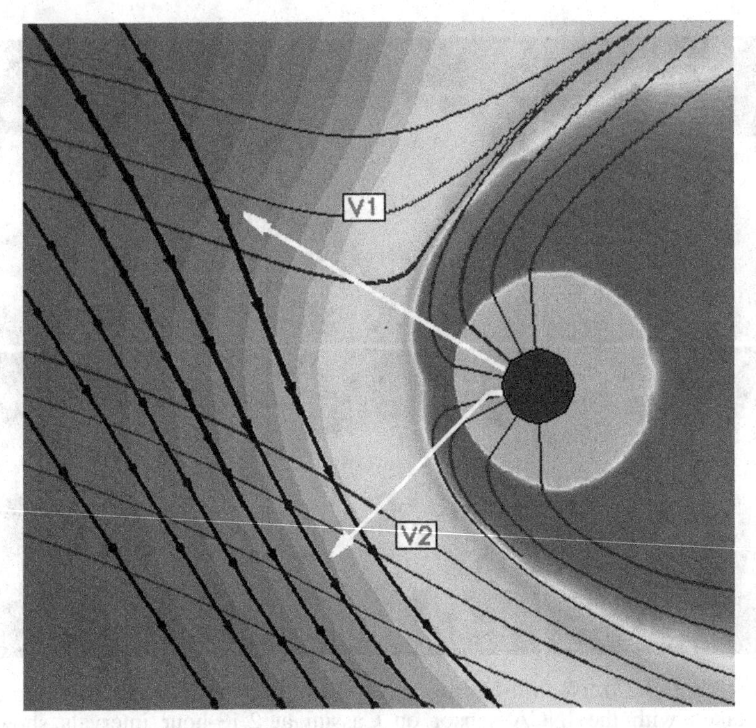

Plate 3 (Fig. 7.4). Meridional cut from a heliosphere simulation including the plasma and the neutral H atoms (Opher, 2009). The contours are the plasma temperature. The blue region is the region beyond the heliopause; the red, the heliosheath; and the central green area is the region upstream of the solar-wind termination shock. The black lines are the interstellar magnetic field and the grey lines are the plasma streamlines. The (projected) trajectories of the Voyager 1 and 2 spacecraft are also indicated.

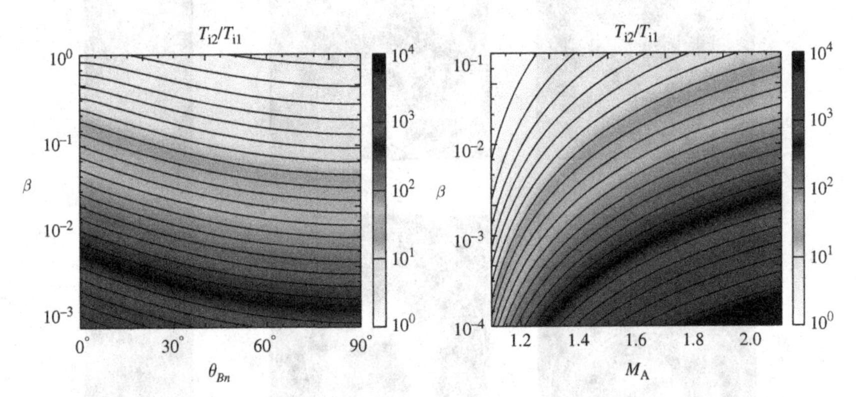

Plate 4 (Fig. 8.1). Iso-contours of shock heating, expressed as the ratio between downstream to upstream ion temperature T_{i2}/T_{i1}, as a function of shock-normal angle θ_{Bn} (fixed $M_A = 2$) and Alfvén Mach number M_A (fixed $\theta_{Bn} = 45°$) for low β plasmas. Derived from standard Rankine–Hugoniot conditions for fast shocks, assuming a specific heat ratio $\gamma = 5/3$. The graphs show that for a wide range of angles there can be very substantial downstream heating at sufficiently low plasma β, as present in much of the solar corona. Such extreme heating may help form a seed population for further acceleration.

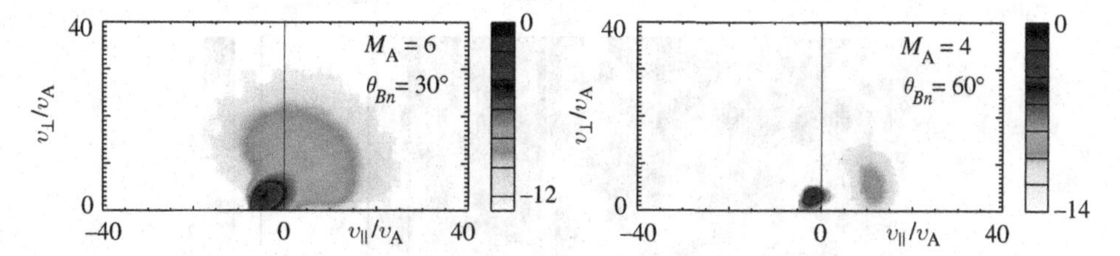

Plate 5 (Fig. 8.5). Sketch of upstream proton distributions (perpendicular and parallel to the ambient magnetic field) in the shock frame from planar, 2D hybrid shock simulations at quasi-parallel ($\theta = 30°$) and oblique ($\theta = 60°$) angles. As in many documented observations of the Earth's bow shock and at sufficiently high Mach number IP shocks, at quasi-parallel shock-normal angles, protons can not only easily travel upstream and generate waves, but they also easily scatter in these self-generated waves to form a diffuse distribution that forms a contiguous cloud of both upstream ($v_\parallel > 0$) and downstream ($v_\parallel < 0$) directed particles. Conversely, at oblique shocks, only a highly dilute upstream-propagating beam with enhanced perpendicular energy is found, and even that can only be seen with very good statistics, in simulations. Unlike the quasi-parallel shock, a higher Mach number does not help initially, but typically makes it more difficult for ions to make it upstream in the first place.

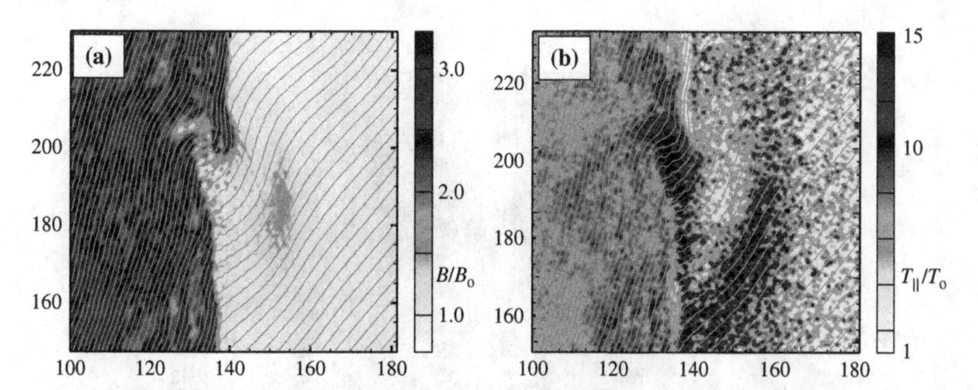

Plate 6 (Fig. 8.6). Magnetic field line contours and (a) total magnetic field, and (b) parallel temperature $T_\parallel$ normalized to upstream in a subset of a 2D hybrid simulation of an oblique shock ($\theta = 50°$). (From Krauss-Varban *et al.*, 2008.) It can be seen how compressional waves generated by dilute beams disrupt the shock and change the local θ_{Bn}, in turn allowing more upstream wave and particle production than expected at the oblique shock. This process appears to enhance upstream energetic proton fluxes by two to three orders of magnitude.

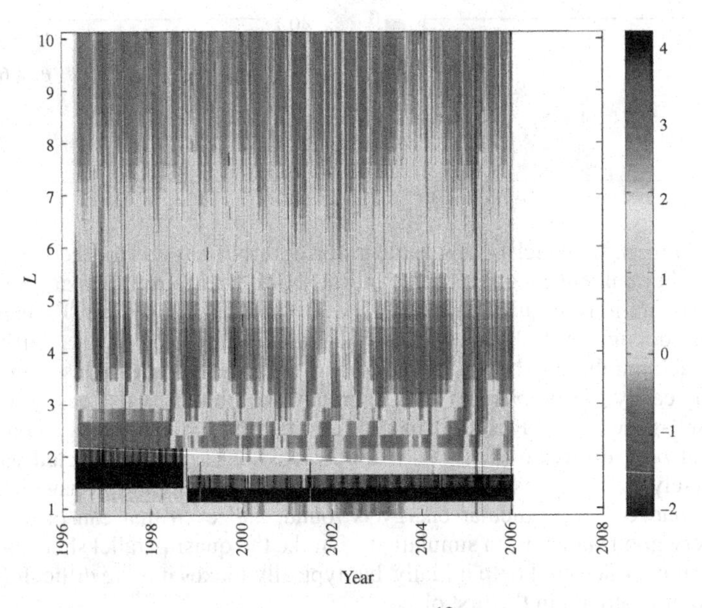

Plate 7 (Fig. 11.5). Radiation belt electron flux (^{10}log(counts/s)) as measured by the Proton Electron Telescope (PET) Elo channel that measures electrons with energies $>$ 1.5 MeV on the SAMPEX satellite. The data are averaged in 0.25L and 1 day bins.

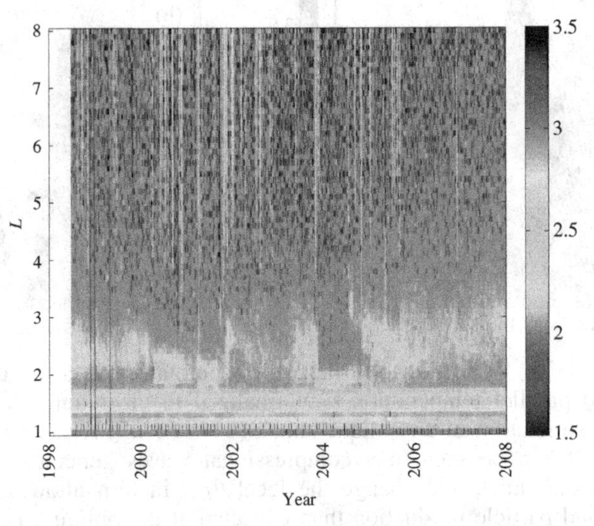

Plate 8 (Fig. 11.6). Radiation belt proton flux (number per cm^2 s str on a logarithmic scale) from the SEM-2 instrument that measures protons with energies between 2.5 and 6.9 MeV on the NOAA-15 satellite. The data are averaged in 0.2L and 1 day bins.

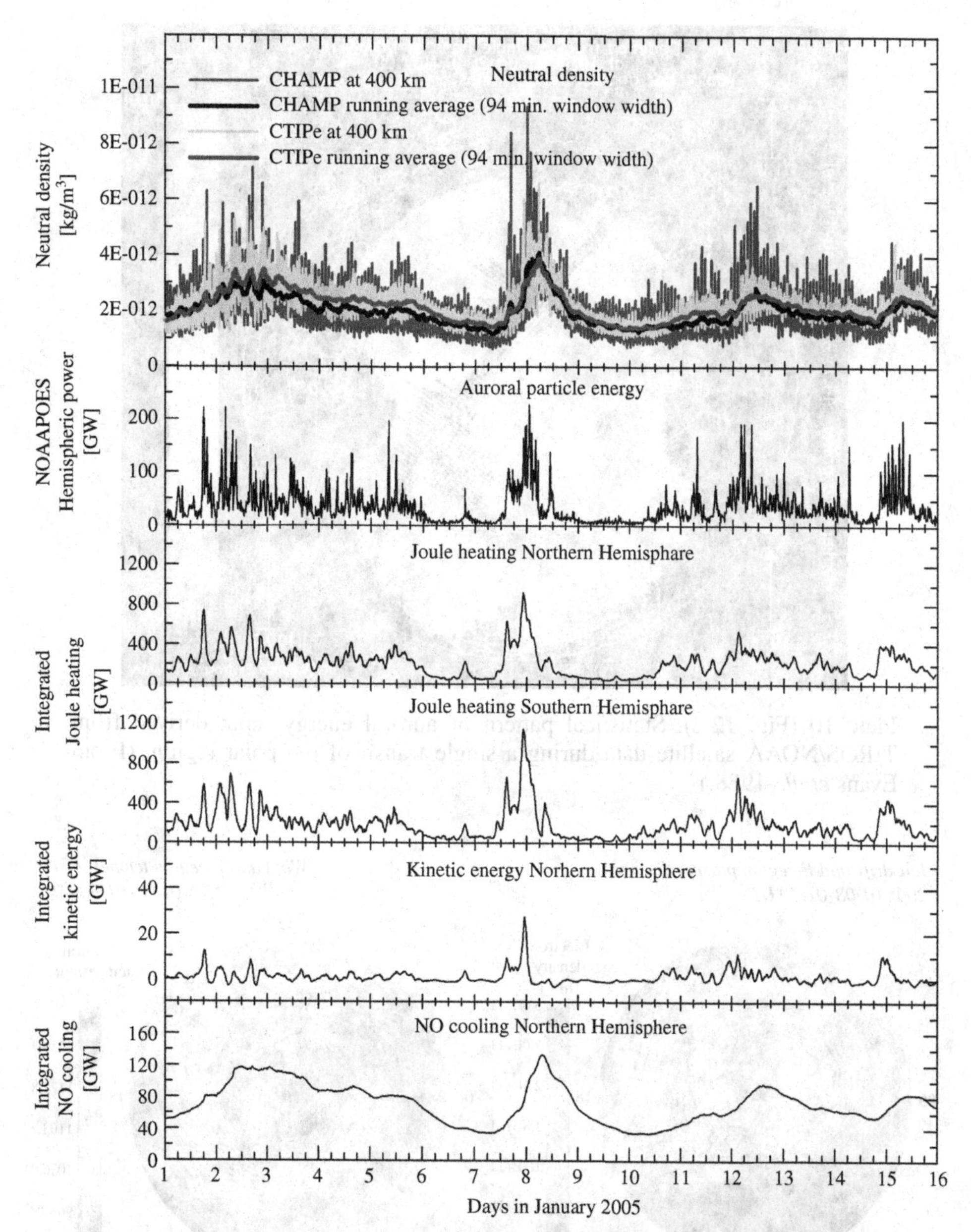

Plate 9 (Fig. 12.1). Ionospheric properties during a geomagnetic storm. The upper panel shows a comparison of CHAMP neutral density measurements at 400 km altitude with a numerical simulation, for a stormy period in January 2005. The lower panels show, from top to bottom, estimates of auroral power, Joule heating in the Northern and Southern Hemispheres, kinetic energy deposition, and nitric oxide infrared cooling rates. (Courtesy of M. Fedrizzi.)

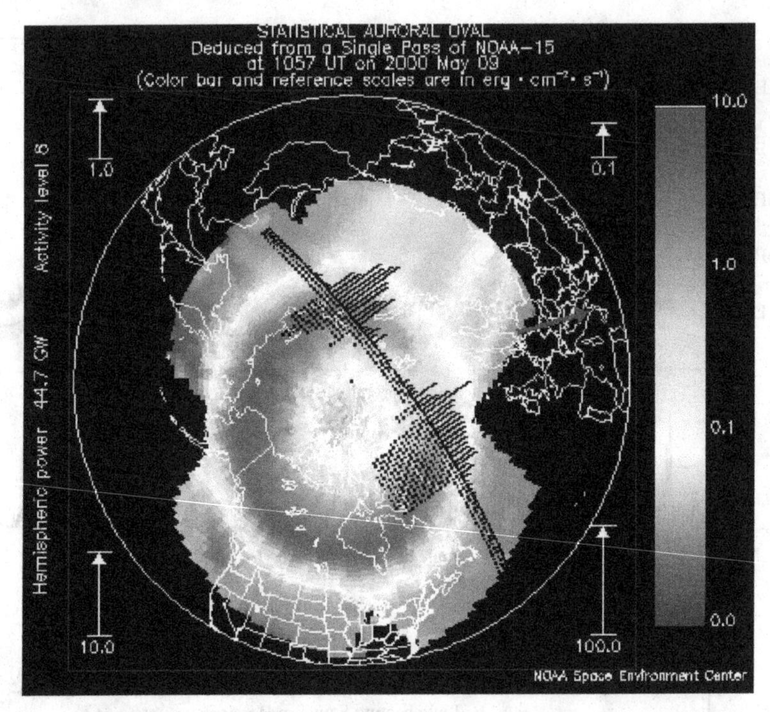

Plate 10 (Fig. 12.2). Statistical pattern of auroral energy input derived from TIROS/NOAA satellite data during a single transit of the polar region. (From Evans *et al.*, 1988.)

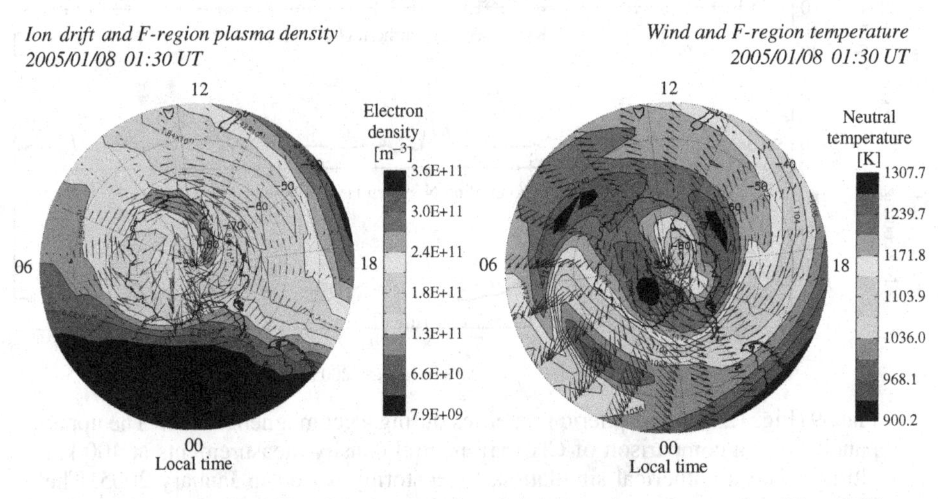

Plate 11 (Fig. 12.3). Simulated response of the F-region plasma densities (left) and neutral winds and temperature (right) at the peak of the storm at 1:30 UT on January 8, 2005, in the Southern Hemisphere. Both represent the response in the upper thermosphere and ionosphere at about 300 km altitude. Peak neutral winds are in excess of 800 m/s. (Courtesy of M. Fedrizzi.)

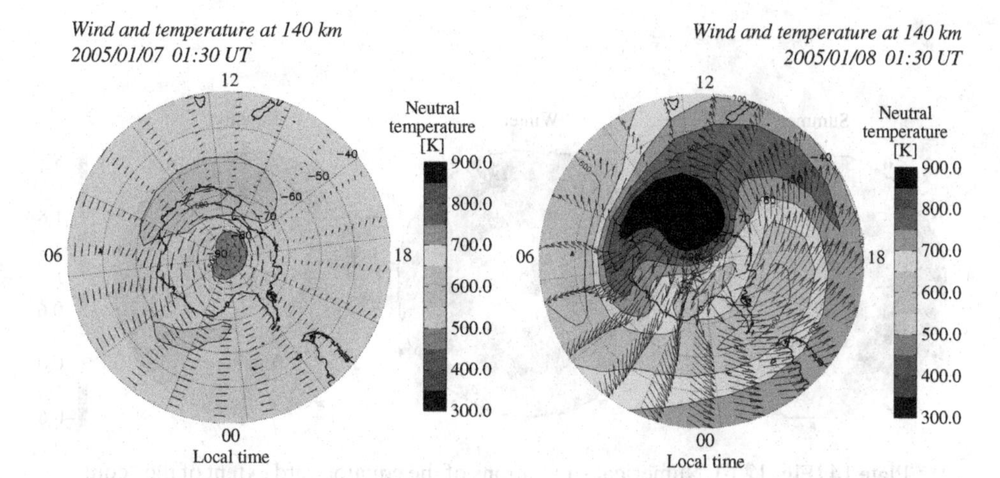

Plate 12 (Fig. 12.4). Neutral winds in the lower thermosphere at around 140 km altitude at the peak of the storm at 1:30 UT on January 8, 2005, in the Southern Hemisphere (right), and at the same UT on the quiet day preceding the storm (left). Winds in the lower thermosphere increase dramatically in response to the storm, but peak magnitudes are about half those at 300 km. Lower thermosphere winds driven by the storm also tend to be slower to dissipate, sometimes acting as a "flywheel" driving Poynting flux upward from the thermosphere/ionosphere to the magnetosphere. (Courtesy of M. Fedrizzi.)

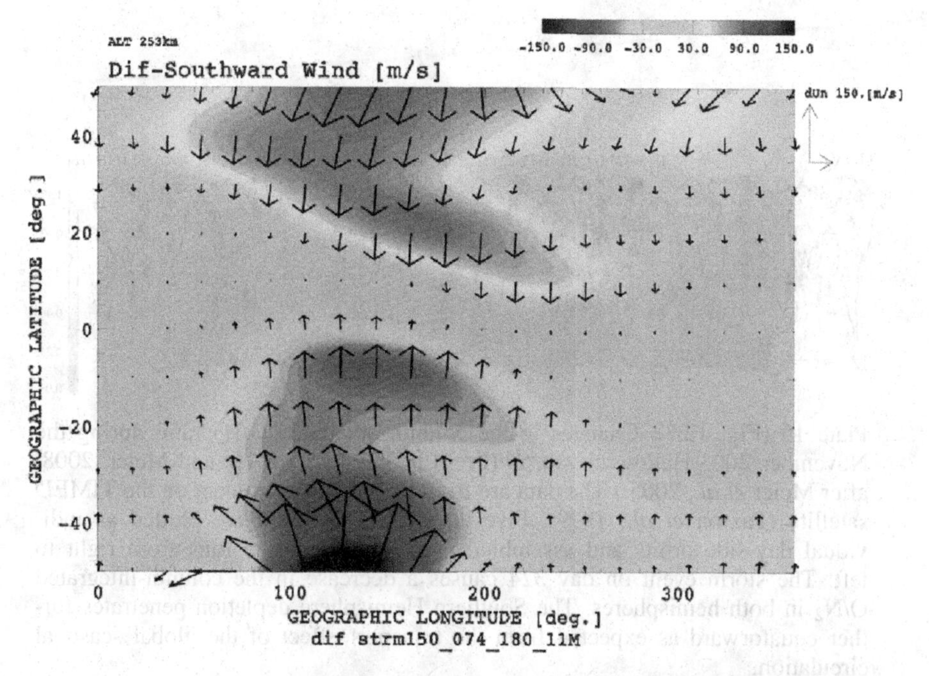

Plate 13 (Fig. 12.5). Simulation of the response of the neutral winds at mid and low latitudes at 250 km altitude, shortly after a sudden increase in high-latitude Joule heating. The region within 50° of the geographic equator is shown at 15 UT, three hours after the increase in high-latitude magnetospheric forcing, equivalent to a $K_p \sim 7$. Wind surges of ~150 m/s are produced, mainly on the night side.

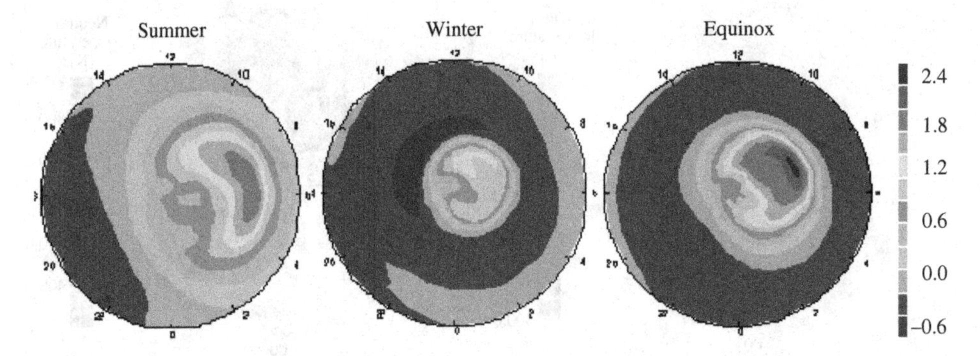

Plate 14 (Fig. 12.6). Numerical simulations of the equatorward extent of the "composition bulge" for equivalent storms in the Northern Hemisphere for summer (left), winter (middle), and equinox (right). The seasonal circulation assists the transport to low latitudes in the summer hemisphere and inhibits the transport in winter.

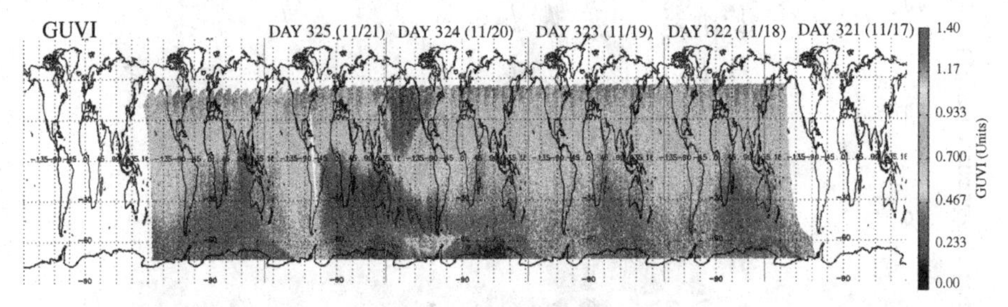

Plate 15 (Fig. 12.7). Changes in the column-integrated O/N_2 ratio during the November 2003 Halloween storm. (From review by Crowley and Meier, 2008; after Meier *et al.*, 2005.) The data are from the GUVI instrument on the TIMED satellite (Paxton *et al.*, 1999). Five days of GUVI data are plotted as individual day-side orbits and assembled as a montage; time runs from right to left. The storm event on day 324 causes a decrease in the column-integrated O/N_2 in both hemispheres. The Southern Hemisphere depletion penetrates further equatorward as expected from the transport effect of the global seasonal circulation.

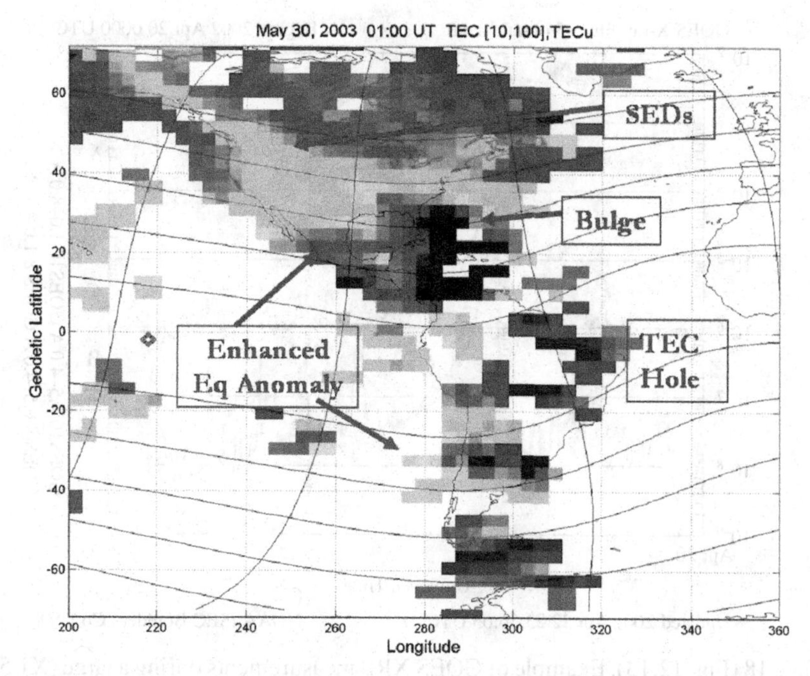

Plate 16 (Fig. 12.9). Illustration of the large enhancement "bulge" in TEC at mid-latitudes during a geomagnetic storm, and showing the plume of plasma (storm-enhanced density, or SED) connecting the bulge to the high latitudes. (Courtesy of J. Foster.)

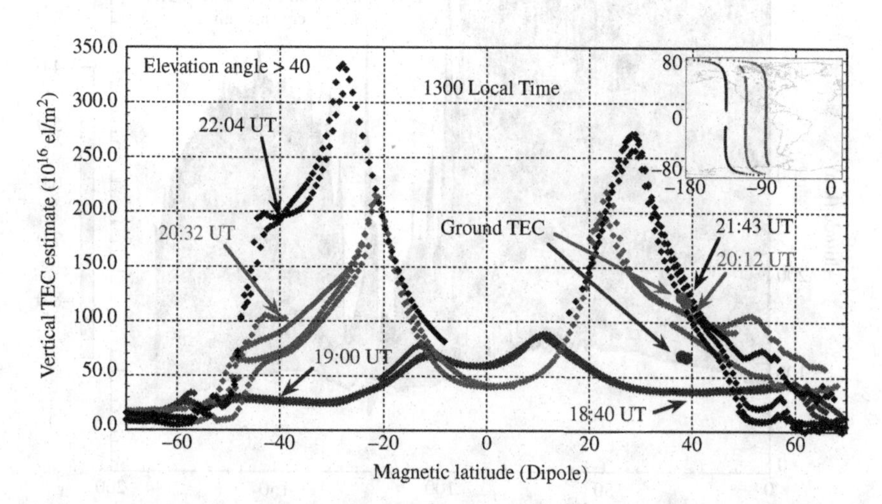

Plate 17 (Fig. 12.10). Order of magnitude increases in over-the-satellite electron content (OSEC) above 400 km during the Halloween storm of October 28, 2003, as measured by the CHAMP satellite. (From Sparks *et al.*, 2005; figure updated by A. Mannucci.)

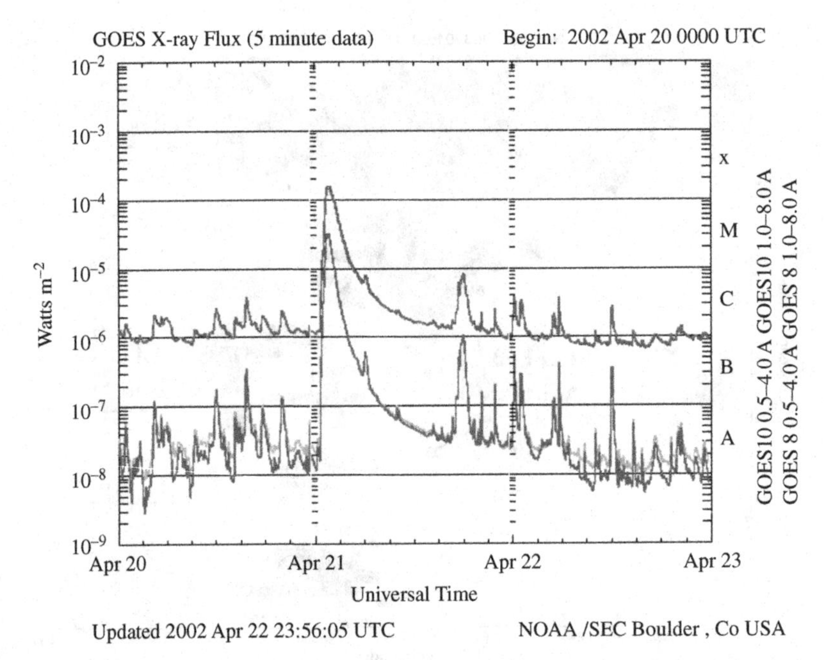

Plate 18 (Fig. 12.13). Example of GOES XRS measurements during a large (X1.5) solar flare.

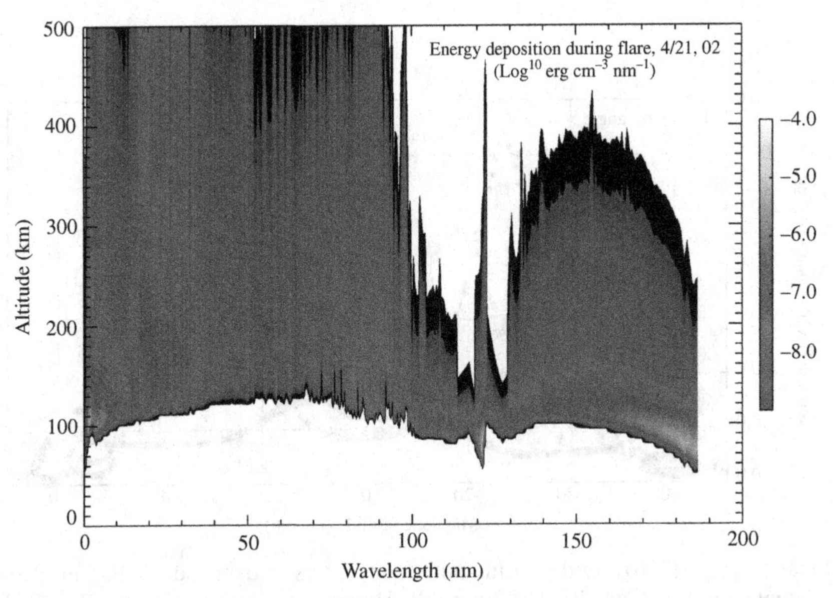

Plate 19 (Fig. 12.17). Energy deposition in the upper atmosphere as a function of wavelength and altitude during a solar flare.

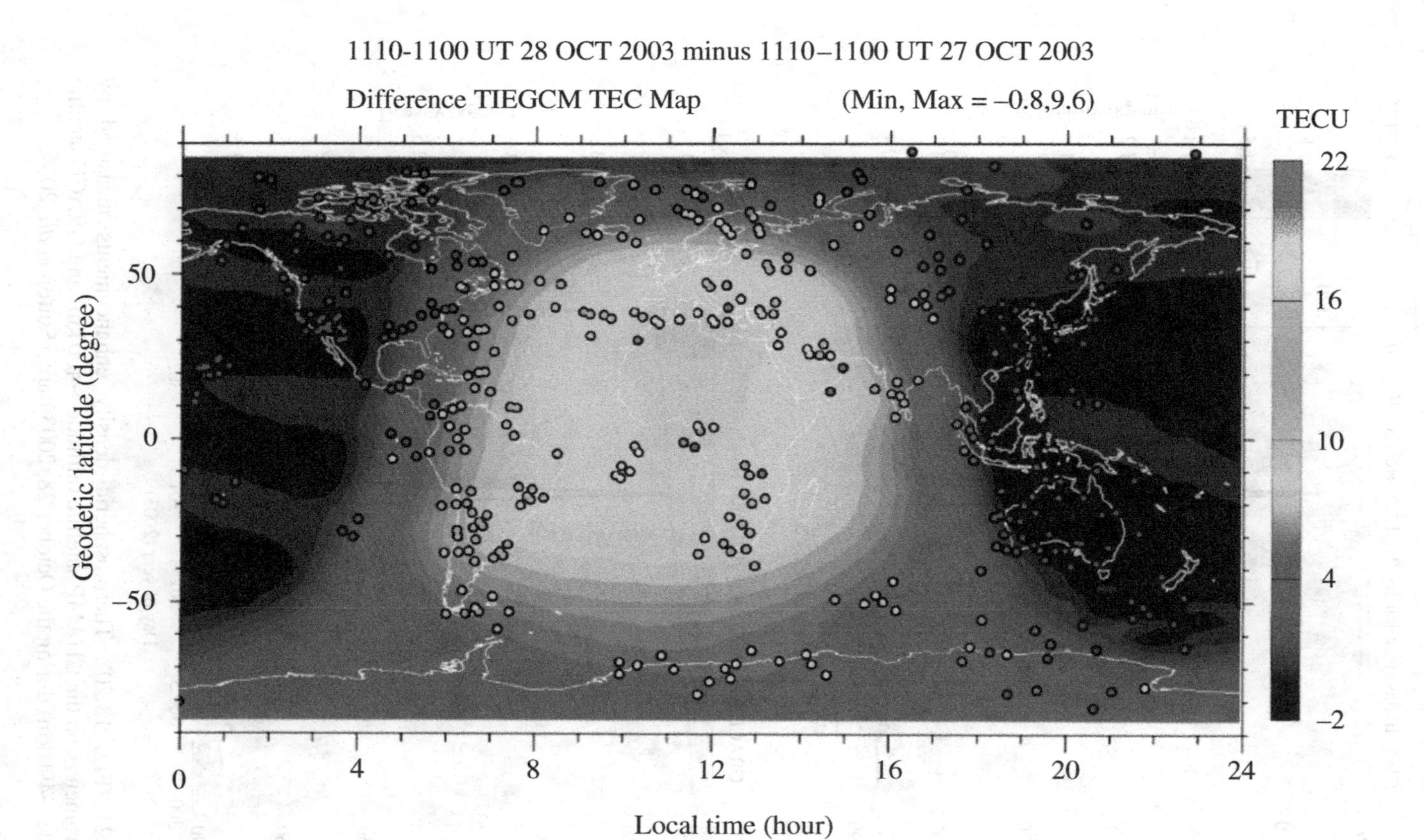

Plate 20 (Fig. 12.19). Comparison of total electron content enhancements during the October 28, 2003 flare, observed by the global network of differential GPS stations, and modeled using the NCAR Thermosphere-Ionosphere-Electrodynamics General Circulation Model (TIE-GCM). Total electron content is the vertically integrated column electron content in units of 10^{16}m^{-2}.

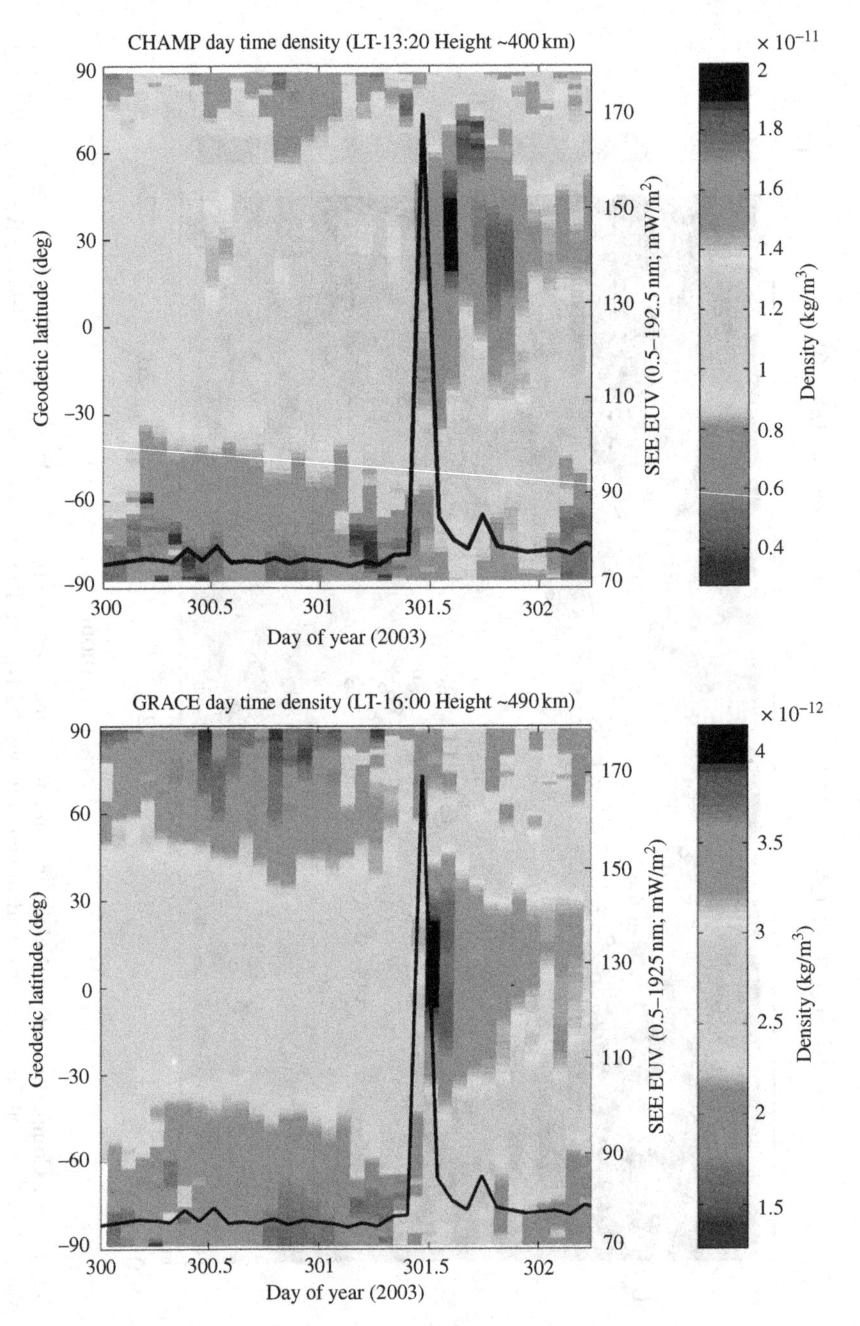

Plate 21 (Fig. 12.20). Thermospheric density enhancements measured by accelerometers on the CHAMP satellite (altitude ~400 km) and GRACE satellite (altitude ~490 km) during the October 28, 2003 flare. (Sutton *et al.*, 2006.)

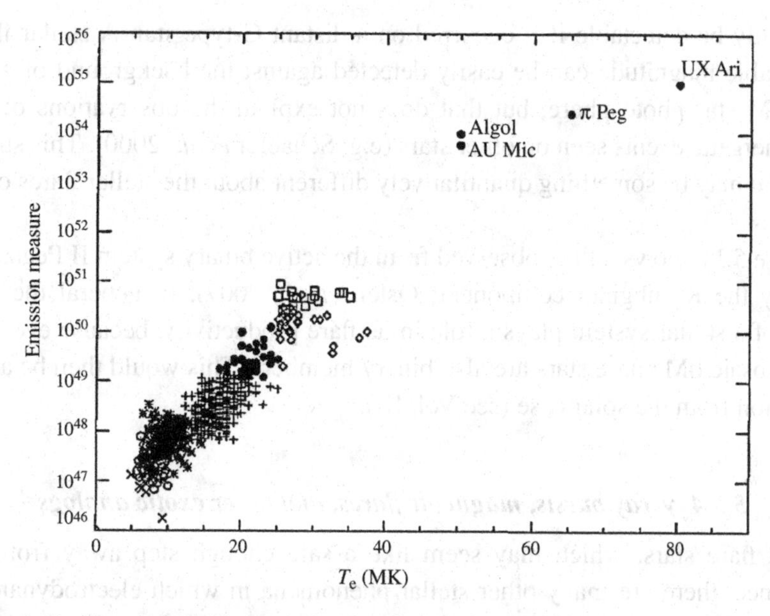

Fig. 5.13. Correlation of temperature and emission measure for solar and stellar flares. The various symbols refer to the original publications, as identified in the paper by Feldman *et al.* (1995) from which this figure is taken. (Reproduced by permission of the AAS.)

5.5.3 Stellar flares

Many observed light-curve properties of solar and stellar flares resemble one another. There is a tendency for the same fast-rise/slow-decay pattern, a similar relationship between hard and soft X-rays (Isola *et al.*, 2007; Osten *et al.*, 2007) and even a stellar Neupert effect visible in comparisons of white light (Hawley *et al.*, 1995) and microwaves (Guedel *et al.*, 1996) with soft X-ray time profiles. In soft X-rays there is a clear statistical relationship between the emission measure ($\int n_e^2\, dV$) and the temperature, as shown in Fig. 5.13 (Feldman *et al.*, 1995; Aschwanden *et al.*, 2008). This correlation, first noted by Feldman *et al.* (1995), while apparently significant, necessarily compares very different kinds of observations. The relatively poor correlations seen in individual data sets suggest that systematic biases play an important role, as yet not well understood, independent of the overall correlation. From the point of view of "universal physical processes", it has been argued that this broad correlation results from a universal kind of magnetic reconnection (Shibata and Yokoyama, 1999). This may be an over-interpretation of effects explainable in other ways (Aschwanden *et al.*, 2008), but at a minimum it suggests the importance of the Alfvén speed as a parameter.

The most easily observable stellar flares are found on the traditional dMe flare stars. These stars are cooler and fainter than the Sun (G2 V; see Vol. III), making it easier to detect brightenings. Indeed, the powerful Carrington flare of 1859 would not readily be detectable if it occurred on a distant G-type star. A stellar flare of comparable magnitude can be easily detected against the background of a much fainter M star photosphere, but that does not explain the observations of much more energetic events seen on other stars (e.g. Schaefer *et al.*, 2000). This suggests that there may be something quantitatively different about the stellar flares or their causes.

Figure 5.11 shows a flare observed from the active binary system II Pegasi (presumably the K subgiant component; Osten *et al.*, 2007). In general the binary nature of a stellar system plays a role in its flare productivity, because even many more prosaic dMe flare stars are also binary members. This would then be another distinction from the solar case (see Vol. III).

5.5.4 *γ*-ray bursts, magnetar flares, and other exotic analogs

Beyond flare stars, which may seem like a safe enough step away from solar experience, there are many other stellar phenomena in which electrodynamics is invoked to explain the observations. Figure 5.14 shows sketches of two examples, which we discuss briefly here. The left panel shows twisted field lines hypothesized to develop in the atmosphere of a "magnetar", a neutron star thought to have interior magnetic-field strengths as large as 10^{15} G (Thompson and Duncan, 1995). The rough idea is that magnetic energy can build up in these twists, maintained by the rigidity of the neutron-star crust, until a giant flare releases it. The right panel is a representation of the "X wind" model of Cai *et al.* (2008),

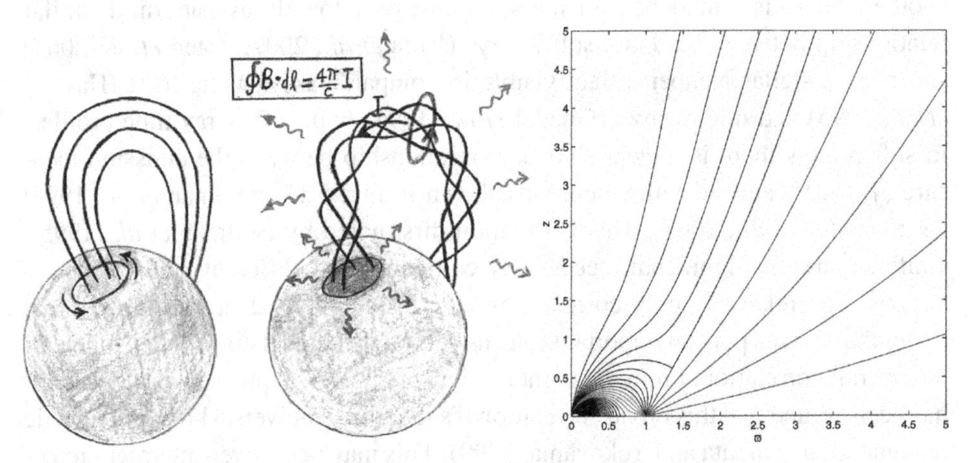

Fig. 5.14. *Left*: Cartoon showing energy storage in the "corona" of a magnetar, a neutron star magnetized to $\sim 10^{15}$ G and capable of giant flares (Duncan and Thompson, 1996; Duncan, 2005). *Right*: Cartoon showing "X-wind" model of magnetic fields involved in the accretion of matter onto a young star (Cai *et al.*, 2008, reproduced by permission of the AAS).

which generalizes the solar ideas by involving the accretion disk of a young star (T Tauri) in the stressing of the field and its release as a flare. The high activity of young stars presumably results from rapid rotation and the presence of an accretion disk.

5.6 Observational aspects of magnetic reconnection

5.6.1 Connectivity

Although magnetic reconnection is only one possible way to extract energy from a magnetic field, this idea dominates most research in flare theory. Most observers therefore try to understand their observations in this way. I take this opportunity to discuss how we detect magnetic reconnection by remote-sensing techniques. In the laboratory or in the magnetosphere it is possible to make measurements on the scale at which particles "demagnetize" so that reconnection can happen. This is unlikely ever to be the case for astronomical observations, unfortunately; the proton inertial length for high-energy protons, c/ω_{pi}, is tiny, approximately 10^3 cm at the top of the VAL-C (Vernazza *et al.*, 1981) chromosphere (Hudson, 2007a).

The best evidence for the occurrence of reconnection, therefore, must come from tracking connectivity, via the identification of magnetic domains (see the discussion in Vol. I, Chapter 4). A flare driven by reconnection would involve the transfer of flux between two domains of different connectivity, in such a manner as to release some fraction of the stored energy. But how to identify the domains? This can really only be done in the context of a coronal magnetic-field model at a level of approximation that permits stresses to remain in the field, and hence separate domains to exist. A field model derivable from a scalar potential will not serve perfectly, but if the coronal field is only weakly non-potential, the separatrices between the domains may be in about the right places.

In a large-scale magnetic reconnection model, one might expect the flare brightenings to appear at the intersections of coronal magnetic separatrix surfaces with the photosphere. The separatrices show the location of sudden changes in the connectivity maps (Titov and Démoulin, 1999). Mandrini *et al.* (1991) observed Hα brightenings at such locations, and more recently Metcalf *et al.* (2003) presented an excellent example of this for the flare shown in Fig. 5.8 (right). Figure 5.15 shows the mapping for this event. Such indirect correspondences provide some of the best evidence to date of the large-scale reconnection picture, but note that the observation is still quite remote from the microphysics of reconnection, and that there are necessarily ambiguities in the interpretation.

Many other flare observations have been interpreted in terms of large-scale magnetic reconnection. The "Masuda flare" (Section 5.5.1) is often cited as conclusive

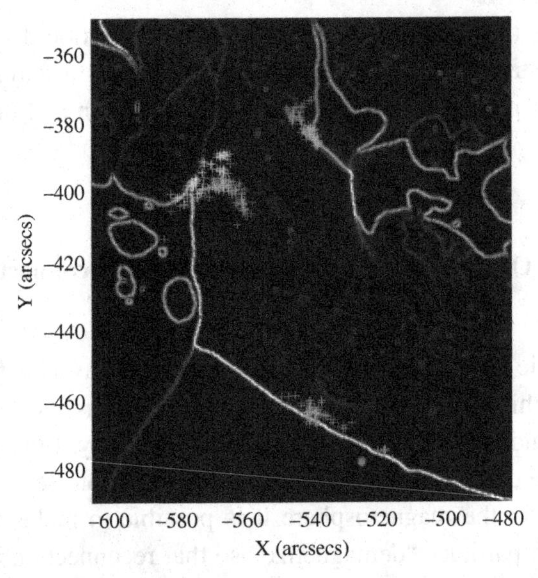

Fig. 5.15. Mapping of separatrices to the photosphere (heavy lines) compared with hard X-ray footpoint locations (crosses) for the flare of August 25, 2001 (Metcalf *et al.*, 2003), also the subject of Fig. 5.8b. The image dimensions are 120 × 150 arcsec in x and y, respectively. The main flare ribbons are at the upper part of the figure, but note how faithfully the remote brightenings follow the projected separatrix at the bottom of the figure as well. (Reproduced by permission of the AAS.)

evidence for such a picture, and a more recent Yohkoh observation of apparent reconnection inflow (Yokoyama *et al.*, 2001) also fits the picture. However, each of these events was quite unusual and may have drawn attention not so much because they were in any sense typical, but simply because they evoked the cartoon. In the Yohkoh era, probably the best circumstantial evidence for reconnection dynamics was in the observation of the "supra-arcade downflows" (McKenzie and Hudson, 1999; McKenzie, 2000; Asai *et al.*, 2004b). Recently, Hara *et al.* (2008) have applied the much better observational material of the Hinode/EIS instrument (Culhane *et al.*, 2007) to a well-observed gradual flare. In principle, such observations would show the reconnection flow fields and thus be a step closer to confirming the reality of the picture. Such an observation appears to have been too difficult even for this instrument, at least for this flare, and no unambiguous results could be obtained.

In general, the observations most strongly suggestive of the reconnection picture apply mainly to the later phases of a flare. Asai *et al.* (2004b) also observed supra-arcade downflows in coincidence with the impulsive-phase hard X-ray bursts of the flare of July 23, 2002. We do not know how this phenomenon applies to flares without eruptions.

5.6.2 Current sheets

Magnetic reconnection requires the existence of a current sheet on a scale fine enough for particle demagnetization to occur. Given the small values of the ion inertial length ($c/\omega_{\mathrm{pi}} \ll 1$ km in the corona), the detectability of a current sheet would be indirect by any known remote-sensing technique. Enhanced density or temperature could be clues, for example, or simple image morphology based on theoretical expectation.

For some CMEs there is clear evidence for the re-formation of a coronal helmet streamer following the event (Kahler and Hundhausen, 1992; Hiei *et al.*, 1993). We interpret this to mean that the juxtaposed open fields of opposite polarities do form an active current sheet during a reconnection process. Webb *et al.* (2003) discuss this CME morphology in detail.

Temperature and density signatures might also be expected in the EUV or soft X-ray ranges, given the dynamics of the reconnecting magnetic field, especially in flares for which the process might be faster and more energetic. These physical parameters translate into an emission measure $n_e n_i L$, where L represents the width of the source in the line of sight. Analytical work or numerical modeling do not give us good predictions for any of these parameters, but UV observations of several linear features behind CMEs strongly suggest that they are in fact the expected current sheets, or else plasma structures closely related to them (see Ciaravella and Raymond, 2008, and other papers cited therein). One distinguishing feature of most of the handful of events detected in this manner is the presence of the high-temperature Fe^{17+} ion.

X-ray observations of flares with RHESSI have also provided indirect evidence for the presence of current sheets in the impulsive phase of a flare, where reconnection models would expect them (Sui and Holman, 2003; Sui *et al.*, 2004).

5.6.3 Coronal motions

The plasma in the core of an active region has a low plasma β; for reasonable values of the physical parameters we find $\beta = 2nk_{\mathrm{B}}T/(B^2/8\pi)$ to be of order 10^{-4} or lower. This means that any detectable features – any emission at any wavelength – will serve mainly as a "leaf in the wind" (Sheeley *et al.*, 1999), helping us to determine the geometry of the flows but not having much physical significance. The bright features are not important physical objects, because they are embedded in a much stronger and pervasive magnetic field that determines the forces dictating the flow.

The discovery of the "supra-arcade downflows" (McKenzie and Hudson, 1999) offers one of the main possible links between the observations of plasma motions in the flare and the idea of large-scale magnetic reconnection. This phenomenon is best appreciated in movie format; although it was discovered with Yohkoh/SXT soft X-ray observations, in fact the higher resolution available in the TRACE 195 Å data make it more visible. The data show a downward flow toward the surface of the Sun from above the developing arcade. The flow speeds are sub-Alfvénic and show deceleration as they approach the arcade loops (McKenzie, 2000; Sheeley *et al.*, 2004). It would be attractive to interpret these motions as confirmation of the standard reconnection model, but the (apparently) sub-Alfvénic speeds provide a major obstacle to this interpretation.

Soft X-ray dimming offers another signature. These were another Yohkoh discovery (Hudson *et al.*, 1995b; but see Rust, 1979, for earlier Skylab observations and Hansen *et al.*, 1974, for still earlier observations from a ground-based coronagraph). The dimming coincides with the impulsive phase of the flare (Zarro *et al.*, 1999) and hence with the acceleration phase of the associated CME (Zhang *et al.*, 2004). It is thus reasonable to associate the dimming signature with the outward flow of mass required by a CME. Although there are many observations now of expanding loops, seen in many wave bands, the dimming signature is more profound and often can be seen in diffuse or unresolved corona. This signature is important for the reconnection models because it may identify newly opened field that can then reconnect.

Some of the dimming may also be related directly to the inflow expected of large-scale reconnection (Yokoyama *et al.*, 2001). In this case the flow field would be essentially horizontal, rather than radially outward as in the "transient coronal hole" interpretation of dimming as the mass being lost to the CME. Measuring the orientation of the velocity field should therefore be a high priority for future spectroscopic observations (via the Doppler effect) and for future high-resolution imaging observations (via "leaves in the wind"). Other examples of this type have been found, but they are rare and do not provide good evidence for a well-understood reconnection scenario in the impulsive phase.

5.6.4 Ribbon motions

The expanding motions of flare ribbons provided one of the first clues to what we think of as the standard reconnection model of a flare (see the sketch of Fig. 5.4 in Vol. I, or others in Chapter 6 of this volume). As pointed out by Poletto and Kopp (1986), these motions can be interpreted as an electric field. This is a motional or "convective" electric field given by $\mathbf{E} = \mathbf{v} \times \mathbf{B}$, and it is often taken as a measure of the reconnection rate. Fletcher and Hudson (2001) point out that the rate at which the ribbons sweep out the field should correspond in some sense to the rate at which energy is released during reconnection, and that at the same time the field guides

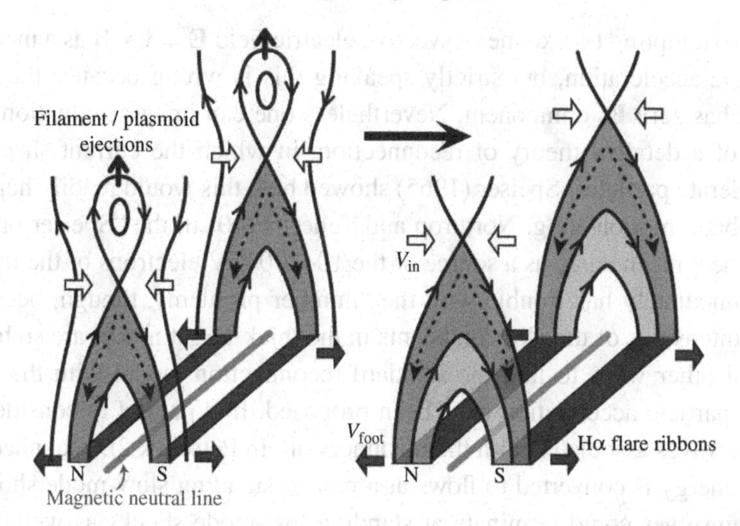

Fig. 5.16. How the ribbon motion sweeps out magnetic field during the reconnection process in the standard model. (From Asai *et al.*, 2004a; cf. Fig. 6.1. Reproduced by permission of the AAS.)

the particle or heat flux responsible for the ribbon excitation. Figure 5.16 shows the geometry.

The actual magnitude of the convective field may be quite large. We can estimate it (in SI units for convenience) for a reconnection flow speed $|\mathbf{v}| = 0.1 v_A$, where v_A is the Alfvén speed, which would plausibly be $v_A = 10^7$ m/s in the core of an active region. Then for $|\mathbf{B}| = 0.1$ T, $|\mathbf{E}| = 10^6$ V/m. Similarly, the Poynting flux can be estimated at 10^5 W/m^2 (e.g. Asai *et al.*, 2004a), approximately the level needed to power a flare with plausible assumptions about the geometry. Asai *et al.* (2004a) also showed that the local Poynting flux appeared to correlate in time with the temporal variations of impulsive-phase signatures, consistent with expectation from the standard reconnection model.

5.6.5 Particle acceleration

How does one understand particle acceleration in the context of magnetic reconnection, and can the particles be a guide to understanding the reconnection physics? At first glance this may seem implausible, because one frequently appeals to reconnection within an MHD framework, as in Fig. 5.16. MHD is a fluid theory and therefore has no particles at all, and so any theory of particle acceleration needs to be grafted on in a non-self-consistent manner as a "test-particle" theory. This would be satisfactory theoretically if the particles were energetically unimportant, but as we have seen (Section 5.2.2) this is not true in the impulsive phase at least.

It is also tempting to take the convective electric field $\mathbf{E} = \mathbf{v} \times \mathbf{B}$ as a mechanism for particle acceleration, but strictly speaking this is wrong because the convective field has zero $\mathbf{E}_{\parallel}$ component. Nevertheless one can imagine situations, in the absence of a detailed theory of reconnection, in which the current sheet can in fact accelerate particles. Speiser (1965) showed how this would readily happen via non-adiabatic motions (e.g. Northrop and Teller, 1960) in the "Speiser orbits". A current-sheet mechanism as a source of the 10–100 keV electrons of the impulsive phase immediately has trouble with the "number problem", though, because the inferred intensities of the electron beams in the thick-target model are so high.

Several other ways to link the standard reconnection model with the requirement for particle acceleration have been proposed. It is natural to consider a role for shock waves associated with the reconnection. In Petschek 2D reconnection, in fact, the energy is converted to flows at a pair of standing slow-mode shocks; the flows themselves could terminate at standing fast-mode shocks as well. Tsuneta and Naito (1998) used the latter for acceleration and the former for trapping. Unfortunately there is no clear evidence for fast reconnection outflows and their attendant fast-mode shocks (see Section 5.6.3). This scenario may then fail as a result of the 2D reconnection picture not providing a good approximation to the 3D situation.

Recently, Fletcher and Hudson (2008) have introduced ideas carried over from the terrestrial aurora and somewhat new to solar physics. These ideas, sketched in Fig. 5.17, make use of the Poynting flux of Alfvén waves generated in the restructuring of the coronal magnetic field (Emslie and Sturrock, 1982). The particle acceleration would hypothetically result from the development of structure on small scales, generating the necessary $|\mathbf{E}_{\parallel}|$ either via kinetic effects in the wave propagation (e.g. Kletzing, 1994) or via the development of turbulence (e.g. Larosa

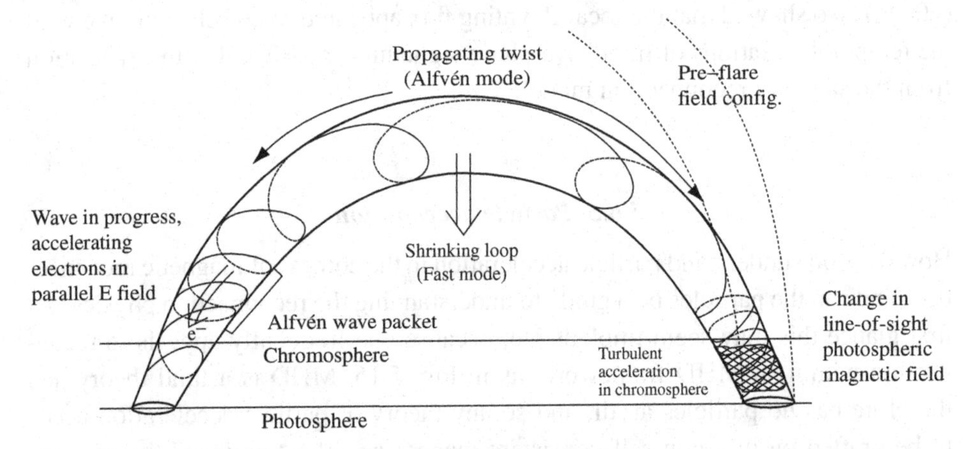

Fig. 5.17. Model put forth by Fletcher and Hudson (2008), showing the extraction of stored coronal magnetic energy via the Poynting flux of waves excited by the restructuring that produces the flare. Particle acceleration in this picture, as in other pictures, remains problematic.

et al., 1994; Petrosian and Liu, 2004). In either case, the actual particle acceleration could take place near the chromosphere and thus have a better chance to avoid the number problem.

5.7 Conclusions

This chapter has reviewed the status of our observational material on solar flares, including other forms of solar activity and some of their analogs in other environments. Observations of flares now span one and a half centuries, and many things have been learned. Nevertheless major questions remain unanswered, and so the observations must be improved. What do we not understand, and how should we proceed to learn more? I will start this conclusion with some general remarks and then get specific about the observations.

Flares are a clear example of a "stick and slip" process, whereby energy builds up slowly and then converts suddenly into other forms. In this case the storage is in the inductive magnetic field of currents driven into the solar atmosphere by convective motions in the solar interior. These currents can find quasi-stable equilibria that evolve until a loss of equilibrium takes place. The energy release in the resulting development of the system is nonlinear and involves a range of scales in the plasma that cannot be described quantitatively at the present time. We thus do not have a predictive theory of the restructuring that releases this stored coronal energy and results in a flare. The paramount problem of flare physics therefore is to understand the transformation of energy in this interesting physical system.

The physical essence of flare physics, regarded most generally, would be in the behavior of the interface between a stellar interior and its atmosphere. The Sun shows us that this interface reacts in quite striking ways to what should be an orderly flow of stellar energy away from its interior sources. Electrodynamic effects dominate the interaction between this flow and the exterior space. For the Sun, the flare is the most common of these effects in terms of coronal signatures. For the most energetic flares the simultaneous occurrence of a CME and its concomitant particle acceleration leads to physically (and perhaps biologically) important interactions with planetary environments. We do not know enough, in spite of a long history of observation. Following the premise of Harwit (1981), we should note that there is a vast unobserved parameter space in the UV and EUV wavelength ranges covering these regions of the solar atmosphere. Harwit argued that cosmic discovery follows almost directly from the opening-up of new parameter domains. The most striking omission in the case of solar flares might be the almost complete lack of hydrogen Lyα observations with sufficiently high spectral, spatial, and temporal coverage. Note that this is the most basic spectral line of the most abundant element on the nearest star!

Other important omissions include sensitive observations, at high resolution, of X- and γ-rays. The most direct insight into flare energy release necessarily must follow from observation of the accelerated particles. In this context radio techniques also have great sensitivity and a parameter space that has not been exploited. Specifically in the microwave band, we have never had sensitive broad-band spectral coverage. Almost all of the observations to date have been at widely spaced fixed frequencies that provide only limited information about the physical properties of the sources, including the all-important coronal magnetic field (e.g. Brosius and White, 2006).

6

Models of coronal mass ejections and flares

TERRY FORBES

Solar flares and coronal mass ejections (CMEs) are closely related phenomena (see Section 5.3.5), and it now seems very likely that they are simply different manifestations of a single, underlying physical process, namely, the release of magnetic energy stored in the magnetic field of the solar atmosphere. In the past there has been considerable controversy about the relation between CMEs and flares. Some authors have argued that flares cause CMEs by creating high enough temperatures to eject both plasma and magnetic field into the interplanetary medium. However, most CMEs are not associated with what is normally considered a flare (Gosling, 1993), and even in those cases that are, the thermal pressure is never enough to force the field open (Low, 2001).

As discussed in the previous chapter, flares occur over a span of energy scales that ranges from very small (microflares at the observable limit) to very large ($>10^{32}$ ergs). Some time ago, Švestka and Cliver (1992) suggested that the main factor that determines whether a CME will be associated with a flare or not, is the strength of the magnetic field in the erupting region. If the ambient magnetic field strength is weak, then the emitted radiation, although still present, is just too faint to be considered a flare according to the traditional definition (Zirin 1988). According to some models, it is possible to have two CMEs with nearly the same trajectories and speeds but with an order of magnitude or more difference in the peak intensities of their light curves (Reeves and Forbes, 2005). A low-mass CME in a weak-field region can experience the same acceleration as high-mass CME in a strong-field region, but the reduced magnetic energy density of the former case leads to much weaker emissions.

6.1 Recapitulation of key observational features

6.1.1 Morphology

Historically, a flare has been defined as a localized brightening of the chromosphere, observed in Hα, over a time scale ranging from a few minutes

Heliophysics: Space Storms and Radiation: Causes and Effects, eds. Carolus J. Schrijver and George L. Siscoe. Published by Cambridge University Press.

to an hour (Zirin, 1988). Nowadays, the definition has been extended to include the rapid onset of X-ray and UV emissions in the corona (Tandberg-Hanssen and Emslie 1988; see also Chapter 5). The Hα brightening in the chromosphere typically occurs in the form of flare ribbons, while the X-ray and UV emissions appear in the form of loops whose feet map to the ribbons (see Figs. 6.1 and 6.2). During the course of the flare, the separation between the ribbons typically increases, and the flare loops grow in size. However, the apparent motions of the loops and ribbons do not correspond to plasma flows in the solar atmosphere. Instead, they correspond to the sequential energization of the plasma in a continuum of nearly stationary magnetic loops (see Fig. 5.1 in Vol. I).

CMEs typically have a three-part structure consisting of an outer bright shell, an interior dark cavity, and filament material located near the center of the cavity as shown in Fig. 5.5 and in the upper left panel of Fig. 6.1 (Low, 1996). The density of the shell is about a factor of ten higher than the ambient coronal density. This value is significantly larger than the maximum value of four that can be produced by shock compression (Ferraro and Plumpton, 1966), so the shell is not likely to be a shock wave. As pointed out in the previous chapter, it most likely results from the pile-up of the material of the helmet streamer that typically overlies the erupting region (Hundhausen, 1988). Outward propagating CMEs do produce shocks, but these shocks are difficult to observe. In those cases where the shock has been detected, the density compression across it is typically in the range from 1.2 to 2.5 (Vourlidas, 2006) – a range that is consistent with the predictions of MHD theory.

More than half of all CMEs are associated with the eruption of filaments (upper right panel of Fig. 6.1). Filaments (called prominences if seen above the solar limb) are filamentary, cloud-like structures consisting of plasma that is about 100 times cooler and denser than the plasma in the surrounding corona, and they occur at altitudes as high as 10^5 km in their quiescent state (Tandberg-Hanssen, 1995). Because of their high density, the filament material is heavy, so some force must act on them to keep them suspended in the corona. It is usually assumed that this force must be magnetic in nature, at least in part (Martens and Zwaan, 2001). The issue of support is complicated by the fact that complex flows with magnitudes on the order of 5 to 20 km/s typically exist within filaments (Zirker *et al.*, 1998). Part of the filament material appears to drain slowly out of the filament towards the surface, while at the same time upflows may occur in other parts. At high altitude, counter-streaming horizontal flows are often observed in the region known as the filament *spine* (Zirker *et al.*, 1998). Some of these flows may be generated by the continual reshuffling of the feet of the filament's magnetic field lines by magnetic reconnection and turbulent flows at the solar surface (Berger *et al.*, 2008).

Large quiescent filaments outside active regions tend to create slow to moderate speed CMEs (Low, 2001). High-speed CMEs are typically produced in active

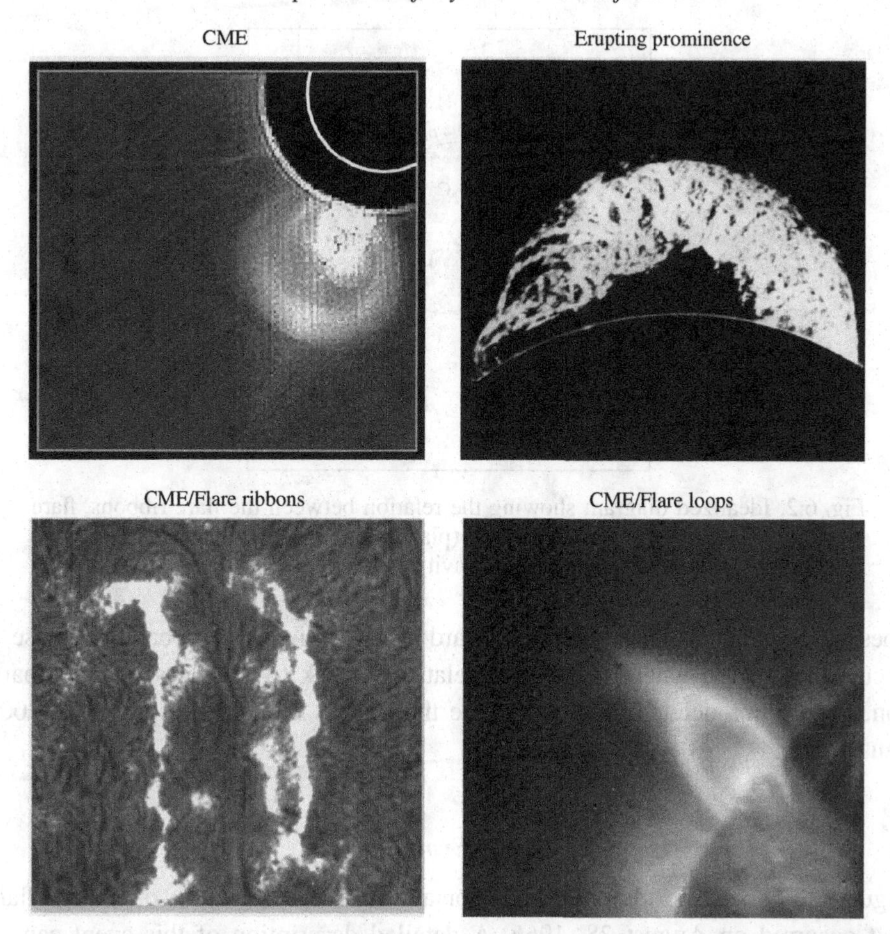

Fig. 6.1. Four different images of solar eruptions obtained by different types of telescopes. The upper left panel shows a coronagraph image from the Solar Maximum Mission satellite, while the upper right panel shows an Hα image of an eruption at the limb of the Sun (both images courtesy of the High Altitude Observatory). The lower left panel shows an Hα image of an eruption seen at disk center (courtesy of the Big Bear Solar Observatory), while the lower right panel shows a soft X-ray image of an eruption at the limb of the Sun (courtesy of the Institute of Space and Astronautical Sciences of Japan).

regions, and they are generally well correlated with flares. As observations have improved, it has become increasingly clear that erupting filaments outside active regions have many features typical of large flares. Like large flares, erupting filaments produce loops and ribbons that move apart in time, but, unlike large flares, the ribbons are usually too faint to be seen in Hα. However, the ribbons can often be seen in the He 10 830 Å line that is a more sensitive indicator of chromospheric excitation (Harvey and Recely, 1984). The eruption of a large quiescent filament

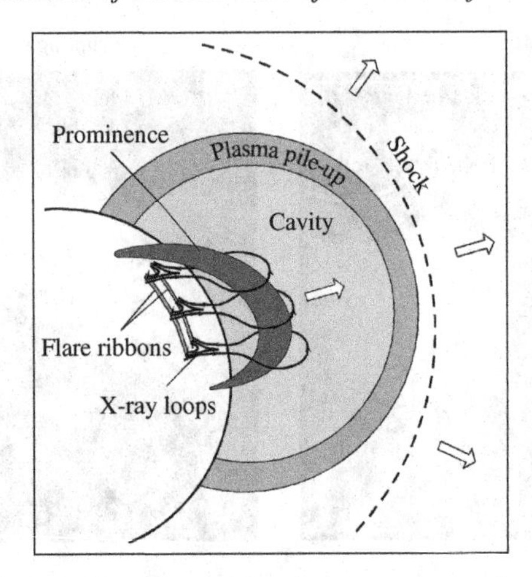

Fig. 6.2. Idealized diagram showing the relation between the flare ribbons, flare loops, the CME shock, the CME shell (plasma pile-up region), the CME cavity, and the filament contained within the cavity.

does not usually produce significant hard X-ray emissions, probably because it occurs in a region where the field is relatively weak (3–10 gauss). By comparison, large flares occur in regions where the field is relatively strong (100–1000 gauss).

6.1.2 Evolution

Figure 6.3 shows the temporal behavior at various wavelengths of a large flare that occurred on August 28, 1966. A detailed description of this event can be found in Švestka and Simon (1969). This event had intense Hα, X-ray, and radio emissions, and it produced a high-speed Moreton wave in the chromosphere. The Moreton wave signal is thought to be due to the downward displacement of the chromosphere caused by the enhanced pressure downstream of the CME shock (Dodson and Hedeman, 1968; Zirin and Lackner, 1969; Uchida, 1970, 1974). Chromospheric Moreton waves are usually seen only in very energetic events. More common are the so-called EIT waves observed in the transition region between the chromosphere and the corona (Dere *et al.*, 1997a). The nature of these waves remains unknown at the present time. Some authors argue that they are the transition region counterpart of the Moreton waves (Thompson *et al.*, 1999; Veronig *et al.*, 2006b), but other authors argue that they are a completely different phenomenon (Shibata *et al.*, 2002; Delannée *et al.*, 2008).

The Hα emission shown in Fig. 6.3 comes from the two chromospheric ribbons whose appearance is the classical signature of flare onset. The Hα emission becomes quite intense within five minutes after onset, but takes several hours to

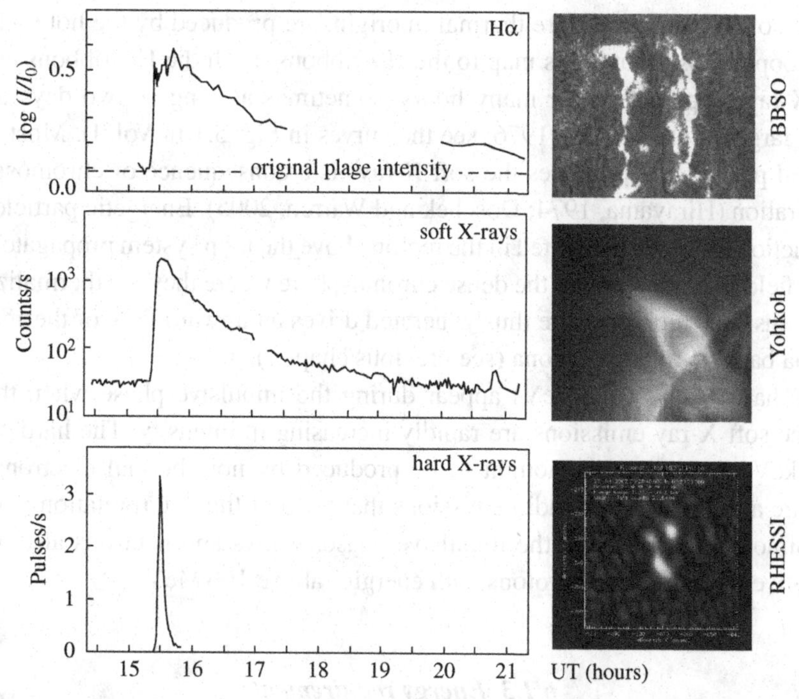

Fig. 6.3. *Top*: Hα (1.902 eV); *middle*: soft X-ray (1.0–6.3 keV); and *bottom*: hard X-ray (10–50 keV) emissions for the large, two-ribbon flare of August 28, 1966 (from Forbes, 2003). The Hα light-curve data are from the McMath–Hulbert Observatory and show the logarithm of the intensity of one of the Hα ribbons in units of the undisturbed Hα background intensity (Dodson and Hedeman, 1968). The soft X-ray data are from Explorer 33 (Van Allen and Krimigis, as published in Zirin and Lackner, 1969). The hard X-rays were measured by an ion chamber on the ATS-6 satellite (Arnoldy *et al.*, 1968). Also shown are images (for different events) of the flare features that give rise to these types of emissions.

fade. Even after six hours, it still exceeds the pre-flare emission by almost a factor of two. During the rapid rise phase of the Hα emission, the flare ribbons move apart at a rate of more than 100 km/s, but as soon as the peak is reached they quickly slow to a speed on the order of a few km/s. The ribbon motion is an apparent one produced by the progressive brightening and fading of the chromosphere at the outer edges of the ribbons (cf. Fig. 5.16; also Švestka, 1976, Fletcher and Hudson, 2001).

An event that lasts many hours, like the one in Fig. 6.3, is known as a *long duration event* (LDE). The long duration of the event is a consequence of the huge geometrical scale created by the stretching of the magnetic field by the outgoing CME, and the long time it takes for reconnection to relax the stretched configuration.

The soft X-rays, which are thermal in origin, are produced by the hot ($>10^7$ K) flare loops whose footpoints map to the Hα ribbons. Both the Hα ribbons and the soft X-ray loops persist for many hours, sometimes as long as two days after a really large event (Švestka, 1976; see the curves in Fig. 5.1 in Vol. I). Most of the thermal plasma that produces the soft X-rays is a consequence of chromospheric evaporation (Hirayama, 1974; Doschek and Warren, 2005). Energetic particles and conduction electrons generated in the region above the loop system propagate down along field lines mapping to the dense chromosphere where they are thermalized by collisions. The high pressure thus generated drives an upward flow of thermalized plasma back up into the corona (see previous chapter).

The hard X-rays (>20 keV) appear during the impulsive phase when the Hα and the soft X-ray emissions are rapidly increasing in intensity. The hard X-rays (>20 keV) are generally thought to be produced by non-thermal electrons, and they are accompanied by radio emissions that support this interpretation (Švestka and Simon, 1969). During the impulsive phase, γ-rays and neutrons also appear indicative of high-energy protons with energies above 100 MeV.

6.1.3 Energy requirements

Magnetic energy is the only source of energy in the solar atmosphere that exists in sufficient quantity to account for the radiative and kinetic energy output of large flares (Forbes, 2000). As shown in Table 6.1, the magnetic energy density of a 100 gauss (10^{-2} tesla) coronal field is about 40 J m^{-3}. By comparison, the thermal energy density is about 0.01 J m^{-3}. The kinetic energy density in the corona is about 10^{-6} J m^{-3}, assuming that the coronal velocity is on the order of the convective velocity imparted by flows at the photospheric level. Finally, the gravitational energy density is on the order of 0.04 J m^{-3} for mass to fall to the surface from a height of 10^8 m. Thus, the magnetic energy density is about three orders of magnitude greater than any of the other types. Since large flares and high-speed CMEs typically have an energy of 10^{25} J (10^{32} ergs) and a volume in the range from 10^{24} m^3 to 10^{25} m^3, an average energy density of 1 to 10 J m^{-3} is required. Only the

Table 6.1. *Characteristic coronal energy densities*

Energy type	Formula	Value (J/m^3)	Parameter values
Magnetic	$B^2/2\mu$	40	$B = 100$ gauss
Thermal	nkT	0.01	$n = 10^{15}$ m^{-3}, $T = 10^6$ K
Bulk kinetic	$m_p n v^2/2$	10^{-6}	$n = 10^{15}$ m^{-3}, $v = 1$ km/s
Gravitational	$m_p n g h$	0.04	$n = 10^{15}$ m^{-3}, $h = 10^8$ m

magnetic energy density is in this range, so non-magnetic models of flares are ruled out for the large events. This conclusion does not rule out the possibility that other forms of energy, such as gravitational, play a role in triggering eruptions. If the energy stored in the magnetic field reaches a critical state for instability, or a loss of equilibrium, then even an energetically weak process can trigger an eruption.

In considering the energy of the coronal magnetic field it is important to distinguish between the potential and the non-potential components of the field. The potential component is produced by current sources located within the interior of the Sun, for example the currents that create the magnetic field within sunspots. The non-potential component is produced by currents located within the corona, such as the currents associated with sheared arcades and coronal flux ropes. The reason it is important to distinguish between these components is because the potential magnetic field is governed by the dynamics of the convection zone, whereas the non-potential magnetic field is governed by the dynamics of the corona. The Alfvén speed at the top of the convection is about 2 km/s, so the time scale for an active region of size 10^5 km is about 14 h (Priest, 1982). By contrast, the Alfvén speed at the base of the corona is about 10^3 km/s or more, and the corresponding dynamic time scale is about 1 to 2 min. Thus, the rapid onset of flares and CMEs implies that these phenomena are a consequence of rapid changes in the coronal magnetic field. The difference between the magnetic energy of the non-potential field and the potential field is often referred to as the *free magnetic energy* since this is the magnetic energy that is obtained if the coronal currents are completely dissipated.

Some authors (e.g. Dryer *et al.*, 1979) have considered the possibility that CMEs might be driven by the high gas pressure produced by a flare. Although the gas pressure is much higher after the flare than before, it is still about an order of magnitude too weak to open the magnetic field (Švestka, 1976; Low, 2001). Typically, only about 20% of the total magnetic energy is estimated to be released as thermal energy during a flare (Reeves and Forbes, 2005), so even after a flare has occurred, the magnetic field is still sufficiently strong to contain the high gas pressure that is produced.

6.1.4 Statistical properties

Models of CMEs and flares need to explain not only the features of individual eruptions, but also the statistical distribution of CME properties such as speed, mass, size, etc. Figure 6.4 shows the distribution of the apparent speeds and angular widths of more than 8000 CMEs observed by the Solar and Heliospheric Observatory (SOHO). Figure 6.5 shows the distribution of CME mass and kinetic energy (Schwenn *et al.*, 2006). The mass is obtained from the polarized brightness images taken by the on-board coronagraph (LASCO). These images record the intensity

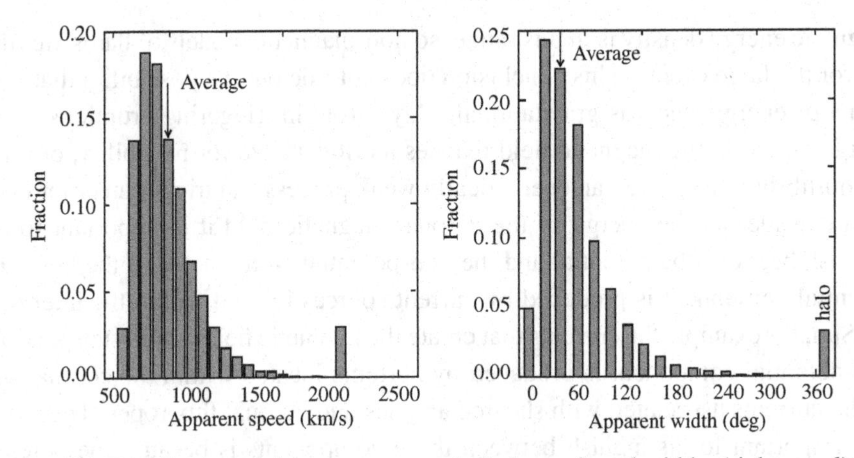

Fig. 6.4. Distribution of the apparent speeds (left panel) and widths (right panel) of CMEs observed by the SOHO coronagraph (LASCO) between 1996 and 2004. The bar in the right panel labeled "halo" refers to CMEs traveling directly toward or away from the Earth. For such events the apparent angular width tends to be near 360 degrees regardless of the actual angular width. (From Schwenn *et al.*, 2006.)

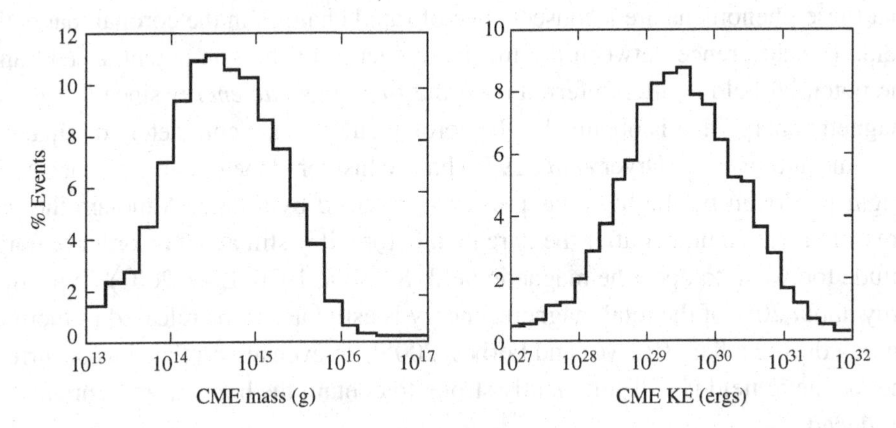

Fig. 6.5. Distribution of mass (left panel) and kinetic energy (right panel) of CMEs observed by the SOHO coronagraph (LASCO) between 1996 and 2004. The decrease in the number of events at low mass and energy is at least partly due to the difficulties encountered in observing small events. (After Schwenn *et al.*, 2006.)

of radiation produced by Thompson scattering and provide a measurement of the column density of the plasma along the line of sight to within an accuracy of 10% or better (Hayes *et al.*, 2001).

The average CME speed and angular width for the events shown in Fig. 6.4 are 487 km/s and 47°, respectively. These average values are somewhat misleading,

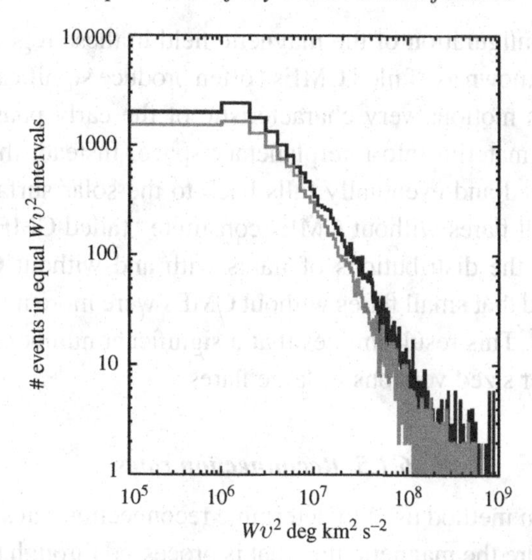

Fig. 6.6. Kinetic energy distribution of CMEs estimated by Yashiro *et al.* (2008) using the CME angular width, W, times the CME velocity, v, squared as a proxy. The black shaded region uses values obtained by hand from the SOHO (LASCO) data base, while the grey shaded region uses values from an automated process (CACTus) of the same database. The approximately straight-line shape of the curve between 10^6 and 10^9 in $\log(Wv^2)$ implies a power-law distribution for CME kinetic energy. (From Yashiro *et al.*, 2008. Source: IAU.)

however, because they are strongly influenced by the inability of the instrument to detect small events. In fact is not clear if any physically real lower limit has been observed for the speed, size, mass, and kinetic energy of CMEs. Figure 6.6 illustrates this point (Yashiro *et al.*, 2008). This figure shows a log–log plot for the distribution of a proxy parameter, Wv^2, for the kinetic energy using two different methods. Since this parameter does include CME mass, there is significant scatter caused by the fact that some events may have the same size and speed but different masses. Nevertheless, the nearly straight-line shape of the curve suggests that the distribution of CME kinetic energy roughly follows a power law. Thus, like flares (Hudson, 1991), CMEs have a wide range in size and energy that extends below the observable threshold.

From a modeling perspective this distribution raises some interesting questions concerning the nature of the trigger mechanism, or mechanisms, for flares and CMEs. One possible way to account for the wide range of energies is to invoke a range of spatial scales and to assume that the eruptive behavior of the field is self-similar. However, various studies (see the review by Schrijver, 2009) show that the distribution is not simply the result of the scale size of the eruptive region. Two regions of the same size may have quite different behavior, especially if either the strength or configuration of the magnetic field in these regions is significantly different. Events known as "failed CMEs" often produce significant flares and show the onset of mass motions very characteristic of the early phase of a CME. Yet they fail to eject material into interplanetary space. Instead, the upward rushing material is confined and eventually falls back to the solar surface (Gilbert *et al.*, 2007). Possibly all flares without CMEs constitute "failed CMEs" in some sense. A comparison of the distributions of flares with and without CMEs by Yashiro *et al.* (2006) found that small flares without CMEs were more numerous than small flares with CMEs. This result implies that a significant number of small flares are not simply smaller sized versions of large flares.

6.1.5 Reconnection rates

The most common method used to determine reconnection rates in solar flares and CMEs is to measure the magnetic flux that is processed through the flare/CME ribbons as they move across the chromosphere (Forbes, 2000; Fletcher and Hudson, 2001; Lin *et al.*, 2005; Longcope *et al.*, 2007; Qiu, 2007; Ning, 2008). This method assumes that the magnetic flux swept out by a ribbon is equal to the flux processed through the reconnection site. This method provides a global measure of the reconnection in terms of webers per second (or maxwells per second in cgs). The rate of change of the reconnected flux in time can also be expressed as a voltage. Typically for solar flares and CMEs the voltage is in the range of 10^{10} to 10^{12} volts. Figure 6.7 shows an example for a relatively small M1 flare (Saba *et al.*, 2006), but even in this case the equivalent voltage exceeds 10^{10} volts. This value corresponds to the potential drop along the X-line (separator line or quasi-separator line) in

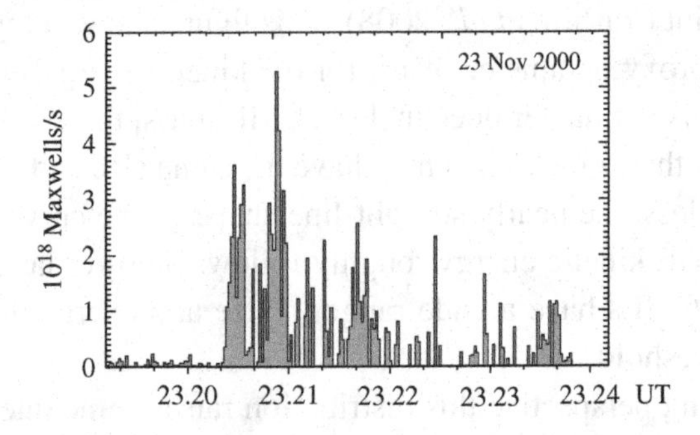

Fig. 6.7. Reconnection rate as a function of time for an M1 class flare observed on November 23, 2000. The rate is determined by calculating the rate at which the line-of-sight magnetic flux measured by the Michelson Doppler Interferometer on SOHO passes through the outer edges of the chromospheric flare ribbons observed by TRACE. A flux rate of 10^{18} maxwells/s (gauss cm^2/s) corresponds to a potential drop of 10^{10} volts. (From Saba *et al.*, 2006. Reproduced by permission of the AAS.)

the corona. The corresponding value for the Earth's magnetotail during a magnetic substorm is only about 5×10^4 volts – five orders of magnitude smaller than the solar value.

A less common and more difficult procedure is to measure the plasma flow into the current sheet where the field is reconnecting. The principal difficulty with this method is that the inflowing plasma is faint and difficult to observe. Such inflows, however, have been observed for a few events, but only during the gradual and late phases of the event (Yokoyama *et al.*, 2001; Lin *et al.*, 2005; Narukage and Shibata, 2006; Nagashima and Yokoyama, 2007). Typically the observed values correspond to inflow Alfvén Mach numbers on the order of 0.001 to 0.1, values that are consistent with the results obtained from the ribbon motion. The highest speed flows and Alfvén Mach numbers occur during the impulsive phase shortly after onset.

6.2 Models

6.2.1 Storage models

Although it is generally agreed that flares and CMEs derive their energy from the Sun's magnetic field, exactly how the magnetic energy is extracted remains uncertain. One possibility is that a flare or CME occurs when a slowly evolving coronal magnetic field reaches a point where a stable equilibrium is no longer possible. The slow evolution of the corona is driven by the changes continually occurring in the photospheric field as a result of solar convection. The equilibrium may disappear altogether or, alternatively, a stable equilibrium may simply become unstable (see Low, 1996). The continual emergence of new flux from the convection zone and the shuffling of the footpoints of closed coronal field lines increase the free magnetic energy in the corona. Eventually, these stresses may exceed a threshold beyond which a stable equilibrium cannot be maintained, and the field erupts. Since the eruption releases the free magnetic energy stored in the corona, models based on this principle are often referred to as *storage models*.

Storage models typically assume that the slow evolution of the photospheric magnetic field can be ignored once an eruption occurs. During the rapid evolution of the corona that occurs after onset, the photospheric field is essentially static, except for changes induced by the rapid variations occurring in the corona. Because the plasma in the photosphere is almost 10^9 times denser than the plasma in the corona, it is difficult for disturbances in the tenuous corona to have much effect on the photosphere and the deeper layers below it. Field lines mapping from the corona to the photosphere are thus said to be "inertially line-tied", which means that the footpoints of coronal field lines are essentially stationary over the time scale of the eruption. In actuality, line-tying at the photospheric level of the atmosphere is not perfect (van der Linden *et al.*, 1994), and photospheric magnetic disturbances have been observed to occur in response to flares (Kosovichev and Zharkova, 1998; Sudol and Harvey, 2005). However, for most storage models it does not really matter whether line-tying occurs at the photospheric level or some deeper level. The important thing is that it exists at some level so that the coronal field lines are well anchored.

Unlike models of confined flares, models of CMEs must be able to explain not only the release of magnetic energy, but also how mass is ejected into interplanetary space. During a CME, magnetic field lines mapping from the ejected plasma to the photosphere are stretched outwards to form an extended, open field structure. This opening of the field creates an apparent paradox for storage models since the stretching of the field lines implies that the magnetic energy of the system is increasing, whereas storage models require it to decrease (Sturrock *et al.*, 1984). Using quite general arguments, Aly (1991) and Sturrock (1991) have argued that a fully opened field configuration must always have a higher magnetic energy than the corresponding force-free magnetic field, if the field is simply connected. Their arguments seem to imply that CMEs are energetically impossible, but, as Aly and Sturrock have noted, there are several ways to avoid this predicament. For example, the magnetic fields may not be simply connected but contain X and O points. Also, an ideal-MHD eruption can quickly change into a non-ideal one by forming a current sheet where reconnection can occur. Alternatively, no paradox occurs if only some of the closed field lines are opened (Wolfson and Low, 1992; Low and Smith, 1993).

6.2.2 Directly driven models

Since the discovery of flares by Carrington in 1859, various researchers have considered the possibility that they are a photospheric or sub-photospheric phenomenon rather than a coronal one. Sen and White (1972), Heyvaerts (1974), Kan *et al.* (1983), and Hénoux (1986), among others, have proposed models that produce a sudden energy release in the corona by means of a surface or sub-surface current generator. In contrast to storage models, there is no buildup of magnetic energy in the corona prior to onset. Instead, there is a sudden injection of current or magnetic flux into the corona from below. As a rule, the models do not address the mechanism that leads to the sudden injection of current or flux. They simply posit that such an injection occurs, and then model the consequences of such an injection for the corona.

Many of the directly driven models are based on the fact that the photosphere is weakly ionized, having less than 10^{-4} charged carriers per neutral particle, compared to the fully ionized corona. They typically invoke the same process that occurs in laboratory MHD generators and the Earth's ionosphere when weakly ionized plasma flows across a stationary magnetic field. However, Melrose and McClymont (1987) have argued that the concept of a photospheric dynamo of this type is inconsistent with the observed properties of the photosphere and the way it is coupled to the regions above and below it. Specifically, the conductivity of the photosphere is too high to allow rapid diffusion of the field, nor is it decoupled from the interior by a non-conducting region of gas as is the case for the Earth's ionosphere.

An alternate subsurface model is shown in Fig. 6.8c. This model, proposed by Chen (1989), impulsively injects magnetic flux and power from the convection zone into the corona at CME onset. It requires a rapid increase in the magnetic energy of the corona during the eruption, rather than a decrease as in the storage models. Also, it does not address the reason why the convection zone should suddenly inject flux into the corona on a time scale that is on the order of the coronal Alfvén time scale, but more than a thousand times shorter than the Alfvén time scale of the photosphere and the regions below it (Priest, 1982).

A flux-injection model requires large-scale surface motions to exist at the photosphere. Although the photosphere is only weakly ionized, it is still an excellent conductor, and field lines there are frozen to the plasma. Thus, any sudden injection of flux from the convection zone into the corona at the start of the eruption must necessarily move the photospheric plasma. To estimate the size of such flows, let us consider the Poynting flux, $\mathbf{S}$, through the surface area, A, during the injection related to the surface flow, $\mathbf{v}$, and the surface field, $\mathbf{B}$, by

$$\mathbf{S} = -[(\mathbf{v} \times \mathbf{B}) \times \mathbf{B}]/(4\pi) \tag{6.1}$$

for an ideal-MHD plasma. This expression can be rewritten as

$$\mathbf{v}_\perp = \frac{4\pi \mathbf{S}}{B^2}, \tag{6.2}$$

where $\mathbf{v}_\perp$ is the flow perpendicular to the magnetic field. For simplicity, the flow parallel to the magnetic field has been set to zero since this flow does not affect the value of $\mathbf{S}$. The magnitude, S, of the Poynting flux is related to the energy released, ΔW, during the flare or CME by

$$S = \frac{\Delta W}{\Delta t A}, \tag{6.3}$$

where Δt is the time scale over which the major portion of the energy release occurs. Therefore, the flow speed at the surface averaged over the area A is

$$v_\perp = \frac{4\pi}{B^2} \frac{\Delta W}{\Delta t A} \tag{6.4}$$

and the corresponding distance, ΔD, traveled by the plasma is

$$\Delta D = \frac{4\pi}{B^2}\frac{\Delta W}{A}. \tag{6.5}$$

For a moderately large flare, $\Delta W \approx 10^{32}$ ergs, $B = 100$ gauss, $A \approx (10^5\ \text{km})^2$, and $\Delta t \approx 10^3$ s or less. These values yield $v \approx 13$ km/s and $\Delta D \approx 1.3 \times 10^4$ km. Flows and displacements of this magnitude are easily detectable, especially as they occur coherently over an area on the order of $10^{10}\ \text{km}^2$. However, if B is set to 1000 gauss, then the flow and displacement reduce to 0.13 km/s and 130 km. These values are probably too small to be detected, but a value of 1000 gauss implies a rather large increase in the coronal magnetic field during the eruption process.

It is possible that some very-low-energy events might be directly powered by flux injection. For example, some CMEs have speeds less than 50 km/s and undergo acceleration over a period lasting many hours rather than seconds. Such events imply speeds and displacements that differ little from the observed speeds and displacements produced by solar convection and flux emergence (Krall *et al.*, 2000).

6.2.3 Pre-eruption current sheet models

Because the magnetic energy in the corona is much larger than the thermal and gravitational energies, the magnetic force ($\mathbf{j} \times \mathbf{B}$) cannot, in general, be balanced by gravity or by a gas pressure gradient. Thus, as a rule, the coronal field will tend to be force-free, meaning that the current will flow along the direction of the magnetic field (see Fig. 6.8a). An exception to this rule occurs when a current sheet

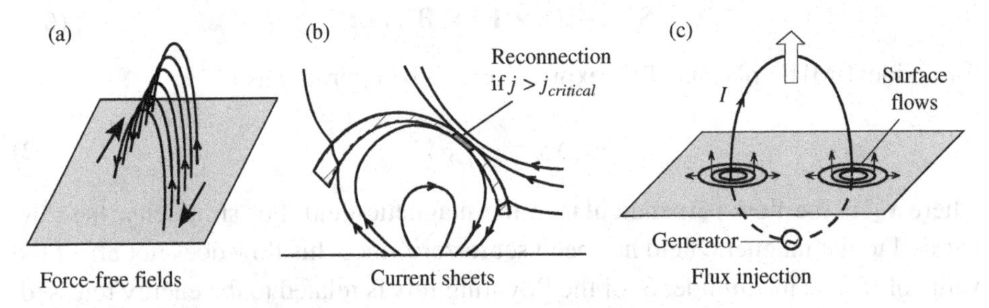

Fig. 6.8. Schematic illustration of three different types of models that use magnetic energy to power a flare or CME. (a) Magnetic energy is stored in the corona in the form of field-aligned currents that eventually become unstable. (b) Magnetic energy is stored in the corona in the form of a thin current sheet that is suddenly dissipated when a micro-instability is triggered within the sheet. (c) An example of a directly driven flare model. Here magnetic flux is suddenly injected from the convection zone into the corona at the onset of the flare or CME. Such a model produces a well-organized flow pattern during the impulsive phase (small arrows at surface in panel c).

is present. In this case gas pressure within the sheet balances the strong magnetic field outside. If the current sheet is sufficiently thin, then the high temperature or density within the sheet may not be detectable. Thus the corona could still have the appearance of a plasma with a low gas to magnetic pressure ratio (i.e. plasma $\beta \ll 1$). Figure 6.8b shows a flare model with such a current sheet, where a micro-instability within the sheet triggers an eruption.

Prior to onset, the current sheet grows as a consequence of the emergence of new magnetic flux into a pre-existing magnetic loop as shown in Fig. 6.8b. As the current sheet grows, it eventually reaches a point where a micro-instability is triggered because the current density exceeds some critical value (Heyvaerts *et al.*, 1977). Once the micro-instability occurs, the electrical resistivity of the plasma in the sheet dramatically increases, and rapid reconnection ensues.

For the model shown in Fig. 6.8b a thermal instability is postulated that creates strong turbulence within the sheet. This turbulence supposedly leads to a dramatic increase in the resistivity. Since no precise quantitative analysis is provided for this process, it is difficult to determine its viability. More recently Cassak *et al.* (2005, 2007) have numerically demonstrated a mechanism for the onset of rapid reconnection in a current sheet that uses the Hall current. Their mechanism relies on a bifurcation in the reconnection rate when the Hall term is important in the generalized Ohm's law (Eq. 5.4 in Vol. I). This allows the current sheet to rapidly reconfigure itself from a slowly reconnecting Sweet–Parker-like configuration to a rapidly reconnecting Petschek-like configuration (see Chapter 5 in Vol. I). Longcope (1996, 2001) has developed a method for analyzing flares in highly complex magnetic fields that uses pre-eruption current sheets to calculate the magnetic energy stored in the corona.

6.2.4 Two-dimensional force-free models

As mentioned previously, many storage models use configurations that have currents flowing parallel to the magnetic field in the pre-eruption state. Thus, there is no magnetic force anywhere in the configuration prior to eruption. To explain an eruption, such models need to show how a strong magnetic force can rapidly appear as a result of the slow evolution of the photospheric boundary conditions.

To illustrate the basic principles, we first consider a relatively simple flux-rope model developed by van Tend and Kuperus (1978), van Ballegooijen and Martens (1989), Forbes and Isenberg (1991), and Forbes and Priest (1995), among others. The particular version we use here is from Lin and Forbes (2000). The model's field is prescribed by

$$B_y + \mathrm{i}B_x = \frac{2\mathrm{i}A_0\lambda(h^2+\lambda^2)\sqrt{(\zeta^2+p^2)(\zeta^2+q^2)}}{\pi(\zeta^2-\lambda^2)(\zeta^2+h^2)\sqrt{(\lambda^2+p^2)(\lambda^2+q^2)}}, \tag{6.6}$$

where $\zeta = x + \mathrm{i}y$ and A_0 is the photospheric magnetic flux, or, equivalently, the magnetic vector potential at the origin (see Reeves and Forbes, 2005). In this expression h is the height of the flux rope above the surface and p and q are the lower and upper tips of a vertical current sheet below the flux rope as shown in Fig. 6.9. The parameter λ is the half-distance between two photospheric field sources located at $\zeta = \pm\lambda$ on the surface. The above expression applies only in the region outside the flux rope. Inside the flux rope the solution is a modified version of one obtained by Parker (1974). (See also Isenberg *et al.*, 1993.)

Application of the frozen-flux condition at the surface of the flux rope determines the current in the rope. This condition keeps the magnetic flux between the flux

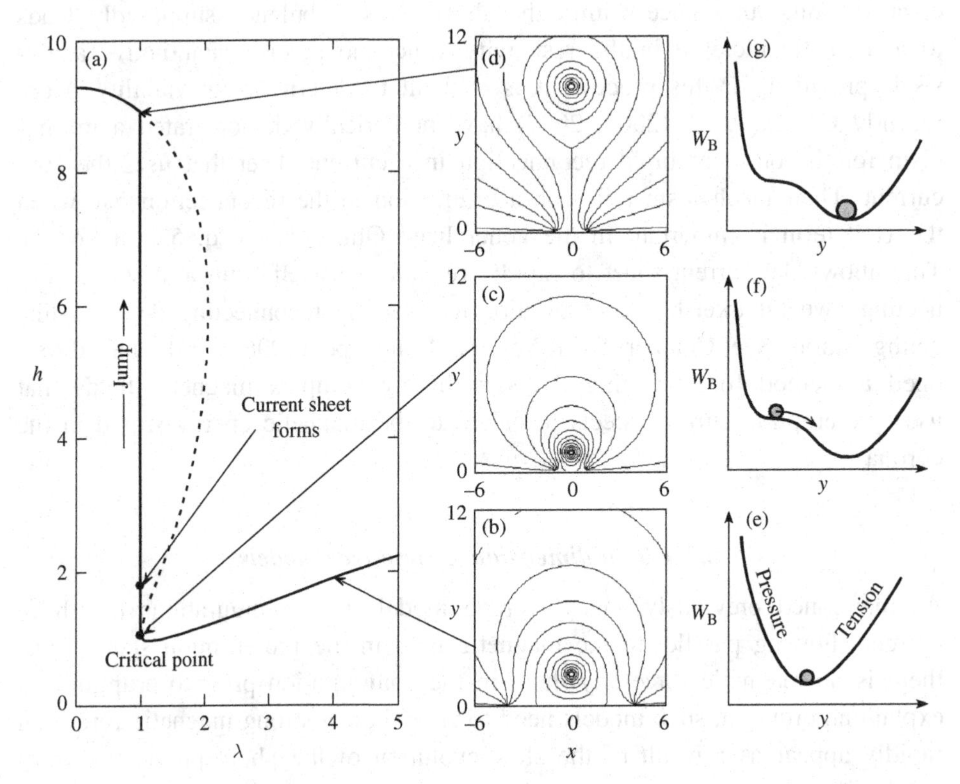

Fig. 6.9. Ideal-MHD evolution of a two-dimensional arcade containing a magnetic flux rope. (a) shows the equilibrium curve for the flux rope height, h, in normalized units, as a function of the source separation half-distance λ; (b), (c), and (d) show the magnetic field configuration at three different locations on the equilibrium curve; and (e), (f), and (g) show the corresponding energy schematic for each configuration. The case shown is for a flux rope radius of 0.1 in normalized units. (After Forbes and Priest, 1995.)

rope and the surface constant in time. It also ensures that during an eruption there is no flow of energy into the corona if the normal component of the field at the base remains invariant. Consequently, the current in the flux rope is prescribed by (Lin and Forbes, 2000)

$$I = \frac{c\lambda A_0}{2\pi h} \frac{\sqrt{(h^2 - p^2)(h^2 - q^2)}}{\sqrt{(\lambda^2 + p^2)(\lambda^2 + q^2)}}. \tag{6.7}$$

This current decreases with time during an eruption as magnetic energy is converted into kinetic energy. This decrease becomes apparent only when the formula giving the dependence of q upon h and p is incorporated into the above expression. Since this formula is rather complex (involving elliptical integral functions), we refer the reader to the paper by Lin and Forbes (2000).

The magnetic field configuration is shown in Fig. 6.9 for three different sets of parameters. The surface at $y = 0$ corresponds to the photosphere, and the boundary condition at this surface is

$$A(x, 0) = A_0 \mathcal{H}(\lambda - |x|), \tag{6.8}$$

where $\mathcal{H}$ is the Heaviside step-function and A_0 is the value of A at the origin. This boundary condition corresponds to two sources of opposite polarity located at $x = \pm\lambda$.

Force-free equilibria are calculated by determining the net force acting on the flux rope and then setting it to zero. Depending on the choice of model parameters, there may be three equilibria, one equilibrium, or no equilibrium for a given set of parameters. In the situations with three equilibria the magnetic energy of each equilibrium is different. For the isolated equilibrium shown in Fig. 6.9b the flux rope sits in an energy well as shown in Fig. 6.9e. If the flux rope is pushed downward toward the surface, compression of the magnetic field between the flux rope and the surface creates an upward force. If the flux rope is pulled upward away from the surface, magnetic tension from the overlying arcade creates a downward force. Line-tying plays a key role in creating the equilibrium because it prevents field lines from being pushed into, or pulled out of, the surface when the flux rope is perturbed.

An evolutionary sequence is created by assuming that the distance between the two sources at $\pm\lambda$ decreases at a rate that is much slower than the Alfvén time scale in the corona. A flux rope located on the lower portion of the equilibrium curves shown in Fig. 6.9a will erupt when the distance between the line sources becomes less than the height of the flux rope. When this location is reached, the unstable and stable equilibria coincide as shown in Fig. 6.9g. Once equilibrium is lost, the flux rope rapidly moves upwards. In the absence of reconnection ($p = 0$) the flux rope does not escape, but, instead, reaches a new equilibrium position with a vertical current sheet, as shown in Fig. 6.9d.

In the absence of any reconnection the amount of energy released by the loss of equilibrium is quite small, less than 5% as shown in Fig. 6.10. Thus, while the loss of equilibrium can account for the rapid onset of an eruption, it cannot, by itself, account for the large amount of energy released. For this, magnetic reconnection is needed.

A key assumption of this model is that the flux-rope radius, a, is much smaller than the flux-rope height, h. This assumption allows one to take advantage of the fact that the magnetic field due to all sources external to the flux rope vanishes at the equilibrium location of the flux rope (Forbes and Isenberg, 1991). For this flux-rope model there are three external sources. The first is the photospheric source current that creates the background field, the second is the current sheet, and the third is the surface current that arises from line-tying. In the absence of the flux rope, the combined field of these three sources creates an X-line at $x = 0, y = h$. Thus, if $a \ll h$, the equilibrium inside the flux rope is approximately the same as it is when there are no external sources, and it can be solved independently of the global equilibrium. As shown in Forbes and Priest (1995), the error is of order $(a/h)^2$.

In the absence of magnetic reconnection, the flux rope in this two-dimensional model cannot escape. If one places the flux rope at the critical point, it will start to move upward, slowly at first, but with an ever-increasing speed until it reaches a height corresponding to about $h = 9$. At this height, it starts to slow down,

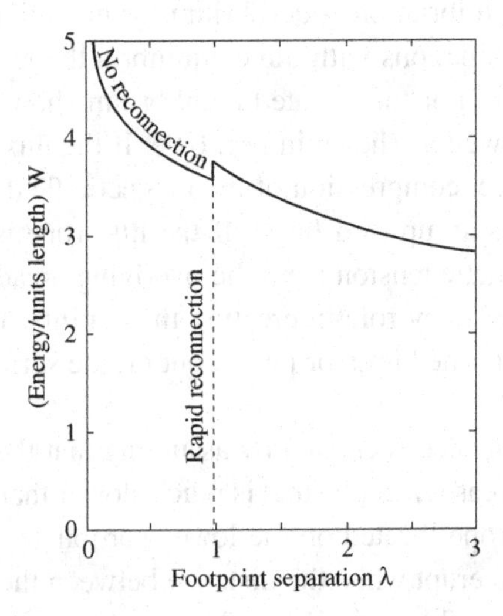

Fig. 6.10. Free magnetic energy released by the loss of equilibrium in the two-dimensional flux-rope model. The solid curve is the ideal-MHD case, and the dashed vertical line is the rapid-reconnection case.

and it will eventually come to rest at this location if its kinetic energy is dissipated. However, if reconnection is allowed, the flux rope will escape provided that the reconnection is fast enough. Lin and Forbes (2000) have shown that for typical coronal conditions a very modest rate of reconnection is sufficient to allow escape. For reconnection rates corresponding to an inflow Alfvén Mach number, M_A, greater than 0.05 (at the midpoint of the current sheet sides) the flux rope can escape without any deceleration. Escape with deceleration occurs as long as M_A is larger than about 0.006. Figure 6.11 shows an example of a non-decelerating trajectory obtained for $M_A = 0.1$. As the flux rope moves upward, the current passing through it decreases causing the flux rope to expand. The lower tip of the current sheet at p moves upward very slowly because its motion is controlled by the slow rate at which reconnection occurs. By contrast, the upper tip of the current sheet at q moves upward quite rapidly at a speed that is only about a factor of two smaller than the speed of the flux rope. Because a loss-of-equilibrium is an ideal-MHD process, the upward speed of the flux rope is close to the ambient Alfvén speed in the region where the loss of equilibrium develops.

The curves shown in Fig. 6.11 are derived from an analytical model that treats the flux rope as a ballistic projectile, and, therefore, they do not account for the

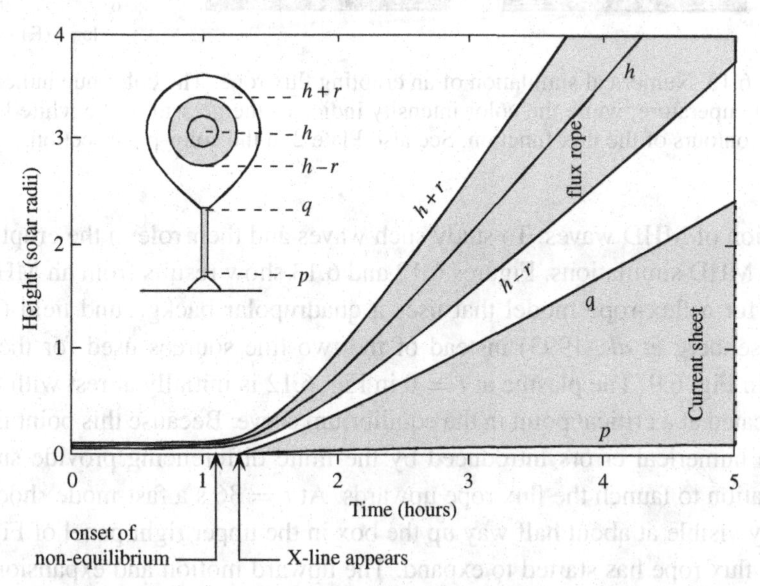

Fig. 6.11. Flux rope and current sheet trajectories obtained from the model in Fig. 6.9 by assuming a constant inflow Alfvén Mach number of 0.1 in the inflow region at the midpoint of the current sheet. The parameters h, r, q, and p are the flux rope's height and radius and the current sheet's upper and lower tips, respectively. The ambient Alfvén speed is calculated from the model magnetic field and the empirical coronal density model of Sittler and Guhathakurta (1999).

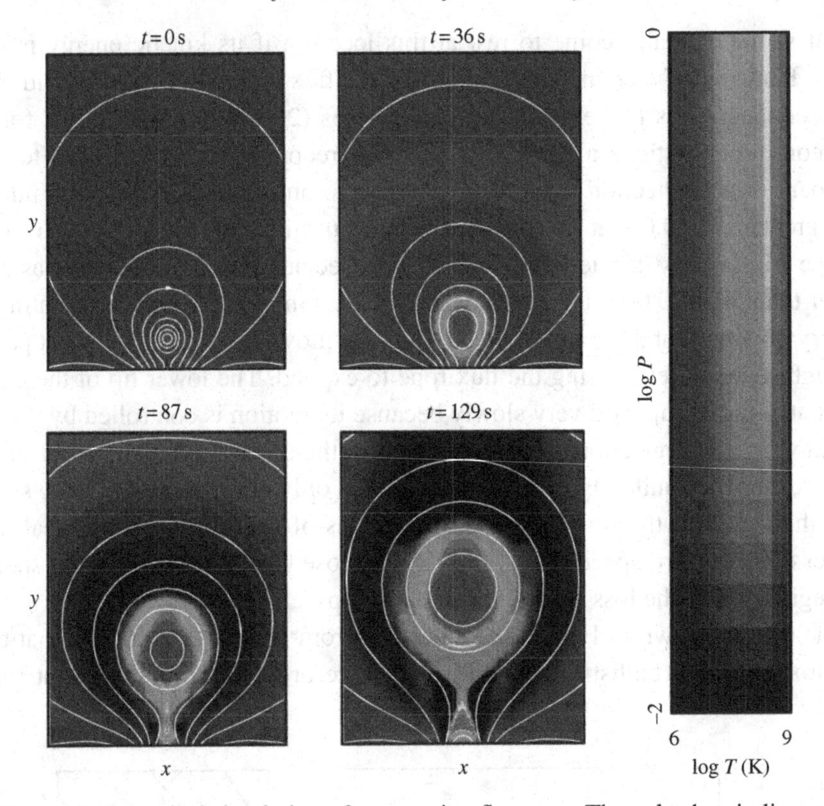

Fig. 6.12. Numerical simulation of an erupting flux rope. The color hue indicates the temperature, while the color intensity indicates the pressure. The white lines are contours of the flux function. See also Plate 2 in the color-plate section.

generation of MHD waves. To study such waves and their role in the eruption one can use MHD simulations. Figures 6.12 and 6.13 show results from an MHD simulation for a flux-rope model that uses a quadrupolar background field (Forbes, 1991; Isenberg *et al.*, 1993) instead of the two line sources used for the model shown in Fig. 6.9. The plasma at $t = 0$ in Fig. 6.12 is initially at rest with the flux rope located at a critical point in the equilibrium curve. Because this point is unstable, the numerical errors introduced by the finite differencing provide sufficient perturbation to launch the flux rope upwards. At $t = 36$ s a fast-mode shock wave is barely visible at about half way up the box in the upper right panel of Fig. 6.12, and the flux rope has started to expand. The upward motion and expansion of the flux rope leads to the formation of a low-density cavity with a moderately strong magnetic field at $t = 87$ s (see Gibson and Low, 1998, for more discussion of why a cavity forms). Finally, by $t = 129$ s, reconnection of field lines below the flux rope has heated the plasma in the closed loop region to a temperature in excess of 10^8 K – higher than expected in reality because the simulation does not include

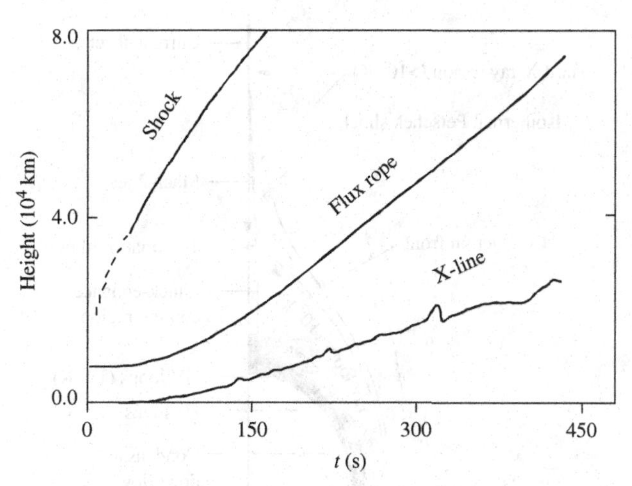

Fig. 6.13. The height of the shock, flux rope, and X-line as a function of time for the numerical simulation shown in Fig. 6.12. The dashed line indicates the compressive wave that eventually steepens into the shock.

cooling due to radiation and thermal conduction. Plasma in the outflow region above the reconnection region is also heated, but the temperature is not as high because of the ongoing expansion.

The trajectories of the flux rope, the shock wave, and the x-line are shown in Fig. 6.13. Because the initial numerical perturbation is very small, the initial upward motion of the flux rope is almost imperceptible. However, as it moves upwards, the flux rope gradually gains speed. The slow upward motion creates a compressive wave that steepens into a fast-mode shock propagating ahead of the rising flux rope. This shock accelerates the plasma through which it passes, so that the amount of material that is being accelerated continually increases with time. Even though the shock plays an important role in accelerating the plasma, the resulting trajectories are quite similar to those predicted by the analytical model. (For another comparison between the analytical two-dimensional flux rope model and a numerical simulation see Webb *et al.*, 2003.)

6.2.5 *Modeling flare emissions*

The Poynting flux passing into the current sheet of the two-dimensional models can be used as an input for a multi-loop numerical model of the flare loop system. At any given time there is only a small number of field lines that map to the current sheet (including any slow shocks). Conduction electrons and energetic particles travel along these field lines down to the chromosphere and result in an impulsive heating of the plasma as shown in Fig. 6.14. Some of the heated plasma then flows

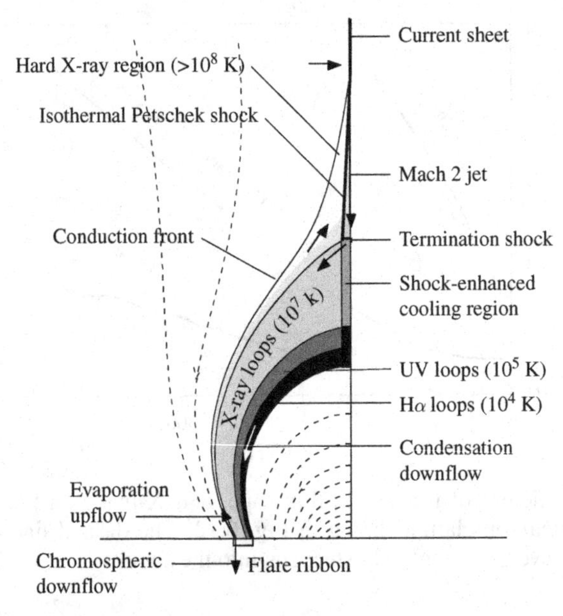

Fig. 6.14. Diagram of the temperature structure and flows predicted by a reconnection model of flare loops. Strong thermal conduction channels the energy released by reconnection to the chromosphere where it heats the plasma. The high pressure thereby created drives plasma upwards into the corona and downwards into the chromosphere. (Compare with Fig. 5.16.)

back up into the corona to produce the system of flare loops shown in the figure (Forbes and Acton, 1996). Once a magnetic field line is disconnected from the current sheet, the plasma on it cools, first by thermal conduction and later by radiation (Cargill *et al.*, 1995). This process leads to the formation of a system of flare loops with different temperatures, densities, and flows.

Because the amount of time that a given field line is connected to the reconnection region is short, it is possible to model the overall emission of the loop system as a collection of many small loops heated at different times by a continually evolving source. Models of this type have been constructed by Hori *et al.* (1997, 1998), Reeves and Warren (2002), Warren and Doschek (2005), and Warren (2006). These models have used either an empirical energy input derived from data or an arbitrary heating function specifically tailored to match the observations. The models show that a sustained energy input is required to account for the emissions produced in LDEs. Reeves *et al.* (2007) have used the reconnection energy output predicted by the two-dimensional model shown in Figs. 6.9, 6.10, and 6.11 to drive a multi-loop model. With the use of the SOLFTM one-dimensional flare code (Mariska, 1987) they obtained predictions for the density, temperature, and bulk flow of the entire flare loop system as functions of space and time. This information was then

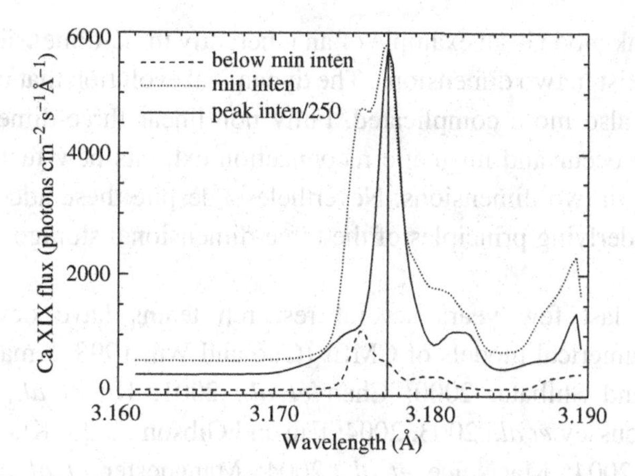

Fig. 6.15. The Ca XIX line profile predicted by the two-dimensional flux-rope model for typical coronal values. Three different times are shown corresponding to the peak intensity (re-scaled solid line), an earlier time when the intensity would first be observed by the Yohkoh BCS (dotted line), and an even earlier time when the intensity would be too low to be observed (dashed line). The vertical line marks the rest wavelength of the resonance line. (From Reeves *et al.*, 2007.)

processed through a numerical routine (bcs-spec, part of the standard SolarSoft package) developed at the US Naval Research Laboratory for the Bragg Crystal Spectrometer (BCS) on the Japanese satellite Yohkoh. Figure 6.15 shows the resulting spectrum for the Ca XIX line at three different times.

The model's parameters were chosen to simulate a B2.5 flare that was a long duration event associated with a CME. Even though the net emission generated by this model consists of many loops containing flowing plasma at different temperatures, the combined spectrum that is generated is similar to the spectrum generated by a single loop at a temperature of 8 MK at the time of the peak emission in Ca XIX. The model predicts that during the early phase of the flare a strong blue shift occurs, but only when the intensity of the line is still below the intensity threshold of the instrument. By the time emission becomes detectable, the spectrum consists of a stationary component with an enhanced blue wing. At the time of the peak emission, only the stationary component is evident. These results nicely account for the observations of actual flares by the BCS (Antonucci *et al.*, 1987; Doschek and Warren, 2005).

6.2.6 Three-dimensional force-free models

It will probably come as no surprise that three-dimensional models are considerably more complex than two-dimensional ones. Three-dimensional field configurations are subject to a much greater number of instabilities. The helical ideal-MHD kink mode is an example of an inherently three-dimensional instability that does not exist in two dimensions. The dynamical evolution that occurs in three-dimensions is also more complicated. Fully non-linear three-dimensional MHD turbulence can occur and magnetic reconnection exhibits new features that have no counterpart in two dimensions. Nevertheless, despite these additional complications, the underlying principles of the three-dimensional storage models remain the same.

During the last few years several research teams have developed three-dimensional numerical models of CMEs (Guo and Wu, 1998; Amari *et al.*, 2000, 2003; Chen and Shibata, 2000; Chen *et al.*, 2001; Hu *et al.*, 2003; Linker *et al.*, 2003; Roussev *et al.*, 2003, 2004; Fan and Gibson, 2004; Kliem *et al.*, 2004; Török *et al.*, 2004; MacNeice *et al.*, 2004; Manchester *et al.*, 2004b; Lugaz *et al.*, 2005a; Wang *et al.*, 2005; Williams *et al.*, 2005; Birn *et al.*, 2006; Dubey *et al.*, 2006; Forbes *et al.*, 2006; Inoue and Kusano, 2006; Jacobs *et al.*, 2006; Lynch, 2006; Fan and Gibson, 2007; Jacobs *et al.*, 2007a, b; Riley *et al.*, 2007; Roussev *et al.*, 2007; Wu *et al.*, 2007; Ye *et al.*, 2007). In order to show the relation of the relatively simple two-dimensional model of the previous section with these three-dimensional models, we take a reductionist approach. That is, we start with a very simple three-dimensional configuration and then sequentially add new features that increase its complexity. We start with the simple toroidal flux rope shown in Fig. 6.16. The anti-parallel orientation of the current flowing on the opposite sides of the ring produces a repulsive force similar to the force between two parallel wires with anti-parallel currents. For a small minor radius, a, this force, sometimes referred to as the hoop force, is approximately

$$F \propto \frac{I^2}{R}\ln(R/a), \tag{6.9}$$

where I is the flux-rope current, R is the major radius, and a is the minor radius of torus. The right-hand side of the above expression is the lowest order term of an expansion in the parameter a/R, so the expression is only valid for $a \ll R$

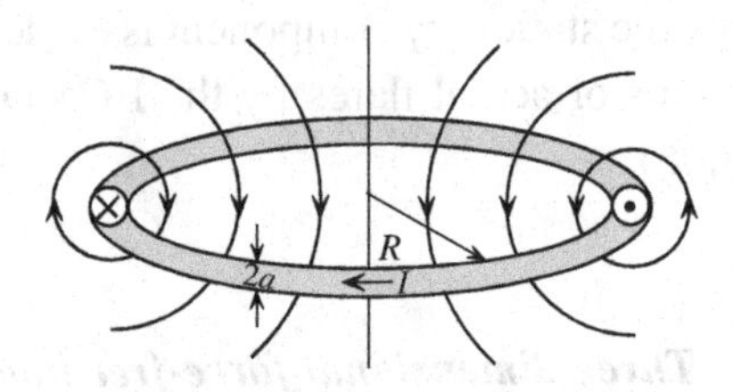

Fig. 6.16. An isolated toroidal flux rope. The flux rope has a major radius, R, a minor radius, a, and carries a net toroidal current, I. The anti-parallel orientation of the current flowing on the opposite sides of the torus creates an outward force in the radial direction.

(Shafranov, 1966). If we had two straight wires, the force between them would be proportional to I^2/R but because we are dealing with a circular ring there is an additional factor of $\ln(R/a)$. This factor is due to the effect, at any given point, of the nearby portions of the ring. Note that as a goes to zero, the repulsive force becomes infinite. Thus, when a is small relative to R, a rather large force is needed to bend the flux rope.

Just as for two-dimensional storage models, the three-dimensional models assume that the time scale of the eruption is so fast that any additional input of magnetic energy after the eruption starts is completely negligible. Therefore, the flux associated with the flux-rope current is conserved. In the limit that a/R tends to zero, the flux-rope current is roughly

$$I \approx \frac{I_0 R_0}{R \ln(R/a)}, \tag{6.10}$$

where I_0 and R_0 are initial values. If one considers the torus configuration as an initial state that subsequently evolves in response to the force, then R will increase to infinity, but as it does, so I will decrease to zero. In the process the magnetic energy associated with the flux rope's initial current is converted into the kinetic energy of the expanding plasma ring.

To create an equilibrium one must add an additional magnetic field of the proper orientation and strength. In tokomak terminology such a field is called a strapping field (Wesson, 1987). Figure 6.17 shows an example of a strapping field obtained by placing a properly oriented dipole at the center of the torus. With the addition of the dipole the force is now

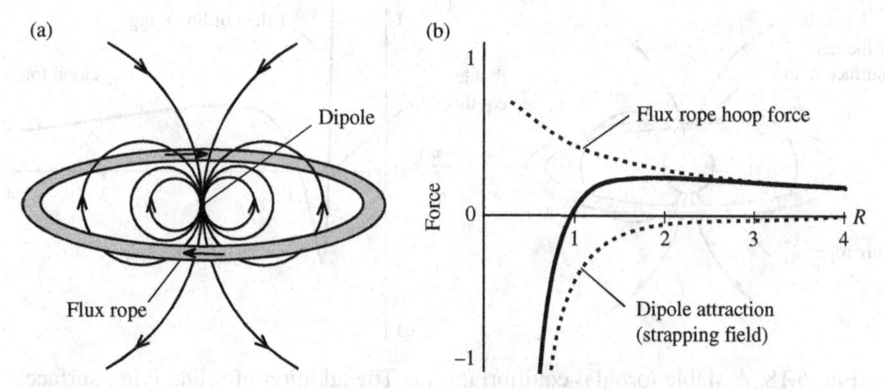

Fig. 6.17. An unstable toroidal equilibrium. (a) The outward force of the curved flux rope is balanced by a properly oriented dipole magnetic field. (b) Schematic diagram showing the forces acting on the flux rope as a function of radial distance. The single equilibrium that exists is unstable because displacements away from it produce forces that act to increase the displacement.

$$F \propto \frac{I^2}{R}\ln(R/a) - \frac{Im}{R^3}, \tag{6.11}$$

where m is the relative strength of the dipole. Although we can now have an equilibrium, the equilibrium will be unstable because a small displacement of the flux rope outward creates a force that acts to increase the displacement. Similarly, a small displacement inward creates a force that pushes the flux rope closer to the dipole. In laboratory plasmas this instability is known as the horizontal tokomak or torus instability (Bateman, 1978). Since this instability is a current-driven instability it is also technically a kind of kink instability (see Bateman, 1978). However, it is distinct from the usual helical instability that most people associate with the term "kink".

Although it is possible to create a stable equilibrium by altering the form of the strapping field, an alternative possibility that is more appropriate for a storage model is to introduce a line-tying surface as shown in Fig. 6.18. The effect of line-tying can be modeled by introducing a fictitious image current below the surface (Lin *et al.*, 1998). With the introduction of this additional current, a new equilibrium appears which, unlike the previous one, is stable. Stabilization is achieved because line-tying prevents field lines from being pushed into, or pulled out of, the surface. One can construct a spherically symmetric analog of the two-dimensional model discussed previously by letting the strength of the dipole field vary in time. The resulting set of equilibria contains a critical point (i.e. a nose point) that is reached when the dipole field becomes too weak.

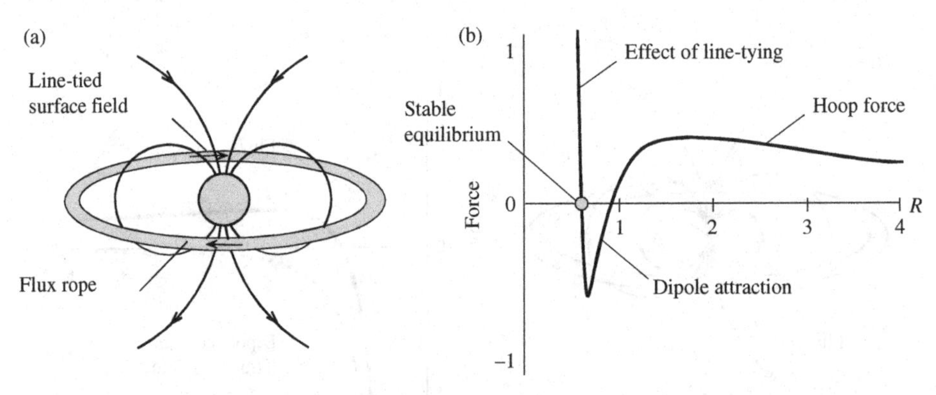

Fig. 6.18. A stable toroidal equilibrium. (a) The addition of a line-tying surface representing the surface of the Sun creates the possibility of a stable equilibrium. Surface currents (which can be modeled using an image current) create an additional magnetic field component that gives rise to a second equilibrium position as shown in (b). The new equilibrium is stable because displacements away from it produce a restoring force.

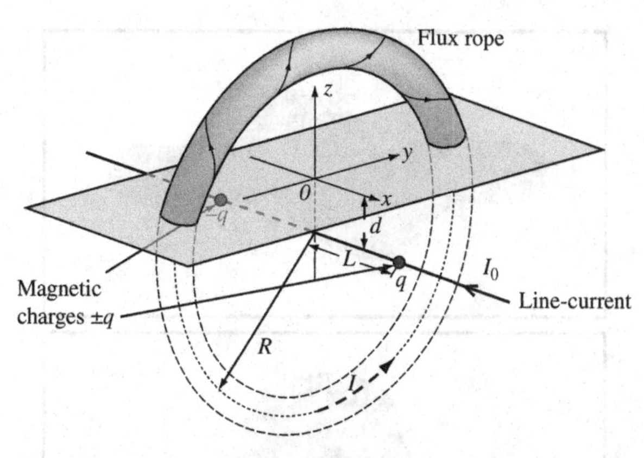

Fig. 6.19. The three-dimensional flux-rope model of Titov and Démoulin (1999). The coronal magnetic field is produced by three different sources consisting of a flux-rope current, a pair of magnetic charges, and a line current. The source regions located below the surface are fictitious constructs used to create the coronal field. The model does not prescribe the form of the subsurface field.

Although we now have an eruptive model with some degree of three dimensionality, it still has the drawback that the flux rope is not itself anchored to the solar surface. An analytical configuration that does have this property is shown in Fig. 6.19. This configuration was proposed by Titov and Démoulin (1999), and it consists of a toroidal flux rope that intersects the photospheric surface. The flux rope, with current I, is held in equilibrium by an overlying arcade (not shown in the figure), which is produced by subsurface magnetic charges $\pm q$ located along the centerline at a depth d below the photospheric surface at $z = 0$. Finally, there is a subsurface line current lying along the centerline. The strength of the current, I_0, flowing in this subsurface line controls the pitch of the coronal magnetic field. When I_0 is varied from small to large values, the configuration changes gradually from a highly twisted flux rope resembling a slinky to one that resembles a sheared arcade without a flux rope.

Although the magnetic field of the Titov and Démoulin (1999) configuration is still azimuthally symmetric about the center line of the torus, the solar surface no longer shares this symmetry. Instead the surface is a flat plane that intersects the flux rope torus at some arbitrary position without influencing the field structure. Thus, any line-tied evolution of this configuration away from the initial state necessarily creates a highly asymmetrical configuration. An example of what such a configuration looks like is shown in Fig. 6.20. This figure shows two different views of an iso-current surface of the current density obtained from a simulation by Török *et al.* (2004). This simulation starts with an unstable Titov and Démoulin configuration that is given a small perturbation. Within a few Alfvén scale times

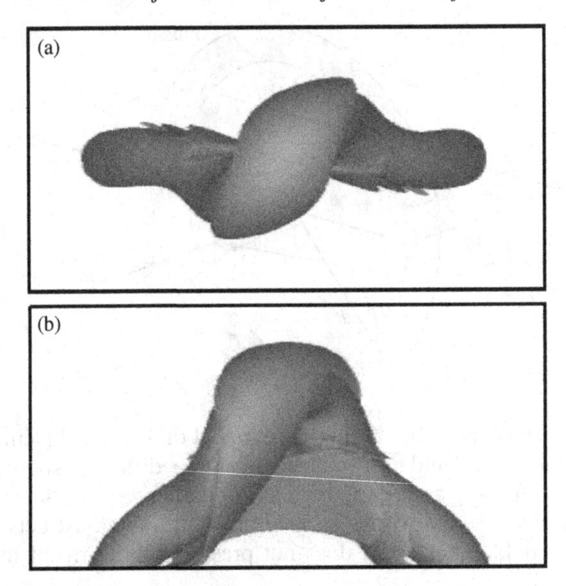

Fig. 6.20. Top view (a) and side view (b) of constant current density surfaces from the simulation by Török *et al.* (2004) for an unstable Titov and Démoulin equilibrium.

the configuration evolves into the kinked, omega-shaped flux rope shown in the figure. For this particular case, the initial instability is actually a helical kink instability rather than the torus instability discussed previously. However, it is possible to construct unstable Titov and Démoulin configurations that are unstable to the torus instability rather than the helical kink (Kliem and Török 2006). Similar simulations using the Titov and Démoulin configuration as a starting point have also been carried out by Roussev *et al.* (2003). Isenberg and Forbes (2007) have used the Titov and Démoulin configuration as the starting point for a fully three-dimensional analytical model that can be evolved in time while still satisfying the line-tying condition that the normal magnetic field at the solar surface remain fixed.

6.2.7 Formation of the pre-eruption field

An important issue that the above flux-rope models do not address is the creation and growth of the magnetic stress that causes the field to erupt. It could be that most of the stress buildup occurs in the convection zone before the field emerges into the corona. Alternatively, it may be that the field emerges in a nearly unstressed, current-free state, and that the stress subsequently develops in response to the observed surface flows. In practice both possibilities are likely to occur at least at some level.

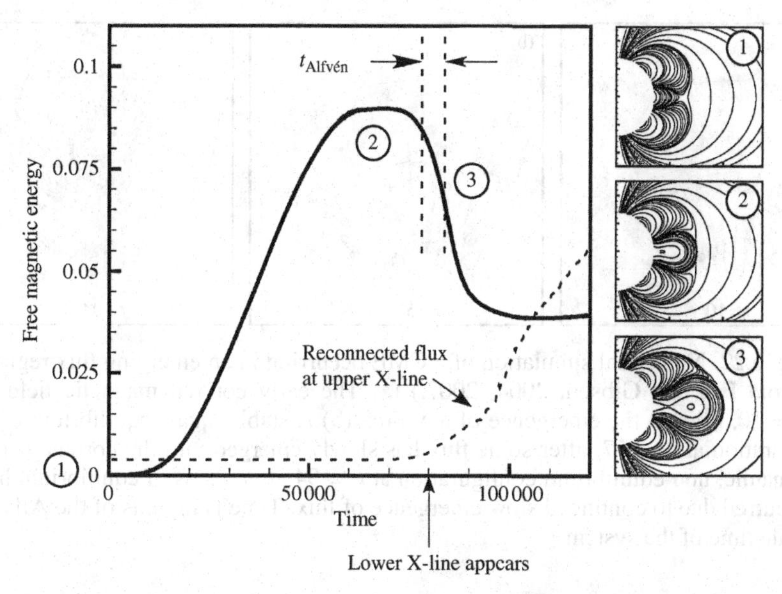

Fig. 6.21. Numerical simulation of a storage model proposed by Antiochos *et al.* (1999). The panel at left shows the free magnetic energy as a function of time, while the three panels at right show contours of the magnetic flux surfaces at three different times. (After MacNeice *et al.*, 2004.)

Several three-dimensional simulations have been carried out that address this issue. Perhaps the best known at the present time is the simulation of what is called the *breakout model* (Antiochos *et al.*, 1999; MacNeice *et al.*, 2004; Lynch, 2006). The evolution of this model is shown in Fig. 6.21. The initial state consists of a quadrupolar magnetic field that carries no current, so it contains no free magnetic energy. Slowly shearing the central arcade around the equator gives rise to a set of stressed loops that push outward against the overlying arcade. As this happens, a curved, horizontal current sheet forms at high altitude at the pre-existing X-line. Eventually, the stresses build up to a level that causes an eruption. The nature of the mechanism that triggers the eruption has not yet been fully resolved, but it is likely that it consists of some kind of combination of both ideal and non-ideal processes (Zhang and Wang, 2007).

A three-dimensional storage model that uses the emergence of a pre-stressed field to model a CME has been carried out by Fan and Gibson (2004, 2007). They start with a flux rope below the surface and then slowly emerge it into the corona as shown in Fig. 6.22. Depending on the various assumptions made about the parameters that characterize the flux rope one may, or may not, have an eruption. Generally, the flux rope will tend to erupt once there are one or two turns in the portion of it that has emerged into the corona. However, if the flux rope emerges into a pre-existing arcade, the strength and orientation of this arcade also

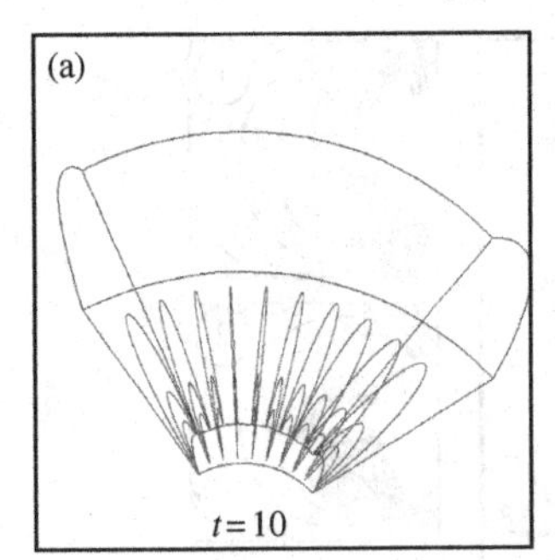

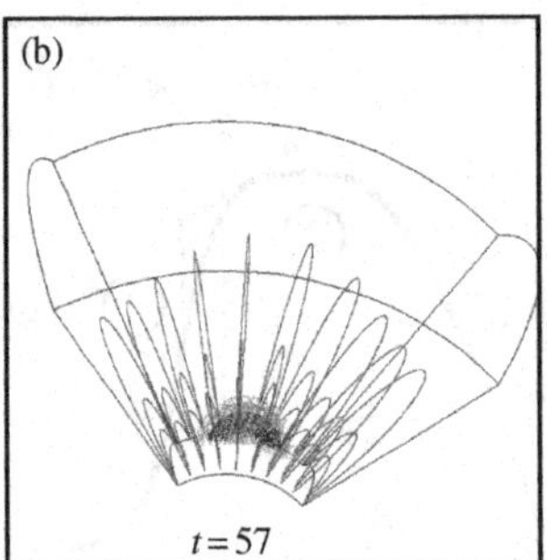

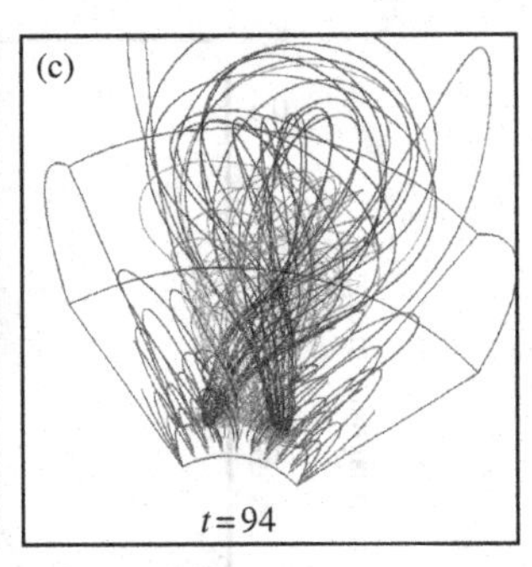

Fig. 6.22. Numerical simulation of a CME occurring in an emerging flux region. (From Fan and Gibson, 2004, 2007.) (a) The early coronal magnetic field at $t = 10$, prior to the emergence of any flux. (b) A stable, quasi-equilibrium configuration at $t = 57$, after some flux has slowly emerged into the corona. (c) A dynamic, non-equilibrium configuration at $t = 94$ after a loss of equilibrium has occurred due to continued slow emergence of flux. Time is in units of the Alfvén scale time of the system.

has a strong effect on whether an eruption occurs or not. Thus, their model shares many of the behavior features found by Kliem and Török (2004) for the model of Titov and Démoulin (1999).

Many studies of flux emergence (Fan *et al.*, 1999; Magara, 2001; Abbett and Fisher, 2003; Manchester, 2004b; Archontis *et al.*, 2007, among others) have been carried out in order to obtain a better understanding of what type of coronal field configuration is likely to form in active regions (for a review see Schrijver, 2009). One of the important issues that these studies address is the effect of mass loading on the emergence of a flux rope into the low-density corona. Most of the CME models discussed in the previous section are based on the supposition that a flux rope exists in the corona prior to onset (Filippov, 2001), but it is not obvious how such a structure could be formed. Formation of the flux rope within the convection zone followed by its buoyant rise into the corona immediately encounters the problem that mass cannot easily drain out of concave-upward portions of the magnetic field. Unless there is a way for the mass to drain out of the flux rope, the rope will remain half buried in the solar surface. One way around this difficulty is to suppose that the flux rope does not exist prior to the emergence of magnetic flux, but instead forms in the corona by a combination of converging flows and slow reconnection (Priest *et al.*, 1996; Martens and Zwaan, 2001). Most dynamo models, however, predict that large-scale flux rope will form near the base of the convection zone and then rise buoyantly to the solar surface to form an active region. Thus, this solution to the mass-loading problem involves both the destruction and reformation of the flux rope below and above the surface.

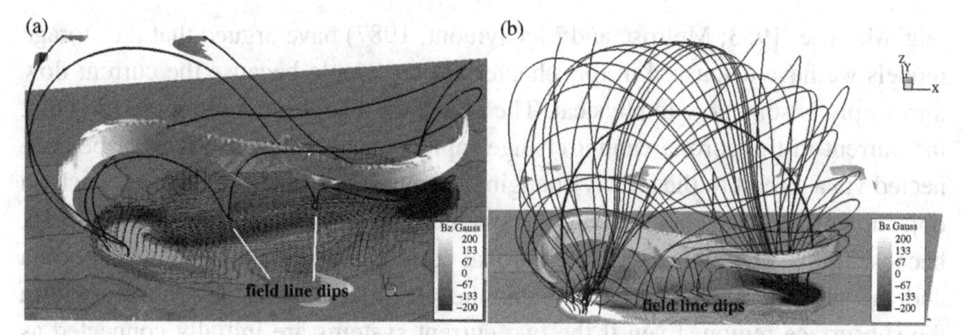

Fig. 6.23. Two views of the magnetic field configuration and current density distribution at $t = 72.8$ in the simulation of Manchester *et al.* (2004b). (a) The view from above; (b) the view from an angle. The base surface is shaded according to the value of the normal magnetic field component. The black and grey curves show two magnetic field lines. The bright grey, ribbon-like structure shows the position of the current sheet that develops during the slow emergence of the flux into the corona. This sheet occurs prior to any eruption and plays a critical role in allowing mass to drain along field lines. (After Manchester *et al.*, 2004b. Reproduced by permission of the AAS.)

A reasonable away around these difficulties has been demonstrated by Manchester *et al.* (2004b) using a fully three-dimensional MHD simulation. Starting with a buoyant flux rope below the surface, they find that short sections of the flux rope containing a single turn of the field line can successfully rise through the solar surface. As shown in Fig. 6.23, the result is a series of arched flux-rope segments that are reminiscent of a hedgerow filament. Slow reconnection plays a key role in transforming the flux rope and allowing mass to drain out of the field as it is transported into the corona. Furthermore, the simulation also shows that the shear flows that are observed near the polarity inversion line of filaments are a natural consequence of the force imbalances generated near the photosphere during the emergence of a flux rope (Manchester, 2003, 2007; Manchester *et al.*, 2004b).

More global simulations of flux emergence carried out by van Ballegooijen and MacKay (2007) confirm the importance of reconnection in transforming rising flux ropes (sometimes called *omega loops*) into weakly twisted coronal flux ropes. These simulations show that many of the large-scale features of filaments such as *switchbacks* are nicely accounted for by the emergence and subsequent reconnection and diffusion of the magnetic field (Mackay and van Ballegooijen, 2005). The process of flux emergence eventually leads to a loss of equilibrium and the ejection of the flux rope into space.

The role of current in the formation and evolution of the flares and CMEs is an issue that has been discussed by several authors. Some of these authors (e.g. Melrose, 1983; Melrose and McClymont, 1987) have argued that the storage models we have discussed in this chapter are not viable because the current flow they imply is somehow unphysical. The objection that is most often raised is that the current in the corona cannot change rapidly in time because it must be connected via a circuit to the slowly changing current in the convection zone. Such an objection clearly does not apply to the symmetrical ring model shown in Fig. 6.18, because in this symmetric case the coronal current closes entirely with the corona and the subsurface current that creates the potential field closes entirely within the subsurface region. Even if the two current systems are initially connected as in the Titov and Démoulin model of Fig. 6.19, inertial line-tying of the magnetic field decouples the currents in the two regions during an eruption. As the current decreases in the corona, surface currents appear that ensure proper closure of the current. These surface currents are clearly seen in the MHD simulations that include the subsurface region within the numerical domain (e.g. Roussev *et al.*, 2003). Although inertial line-tying may not be as perfect as these models assume, it is sufficiently effective to ensure decoupling of the two current systems (van der Linden *et al.*, 1994).

Another, related, objection that is sometimes raised is that any current flow created in the corona by flux emergence or by shearing will produce a shielding current that flows in the opposite direction to the current within the rope or arcade. The existence of a shielding current is a natural consequence of the intrusion of new magnetic flux into a highly conducting plasma. However, those who raise this objection usually go on to assume that the shielding current is located adjacent to the main current and prevents this current from interacting with any other field. MHD simulations carried out by Amari *et al.* (2005) and Delannée *et al.* (2008) have demonstrated that this objection is not valid for flux emergence models. These studies show that the shielding current propagates outward to very large distance due to the expansion of the plasma as it transforms from a high-β plasma to a low-β plasma on its passage through the surface. The effect of the shielding current on the equilibria and their stability is therefore relatively minor.

It is perhaps not too surprising that objections based solely on current flow arguments typically turn out to be spurious and inconsistent with the results of MHD simulations. Both Parker (1996, 2007) and Vasyliunas (2005) have emphasized that, in astrophysical plasmas, current and electric field are secondary quantities and are mathematically insufficient to prescribe the plasma's behavior. Assuming a priori how the current will flow in a given system is likely to give incorrect results because it is the magnetic field and velocity that determine the current and the electric field and not the other way around.

6.3 Some topics for future research

This chapter has focused primarily on the contemporary understanding of the origin of flares and CMEs with the emphasis primarily on CMEs. At the present time there is a very strong consensus in the solar research community that these phenomena derive their energy from the Sun's magnetic field. It is also generally accepted by most researchers that it is the free magnetic energy associated with coronal currents that is available to drive flares and CMEs. However, beyond these generalities, there is a great deal of uncertainty about the specific mechanism, or mechanisms, that trigger eruptions. How stresses actually build up in the corona remains difficult to assess, as does the relative role of ideal-MHD processes (e.g. the kink instability) and non-ideal-MHD processes (e.g. magnetic reconnection).

Inertial line-tying is an important feature for most models of CMEs and flares, yet very little is known about the altitude in the solar atmosphere where it becomes effective. Van der Linden *et al.* (1994) have argued that the ability of the photosphere and surrounding region to stabilize the coronal field depends on more than just the inertia or density of the plasma. The gravitational and pressure forces play a role as does the structure of the magnetic field. Reports of sustained changes in the normal component of the photospheric magnetic field (e.g. Sudol and Harvey, 2005; Wang, 2006) imply that the field is not perfectly line-tied at the photospheric level.

One area of research that is particularly important for NASA's Living With a Star program, but has received very little attention so far, is the theoretical limits on predicting the onset of solar eruptions. It seems almost certain that solar eruptions share the same property as other catastrophic phenomena, such as avalanches and earthquakes, that very small, nearly imperceptible effects can trigger an eruption once stresses build up to a critical level. Even if one had perfect measurements, it is unlikely that perfect predictions could be made. What would really be useful to know is what degree of predictability could be achieved given a certain level of observations and numerical modeling.

7

Shocks in heliophysics

MERAV OPHER

This chapter describes several types of shocks, focusing on the ones that prevail in the heliosphere. The chapter addresses why shocks happen, describes the Rankine–Hugoniot jump conditions, reviews the classification of shocks, discusses contact and tangential discontinuities, and closes with a discussion of the physical processes yet to be explored for shocks. The sections contain specific examples such as coronal shocks, shocks driven by coronal mass ejections, planetary shocks, and the termination shock and heliopause. For further reading, we refer to Burlaga (1995), Kallenrode (2004), Gurnett and Bhattacharjee (2005), Goedbloed and Poedts (2004), Kulsrud (2005), and Opher (2009), upon whose work much of this chapter is based.

7.1 Introduction

Shock waves are an important manifestation of solar activity. They play an important role in space weather because they can accelerate particles to high energies, creating solar energetic particle (SEP) events, and produce storms at Earth (Gopalswamy *et al.*, 2001). They also produce radio emission at various distances from the Sun, which allows us to track shock propagation throughout the corona and heliosphere.

Near the Sun, shocks are believed to be mainly driven by solar disturbances such as coronal mass ejections (CMEs). The CMEs and the SEP events associated with them are of particular importance for space weather because they endanger human life in outer space and pose major hazards for spacecraft. High-energy solar protons ($>$100 MeV) can be accelerated within a short period of time ($\sim$1 h) after the initiation of CMEs, which makes them difficult to predict, and therefore they pose a serious concern for the design and operation of both manned and unmanned space missions.

Heliophysics: Space Storms and Radiation: Causes and Effects, eds. Carolus J. Schrijver and George L. Siscoe. Published by Cambridge University Press.

Measurements by Haggerty and Roelof (2002), Simnett *et al.* (2002), and Falcone *et al.* (2003), Mewaldt *et al.* (2003), Tylka *et al.* (2003) are consistent with SEP production by CME-driven shocks beginning very near the Sun, at distances of only a few solar radii. The most efficient particle acceleration takes place near the Sun (at distances of 2–15 $R_\odot$), and the fastest particles can escape upstream of the shock, reaching the Earth shortly after the initiation of the CME (~1 h). This mechanism, also known as diffusive shock acceleration, is well supported by both theory (Lee, 1997; Ng *et al.*, 1999, 2003b; Zank *et al.*, 2000) and observations (Kahler, 1994; Tylka *et al.*, 1999, 2005; Cliver *et al.*, 2004). The CME-driven shock continues to accelerate particles, and the shock passage is often accompanied by an enhancement of the energetic-particle flux. These theories, however, have been debated within the community (Reames, 1999, 2002; Tylka, 2001), since very little is known from observations about the dynamic properties of CME-driven shock waves in the low corona soon after the onset of the eruption (Gopalswamy *et al.*, 2001). How soon after the onset of a solar eruption a shock wave forms, and how it evolves over time, depends largely on how this shock wave is driven by the erupting coronal magnetic field and by the background solar wind into which it propagates. As proposed by Tylka *et al.* (2005), the shock geometry plays a significant role in the spectral and compositional variability of SEPs above ~30 MeV/nucleon. The acceleration mechanism depends on whether the shock is parallel or perpendicular. As we will describe in Section 7.4.2, shocks are considered parallel if the angle θ_{Bn} between the normal of the shock and the magnetic field is 0° and perpendicular if $\theta_{Bn} = 90°$.

Shocks also can be driven in the corona by flares (e.g. Vršnak and Cliver, 2008). Signatures of coronal shock waves are radio type II bursts and Moreton waves. The type II burst is a narrow-band radio emission excited at the local plasma frequency by a fast-mode MHD shock. As the shock propagates outwards through the corona, the emission drifts slowly towards lower frequencies due to decreasing ambient density. Radial velocities, inferred from the emission drift rates by using various coronal density models, are found to be on the order of 1000 km/s. The Moreton wave is a large-scale wave-like disturbance of the chromosphere, observed in Hα, which propagates out of the flare site at velocities also on the order of 1000 km/s (Vršnak and Cliver, 2008). A shock could be driven over large distances by the eruption of structures that evolve into CMEs, or can be ignited by a smaller-scale process associated with the flare energy release, e.g. by the expansion of hot loops or by small-scale eruptions.

Shocks are also present in the interplanetary medium and have been detected at 1 AU and beyond (von Steiger and Richardson, 2006; Richardson *et al.*, 2006). Shocks in the interplanetary medium consist of interplanetary coronal mass ejections (ICMEs) driving a shock. While ICMEs are sometimes obvious features in *in-situ* observations at 1 AU, many ICMEs are difficult to identify with certainty. At larger distances, the effects of ICME and solar wind evolution and ICME interactions with the ambient solar wind and other ICMEs complicate their identification even more (Richardson *et al.*, 2006). Several ICMEs have been traced

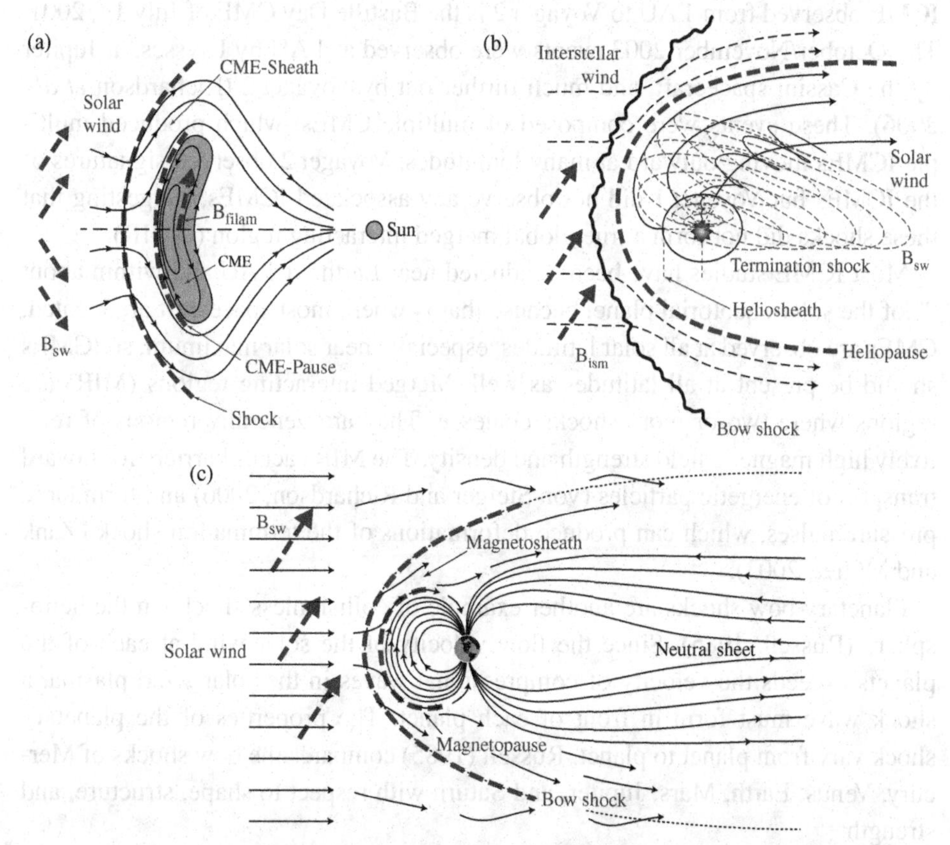

Fig. 7.1. Schematic comparison of shocks around CMEs, the heliosphere, and the magnetosphere. The figure shows some of the types of shocks and sheaths that exist in the heliosphere and their universal basic structures: (a) a CME, (b) the outer heliosphere, and (c) Earth's magnetosphere. The same basic structures appear: shocks where the solar wind becomes subsonic; the sheaths that separate the subsonic solar wind from the obstacle ahead; and the "pause" where there is a pressure equilibrium between the subsonic solar wind and the obstacle's environment. In the case of a CME these three structures are the shock, CME-sheath, and CME-pause, and the obstacle is the magnetic filament that drives the CME. In the case of the outer heliosphere the structures are the termination shock, heliosheath, and heliopause. The obstacle is the interstellar wind and the magnetic field it is carrying. If the interstellar wind is supersonic there is an additional shock, the bow shock. In the case of the Earth's magnetosphere the structures are the shock, the magnetosheath, and the magnetopause and the obstacle is the Earth's dipolar magnetic field.

from the inner to the outer heliosphere using enhanced He/H ratios as tracer (Paularena *et al.*, 2001; Richardson *et al.*, 2002), observed at 1 AU by the Wind spacecraft, at 5.2 AU by Ulysses, and at 58 AU by Voyager. These spacecraft were separated by 200° longitude so perhaps they observed different parts of the ICME or different ICMEs from the same active region. Another example of an ICME observed from 1 AU to Voyager 2 is the Bastille Day CME of July 14, 2001. The October/November 2003 events were observed at 1 AU by Ulysses, at Jupiter by the Cassini spacecraft, and much further out by Voyager 2 (Richardson *et al.*, 2006). These events were composed of multiple CMEs, which produced multiple ICMEs moving outward at many longitudes. Voyager 2 observed signatures of the ICMEs but Voyager 1 did not observe any associated ICMEs, suggesting that these shocks did not form a true global merged interacting region (GMIR).

Most ICME studies have been conducted near Earth, at 1 AU and within about 7° of the solar equatorial plane, because that is where most spacecraft are located. CMEs are observed at all solar latitudes, especially near solar maximum, so ICMEs should be present at all latitudes as well. Merged interacting regions (MIR) are regions where two or more shocks coalesce. They are generally regions of relatively high magnetic field strength and density. The MIRs act as barriers for inward transport of energetic particles (von Steiger and Richardson, 2006) and form large pressure pulses, which can produce deformations of the termination shock (Zank and Müller, 2003).

Planetary bow shocks are another example of collisionless shocks in the heliosphere (Russell, 1985). Since the flow velocity of the solar wind at each of the planets exceeds the velocity of compressional waves in the solar wind plasma, a shock wave must form in front of each planet. The properties of the planetary shock vary from planet to planet. Russell (1985) compares the bow shocks of Mercury, Venus, Earth, Mars, Jupiter, and Saturn with respect to shape, structure, and strength.

The farthest shock in the heliosphere is the termination shock where the solar wind slows down in the interaction with the interstellar medium. Voyager 2 crossed the undulating termination shock several times (Richardson *et al.*, 2008; Stone *et al.*, 2008).

7.2 Why shocks happen: non-linear steepening and shocks

In the small-amplitude limit, the profile of a magnetohydrodynamic (MHD) wave does not change as it propagates, but even a small-amplitude wave will eventually distort due to *wave steepening*. The wave steepening happens when gradients of pressure, density, and temperature become so large that dissipative processes (e.g. viscosity, thermal conduction) are no longer negligible. In the steady state, a steady wave-shape – a *shock wave* – is formed in which the steepening effect of non-linear convective terms balance the broadening effects of dissipation. The shock waves move at speeds larger than the ambient intrinsic speed, which for magnetized ionized matter in the heliosphere is the magnetosonic speed. If the shock moves much faster than the magnetosonic wave, it is called a strong shock; if it moves just slightly faster, it is called a weak shock. The dissipation inside the shock front leads to a gradual conversion of the energy being carried by the wave into heat. In the heliospheric plasma, we have collisionless shocks in which the thermalization happens through wave–particle interactions.

The strength of a shock is given by the Mach number $M = v_1/\gamma^{1/2}v_s$, where v_1 is the shock velocity with respect to the flow and $\gamma^{1/2}v_s$ is the sound speed ahead of the shock.

An example is the adiabatic propagation of sound waves. The propagation velocity of a sound wave is given by $v_s^2 = \mathrm{d}P/\mathrm{d}\rho$. For an adiabatic equation of state P/ρ^γ = constant, so that $v_s \propto P^\alpha$, where $\alpha = (\gamma+1)/2\gamma$. Figure 7.2 shows a sketch of the steepening of a pressure (sound) wave.

A propagating wave described by the ideal fluid equations leads to infinite gradients in a finite time. There is no solution for the ideal MHD equations. This is not surprising: ideal equations are valid when scales of variations are larger than the mean-free path. The breakdown in ideal equations occurs in a very thin region, while the fluid equations are valid everywhere else. In this very thin region, it is difficult to describe the plasma in detail. The simple picture is a discontinuity dividing two roughly uniform fluids. An important aspect is that the simple picture of a discontinuity dividing two roughly uniform fluids is not usually applicable in a plasma. Shocks can involve turbulence for example. For this initial discussion, we make the simplifying assumption that there is a planar discontinuity of zero thickness that separates two uniform fluids, as depicted in Fig. 7.3. We also assume that the shock is stationary.

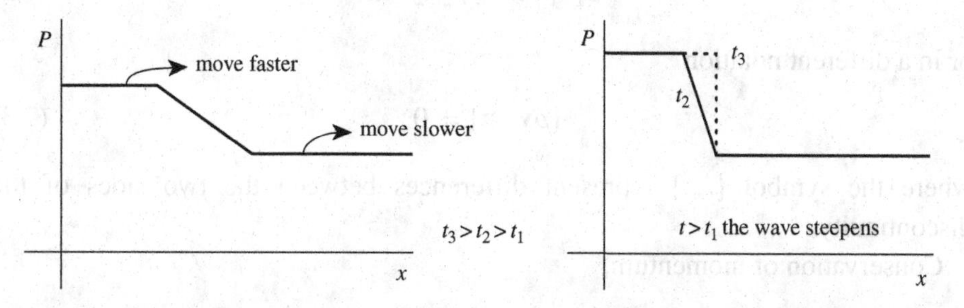

Fig. 7.2. Diagram of the steepening of a wave. Three phases are shown: t_1 in the left panel, and $t_{2,3}$ in the right panel, with $t_3 > t_2 > t_1$.

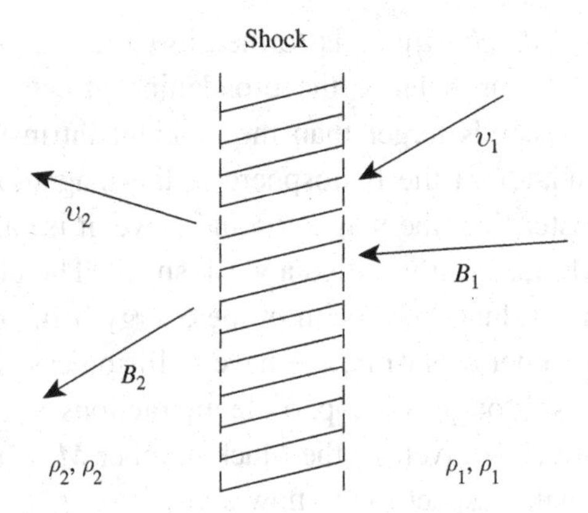

Fig. 7.3. Diagram showing the region upstream (right) and downstream (left) of a shock.

Figure 7.3 sketches the regions upstream and downstream of a shock. Region 2 is downstream and region 1 is upstream of the shock. The transition must be such as to conserve mass, magnetic flux, and energy. The MHD jump conditions are independent of the physics of the shock itself and are known as the *Rankine–Hugoniot jump conditions.*

7.3 Rankine–Hugoniot jump conditions

It is straightforward to obtain the Rankine–Hugoniot jump conditions from the MHD equations. Assuming steady state in the frame of reference of the shock, the equation for the conservation of mass,

$$\frac{\partial \rho}{\partial t} + \nabla \cdot (\rho \mathbf{v}) = \mathbf{0}, \tag{7.1}$$

gives

$$\rho_1 \mathbf{v_1} \cdot \mathbf{n} = \rho_2 \mathbf{v_2} \cdot \mathbf{n}, \tag{7.2}$$

or in a different notation

$$\{\rho \mathbf{v} \cdot \mathbf{n}\} = 0, \tag{7.3}$$

where the symbol $\{\ldots\}$ represent differences between the two sides of the discontinuity.

Conservation of momentum,

$$\frac{\partial(\rho \mathbf{v})}{\partial t} + \nabla \cdot \left[\rho \mathbf{v}\mathbf{v} + \left(p + \frac{B^2}{2\mu_0}\right)\mathbf{I} - \frac{\mathbf{B}\mathbf{B}}{\mu_0}\right] = 0, \tag{7.4}$$

yields

$$\left\{\rho\mathbf{v}(\mathbf{v}\cdot\mathbf{n}) + \left(\mathbf{p} + \frac{\mathbf{B^2}}{\mathbf{2}\mu_0}\mathbf{n} - \frac{\mathbf{B}}{\mu_0}(\mathbf{B}\cdot\mathbf{n})\right)\right\} = 0. \tag{7.5}$$

Conservation of energy,

$$\frac{\partial}{\partial t}\left(\frac{1}{2}\rho v^2 \frac{P}{\gamma - 1} + \frac{B^2}{2\mu_0} + \nabla\cdot\left(\frac{\mathbf{1}}{\mathbf{2}}\rho\mathbf{v^2v}\right) + \frac{\gamma\mathbf{P}}{\gamma - \mathbf{1}}\mathbf{v} + \frac{\mathbf{1}}{\mu_0}\mathbf{E}\times\mathbf{B}\right) = 0, \tag{7.6}$$

results in

$$\left\{\left(\frac{1}{2}\rho v^2 + \frac{\gamma P}{\gamma - 1}\right)(\mathbf{v}\cdot\mathbf{n}) + \frac{\mathbf{1}}{\mu_0}(\mathbf{E}\times\mathbf{B})\cdot\mathbf{n}\right\} = 0. \tag{7.7}$$

Conservation of magnetic flux,

$$\nabla\cdot\mathbf{B} = \mathbf{0}, \tag{7.8}$$

gives

$$\{\mathbf{B}\cdot\mathbf{n}\} = \mathbf{0}. \tag{7.9}$$

The equation

$$\nabla\times\mathbf{E} = -\frac{\partial\mathbf{B}}{\partial\mathbf{t}} \tag{7.10}$$

can be written as

$$\{\mathbf{E}\times\mathbf{n}\} = \mathbf{0}. \tag{7.11}$$

Let us consider, now, the normal n and the tangential t components relative to the shock surface so that the jump conditions can be written as

$$\left\{\rho v_n^2 + P + \frac{B_t^2}{2\mu_0}\right\} = 0, \tag{7.12}$$

$$\left\{\rho\mathbf{v_t v_n} - \frac{\mathbf{B_t}}{\mu_0}\mathbf{B_n}\right\} = 0, \tag{7.13}$$

$$\left\{\left(\frac{1}{2}\rho v^2 + \frac{\gamma P}{\gamma - 1} + \frac{B^2}{\mu_0}\right)u_n - (\mathbf{v}\cdot\mathbf{B})\frac{B_n}{\mu_0}\right\} = 0, \tag{7.14}$$

$$\{B_n\} = 0, \tag{7.15}$$

$$\{\mathbf{v_n}\times\mathbf{B_t} + \mathbf{v_t}\times\mathbf{B_n}\} = \mathbf{0}. \tag{7.16}$$

Equations (7.12)–(7.16) are the Rankine–Hugoniot jump conditions that describe all types of shocks.

7.4 Definition and classification of shocks

From the Rankine–Hugoniot jump conditions, discontinuities and shocks are classified as summarized in Table 7.1. In Sections 7.4.1 and 7.4.2 we describe discontinuities and shocks, respectively.

7.4.1 Contact and tangential discontinuities

Discontinuities can be classified as either contact or rotational discontinuities. Contact discontinuities happen when there is no flow across the discontinuity, i.e. $v_n = 0$ and $\{\rho\} \neq 0$. A classic example is the contact discontinuity of a mix of vinegar and olive oil. If $\{B_n\} \neq 0$ at a contact discontinuity then only the density changes across the discontinuity, which is rarely observed in plasmas. A tangential discontinuity occurs when $\{B_n\} = 0$, then $\{v_t\} \neq 0$ and $\{B_t\} \neq 0$ and $\{p + B^2/2\mu_0\} = 0$. This means that the fluid velocity and magnetic field in this case are parallel to the surface of the discontinuity but change in magnitude and direction, and that the sum of thermal and magnetic pressures is constant.

An example of a tangential discontinuity is the heliopause. The motion of the solar system through the interstellar medium with a velocity of ~26 km/s compresses the heliosphere, producing a comet-like shape with an extended tail (Fig. 1.4 in Vol. I). The heliosphere is created by the supersonic solar wind, which abruptly slows, forming a termination shock as it approaches contact with the interstellar medium at the heliopause. Beyond the termination shock, the solar wind is gradually deflected tailward. As the Sun rotates, the solar magnetic field carried outward by the solar wind forms a spiral, becoming almost completely azimuthal in the outer heliosphere. Beyond the heliopause, the interstellar wind contains neutral atoms, mainly hydrogen and helium, and ions that carry the frozen-in interstellar magnetic field. Figure 7.4 shows a meridional cut from a simulation of the heliosphere.

Another example of tangential discontinuities is planetary magnetospheres (e.g. Fig. 10.3 in Vol. I). If there is not much reconnection (i.e. for a "closed magetosphere" as in Fig. 10.1 in Vol. I) $\{v_n\} \sim 0$ and $\{B_n\} \sim 0$, so that solar wind plasma and magnetic field do not penetrate into the magnetosphere.

Table 7.1. *Classification of shocks based on the Rankine–Hugoniot jump conditions*

	$v_n = 0$	$v_n \neq 0$
$\{\rho\} = 0$	—	rotational discontinuity
$\{\rho\} \neq 0$	contact discontinuitiy	shock wave

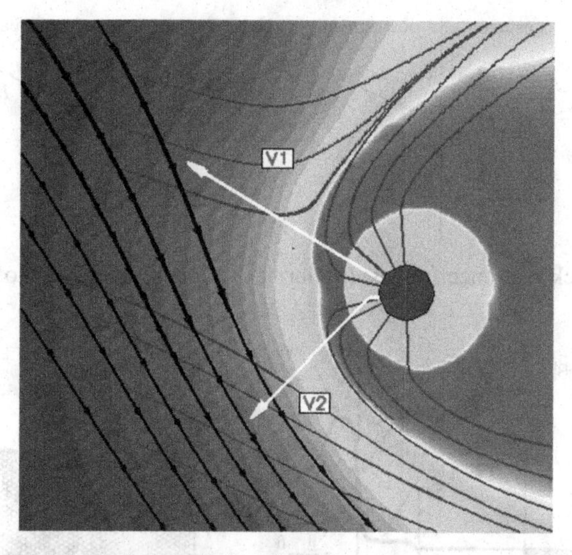

Fig. 7.4. Meridional cut from a heliosphere simulation including the plasma and the neutral H atoms (Opher, 2009). The contours are the plasma temperature. The blue region is the region beyond the heliopause; the red, the heliosheath; and the central green area is the region upstream of the solar-wind termination shock. The black lines are the interstellar magnetic field and the grey lines are the plasma streamlines. The (projected) trajectories of the Voyager 1 and 2 spacecraft are also indicated. See also Plate 3 in the color-plate section.

A rotational discontinuity occurs when $\{v_n\} \neq 0$ and $\{\rho\} = 0$. From the jump conditions this implies that $\{v_n\} = 0$ and $\{p + B_t^2/2\} = 0$ so $\mathbf{v}_1 \cdot \mathbf{n} = \mathbf{v}_2 \cdot \mathbf{n} = v_n$ and $\rho_1 = \rho_2$. After some math, we find that $v_n^2 = B_n^2/\mu_0\rho$, and that B_t remains constant in magnitude but rotates in the plane of the discontinuity. A rotational discontinuity occurs, for example, if the reconnection rate between the solar wind magnetic field and the planetary magnetic field is substantial, so that the plasma can penetrate significantly into a magnetosphere. In this case, the magnetopause becomes a rotational discontinuity.

7.4.2 Shock waves

Shock waves are characterized by fluid flows across the discontinuity $\{v_n\} \neq 0$ and a non-zero jump discontinuity in at least the plasma density. There are two frames of reference for MHD shocks: (i) normal incident frame (coordinate system moving along the shock front with speed v_t) and (ii) the de Hoffman–Teller frame (coordinate system in which the plasma moves parallel to the magnetic field on both sides and the reference frame moves parallel to the shock front with the de Hoffman–Teller speed). Figure 7.5 shows these two reference frames.

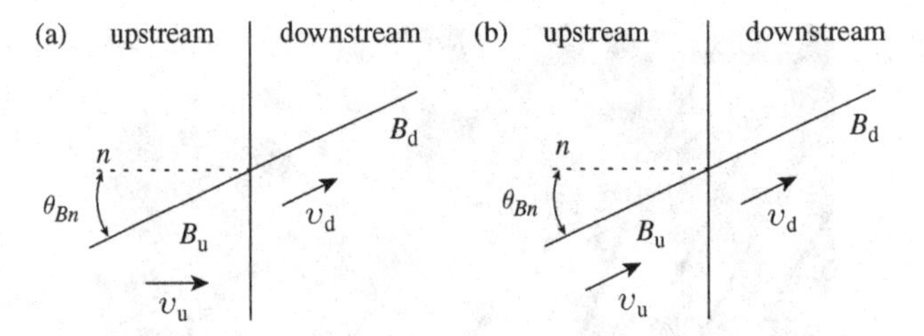

Fig. 7.5. Shock reference frames: (a) normal-incident and (b) de Hoffman–Teller frame.

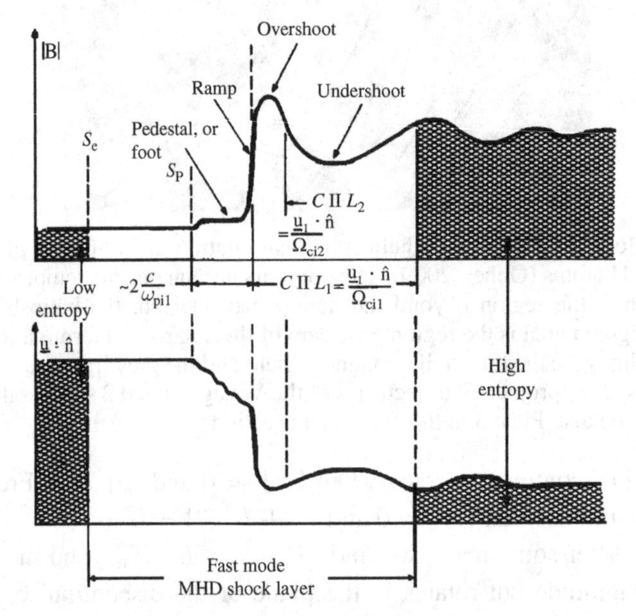

Fig. 7.6. Substructure terminology of supercritical, fast mode, collisionless shock layer. (From Scudder *et al.*, 1986a.)

The Rankine–Hugoniot jump conditions have 12 unknowns. Four upstream parameters are specified: ρ, v_s, B_t, and B_n, so we have seven equations for eight unknowns. Therefore we need to specify one more quantity, namely the strength of the shock $\delta = \rho_2/\rho_1$.

Figures 7.6 and 7.7 show observations of MHD shocks. Figure 7.6 shows the Earth's bow shock (shown at a distance 15.4 R_E upstream from Earth). In this example $\theta_1 = 76°$ (between **B** and **n**), $v_1 = 294$ km/s $> v_A = 37.8$ km/s. Figure 7.7 shows the crossing of the termination shock by Voyager 2 in August of 2007.

We differentiate between three types of shocks: perpendicular, parallel, and oblique shocks. They are classified accordingly with the angle θ_{Bn} between the magnetic field direction and the shock normal. A perpendicular shock has $\theta_{Bn} \equiv 90°$, a parallel, $\theta_{Bn} \equiv 0°$, and an oblique shock propagates with θ_{Bn} between 0° and 90°. We describe below several examples of shocks in the heliosphere.

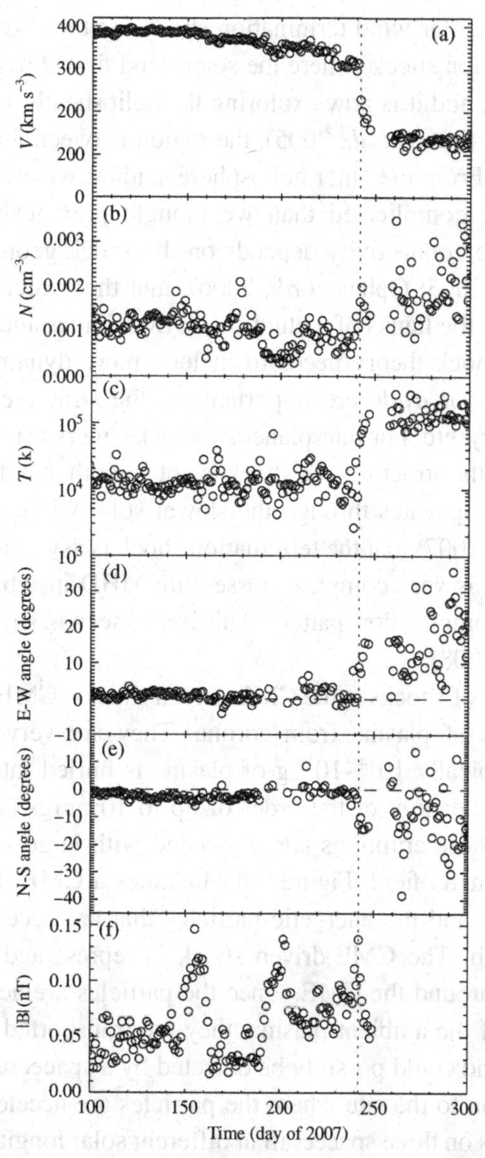

Fig. 7.7. Crossing of the termination shock by Voyager 2. Daily averages of solar wind speed V (a), density N (b), temperature T (c), east–west flow angle (d), north–south flow angle (e), and magnetic field magnitude (f). Flow angles are in the RTN coordinate system, where R is radially outwards, T is parallel to the plane of the solar equator and positive in the direction of the Sun's rotation, and N completes a right-handed system. The east–west angle is the angle in the R–T plane and the north–south angle is the angle out of the R–T plane. The dashed line shows the termination shock crossing, where the speed decreases by a factor of about two, the density increases by a factor of two, the proton temperature increases to near 100 000 K, and the flow is deflected consistent with flow away from the nose direction of the heliosphere, that is, the direction toward the local interstellar medium flow. (From Richardson *et al.*, 2008.)

We start with the solar wind termination shock. At the end of 2004, Voyager 1 crossed the termination shock, where the solar wind first slows its approach to the interstellar medium, and it is now exploring the heliosheath (Burlaga *et al.*, 2005; Decker *et al.*, 2005; Stone *et al.*, 2005), the region between the termination shock and the heliopause. From the outer heliosphere studies, we are learning that shock structures are more complicated than we thought previously: the observations reveal that magnetic connectivity depends on the shock geometry (Jokipii *et al.*, 2004; Stone *et al.*, 2005; Opher *et al.*, 2006), that the source of anomalous cosmic rays could be at the flanks of a blunt shock (McComas and Schwadron, 2006), that the diffusive shock theory needs to include more dynamic effects, and that the details of the solar cycle are important for the structure of the termination shock, its asymmetry, etc. For interplanetary shocks, works such as by Manchester (2003) showed that the structure of a shock is not smooth, but develops dimple-like indentations as it propagates through the slower solar wind. Similarly, we found (Opher *et al.*, 2006, 2007) that the termination shock is asymmetric. We also found that the heliosheath is very complex, possessing MHD instabilities (Opher *et al.*, 2003, 2004) and a complex flow pattern. This is also seen in sheets of interplanetary shocks (Liu *et al.*, 2008).

Another category of shocks is the CME-driven shocks. CMEs have been defined as large explosions of plasma from corona. They are very-large-scale coronal events in which typically 10^{14}–10^{16} g of plasma is hurled into the interplanetary space with a kinetic energy of the order of up to 10^{32} ergs (see Chapter 6). It is also believed that these eruptions are associated with large-scale reconfiguration of the coronal magnetic field. Figure 7.8 illustrates a CME, the geometry of the CME-driven shock, and the energetic particles that are accelerated at the shock that arrives at Earth. The CME-driven shock is represented by the thick black line, which wraps around the CME. Once the particles are accelerated out of the suprathermal tail of the ambient plasma, they are transported along the magnetic field line to 1 AU and could possibly be detected by a spacecraft if that has a good magnetic connection to the site where the particles are accelerated. The detected time profile of SEPs on three spacecraft at different solar longitudes are also shown at the corresponding locations in Fig. 7.8.

The propagation of a CME from the inner corona to 1 AU has been numerically modeled (e.g. Usmanov and Dryer, 1995; Wu *et al.*, 1999; Groth *et al.*, 2000; Odstrcil *et al.*, 2002; Riley *et al.*, 2002; Manchester *et al.*, 2004a) with increasing sophistication. Manchester (2003), for example, showed that a CME shock is distorted in its interaction with the stratified solar wind forming a dimple in the slow solar wind. Studies such as by Manchester (2003) (see also Manchester *et al.*, 2004a,b; Lugaz *et al.*, 2005b) indicate how the shock is distorted as it propagates from the Sun towards the Earth.

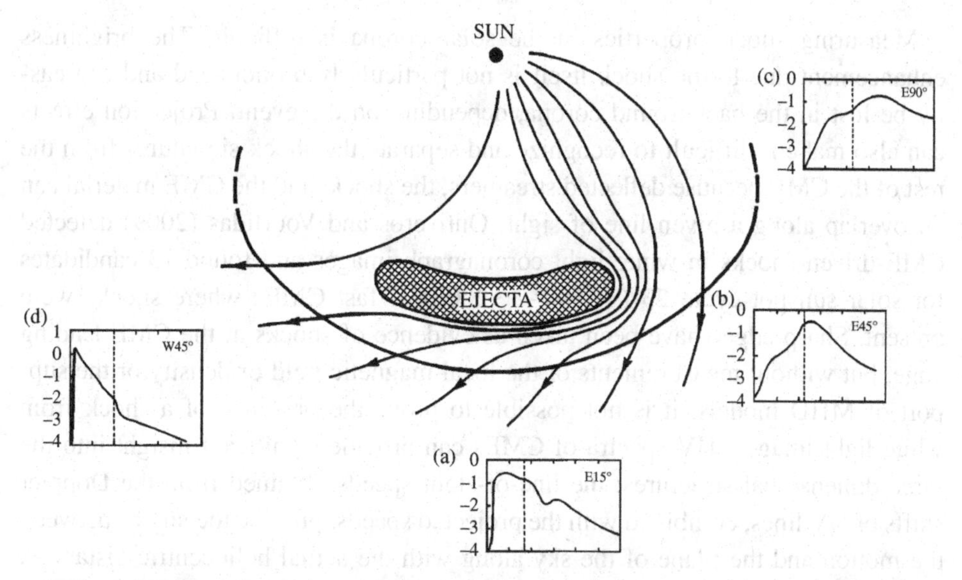

Fig. 7.8. Representative profiles of 20 MeV proton events for different positions of the observatory with respect to a shock. The draping of the field lines around the ejecta is only a suggestion. (From Cane *et al.*, 1988.)

The CME-driven shock is very effective in generating strong geomagnetic activity at Earth when there is (a) a strong sustained southward B_z (lasting more than ~10 h) and (b) a substantial pressure increase associated with the CME-driven shock that compresses the magnetosphere. Manchester *et al.* (2005) studied the evolution of CME-driven shock in a modeled ambient solar wind. Their result showed that the range of θ_{Bn} (the angle between the magnetic field and shock normal) changes for magnetic field lines at different solar latitude. They also conclude that the magnetic field line bends around the ejecta, which could also contribute to the acceleration of energetic particles (Kóta *et al.*, 2005). Their solar wind was based on the model developed by Groth *et al.* (2000), a highly simplified model based on a dipolar solar magnetic field and ad-hoc heating function.

Type II radio emission has long been associated with propagating shock waves in the solar corona and interplanetary medium (Wild and Smerd, 1972; Cane *et al.*, 1981; Nelson and Melrose, 1985; Lengyel-Frey *et al.*, 1997). There is strong evidence that at least some type II radio emission is generated in electron foreshock regions upstream of the shock waves ahead of the CMEs (Cane *et al.*, 1981; Reiner *et al.*, 1998, 1999; Bale *et al.*, 1999; Reiner and Kaiser, 1999). As summarized by Gopalswamy *et al.* (2003), type II radio bursts are thought to be produced by shocks that accelerate not only electrons but also ions, so we should expect a high degree of association between type II bursts and SEP events (Kahler *et al.*, 1984, Cane *et al.*, 1990).

Measuring shock properties in the solar corona is difficult. The brightness enhancement due to the shock itself is not particularly pronounced and can easily be lost in the background corona, depending on the event. Projection effects can also make it difficult to recognize and separate the shock signatures from the rest of the CME because deflected streamers, the shock, and the CME material can all overlap along a given line of sight. Ontiveros and Vourlidas (2009) detected CME-driven shocks in white-light coronagraph images and found 13 candidates for solar sunspot cycle 23 (~1997 – ~2008) of fast CMEs where shocks were present. Sharp edges have been taken as evidence of shocks at the CME leading edge, but without measurements of the local magnetic field or density or the support of MHD models, it is not possible to prove the presence of a shock from white-light images. UV spectra of CMEs can provide significant insight into the three-dimensional structures: the line-of-sight speeds obtained from the Doppler shifts of UV lines, combined with the projected speeds, provide the angles between the motion and the plane of the sky along with the actual heliocentric distances. Spectra can also diagnose the presence of shocks at the CME fronts: line profiles carry information on the bulk expansion and thermal status of the CME material. The number of UV observations of CMEs is not as large as those by white-light coronagraphs because the Ultraviolet Coronagraph Spectrometer (UVCS) can only observe CMEs through its long narrow slit. In another study, Ciaravella *et al.* (2006) looked at 22 halo or partial halo CMEs observed by UVCS and found signatures of shocks in front of halo CMEs.

7.5 Physical processes in shocks and future work

Some aspects of shocks for which continued exploration is necessary involve questions such as these: Which kinds of flows exist in shocks and their accompanying sheaths? How do magnetic effects influence shocks? What is the role of MHD instabilities? How are shock geometries affected by their interaction with plasma and field through which they propagate? Which types of instabilities happen in CME sheaths, in the Earth's magnetosheath, and in the heliosheath? In particular, as we mentioned above, the processes in the sheaths are not well known. As discussed by Siscoe and Odstrcil (2008), the ICME sheaths differ from magnetosheaths. Siscoe and Odstrcil (2008) argue that the ICME sheaths are thinner as compared to magnetosheaths. Their assumptions, however, do not include three-dimensional, magnetic effects or a structured solar wind.

Another important question is how far CME-driven shocks persist. Interplanetary shocks are observed throughout the heliosphere, but it remain a challenge to model their evolution as they move through a realistic three-dimensional solar wind from the Sun to Earth and beyond to the outer heliosphere. Only when we can do that, can we access aspects such as the longitudinal extension of ICME shocks and how their magnetic structure evolves.

It is also crucial to constrain the Alfvén wave profile in the lower corona in order to know where shocks first form. Due to the absence of such observations, studies rely heavily on MHD models, which also need to be advanced (Evans *et al.*, 2008).

Another important question is which non-linear processes dominate shock dissipation. This could explain how far out CME-driven shocks persist, and why. Another question is why some shocks are supposed to merge and others can apparently pass through each other. Lugaz *et al.* (2005b), for example, use a three-dimensional MHD model to analyze how two CMEs interact: two similar CMEs were launched and after an initial phase the two shocks merged. They found that a reverse shock is driven after the collision of the two magnetic clouds.

We still lack the knowledge of how the solar magnetic field is wrapped around the CMEs as it evolves towards Earth. The transport of SEPs will depend on the magnetic connectivity. Acceleration of SEPs by the front shock of a very fast CME has been successfully simulated (Sokolov *et al.*, 2006), following a magnetic field line from the three-dimensional MHD model of Manchester *et al.* (2004a). Similar studies with different velocities and initiation processes of CMEs are needed. It is also not clear where in the shock front radio emission is produced. As commented above, Reames (2002) shows how different spectra of particles can be produced depending on the connectivity of the field lines from the Sun to the shock and to the observer. The connectivity as shown by Opher *et al.* (2006) plays a crucial role to explain the beam of particles seen by Voyagers 1 and 2. This aspect needs to be explored in future studies of shocks, both near the Sun as well as in interplanetary shocks.

Other questions are: How does the large-scale magnetic field modify the shock (shape, properties) in the corona? When, and at what distance, does the shock escape from the driver? How are the CME sheath flows modified by the large-scale magnetic field? Does reconnection affect shock structures? For example, Liu *et al.* (2008) seem to have detected a reverse shock around where reconnection happens in the back of a CME. The presence of reverse shocks around expanding CME and ICME still needs to be explored.

These questions illustrate the richness of the study of shocks in the heliosphere. Much still needs to be explored in order to help us understand the formation and evolution of these shocks.

from the Sun to Earth and beyond to the outer heliosphere. Any knowledge [illegible] that can be [illegible] the longitudinal [illegible] of CME shocks and flows [illegible] the structure [illegible]

It is also crucial to constrain the [illegible] wave [illegible] to [illegible] the [illegible] Due to the absence of [illegible] studies [illegible] MHD models [illegible] (2008).

[illegible] [illegible] non-linear [illegible] [illegible] CME-driven shocks [illegible] and why some [illegible] shocks [illegible] and others can apparently pass through each other [illegible] et al. (2008) [illegible] MHD model [illegible] CME [illegible] CME [illegible] shocks [illegible] reverse shock [illegible] of the two [illegible]

[illegible] how the solar magnetic field is [illegible] around the CMEs as it [illegible] Earth. The transport of SEPs with [illegible] magnetic connectivity. Acceleration of SEPs by [illegible] shocks [illegible] CME [illegible] (Sokolov et al. 2004) [illegible] magnetic field [illegible] MHD model [illegible] et al. (2004). Similar studies with different [illegible] processes of CMEs [illegible] It is also [illegible] emission is produced. As [illegible] (200[illegible]) [illegible] of [illegible] can be [illegible] depending on [illegible] the field lines from the [illegible] Voyagers [illegible] 2 [illegible] to explain the [illegible] by Voyagers 1 and 2 [illegible] captured in [illegible] studies [illegible] near the Sun [illegible] shocks.

Other questions [illegible] how [illegible] magnetic field [illegible] the shock [illegible] When, and at what distance [illegible] modified by the large-scale magnetic field [illegible] affect shock structures? For example, [illegible] [illegible] have detected a reverse shock [illegible] should be expected.

These questions [illegible] the study of shocks in [illegible] still needs to be [illegible] in order to better understand the formation and [illegible]

8

Particle acceleration in shocks

DIETMAR KRAUSS-VARBAN

In this chapter, we review the basic principles and characteristics of shock acceleration. After a brief description of the pertinent kinetic scales at shocks and a discussion of heating versus acceleration, we outline the different mechanisms that contribute to accelerating charged particles at shocks. The main emphasis throughout this chapter is on ions, and more importantly, on protons. Acceleration of other ion species or electrons is mentioned in passing and when contrasting interesting differences. Also, we restrict the discussion to the collisionless and non-relativistic shocks that occur in the heliosphere. Finally, we describe particle acceleration at interplanetary shocks and at the Earth's bow shock in greater detail, and discuss the differences between these two. Throughout the chapter, fundamental, underlying principles, historic results, and current research interests are brought together as much as possible.

8.1 Introduction

More than half a century ago, energetic particle events detected at Earth with energies into the GeV range were for the first time unambiguously associated with activity in the solar corona. While this link was established based on concomitant solar flare observations, in the 1970s and early 1980s evidence accumulated that so-called "gradual" solar energetic particle (SEP) events are actually caused by acceleration at coronal and interplanetary (IP) shocks (Sarris and Van Allen, 1974; Cliver *et al.*, 1982; Mason *et al.*, 1984). The 1970s and early 1980s also saw a rapid development in the theory of charged particle acceleration at shocks, and the realization that virtually all heliospheric shocks carry with them energetic particle populations. At the same time, the theoretical developments made it possible to point to interstellar shocks as the origin of galactic cosmic rays.

To understand the ubiquitous presence of energetic particles it is important to realize that, except for planetary ionospheres and the lowest layers of the Sun's

Heliophysics: Space Storms and Radiation: Causes and Effects, eds. Carolus J. Schrijver and George L. Siscoe.
Published by Cambridge University Press.

corona, most plasmas in the heliosphere are basically collisionless. That is, the mean free path of charged particles is larger than most scales of interest. For example, in the undisturbed solar wind, the mean free path for ions is of the order of 1 AU. The lack of such collisions means that there exists no primary mechanism that forces the particles to assume thermalized Maxwellian distributions. In fact, observed distributions, often on top of thermal (colder) approximate "core" Maxwellians, almost universally contain energetic tails, which usually can be described by power laws. In real-world plasmas, a multitude of processes are responsible for generating such suprathermal and high-energy tails; many times, so-called wave–particle interactions are involved.

As detailed in several chapters of this book, the heliosphere is marked by many different plasma regions, which themselves are bounded by transitions and boundary layers in which energy is continuously, or at times in a bursty fashion, transferred from one form (e.g. magnetic or flow) into another (e.g. thermal and flow). Quite often, these energy transfers are accompanied by the production of energetic particles. Two of the processes that are widely held responsible for the generation of energetic charged particles in the heliosphere, and in astrophysical plasmas in general, are magnetic field line reconnection and acceleration at shocks. It turns out that the two are not necessarily mutually exclusive: even during reconnection events, shocks may play an important role.

In the first section, we follow up on the discussion of the types of MHD shocks from Chapter 7, and illustrate their relation to particle acceleration. Then, we investigate the inherent scales of shocks more closely, and their impact on charged particle motion and heating at shocks. This is followed by a discussion of the pertinent, local and non-local mechanisms that accelerate charged particles at shocks. After that, we put these mechanisms into the wider context of the interaction with their surrounding plasma. We conclude this chapter with a more detailed description of particle acceleration at the Earth's bow shock and at interplanetary shocks, and a brief summary. We should note that the transport of energetic particles, as well as the difficult combined problem of simultaneous acceleration and transport, are described in much more detail in the subsequent Chapter 9 of this book.

8.2 Types of shocks and plasma parameters

In general, in magnetohydrodynamic (MHD) theory, the transition from one region of plasma (with its own density, temperature, magnetic field strength and direction, and flow velocity) to another is accomplished through a number of successive discontinuities, each of them responsible for part of the total change in plasma properties, and each propagating at a peculiar speed. As detailed in the preceding chapter, when the upstream speed in the frame of the discontinuity is larger than one of the characteristic speeds in the plasma, the discontinuity is an MHD shock. MHD shocks convert flow energy into thermal energy, and also change the size and direction of the magnetic field, the density, and the temperature. Because the largest available energy is at the highest speeds, the shocks of most interest to particle acceleration are MHD fast mode shocks, which compress both the density and the magnetic field.

Yet slow mode shocks may also play a role under certain circumstances. It is believed that, at least from a macroscopic viewpoint, magnetic field line reconnection in a collisionless plasma is often achieved via a configuration first described by Petschek (1964; see Vol. I Section 5.3), in which the discontinuities attached to the diffusion region that separate the inflow from the outflow are slow shocks (or more generally perhaps slow-shock-like discontinuities, like so-called time-dependent intermediate shocks (Wu, 1990; Krauss-Varban *et al.*, 1995; Karimabadi *et al.*, 1995). Now, in much of the solar corona, the plasma β is very small. Here, β is the ratio between thermal and magnetic energy; for protons, $\beta = (v_{\text{th}}/v_{\text{A}})^2$, where v_{th} is the thermal particle speed and v_{A} the Alfvén speed. If and when the incoming, low-β plasma thermalizes in the reconnection outflow, a significant fraction of the magnetic energy is converted to thermal energy (Cargill and Priest, 1982). In a collisionless plasma, this process is associated with the collision of ion beams with speed $\sim v_{\text{A}}$, but narrow thermal width $v_{\text{th}} \ll v_{\text{A}}$, and resulting in $v'_{\text{th}} \sim v_{\text{A}}$ after thermalization. In other words, the heating is proportional to the initial $1/\beta$ (Krauss-Varban and Karimabadi, 2006; Krauss-Varban and Welsch, 2007). The turbulence resulting from the beam interaction may then cause further ion and electron acceleration, on top of the already very efficient heating for very low β.

Returning to fast mode shocks, there is a further distinction based on the downstream speed. Although a fast mode shock by definition is super-Alfvénic upstream, the downstream speed may or may not exceed the sound speed. If that speed is exceeded, the fast mode shock is called supercritical (Kennel *et al.*, 1985). While this may bear on the required dissipation mechanisms within the shock, this transition takes place at low Alfvén Mach numbers M_{A} (the ratio of the upstream normal speed to the Alfvén speed), and thus is typically not of interest for shocks that produce any significant amount of energetic particles.

An exception may again be the solar corona, which has very high Alfvén speeds, and for which shocks at initiation are believed to often have quite low Alfvénic Mach numbers. This topic area is still under active investigation, but one reason coronal shocks may be different from those in most other, familiar space plasmas is that, as mentioned above, the plasma β can be exceedingly low, much smaller than unity. In that case, as for the slow shocks, conversion of even a small amount of the flow energy (of the order of v_{A} for $M_{\text{A}} \sim 1$) may increase the thermal energy of the particles by a huge amount (see e.g. Fig. 5 in Kennel *et al.*, 1985) and subsequent processes may lead to considerable high-energetic tails.

On the other end of the spectrum, no matter how high the Mach number, the density and magnetic field compression in fast shocks is limited to $n_2/n_1 = B_2/B_1 = 4$ for a specific heat ratio of $\gamma = 5/3$ (e.g. Kennel *et al.*, 1985). Both of these ratios enter in the shock acceleration processes discussed below.

In addition to the plasma β, which is close to unity in most other common heliospheric environments, and the shock speed or M_A, there is one other variable that plays a crucial role in determining how charged particle acceleration processes unfold: the shock-normal angle θ_{Bn}, which is the angle between the shock normal and the upstream magnetic field. We discuss the impact of θ_{Bn} on acceleration mechanisms in Section 8.4. But first, in the next section, we take a look at the length scales associated with shock transitions and dissipation. In ideal MHD, there is no scale beyond the system size. Scale sizes of discontinuities such as shocks are determined by non-ideal processes. In MHD simulations, these may be explicitly included (e.g. using a defined resistivity and/or viscosity) or may be provided by numeric (grid scale) effects. In more self-consistent, kinetic descriptions of the plasma, several small temporal and spatial scales are present that are associated with the kinetic properties of the charged particles.

8.3 Kinetic shock physics

8.3.1 Scales and particle dynamics

All mechanisms that contribute to the acceleration of charged particles at shocks rely on the particle orbits in the spatial and temporal features of the electric and magnetic field environment of the shock. Roughly speaking, such processes are called kinetic when they go beyond the fluid (MHD) properties of the shock, when they are related to the scales associated with the charged particle motion, and when they require some self-consistent back-reaction between the charged particles and the plasma, e.g. in the form of wave generation. For the highest particle energies, gyroradii are so large that the size of the shock transition and even that of many local waves no longer matters. Conversely, for the thermal and so-called suprathermal particles (just above the thermal energy to several thermal energies), the intrinsic shock scales and locally generated waves do matter. As a consequence, the intrinsic shock scales and associated mechanisms play an important role not only for the general dissipation at the shock (the conversion to thermal energy; see Fig. 8.1 for shock heating as a function of plasma β, shock-normal angle, and Mach number), but also in providing a first, background level of energetic particles from "seed particles" in the thermal and suprathermal energy range.

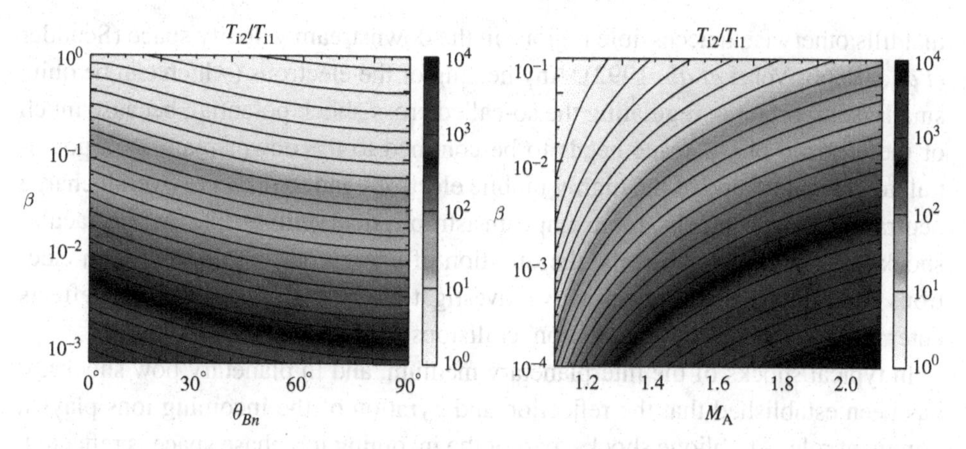

Fig. 8.1. Iso-contours of shock heating, expressed as the ratio between downstream to upstream ion temperature T_{i2}/T_{i1}, as a function of shock-normal angle θ_{Bn} (fixed $M_A = 2$) and Alfvén Mach number M_A (fixed $\theta_{Bn} = 45°$) for low β plasmas. Derived from standard Rankine–Hugoniot conditions for fast shocks, assuming a specific heat ratio $\gamma = 5/3$. The graphs show that for a wide range of angles there can be very substantial downstream heating at sufficiently low plasma β, as present in much of the solar corona. Such extreme heating may help form a seed population for further acceleration. See also Plate 4 in the color-plate section.

The two most important scales in collisionless shocks are the proton inertial length $\lambda_p = c/\omega_p$ and the proton gyroradius $\rho_p = m_p v c/eB$, which are related via the proton β by $\rho_p/\lambda_p = \sqrt(\beta_p)$. Here, c is the speed of light, and $\omega_p = \sqrt{4\pi n e^2/m_p}$ is the proton plasma frequency. The width of the transition for many shocks is either the convected gyro radius v_0/Ω_p or the proton inertial length, whichever is larger (Bale *et al.*, 2003). Here, v_0 is the upstream flow speed in the normal incidence frame (NIF), and Ω_p is the proton gyrofrequency. Exceptions are the almost perpendicular shock, which can be cyclically reforming and steepen to electron scales (Hellinger *et al.*, 2002), and quasi-parallel shocks, which are not only reforming (Burgess, 1989), but at sufficient Mach number have extended regions of steepening upstream waves (e.g. Schwartz *et al.*, 1992), and highly non-linear turbulence downstream (e.g. Krauss-Varban *et al.*, 1994, and references therein).

8.3.2 Heating versus acceleration, and partitioning between ion and electron thermal energy

In most shocks in the heliosphere, the thermalization of the upstream flow is primarily achieved via the ion dynamics, whereas the electrons mostly "just go along for the ride", i.e. they move almost adiabatically, with some subsequent scattering that fills otherwise inaccessible regions in the downstream velocity space (Scudder *et al.*, 1986b; Veltri *et al.*, 1992). Any heating of the electrons (which can be quite small) is important in regulating the so-called cross-shock potential, because much of the electron phase space needs to be confined to the downstream by a potential, to prevent escape of the highly mobile electrons and to preserve overall charge neutrality. In other shocks, for example in astrophysical settings like the spectacular shocks of supernova remnants, the question of energy partitioning between electrons and protons is still under active investigation. Not only do relativistic effects enter at those shocks, but in addition, collisions likely become important.

In typical shocks of the interplanetary medium, and in planetary bow shocks, it has been established that the reflection and gyration of the incoming ions plays a dominant role. At oblique shocks, part of the incoming ion phase space is reflected, but then convected back into the downstream (Gosling *et al.*, 1982; Gosling and Robson, 1985). That is, after reflection, at sufficient Mach number, any upstream-directed parallel velocity of most thermal and even many suprathermal particles is not sufficient to overcome the general plasma drift into the shock. Much of the converted flow energy is initially stored in these gyrating ions, which during this process have attained elevated perpendicular temperatures from the magnetic field jump. Depending on parameters, it may take a while before these protons are thermalized downstream, typically in self-consistently generated Alfvén wave turbulence driven by the temperature anisotropy $T_\perp > T_\parallel$. Generally speaking, the closer to perpendicular the shock, the more difficult it is for both particles and waves to escape upstream.

In contrast, in quasi-parallel shocks reflected (and partially gyrating) ions also play a role, but they can much more easily escape upstream against the flow, because the magnetic field direction is close to the shock normal. There, they generate both obliquely propagating, compressional fast-mode waves, and parallel-propagating Alfvén waves. These waves can grow to large, non-linear amplitudes while convected back towards the shock, where the beam density and growth rate are largest. However, below Alfvén Mach numbers of about $M_\mathrm{A} < 2.8$, the majority of resonantly generated waves are no longer convected back and therefore do not steepen as easily (Krauss-Varban and Omidi, 1991) and do not impact the shock any longer, thus resulting in fewer ions making it upstream to generate waves in the first place. As we will see below, the resulting lower level of turbulence also has a negative impact on ion acceleration to higher energies.

Generally speaking, there is not a clear-cut boundary between particle heating at shocks and particle acceleration. As pointed out in Section 8.1, particle distributions in space plasmas rarely resemble Maxwellians outside some core energy range. Suprathermal and yet more highly energetic tails are an every-day occurrence. Some of the processes that heat particles at thermal energies will also elevate the energy at the upper part of the range. However, such enhancements are often only by a more or less constant, relatively minor factor. An example of this is the adiabatic heating of ions due to the magnetic field compression at the relatively narrow oblique or nearly perpendicular shocks associated with the compression of the plasma ("shock spike event"). While such heating/acceleration is often observed, its simple, approximately adiabatic transformation of the upstream to downstream fluxes distinguishes it from "true" acceleration that may span several orders of magnitude, and will generally also significantly alter the power-law index or general shape of the distribution (e.g. Lario *et al.*, 2005).

8.3.3 Protons, electrons, and minority species

For most heliospheric shocks, proton acceleration is of prime interest. Protons can easily reach energies of tens, if not hundreds of MeV, and as such have a large range of societal consequences such as malfunction or destruction of equipment in space, and posing danger to astronauts or crew and passengers of high-flying aircraft (see Chapters 13 and 14). Electrons, on the other hand, are rarely accelerated to comparable fluxes at these energies, except perhaps at processes well inside the magnetosphere that periodically lead to huge enhancements of trapped populations (see Chapter 11).

Very rarely, IP shocks have such high Mach numbers that the pressure contained in the accelerated charged particles becomes comparable to or even exceeds that of the thermal population (e.g. Terasawa, 1999). These types of shocks are also called "cosmic-ray modified", because the presence of the accelerated particle environment changes the standard Rankine–Hugoniot jump conditions. In such cases, even energetic electrons may become quite important, when the effect of their pressure becomes sizeable in the conservation equations.

Finally, there is significant interest in minority ion species, which can exist in different charge states (see e.g. the excellent review by Reames, 1999). Some ions can have particularly damaging effects, because they are heavy, yet relatively common, and at times selectively accelerated via resonant processes that are still poorly understood. In addition, minority species such as He, O, and Fe, and the ratios of their various charge states, serve as important witnesses of processes occurring during their generation, far away from their points of observation. For example, it is thought that the charge state of solar wind minority ions is at first "frozen in" at particular temperatures and ionization levels in the solar corona. As such, their presence and fluxes provide an important diagnostic of the source plasma, and where in the corona or at what distance from the Sun the material was generated and energized. Also, minority species may be present at vastly different levels, or may be peculiarly enhanced one versus another during particular events, due to acceleration mechanisms that can favor specific charge-to-mass ratios. Finally, some ions such as Fe more easily escape IP shocks, and thus can provide advanced warning of a major SEP event to follow.

In recent years, study of minority ions and their respective, selective acceleration processes has vastly improved our understanding of the origin of energetic ions, and the respective role of solar flares and coronal mass ejection (CME) driven shocks. The reader is referred to the review literature on this topic (Kahler, 1992; Reames, 1999) for more information; in this chapter we will primarily concentrate on protons. The next section discusses the pertinent acceleration processes of protons in supercritical, fast mode shocks in greater detail.

8.4 Particle acceleration mechanisms at shocks

There is a long history of work that attempts to analyze and explain charged particle acceleration at shocks (see e.g. Forman and Webb, 1985; Lee, 2005). For ions, and for the energy range typically observed in the heliosphere, it is well accepted that two distinct acceleration mechanisms are at play.

8.4.1 Kinematic versus kinetic acceleration

The first mechanism, perhaps easier to understand from basic plasma physics principles, is shock-drift acceleration, or one of its variants. This is a "kinematic" process in the sense that the particles simply perform their usual, mostly adiabatic orbits in the given, static or average electric and magnetic fields of the shock transition, neglecting any scattering. From what we know about the shock transition (see Section 8.3.2), it is clear that such an approach only makes sense at oblique shocks. It turns out that one does not have to worry about the complicated quasi-parallel shock transition for this particular process, because it is only effective at highly oblique shocks in the first place. To understand shock-drift acceleration, it is often useful to transform from the (usual) shock frame, in which the upstream magnetic field and flow are in the same plane and the upstream velocity is aligned with the shock normal (normal-incidence frame; NIF), to one in which the upstream flow is fully aligned with the upstream magnetic field (de Hoffmann–Teller frame; HTF; de Hoffmann and Teller, 1950). As explained in more detail below, a particular region of phase space undergoes acceleration simply during the gyro motion in the vicinity of the shock transition.

The second mechanism is of "kinetic" nature, in the sense that wave–particle interactions play the decisive role. As explained in Section 8.3.2, reflected or otherwise energized ions can easily escape into the upstream at quasi-parallel shocks, where they self-consistently generate waves due to the fact that their distributions approximate beams, or ring-beams (beams with an enhanced – or threshold – perpendicular temperature) that carry with them a considerable amount of free energy. Once the waves grow to sufficient amplitudes, the particles scatter in both directions, creating a more "diffuse" distribution that surrounds the entire neighborhood of the shock, from the far upstream to the far downstream. Even in the absence of self-generated waves at the required resonant frequencies, existing turbulence in the medium may be sufficient to achieve some scattering. As explained in more detail below, the particles continue to gain energy by crossing the shock multiple times, until some equilibrium is achieved with escape in either direction. Observations indicate that this process is also operable at oblique and perhaps even at nearly perpendicular shocks. However, how this actually is possible at highly oblique shocks is still under active investigation.

8.4.2 Shock-drift acceleration

As we described in Section 8.3.2, reflection of a portion of the incoming proton phase space, and subsequent convection downstream, is the prime mechanism that eventually provides the dissipation at quasi-perpendicular shocks. Even at highly oblique shocks, a small fraction of these ions will have sufficient parallel speed to make it upstream instead of being convected downstream, but the flux of such ions is strongly diminishing with larger θ_{Bn}, making upstream wave generation increasingly difficult. Although the thermal proton gyroradius is typically comparable to the shock width, and that of suprathermal ions clearly larger than the shock transition, surprisingly, many ions approximately behave adiabatically in simple shock transitions with sufficiently homogeneous upstream and downstream fields. A portion of the ion phase space then gains energy through their gyromotion under consideration of the shock electric fields. The family of such processes is called shock-drift acceleration (SDA).

Consider a steady-state, one-dimensional shock. In this case, in the normal-incidence frame (NIF), there will be an out-of-plane electric field given by the cross-product of the upstream flow and magnetic field $\mathbf{E}_p = -\mathbf{v}_1 \times \mathbf{B}_1/c$. Both the size of the normal component of the flow and the size of the tangential magnetic field aligned with the shock surface change from upstream to downstream, but from Maxwell's equations this necessarily happens such that their cross-product and thus the out-of-plane electric field remain constant. The main difference from the de Hoffmann–Teller frame (HTF) is that in the latter this electric field vanishes (Fig. 8.2). In the NIF, the particles undergo both curvature and gradient drift in the changing magnetic field of the shock transition. These drifts are in opposite directions and perpendicular to the shock plane exactly like the out-of-plane motional electric field, with the gradient drift aligned with $\mathbf{E}_p$ such that the ions

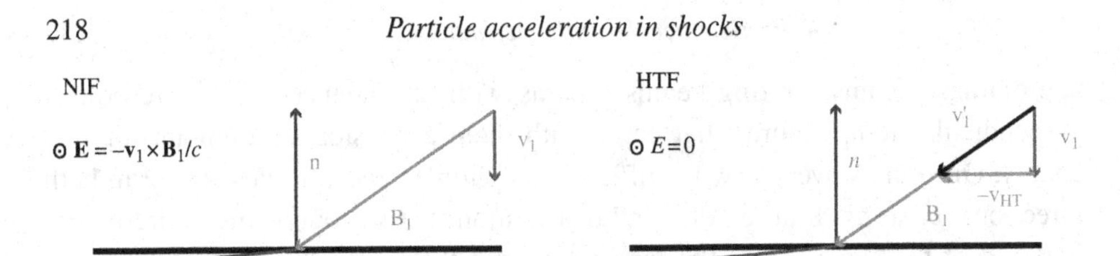

Fig. 8.2. Difference between normal-incidence frame (NIF) and de Hoffman–Teller frame (HTF) at fast mode shocks. The NIF is the shock frame in which the upstream flow is aligned with the shock normal. As a consequence, the upstream out-of-plane motional electric field is non-zero and, from Maxwell's equations in steady state, actually the same downstream. Transformation to the HTF is along the plane shock surface until the upstream flow vector coincides with the magnetic field. Therefore, the motional electric field vanishes, and the description of particle motion simplifies to energy and magnetic moment conservation. When back-transforming to the NIF, one discovers that reflected particles have attained a speed close to twice the transformation velocity VHT, which evidently becomes very large for nearly perpendicular shocks.

gain energy, while they lose energy through curvature drift. It turns out that, at quasi-perpendicular shocks, gradient drift wins out for most ions, which then gain energy proportional to the distance they drift along $\mathbf{E}_\mathrm{p}$ (e.g. Jones and Ellison, 1991).

Alternatively (e.g. Krauss-Varban and Wu, 1989), SDA can be understood by transforming into the de Hoffmann–Teller frame (flow field-aligned upstream; transformation velocity $v_{\mathrm{HT}} = v_1 \tan\theta_{Bn}$). In this frame, the motional electric field vanishes, energy is conserved in the absence of other processes, and the only allowed change in the absence of scattering is between the perpendicular and parallel velocity components. Close to perpendicular shocks, the field-aligned velocity component becomes increasingly larger due to the transformation into the HTF. Because the perpendicular energy gain under magnetic moment conservation is simply a factor based on B_2/B_1, only ions with sufficient initial perpendicular energy may exchange large fractions of their velocity components, while slowing down significantly or reflecting in the magnetic field gradient and in the cross-shock potential. Subsequent back-transformation shows that they have gained energy proportional to the squared transformation velocity (Sonnerup, 1969). While this energy gain can be huge close to $\theta_{Bn} \sim 90°$, an increasingly smaller subset of phase space has sufficient perpendicular energy to effectively participate.

Similar ideas have been applied to the acceleration of electrons at nearly perpendicular shocks (Wu, 1984; Krauss-Varban *et al.*, 1989). However, unlike moderate-energy electrons, the ion gyroradius is comparable to or larger than the shock transition, such that the gyrophase at entry becomes important.

As mentioned above, there is an additional electric field in the shock transition in the normal direction, from the cross-shock potential, which is required to restrain the more mobile electrons that typically have comparable, albeit not necessarily identical temperature to the ions in most heliospheric plasmas. Thus, in a variant of SDA, a subset of ions gains energy in the shock electric field of the NIF not from gradient drift, but because they are "held back" by the cross-shock potential, and thus can accelerate along $\mathbf{E}_p$ for a while, either without gyrating out of the shock transition or via multiple gyration encounters. This mechanism is only effective in thin, nearly perpendicular shocks, and is sometimes called "shock surfing" (SSA; Sagdeev, 1966; Lever *et al.*, 2001). It is a bit different from regular SDA, in that it is mostly initially low-energy particles that spend any appreciable time in the vicinity of the shock transition. There, a portion of the phase space that has a small velocity component into the shock may decelerate or get turned around due to the cross-shock electric field. This extends the time that the particles spend in the shock transition, where they are accelerated in the out-of-plane electric field of the NIF, until the Lorenz force from the newly gained perpendicular velocity suffices to overcome the cross-shock potential. This process is self-restrictive, in the sense that, once the ions have gained any appreciable energy, they are taken "out of resonance" or out of the effective orbit and escape downstream, in nearly perpendicular shocks.

Without additional scattering mechanisms, all these processes are highly limited in the portion of phase space that is affected, and in the amount of energy gain. By themselves, they cannot explain the often substantial observed fluxes of ions at IP shocks, into the hundreds of MeV energy range.

In fact, for the majority of affected ions, both SSA and SDA only increase their energy by a factor of a few, and resultant fluxes at high energy are typically very low. On the other hand, SDA can operate on ions that have previously been accelerated by another mechanism, or that return to the shock due to scattering off waves (e.g. Kallenrode, 1998). Conversely, it is sometimes thought that these mechanisms provide the seed population to get the diffusive acceleration at highly oblique shocks going, where sufficient particle energy is required such that they can make it back upstream against the unfavorable shock-normal angle. Alternatively, highly oblique shocks may accelerate ions via strong scattering across the magnetic field – a process that is still poorly understood (see Chapter 9).

SDA or simple reflection at the shock may enhance a given seed population of the suprathermal energy range to sufficient fluxes such that, even at oblique shocks, upstream wave growth may occur to get the diffusive acceleration process going in the absence of sufficient inherent upstream or downstream turbulence. This appears feasible at oblique shocks in the range of perhaps $\theta_{Bn} \sim 45°$ to $60°$ (Krauss-Varban *et al.*, 2008; see also Section 8.5.4 below). However, between $\theta_{Bn} \sim 60°$ to $90°$, self-consistent upstream wave generation is very difficult, and in addition, there is a "hole" where neither parallel nor perpendicular particle scattering and transport appears to be sufficient to enhance kinematic processes via diffusive shock acceleration (e.g. Zank *et al.*, 2007).

8.4.3 Diffusive shock acceleration

The other main acceleration mechanism is first-order Fermi or diffusive shock acceleration (Axford *et al.*, 1977; Krymskii, 1977; Bell, 1978a,b; Blandford and Ostriker, 1978). It relies on the existence of upstream and downstream scattering centers that are converging due to the difference in the respective flow speeds, thus providing the energy for acceleration. First-order Fermi acceleration produces a power-law distribution and intensities that depend on the shock strength (compression ratio). Power-law distributions are as ubiquitous for SEPs (Ellison and Ramaty, 1985) as they are in cosmic plasmas, in general. However, efficiency, self-consistently generated waves, and the ability to trap the ions decrease with energy and distance to the shock, modifying the expected power-law behavior (e.g. Reames, 1999; Lee, 2005). Eventually, in all real shocks, the restricted temporal and spatial dimensions available lead to an upper cutoff of the spectra at high energies – typically between 10 MeV and 100 MeV for SEPs escaping IP shocks.

The detailed aspects of diffusive shock acceleration have been reviewed many times over the past three decades, see e.g. Axford (1981), Drury (1983), Forman and Webb (1985), and Lee (2005). Here it suffices to outline the basic principles.

For simplicity, assume a planar, parallel shock. Then, a downstream particle with speed v in the downstream (scattering) frame and pitch-angle θ (with respect to the shock normal or magnetic field) will travel upstream if $v \cos\theta > v_2$, where v_2 is downstream velocity. Because scattering preserves the energy in the respective downstream and upstream *wave* frames, *upstream* scattering of this particle means there is an energy gain when transformed back into the *downstream* frame. For particles that are already significantly faster than the flow speed, the associated momentum gain of a returning particle is:

$$\delta p/p = (v_1 - v_2)/v(\cos\theta - \cos\theta'), \tag{8.1}$$

where the prime denotes the new pitch angle. If one now assumes an almost isotropic distribution of particles, one can average over all pitch angles, and the cos terms simply convert into a constant factor. One then proceeds to calculate the probability of escape downstream (which is simply given by the ratio of the downstream to upstream flux) versus the probability of an acceleration cycle. From the calculation it follows that the particle distribution assumes a power law with index q, which depends on the shock compression ratio: $q = 3r/(r-1)$, where from mass continuity in the assumed one-dimensional shock: $r = v_1/v_2 = n_2/n_1$, i.e. the compression ratio between the downstream and upstream densities. To arrive at this, the difference between the respective average wave frames and plasma frames has been neglected (e.g. Lee 2005).

Because waves that make up efficient scattering centers should be generated self-consistently by the energetic ions (Lee, 1982, 1983), must exist for extended regions upstream and downstream of the shock, and should not be convected towards or away from the shock too quickly, diffusive shock acceleration is most efficient and most easily understood for fairly high Mach number, almost parallel shocks. Conversely, it is much less understood how this process can be so efficient at the low-to-medium Mach number, oblique shocks that make up most IP shocks (see below). In particular, at nearly perpendicular shocks, diffusive acceleration may require effective scattering across the magnetic field (Webb *et al.*, 1995; Giacalone and Jokipii, 1999; Zank *et al.*, 2007).

The theory of diffusive shock acceleration still faces several significant challenges. The simplest approach assumes that the existing turbulence is sufficient to create an almost isotropic plasma, which in turn allows a fluid-like description of the energetic ions, replacing velocity-space scattering with spatial diffusion. More realistic and elaborate analytical or numerical methods include a treatment of the self-generated upstream wave field, and its fall-off with distance away from the shock (Ng and Reames, 1994; Ng *et al.*, 2003a; Lee, 2005). In fact, the transition from the region of efficient scattering close to the shock to the almost scatter-free transport (i.e. in the solar wind for IP shocks) is a crucial element that allows the energized ions to escape. This transition region is analytically described by the focused transport equation, which treats the proper pitch-angle scattering in the wave field, rather than spatial diffusion (e.g. Lee, 2005; Chapter 9). Assumptions for the required diffusion coefficients are by necessity elementary and usually simply derived for parallel shocks and parallel propagating waves only, while "real" shocks exhibit a mixture of compressional and Alfvén waves upstream over a wide range of propagation angles. Finally, it may very well be that, in some shocks, second-order Fermi acceleration (i.e. the random scattering in waves propagating in both directions) also contributes. This may in particular be the case in the downstream of coronal shocks, which due to the much larger magnetic field and density close to the Sun have much shorter kinetic scales associated with them. In other words, those shocks have, relatively speaking, much more time available to make use of the otherwise inefficient second-order Fermi mechanism.

Modeling efforts in the past have been hampered by the lack of a proper separation of scales (see e.g. Galinsky and Shevchenko, 2000) and by use of ad-hoc upstream and downstream wave models, or by using the unrealistic Bohm diffusion limit (see e.g. Lee, 2005). Rice *et al.* (2003) improved on this by including an upstream wave model based on the steady-state solution for wave growth (Gordon *et al.*, 1999). However, even in this improved model, the dependence of the turbulence away from the shock is ignored and replaced with a constant value. More recent work attempts to incorporate a sophisticated description of the spatial and temporal evolution of the waves that are self-consistently generated by the streaming particles.

It is important to point out that, when modeling the wave fields surrounding shocks, the quasi-linear approximation is not always appropriate. This is so because the wave–particle interaction is essentially non-linear in the vicinity of shocks that efficiently accelerate particles. Moreover, the quasi-linear approximation typically ignores the presence of multiple wave modes with different propagation directions, which changes the resonance width beyond standard resonance broadening (Karimabadi and Krauss-Varban, 1992; Karimabadi *et al.*, 1992). In both observations and simulations (see below), in addition to parallel-propagating Alfvén waves produced by field-aligned beams, ring-beam type ion distributions are found to generate obliquely propagating fast/magnetosonic waves. As also pointed out by Ng and Reames (1994), effects of this nature cause waves generated by low-energy protons to scatter ions of higher energy more efficiently than otherwise expected, allowing efficient trapping near the shock also for higher-energy ions.

Ng and Reames (1994) have extensively modeled radial transport considering self-consistent wave generation, acceleration, and escape of particles at IP shocks (Reames and Ng, 1998). They found that in the presence of large resonant wave fields, scattering limits the intensities of particles that can stream away (the so-called "streaming limit"). Conversely, intensities rise much higher in the peak near shock passage (ESP event, see Section 8.5.2) when what is observed is the ions trapped in the large wave fields near the shock. The spectrum in this peak observed close to Earth is generally softer because the local IP shock is a less efficient accelerator than the shock when it is in the corona, and the more energetic particles do not scatter as efficiently and so leave the vicinity of the shock promptly. The problem of the combined particle acceleration and transport is discussed in greater detail in the subsequent Chapter 9.

8.4.4 Typical configurations and complicating scenarios

It is expected that the above acceleration mechanisms often work together, which makes their theoretical treatment rather difficult. In particular for CME-driven IP shocks, there are many complicating factors that enter the efficiency of energetic particle production. Therefore, modeling SEP events from shock acceleration and transport theories is a very challenging topic (Zank *et al.*, 2000; Li *et al.*, 2003). As mentioned above, in particular, highly oblique and nearly perpendicular shocks are still poorly understood. Transport of charged particles in those field geometries in principle requires three dimensions (because in 2D, particles are essentially bound to the magnetic field lines, see Chapter 9). Such transport may also be achieved by "meandering field lines" of the intrinsic solar wind turbulence (Giacalone, 2005). For example, such meandering field lines may have sufficient amplitude in the heliospheric termination shock.

It is also known that multiple shocks generate a much more efficient acceleration environment (Gopalswamy *et al.*, 2002). Not only does the first shock leave a much more turbulent and seed-particle rich upstream for the following shock, but particles may scatter multiple times in both shocks. Of course, the upstream seed particle spectrum and background turbulence is highly variable in the solar wind in general, and will have an impact on achieved fluxes.

The overall geometry of the interplanetary environment and the history of the shock are also important considerations. For example, the field lines downstream of IP shocks typically converge towards the Sun. Mirroring in this configuration may prevent the rapid escape of energetic particles, in particular, at oblique shocks that otherwise have too little downstream turbulence to effectively scatter and capture them (Krauss-Varban *et al.*, 2008). Finally, one has to be careful when interpreting local IP shock observations. Not only may the shock on average have very different conditions on a larger scale (e.g. shock jumps and θ_{Bn}) than locally observed, but the energetic particle environment carried with the shock may have originated when the shock was much closer to the Sun, and its properties may have been quite different then. We discuss further details and aspects of IP shocks in Section 8.5.2 below.

8.5 Particle acceleration at the Earth's bow shock and at interplanetary shocks

The two historically best and most convenient laboratories we have had for studying charged particle acceleration at collisionless shocks *in situ* are the Earth's bow shock and interplanetary shocks. In particular, close to solar maximum, CME-driven IP shocks of sufficient Mach number frequently pass the Earth.

Shocks can form in a number of ways in the solar wind; the types of IP shocks and their origin are described in more detail in Chapter 7. Both co-rotating interaction regions and CME-driven shocks are capable of accelerating charged particles; however, not surprisingly, the largest events are associated with the fastest CMEs and can reach Alfvén Mach numbers of 5 to 6, and occasionally even higher. These Mach numbers are comparable to the Earth's bow shock; yet energetic particle energies and fluxes observed at the bow shock are almost dismal compared to those at the largest CME-driven events. While the Earth's bow shock virtually always generates upstream energetic ions, the same cannot be said for IP shocks. As we discuss in detail below, most IP shocks do not accelerate charged particles to any noteworthy degree, at all. This is particularly so from the space weather viewpoint, where the main interest is in protons above 10 MeV. To understand these peculiarities, we need to take a closer look at the similarity and differences between these shocks. Finally, the heliospheric termination shock is also generally viewed as capable of producing highly energized ions.

8.5.1 The Earth's bow shock

Generally speaking, bow shocks form around many objects subjected to "supersonic" flow. In plasmas (versus hydrodynamics), this may happen (see Fig. 7.1) around stars at the boundary of the stellar wind and the interstellar medium, or around planets with an intrinsic magnetic field (and ensuing magnetosphere), providing the boundary between the solar wind and the magnetosheath, which directs the shocked solar wind around the obstacle (the magnetopause). In the latter case, the dipole magnetic field acts as the primary obstacle and the magnetopause is the boundary where the compressed magnetic field and thermal pressure are balanced by the force of the impinging, shocked solar wind. By definition, planetary bow shocks are of finite size, and, as such, any production of energetic particles is both localized and highly non-local: some regions (i.e. the quasi-parallel portion) are much more able to easily generate energetic ions, while any ions propagating upstream, or waves excited upstream of the oblique portion, are quickly convected to a different portion of the finite-size bow shock, or around the obstacle, altogether. The general scale size of the Earth's bow shock is of the order of 20 R_{E} (Earth radii); the stand-off distance is variable, but typically 15 R_E.

The energetic ion environment of the bow shock has been extensively studied over the past 25 years (e.g. Bonifazi and Moreno, 1981; Ipavich *et al.*, 1981; Paschmann *et al.*, 1981; 1982; Thomsen, 1985; Fuselier, 1994; Desai *et al.*, 2000; Freeman and Parks, 2000; Meziane *et al.*, 2002). While the picture that emerged from the ISEE missions (*JGR* special edition; Tsurutani and Rodriguez, 1981) has been improved upon in certain details since then, also with the help of the recent CLUSTER mission, many of the basic ideas have remained. That is, there is an ion foreshock that starts somewhere below $\theta_{Bn} \sim 45°$ and permeates the quasi-parallel domain, while the faster electrons (energized by the "fast-Fermi" process; Leroy and Mangeney 1984; Wu, 1984; Krauss-Varban *et al.*, 1989) form a foreshock boundary close to the perpendicular shock.

Figure 8.3 shows a snapshot of a 2D bow shock simulation to further demonstrate this point. Shown are the parallel temperature (as a proxy for energetic protons) and the magnetic field lines (interplanetary magnetic field (IMF)

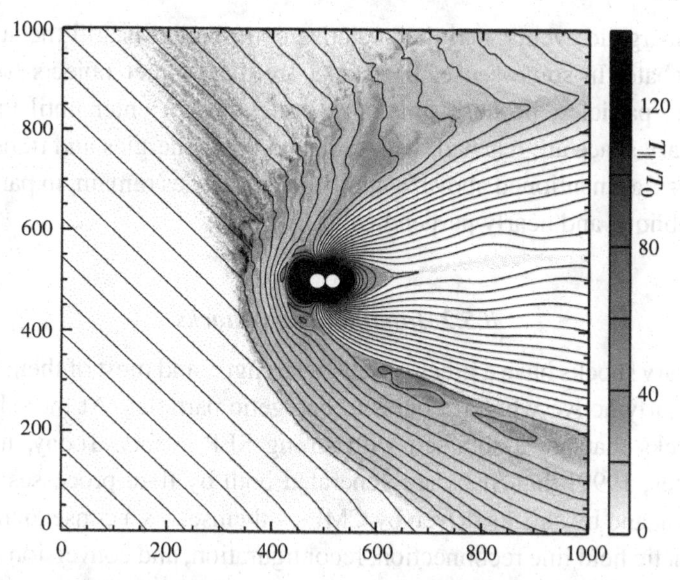

Fig. 8.3. Example of a two-dimensional (2D) hybrid simulation of the solar wind–magnetosphere interaction. (From Krauss-Varban *et al.*, 2008.) Shown are contours of the magnetic field lines (upstream IMF angle $\theta = 45°$) and the normalized parallel ion temperature $T_{\|}$, as a proxy of ion acceleration. As well-documented in many observations of the Earth's bow shock, the ion foreshock starts close to $\theta_{Bn} = 45°$ with energized and backstreaming ions, and simultaneous excitation of waves (visible in the field line undulations). Conversely, at this scale, and with the number of pseudo-particles used in the simulation, there are virtually no accelerated upstream ions at larger shock-normal angles.

direction 45°). The turbulence upstream and downstream of the quasi-parallel portion is clearly visible, as is the large enhancement of upstream-propagating, energetic protons (see also Lin, 2003, and Blanco-Cano *et al.*, 2006). Conversely, there is virtually no upstream activity at or beyond 45°. There are two caveats to this simple interpretation, namely, the limited number of particles per cell, and the fact that the simulations are scaled down with respect to the ratio of the proton gyroradius to the system size. Still, the picture provided by these simulations not only agrees with the observational understanding of the main processes at the Earth's bow shock that has formed since the ISEE missions, but also illustrates why there is so little activity upstream of the oblique portion beyond $\theta_{Bn} \sim 45°$: any ions that manage to make it upstream of the oblique portion, and any waves generated there, are either convected into the quasi-parallel portion of the bow shock, instead, or move past the finite-sized obstacle altogether.

Finally, it should be mentioned that planetary magnetospheres may continue to accelerate ions downstream of the bow shock though a variety of processes. Leakage of such highly energetic magnetospheric particles has been invoked to explain upstream energetic events, and their relative contribution is still the subject of an ongoing debate. In some sense, IP shocks are the simpler objects for the study of energetic particles, because finite-size issues do not enter until the scales of interplanetary space are reached, with corresponding energies and transport times. Nevertheless, as mentioned above, a number of mysteries remain, in particular with regard to oblique and nearly perpendicular IP shocks.

8.5.2 Interplanetary shocks

Interplanetary shocks have a great variety of strength, and most of them are actually not particularly active when it comes to energetic particles. At the other extreme are IP shocks that are associated with strong SEP events. Today, it is thought (e.g. Reames, 1999) that SEPs are generated both by flare processes deep in the solar corona, and by shocks driven by CMEs – themselves a consequence of large-scale magnetic field line reconnection, reconfiguration, and conversion of magnetic energy into flow, thermal, and energetic particle energy (see Chapter 6). In fact, it is estimated that almost all of the magnetic energy released in flares goes into energetic particles, with perhaps approximately an equal share between the ions and electrons. These particles show up as "prompt" events when observed at Earth: extremely energetic ions can traverse the distance from the Sun in minutes, with little delay compared to observed X-ray flare signatures at the Sun.

Conversely, so-called "gradual" solar energetic particle events are generally accepted to be associated with coronal and IP shock acceleration (e.g. Reames, 1999), driven by CMEs. Even in this case, the most energetic particles are produced when the shock is in the corona, with resulting hard spectra that are observed at Earth within tens of minutes. However, production of energetic ions continues to 1 AU and beyond, and peak fluxes, with a softer spectrum, often arrive at Earth with the shock itself – historically called energetic storm particle (ESP) events.

As shown in Fig. 8.4, forward-propagating IP (fast mode) shocks near 1 AU typically have an Alfvén Mach number $M_A < 3$ (Gosling *et al.*, 1984). Using Wind and ACE observations, Echer *et al.* (2003) compiled statistics of IP shocks at both solar minimum and maximum and found similar results. Only in rare cases are IP shocks observed with $M_A > 4$, although evidently those are associated with the strongest CMEs and largest SEP events. The strongest observed IP shocks can be called "cosmic-ray modified", because the pressure of the energized ion and electron populations strongly extends and affects the shock jump (e.g. Terasawa, 1999).

Gosling *et al.* (1984) found that, despite their low Mach number, about one-half of 17 observed shocks had identifiable (albeit relatively low energy) upstream, energized ions and associated waves. Both Gosling *et al.* and Kennel *et al.*

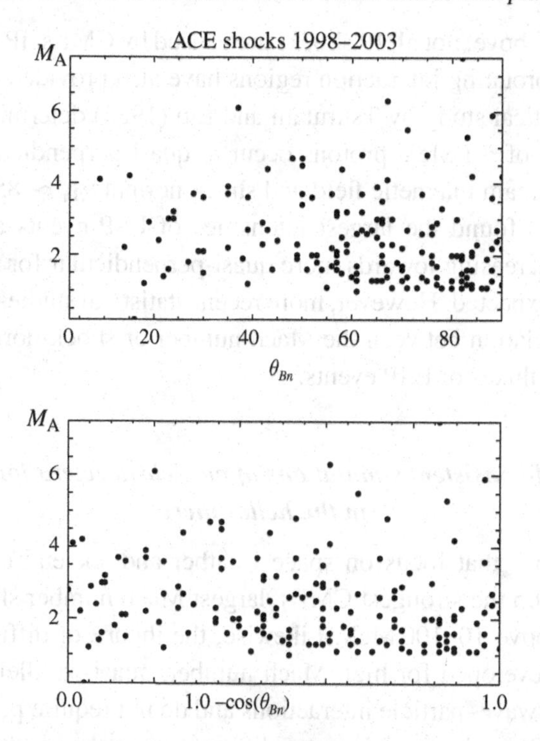

Fig. 8.4. Scatter plot of all IP, forward-propagating fast mode shocks observed with the ACE satellite and in the ACE magnetometer database, in the period 1998 to 2003 (M_A and θ_{Bn} as reported from the ACE magnetometer team database). Top: ordering with shock-normal angle θ_{Bn}. Bottom: ordering with $1.0-\cos(\theta_{Bn})$, which takes into account the solid-angle viewing statistics. Even in this corrected plot, one can see a slight preference for oblique angles, as expected from the solar wind Parker spiral. More importantly, it is evident that most IP shocks are rather slow. And, while they typically will have a detectable energetic particle environment, the associated energy range and fluxes are of little interest in the context of detrimental space weather effects, except for the rare, higher M_A cases. (ACE magnetometer team database, and processing by Yan Li, personal communication.)

(1986) found the distribution of these ions to be fairly isotropic, whereas in only a few cases, upstream beams were observed (Vinas *et al.*, 1984). This behavior also extends to higher energies (Scholer *et al.*, 1983; van Nes *et al.*, 1984; Sanderson *et al.*, 1985; Tsurutani and Lin, 1985; Wenzel *et al.*, 1985) and may be interpreted as a consequence of the large spatio-temporal scales of IP shocks, which rarely allows one to see the initial evolution of wave–particle interactions. While the large scales provide an important clue, and energetic seed particles may play an additional role, currently no scenario self-consistently accounts for the observed energetic ion environment of the weaker and oblique shocks.

As mentioned above, not all IP shocks are caused by CMEs. IP shocks associated with so-called corotating interaction regions have also provided much insight. For example, a statistical study by Tsurutani and Lin (1985) determined that the maximum intensities of ~1 MeV protons occur at quasi-perpendicular shocks (angle between the upstream magnetic field and shock normal $\theta_{Bn} \sim 85°$). Similarly, van Nes *et al.* (1984) found the largest intensities of ESP events at oblique angles, with the angle increasing towards more quasi-perpendicular for irregular or shock spike events, as expected. However, more recent statistical studies have found much less, if any, correlation between the Mach number or shock-normal angle and the energetic proton fluxes of ESP events.

8.5.3 Self-consistent simulations of particle acceleration at shocks in the heliosphere

Theoretical studies that focus on space weather and societal consequences usually concentrate on the strongest CMEs, largest Mach number shocks, and highest energies at or above 10–100 MeV. Likewise, the theory of diffusive shock acceleration is best developed for high Mach number, quasi-parallel shocks that have well-understood wave–particle interactions and do not require perpendicular transport. However, IP shocks come in many different strengths and propagation angles, with the vast majority at relatively low M_A, and with unexpectedly high observed energetic ion fluxes in the oblique to quasi-perpendicular regime.

Understanding of the underlying physics of particle acceleration and wave–particle interactions can best be achieved by considering the entire range of events. A method that does not depend on a-priori assumptions about the wave field and that is not restricted to high M_A or a particular regime of B_n, is particle simulations. In particular, for the problem at hand it usually suffices to only treat the ions kinetically, with a fluid description of the electrons. These so-called hybrid simulations have been used to study shocks and associated ion energization for more than the past two decades (Krauss-Varban and Omidi, 1991; Kucharek and Scholer, 1992; Giacalone *et al.*, 1997; Scholer *et al.*, 2000; and references therein).

Today, such simulations are by design necessarily local. That is, one can calculate the resonant frequency/wavelength of a maximum energy range of interest, and fit just a few such wavelengths into, say, a typically two-dimensional simulation domain. Even using parallel processing, the energy range beyond 10 MeV or so is difficult to address on a routine basis, today. However, past progress in computational resources projected into the future virtually guarantees that eventually the entire history of the shock in the interplanetary medium, from the Sun to the Earth and beyond, will be tackled with kinetic simulations.

8.5.4 Comparison between the Earth's bow shock and interplanetary shocks

While one has to be cautious drawing parallels, it is quite enlightening to compare these findings to observations at the bow shock. Due to its curved geometry and finite extent, the important wave–particle region in front of it cannot fully develop. Moreover, the upstream of the quasi-perpendicular region is convected toward the quasi-parallel shock. It is therefore not surprising that while observations often exhibit back-streaming beams, they generally show a cutoff at a relatively low 200–330 keV, with only rare events extending to 2–3 MeV (Skoug *et al.*, 1996; Meziane *et al.*, 2002). Meziane *et al.* (2002) separated 216 events carefully with respect to the presence or absence of a pre-existing ambient solar wind energetic population. Interestingly, they only found a dependence on θ_{Bn} (and ions up to 2 MeV) in the case of pre-existing ambient energetic ions, and ions above 0.5 MeV only for $\theta_{Bn} > 45°$. Moreover, fluxes of high-energy ions increased with θ_{Bn}.

These results are at least in partial agreement with those at fairly low Mach number IP shocks. Meziane's finding could mean that at the bow shock, SDA efficiently accelerates out of the given (and highly variable) seed population present in the solar wind. Conversely, the larger spatial and temporal scales of upstream and downstream turbulence at IP shocks could provide the seemingly required enhancement to SDA that appears to be lacking at the bow shock.

Local simulations of planar shocks traditionally have replicated ion distributions upstream of the bow shock, at least where those were relatively undisturbed and

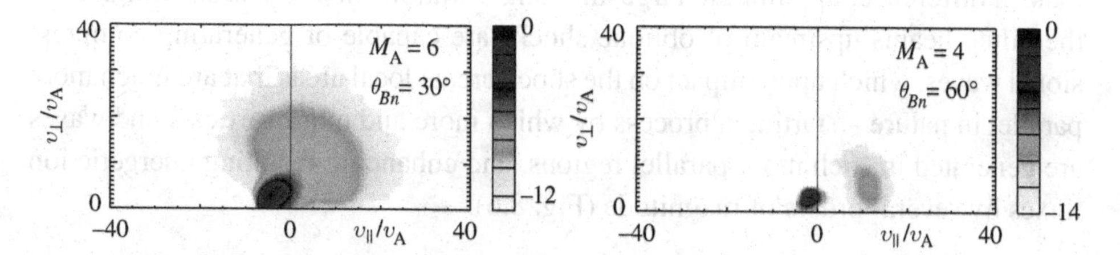

Fig. 8.5. Sketch of upstream proton distributions (perpendicular and parallel to the ambient magnetic field) in the shock frame from planar, 2D hybrid shock simulations at quasi-parallel ($\theta = 30°$) and oblique ($\theta = 60°$) angles. As in many documented observations of the Earth's bow shock and at sufficiently high Mach number IP shocks, at quasi-parallel shock-normal angles, protons can not only easily travel upstream and generate waves, but they also easily scatter in these self-generated waves to form a diffuse distribution that forms a contiguous cloud of both upstream ($v_\parallel > 0$) and downstream ($v_\parallel < 0$) directed particles. Conversely, at oblique shocks, only a highly dilute upstream-propagating beam with enhanced perpendicular energy is found, and even that can only be seen with very good particle statistics, in simulations. Unlike the quasi-parallel shock, a higher Mach number does not help initially, but typically makes it more difficult for ions to make it upstream in the first place. See also Plate 5 in the color-plate section.

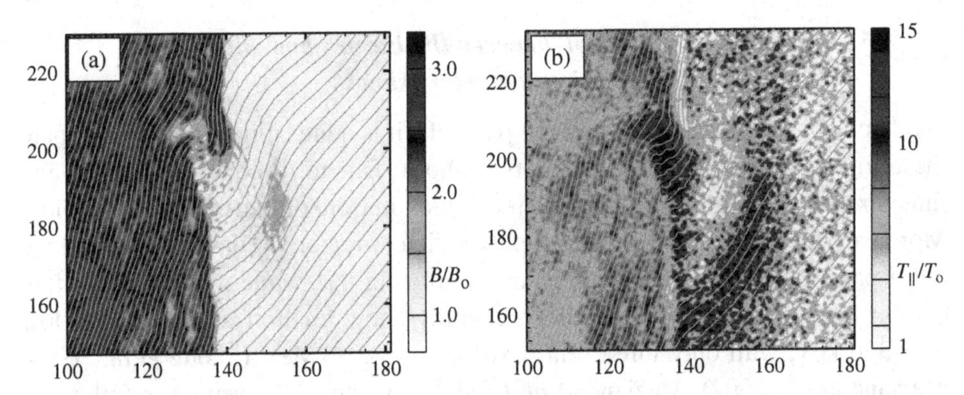

Fig. 8.6. Magnetic field line contours and (a) total magnetic field, and (b) parallel temperature $T_{\|}$ normalized to upstream in a subset of a 2D hybrid simulation of an oblique shock ($\theta = 50°$). (From Krauss-Varban *et al.*, 2008.) It can be seen how compressional waves generated by dilute beams disrupt the shock and change the local θ_{Bn}, in turn allowing more upstream wave and particle production than expected at the oblique shock. This process appears to enhance upstream energetic proton fluxes by two to three orders of magnitude. See also Plate 6 in the color-plate section.

not further processed. That is, both such simulations and bow shock observations show diffuse ions in front of sufficiently strong quasi-parallel shocks, and relatively "un-scattered" beams with enhanced perpendicular energy upstream of oblique and nearly perpendicular shocks (Fig. 8.5).

However, recently we have shown that the large available scales at IP shocks do make a difference: appropriate large and long-duration simulations show that even the dilute beams upstream of oblique shocks are capable of generating compressional waves, which upon impact on the shock create local areas that are much more parallel in nature – starting a process by which more and more particles and waves are generated in such more parallel regions, and enhancing resulting energetic ion fluxes by several orders of magnitude (Fig. 8.6).

8.6 Summary

In this chapter, we approached the question of charged particle acceleration starting from the kinetic description of shocks. This provides the scale sizes of the electric and magnetic field environment, which in turn illuminates both the dissipation at shocks as well as those aspects of particle acceleration that directly result from the finite shock transition. After that, the many different facets of diffusive shock acceleration were touched upon. Finally, we took a look at the Earth's bow shock and at interplanetary shocks in greater detail, emphasizing both their communality and their major differences, highlighting recent research results.

There are still numerous challenges before we arrive at a satisfactory quantitative description of charged particle acceleration, in particular regarding interplanetary shocks. Their environment and strength changes during propagation through the heliosphere, while seed particles, downstream magnetic topology, the presence of other shocks and discontinuities, as well as the relevant solar wind history all have a major impact on the energetic ion fluxes and spectra. Upcoming spacecraft missions and improving computational power will undoubtedly advance our general understanding and modeling capability. In the next chapter, the extraordinary problem of combining the physics of acceleration with that of transport is explained in greater detail.

9

Energetic particle transport

JOE GIACALONE

The purpose of this chapter is to lay out the basic physical foundations of cosmic-ray transport in space. We will discuss the basic physics of the cosmic-ray transport equation, first written down by Parker (1965). This equation is remarkably robust and is widely used to study cosmic-ray transport in the solar system and the interstellar medium. The chapter starts with a general background on cosmic rays in the solar system. The transport equation itself is not formally derived, but the basic physical processes that lead to the various terms in this equation are discussed in detail. We will also address what assumptions are made about this equation and under what conditions it is applicable. At the end of this chapter, we provide a few applications related to specific heliophysics phenomena, such as the propagation of solar-energetic particles in the inner heliosphere, the modulation of galactic cosmic rays, and the drift motions of cosmic rays in the solar system.

9.1 Cosmic rays in the solar system

A fundamental and outstanding problem in astrophysics concerns the origin of high-energy charged particles in space. This problem has been known since the early 1900s when Victor Hess performed his famous electroscope experiments on balloon flights to demonstrate that the excess atmospheric radiation increased with increasing distance from the Earth's surface. This led to the discovery of cosmic rays. In the 1920s Chapman and Ferraro considered the effect on Earth of localized and intermittent streams of corpuscular radiation. It was later determined that the solar corpuscular radiation had two components: one that was steady, now known as the solar wind, and the other intermittent, that we now know to be solar energetic particles.

The problem of the origin of cosmic rays of any source has two key components. The first concerns the problem of acceleration, and the second is how the particles are transported in space to an observer. This chapter deals with the

Heliophysics: Space Storms and Radiation: Causes and Effects, eds. Carolus J. Schrijver and George L. Siscoe. Published by Cambridge University Press.

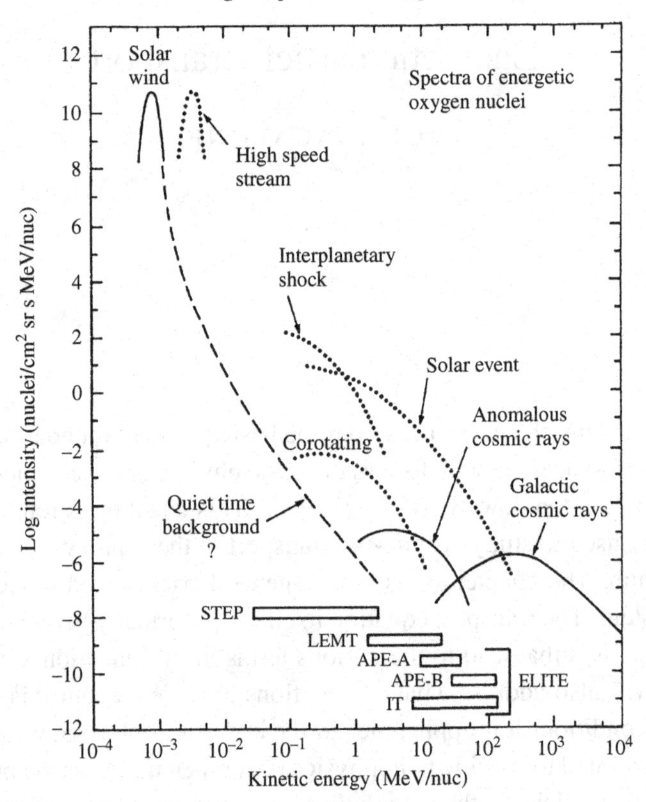

Fig. 9.1. An illustration of the energy spectrum of cosmic rays in the heliosphere based on spacecraft observations. A phenomenological description of the various types of energetic particles indicated in this figure is given in Section 9.1 (see also Fig. 3.1). (From Reames, 1999.)

latter. It is important to note, however, that both problems are coupled by the same basic physical processes. They are both aspects of the more general term "particle transport".

Figure 9.1 shows the energy spectrum† for oxygen nuclei for various particle populations in the space near Earth (Stone *et al.*, 1990). Note that these are not observed distributions from individual events, but instead are representative of those that are typically observed.

The most abundant populations of charged particles at 1 AU are solar wind protons and electrons. Their distribution is approximately Maxwellian at energies of about 1 keV (in the spacecraft frame of reference), but there also exists a suprathermal tail. The combination of these is sometimes modeled as a so-called kappa

† The spectrum here is also known as differential intensity and is strictly defined as $p^2 f$, where p is the magnitude of the momentum and f is the phase-space density (see Eq. 3.1). This is discussed in more detail in Section 9.3.1.

distribution. The nature of the quiet-time suprathermal tail is presently not well understood, although observations suggest that the ion phase-space distribution, remarkably, has a power-law dependence on velocity with a spectral exponent of −5 (Fisk and Gloeckler, 2006). The solar wind is known to have two main components, a high-speed wind, which moves at about 800 km/s and originates from large open-field regions, particularly the polar coronal holes during phases around sunspot minima, and a low-speed wind moving at about 400 km/s (see Vol. I, Chapter 9). At much higher energies the particles come from several distinct sources. We now proceed to discuss these separately.

9.1.1 Galactic cosmic rays

Galactic cosmic rays (GCRs) are charged particles that have energies up to 10^{21} eV. They come from outside our solar system and probably come from a variety of sources (see Vol. III). Those GCRs with energies up to about 10^{15} eV, where there exists a "knee" in the cosmic-ray spectrum, are understood to originate in supernova explosions. The typical time a galactic cosmic ray spends in the solar system is actually rather short, because most of them are swept out of the solar system by the solar wind and its entrained heliospheric magnetic field. However, for those that reach Earth's orbit, where they are observed, the typical lifetime is of the order of a few months to a year (this is discussed further in Section 9.5.3).

Figure 9.2 shows the cosmic-ray spectrum. Note that at low energies, about 1 GeV, the spectrum drops suddenly. This is known as cosmic-ray modulation. This arises because low-energy cosmic rays that come from outside our solar system do not easily penetrate into the inner solar system because of the "sweeping-out" effect of the solar wind. This is discussed further in Section 9.5.2. Note that we cannot directly measure the interstellar cosmic-ray spectrum below about 1 GeV because of heliospheric modulation. In order to directly measure this part of the spectrum a spacecraft would have to be well beyond the outermost edge of the heliosphere.

Cosmic rays with energies well below about 10^{12} eV have small enough gyroradii that they are scattered by irregularities in the turbulent interplanetary magnetic field. Those with energies larger than a few times 10^{12} eV have large enough gyroradii (several tens of AU in the interplanetary field at Earth's orbit) that they move with little scattering in the heliosphere. They can, however, be deflected by the strong fields of the Sun and Earth when they pass close to these objects. These lead to an interesting phenomenon known as the Sun's "shadow" (Amenomori *et al.*, 2000). If a map of the sky is made from the cosmic-ray intensity at these energies, the Sun appears dark because it is opaque to cosmic rays; however, this "shadow" is offset from the actual location of the Sun. This offset is produced by a combination

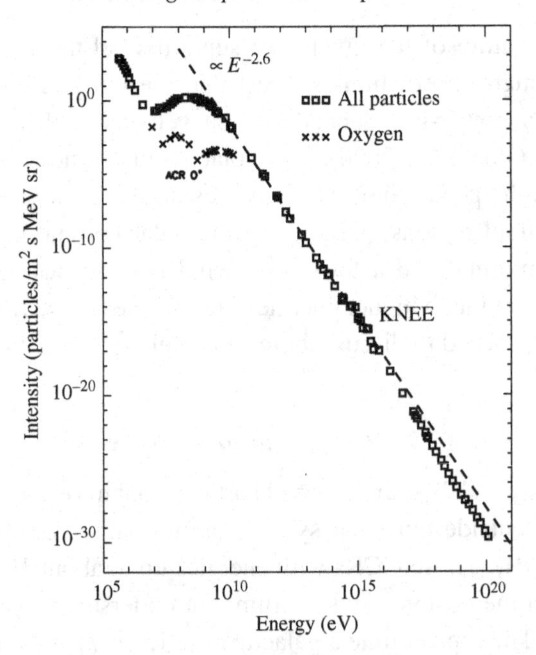

Fig. 9.2. The cosmic-ray spectrum observed at Earth's orbit.

of the effects on particle motions arising from the solar and geomagnetic fields (note that there is also an offset in the Moon's shadow).

9.1.2 Anomalous cosmic rays

The anomalous component of cosmic rays (ACRs) can be seen in Fig. 9.2 as a secondary bump at about 100 MeV (crosses). ACRs consist of helium, nitrogen, oxygen, neon, and protons and are notably lacking in carbon (Klecker, 1995). They are enhanced in a region of the energy spectrum ranging over a kinetic energy of 20–300 MeV per nucleon as shown in Fig. 9.1. ACRs increase in intensity with radial distance from the Sun, indicating that this component probably originates in the interaction of the solar wind with the interstellar medium.

It was first pointed out by Fisk *et al.* (1974) that interstellar neutral atoms, streaming into the solar system due to the Sun's relative motion with the local interstellar gas cloud (e.g. Frisch, 1996), become ionized by either photoionization or charge-exchange with the solar wind. Once they are ionized, electromagnetic forces from the solar wind and interplanetary magnetic field dominate their motion and these particles are swept out of the inner solar system. As they move outward, they encounter the solar wind termination shock (which was crossed by both the Voyager 1 and 2 spacecraft; Stone *et al.*, 2008), where they are accelerated via the mechanism of diffusive shock acceleration (Pesses *et al.*, 1981). It can be readily demonstrated quantitatively that most of the observed characteristics of ACRs can be explained by a standard cosmic-ray transport model including acceleration at a termination shock (Jokipii, 1986). This paradigm has met with some challenges to explain Voyager observations and, as such, other ideas have been presented. These include acceleration at a non-spherical shock (McComas and Schwadron, 2006) and acceleration in the turbulent heliosheath (the region between the termination shock and the heliopause) via a process that is similar to second-order Fermi acceleration (Fisk and Gloeckler, 2008). This issue has yet to be resolved.

9.1.3 Solar energetic particles

Solar energetic particles, or SEPs, constitute a class of energetic nuclei that are of "solar" origin. They are usually classified into different types. Figure 9.1 shows three different types (1) corotating (associated with corotating interaction regions in the outer heliospheric solar wind), (2) interplanetary shock, and (3) "solar event" which has at least two different sub-classifications itself.

Events of the first of these three types are events known to be associated with so-called corotating interaction regions (CIRs). CIRs are structures in the solar wind that arise from the interaction between high-speed and low-speed solar wind. As high-speed wind overtakes the slow-speed wind, an interaction region forms which moves at a speed intermediate to these two (see Pizzo, 1985). At distances beyond Earth's orbit, the leading and trailing edges of this interaction region are bounded by shocks, a forward shock that moves radially outward from the interaction region, and a reverse shock that moves radially inward from the interaction region. These shocks are known to accelerate particles that are then transported to observers at Earth (Fisk and Lee, 1980). It is also known that closer to 1 AU, where the shocks are not yet formed, local particle acceleration occurs via the mechanism of diffusive compression acceleration (Giacalone *et al.*, 2002; Mason, 2000).

The next two types of SEP events indicated in Fig. 9.1 are associated with solar eruptions. When the first observations of large solar cosmic-ray events were made, largely by instruments on balloon flights, and by ground-based neutron monitors, early concepts proposed that the particles originated from solar flares. The particles arrived at Earth by moving along the Parker-spiral magnetic field (compare Fig. 4.7 in Vol. I). Later, this picture was revised to include two separate classes of events: (1) those that are accelerated by solar flares (as in the original picture), and (2) those that are accelerated by a fast-moving shock wave driven by a coronal mass ejection. This is nicely reviewed by Reames (1999) and has been known for some time as the two-class paradigm of SEPs. The label "solar event" in Fig. 9.1 contains both of these classes, although the one labeled "interplanetary shock" refers to those particles that are associated with the shock as it crosses Earth's orbit. These

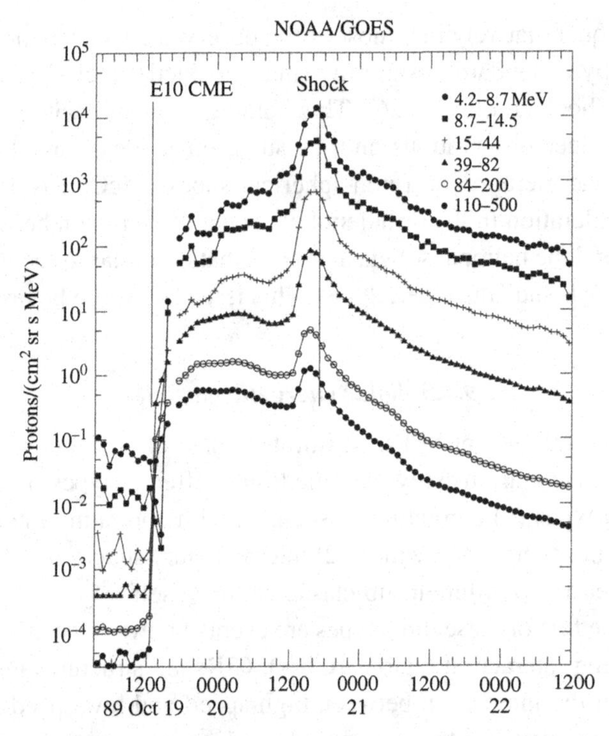

Fig. 9.3. The intensity of energetic protons as a function of time for a solar-energetic particle event associated with a coronal mass ejection on October 19, 1989 (Reames, 1999).

are sometimes referred to as "energetic storm particle" or ESP events in the scientific literature. Although this two-class paradigm is an attractive means of sorting the complex variety of high-energy particle events from the Sun, more sensitive spacecraft measurements by ACE, Wind, and SoHO have indicated that it is probably not correct. There are numerous events that show a mixture of the two classes of events, which is not surprising because most X-class flares have an associated coronal mass ejection (CME) (see Chapter 5). There is likely a combination of physical effects taking place.

Figure 9.3 shows a characteristic SEP time–intensity profile. The sequence of events that leads to this profile is as follows: (1) a CME is seen in a corongraph image at about 1200 UT on October 19, 1989 (the time of this event is indicated in the figure); (2) a few hours later, high-energy particles arrive at the spacecraft; (3) a little over a day later, a shock wave, presumably driven by the CME is detected at the spacecraft and this is coincident with an increase in energetic-particle intensity; (4) the particle event subsides and the intensity of energetic particles declines. The physical picture leading to this sequence of events is that a CME forms low in the solar corona and accelerates particles to high energies very rapidly (minutes or less) as it moves through strong magnetic fields in the corona; these high-energy particles escape upstream of the CME and move through interplanetary space at speeds much greater than the CME itself and are seen to arrive well before the shock; later the slower-moving shock crosses the spacecraft and particles are still being accelerated by it, and trapped by turbulence in its vicinity, leading to an increase in intensity; finally, the combination of diffusion in the interplanetary magnetic field and adiabatic cooling in the solar wind leads to a decline in the particle intensity. The precise details of the resulting time–intensity profile of an SEP event seen at Earth depend on a number of things, including the location of the source on the Sun, the strength of the interplanetary shock, and the nature of the solar wind at the time of the event.

9.2 The motion of individual charged particles

The relevant physics that describes the motion of charged particles in space follows Newton's and Maxwell's equations. Relativistic effects are important for high-energy cosmic rays (protons with energies greater than a few GeV, and electrons with energies greater than a few MeV) and it is straightforward to include these as necessary. We will proceed using the non-relativistic form of the equations.

It is important to note that energetic charged particles exist only when the ambient plasma is very rarefied so that Coulomb collisions, which act to thermalize the distribution, are negligible. Also, since energetic particles have a very small contribution to the number density of the plasma, it is often assumed that they can be treated as test particles. Adding the effects of energetic particles that have a significant contribution to the energy density, which is known to be the case in certain astrophysical plasmas, can be done with certain approximations but is beyond the scope of this chapter.

Before proceeding to discuss the collective motion of an ensemble of charged particles, which is our ultimate goal, it is instructive to examine the forces on individual charged particles. This is also relevant to large-scale computer simulations that integrate the equations of motion of a large number of individual particles. Particularly important in this regard is the fact that certain assumptions about the nature of the electric and magnetic fields that govern the particle motion can lead to restrictions on the particle motion. This is discussed in Section 9.2.5 below.

9.2.1 The Lorentz force

Because we are concerned with the motion of charged particles, the relevant force acting on the particles is the Lorentz force. It is straightforward to show that this force dominates the gravitational force in nearly all relevant applications. It is given by

$$m\frac{d\mathbf{v}}{dt} = q\mathbf{E} + \frac{q}{c}\mathbf{v} \times \mathbf{B}, \tag{9.1}$$

where q, m, and $\mathbf{v}$ are the particle's charge, mass, and velocity, respectively, and $\mathbf{E}$ and $\mathbf{B}$ are the electric and magnetic fields that provide the forces governing the motion of the particles. Note that this equation is written down in cgs units.

9.2.2 Gyromotion

For the simplest case of no electric field and a constant magnetic field in the z direction, the solution to Eq. (9.1) is straightforward. It is given by

$$\begin{aligned} v_x &= v \sin\alpha \cos(\Omega t - \phi), \\ v_y &= -v \sin\alpha \sin(\Omega t - \phi), \\ v_z &= v \cos\alpha, \end{aligned} \tag{9.2}$$

where $\Omega = qB/(mc)$ is the cyclotron (gyro-)frequency, α is called the pitch angle (note that our definition is such that $\alpha = 0$ implies the particle is moving directly along the magnetic field), ϕ is the phase angle, and v is the magnitude of the particle velocity.

It is often convenient to define other quantities such as the cosine of the pitch angle,

$$\mu = \cos\alpha, \tag{9.3}$$

and the components of the particle velocity parallel and perpendicular to the magnetic field,

$$v_\parallel = v\mu, \qquad v_\perp = v\sqrt{1-\mu^2}. \tag{9.4}$$

9.2.3 Particle drifts

There are a number of charged-particle drifts that arise because of crossed electric and magnetic fields, or spatially varying magnetic fields. The simplest of these is due to the presence of a constant electric field that is everywhere normal to the magnetic field. It is straightforward to show that there is a drift in the direction that is normal to both the electric field and magnetic field given by

$$\mathbf{v}_E = \frac{c}{B^2}\mathbf{E} \times \mathbf{B}. \tag{9.5}$$

Note that this drift applies to all charged particles in a plasma. As such, it can be used to derive the fundamental equation of ideal MHD: $\mathbf{E} = -\mathbf{v} \times \mathbf{B}/c$, where $\mathbf{v}$

is the bulk, or average, speed of the ions. To show this, we start by realizing that each particle's motion consists of this drift plus gyromotion about the field. When summing over a large number of particles, the gyromotion averages to zero, leaving only this drift speed. Thus, $\mathbf{v} = \frac{c}{B^2}\mathbf{E} \times \mathbf{B}$ and, with the use of a simple vector identity, it is trivial to derive the ideal MHD equation for the electric field.

Other drifts arise when the magnetic field varies in space on a characteristic scale that is much larger than the gyroradius of the particle. For example, the variation in the magnitude of $\mathbf{B}$ leads to a drift known as the ∇B-drift, and the curvature of the lines of force leads to a curvature drift. For the special case in which $\nabla \times \mathbf{B} = 0$, these are straightforward to derive (see Boyd and Sanderson, 2003) and are given by

$$\mathbf{v}_G = \frac{cW_\perp}{qB^3}\mathbf{B} \times \nabla B, \tag{9.6}$$

$$\mathbf{v}_C = \frac{2cW_\parallel}{qB^3}\mathbf{B} \times \nabla B, \tag{9.7}$$

where $W_\perp = (1/2)mv_\perp^2$ and $W_\parallel = (1/2)mv_\parallel^2$. Note that these expressions are for the case of non-relativistic particles.

However, in most applications of interest $\nabla \times \mathbf{B} \neq 0$. A more-general expression for the particle drift can be derived by expanding the magnetic field about the smallness parameter r_g/L where r_g is the particle gyroradius and L is the characteristic scale of the variation of the magnetic field (Northrop, 1963). The resulting guiding center drift velocity, in the non-relativistic limit, is given by

$$\mathbf{v}_{\text{g.c.}} = \left[v_\parallel + \frac{cW_\perp}{2qB}\mathbf{b} \cdot (\nabla \times \mathbf{b}) \right]\mathbf{b} + \frac{cW_\perp}{2qB^2}\mathbf{b} \times \nabla B + \frac{cW_\parallel}{qB}\mathbf{b} \times (\mathbf{b} \cdot \nabla)\mathbf{b}, \tag{9.8}$$

where $\mathbf{b} = \mathbf{B}/B$. The gradient and curvature drifts are associated with the last two terms in this equation, which are in the direction normal to the magnetic field; however, it is important to note that there exists a component of the drift *along* the magnetic field in addition to these.

When Eq. (9.8) is averaged over an isotropic distribution of particles, one obtains the drift velocity $\mathbf{V}_d = (cmv^2/q)\nabla \times (\mathbf{B}/B^2)$ (Isenberg and Jokipii, 1979), which is commonly used in models of cosmic-ray transport (see Vol. III).

9.2.4 Particle scattering: resonances

To this point we have considered only smoothly varying electric and magnetic fields as compared to the radius of gyration of the particles, $r_g = v/\Omega$. For such cases, the particle speed and pitch angle change very slowly compared to the cyclotron period. However, when the typical scale of the variation in the fields, L, is of the order of r_g, the speed, phase, and pitch angle can undergo more rapid changes over shorter periods of time. This leads to a form of scattering that is loosely analogous to classical scattering, although it differs from that in important ways. For instance, the particles do not actually collide off one another, as in the lower portions of Earth's atmosphere, nor do they collide off large targets, like photons moving through a dense gas, but rather, they scatter off irregularities in the magnetic field. Formally one can solve the equations of motion under the approximation that the amplitudes of the magnetic fluctuations are small and show that there exists a resonance condition,

$$v_{\|} \sim L\Omega, \tag{9.9}$$

for which the equations become undetermined. At such instances, the particle is said to "scatter" and it reverses its pitch angle and its phase angle becomes randomized. Figure 9.4 shows the orbit of a charged particle moving in an irregular magnetic field. It shows the behavior of the pitch angle and position as a function of time in the left two plots and the orbit as projected onto a given plane in the right plot. The magnetic field in this case has a mean that points in the positive z direction, and a fluctuating component that points in the x direction. There is no electric

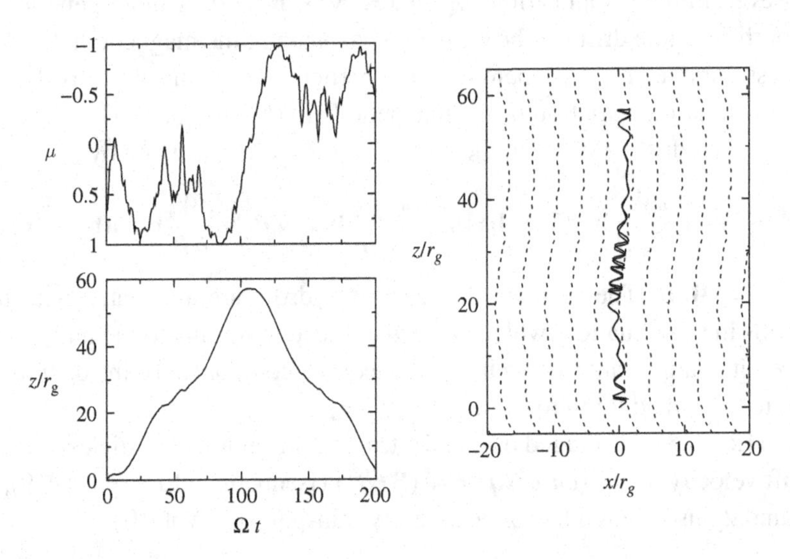

Fig. 9.4. Various representations of the orbit of a single proton moving in an irregular magnetic field, which contains a variety of scales, including those that are comparable to the gyroradius of the proton. The upper left plot shows the cosine of the pitch angle as a function of time, and the lower left plot shows the position along the direction of the average magnetic field (z direction), as a function of time. The right plot shows the position of the particle as projected onto the x–z plane.

field. The particle "scatters" at about $\Omega t = 100$ where it reverses its direction. Note that μ (the cosine of the pitch angle) changes sign.

It is also noteworthy in Fig. 9.4 that, when the particle reverses direction, it traces out essentially the same path in the x–z plane. This is an artificial (and unphysical) aspect of the assumption made about the magnetic field. In this case, it was assumed that the field depends only on the z spatial coordinate (so-called "slab" turbulence). As we will see in the next subsection, this restricts the particle motion.

Since particle scattering is a stochastic process, it is most useful to perform a statistical analysis on a large number, or *ensemble*, of charged particles. The relationship between the average particle motion and the magnetic field can be determined from the quasi-linear theory (Jokipii, 1966). It is found that the dynamical behavior of the distribution function obeys the standard diffusion equation in classical statistical physics. This is discussed further in Section 9.3.

9.2.5 Restricted particle motion in one- and two-dimensional fields

In magnetic and electric fields that depend on only one or two spatial coordinates, certain artificial and unphysical restrictions exist on the motion of individual charged particles (Jokipii *et al.*, 1993; Jones *et al.*, 1998). This is the case even if the model is fully self-consistent, time dependent, and/or contains large-amplitude electromagnetic fluctuation. In such cases, it can been shown rigorously that a charged particle is forever tied to the magnetic line of force on which it begins its motion because of the presence of an ignorable spatial coordinate. This artificial restriction on the particle motion must be realized when doing any calculation in which the model assumes, for simplicity, that the electromagnetic fields are not fully three dimensional. This is particularly relevant to numerical simulation studies in which such restrictions are often overlooked and can lead to unrealistic or misleading conclusions.

The proof of this theorem is straightforward and is given in the references above. A key aspect of it is that the component of the canonical momentum (see any good textbook on electricity and magnetism for a definition of this) of a particle in the direction of any ignorable coordinate is conserved. Since the canonical momentum is a combination of the linear momentum and vector potential, if it is to be constant the particle velocity is bound to a particular value of the vector potential representing the magnetic field. Moreover, the component of the vector potential in the ignored direction is constant along field lines. Using these facts, is can be demonstrated that a charged particle must always remain within one gyroradius of the magnetic field line on which it begins its motion. Other discussions of this can be found in the articles by Thomas and Brecht (1988) and Cowley (1977). It is important to point out that the theorem only applies to the projection onto the plane not

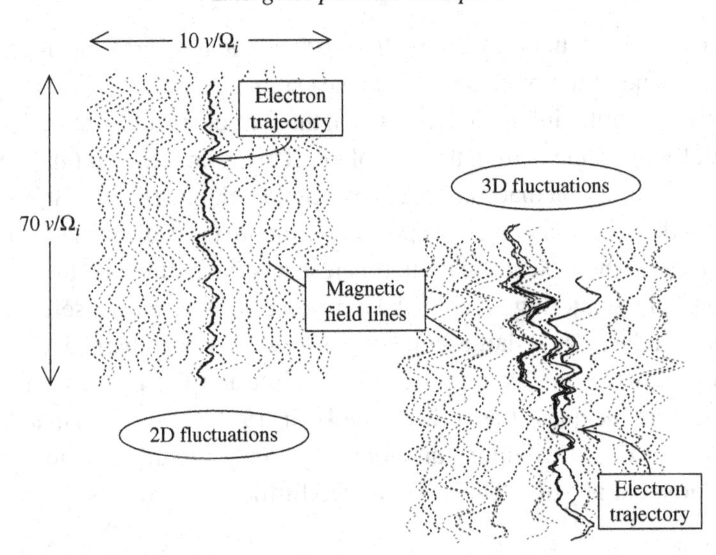

Fig. 9.5. The trajectories of two electrons moving in a spatially irregular (but static in time) magnetic field. In the left plot, the magnetic field depends on only two spatial coordinates, in which case theory requires that the particle remains within one gyroradius of a particular field line, which is the case. In the right panel, the field depends on all three spatial coordinates and the electron is not strictly tied to the same magnetic line of force.

containing the ignorable coordinate, and allows for gradient and curvature drifts in the direction of the ignorable coordinate.

Figure 9.5 illustrates the theorem (see also Giacalone and Jokipii, 2004, and Giacalone, 2004). Shown is the trajectory of an electron in a turbulent two-dimensional (left) and a three-dimensional (right) magnetic field. Also shown are several magnetic field lines. The turbulence was generated using an algorithm similar to that described in Giacalone and Jokipii (1999). As the theory requires, the electron is restricted to move along a particular magnetic field line for the case of the two-dimensional field. This is clearly not physical since the electron can move normal to field for the case of the three-dimensional field.

Including a time-dependent magnetic field does not affect the conclusions of the theorem. This has been verified using similar simulations and self-consistent hybrid simulations (Giacalone, 2004), but is not included here. Additionally, an interesting discovery was made during this confirmation of the theorem. If the time variation is slow compared to the gyroperiod, then the induced electric field is small; however, it *cannot* be neglected, even if the electric force is a few orders of magnitude smaller than the magnetic force. It was discovered that if it is neglected (thereby not satisfying Maxwell's equations), then particles can move normal to the magnetic field, which is in violation of the theory. Therefore, not only must the simulations be fully three dimensional, but they must also exactly satisfy Maxwell's equations in order to accurately describe the motion of the charged particles. We note that this poses a serious challenge to models that simultaneously solve the field equations (e.g. MHD simulations) and the equations of motion for individual charged particles.

9.3 The cosmic-ray transport equation

The evolution of a distribution of charged particles moving in the electromagnetic fields of space is governed by the cosmic-ray transport equation first written down by Parker (1965). This equation is remarkably robust and is widely used to model the propagation and acceleration of cosmic rays in astrophysical plasmas.

The equation is written down in Section 9.3.7. We do not attempt to derive it formally, but rather we will discuss various aspects of charged-particle transport in terms of basic physical principles to see what the various terms represent. These include diffusion, advection, energy change, and particle drifts. These are all aspects of the transport equation that naturally arise from the basic equations of Newton and Maxwell following from certain assumptions (diffusion approximation) and appropriate averaging (the transport equation is averaged over pitch and phase angles). There exist other transport equations, which include, for example, pitch angle information, but we will not discuss these here. The interested reader is directed to the works of Roelof (1967), Ruffalo (1995), Isenberg (1997), and Kota (2000) for further reading on these and other approximations. In this chapter, we deal strictly with the Parker transport equation.

9.3.1 The distribution function

We start with a brief discussion about the distribution function (see also Section 3.2). The number of particles per phase-space volume is known as the phase-space distribution function, f, which is a function of the six dimensions of phase space and time $(\mathbf{p}, \mathbf{r}, t)$, where $\mathbf{p}$ is the particle momentum vector ($\mathbf{p} = m\mathbf{v}$). The number density of particles at a given location at a given time, $n(\mathbf{r}, t)$, is related to the phase-space distribution function by

$$n(\mathbf{r}, t) = \int f(\mathbf{p}, \mathbf{r}, t)\, \mathrm{d}^3\mathbf{p}, \tag{9.10}$$

where $\mathrm{d}^3\mathbf{p}$ is the volume element of phase space. For example, for a Cartesian geometry $\mathrm{d}^3\mathbf{p} = \mathrm{d}p_x\, \mathrm{d}p_y\, \mathrm{d}p_z$ and for a spherical geometry it is $\mathrm{d}^3\mathbf{p} = \mathrm{d}\phi \sin\alpha\, \mathrm{d}\alpha\, p^2\, \mathrm{d}p$, with the same definitions as in the previous section (i.e. α is the pitch angle).

The differential intensity, shown in Figs. 9.1 and 9.2, which is a common representation of distributions of energetic particles in space, is related to the phase-space distribution function by

$$J = p^2 f. \tag{9.11}$$

Sometimes this is written as $\mathrm{d}J/\mathrm{d}E$. This has units of particles per area, per time, per energy, per solid angle. If one integrates J over energy and solid angle (i.e. a spacecraft detector with a given acceptance cone that sums over all energy channels), the result is the *flux density* of particles, or the number of particles crossing per area per time.

9.3.2 The diffusion equation

In Section 9.2.1 we discussed the motion of charged particles in an irregular magnetic field. We mentioned that particles undergo changes in pitch and phase angle that are qualitatively similar to that which one gets from diffusion, as in classical physics. As such, the distribution function of charged particles that undergo such scattering obeys the standard diffusion equation. It is important to keep in mind that this equation is strictly valid only for time scales that are long compared to the time in between scatterings (the scattering time) and spatial scales that are large compared to the distance traveled between scatterings (the mean free path). The diffusion equation in one dimension is given by the second-order partial differential equation:

$$\frac{\partial f}{\partial t} = \frac{\partial}{\partial x}\left(\kappa \frac{\partial f}{\partial x}\right), \tag{9.12}$$

where κ is the diffusion coefficient. For the case of charged particles moving in an irregular magnetic field, κ is related to the statistical properties of the magnetic field, in particular, its power spectrum (Jokipii, 1966; Earl, 1974; Luhmann, 1976). This is discussed more in Section 9.4.1.

We note that for Eq. (9.12) we have assumed that the distribution function varies only in one spatial direction. This should not be confused with the discussion in Section 9.2.5 about the restriction on particle motion arising from fields that vary with only one spatial coordinate. By using Eq. (9.12), we have already assumed that the process is diffusive. If, for example, x is taken to be the direction normal to a mean magnetic field, then the use of this equation implies that the field must be fully three dimensional in order for cross-field diffusion to take place. The key is that the field is fully three dimensional but it is also statistically homogeneous in space.

9.3.3 Advection with a plasma

Since the magnetic field in space exists in a highly electrically conductive plasma, the field moves with the flow of the plasma (it is said to be "frozen in", see e.g. Section 3.2.3.1 in Vol. I). In the limit of ideal magnetohydrodynamics (MHD), which is the limit we are concerned with for energetic-particle transport, there is no electric field in the frame moving with the plasma. Thus, as a charged particle scatters off of a magnetic irregularity, its energy in the frame of reference moving with the plasma remains unchanged.† Stated another way, the magnetic fluctuations, which provide the scattering centers, move with the bulk plasma. Thus, the position coordinate x in Eq. (9.12) refers to the frame moving with the bulk flow v. In the inertial frame, the evolution of f satisfies the diffusion–advection equation, which in one-spatial dimension is given by

$$\frac{\partial f}{\partial t} = \frac{\partial}{\partial x}\left(\kappa \frac{\partial f}{\partial x}\right) - v\frac{\partial f}{\partial x}. \tag{9.13}$$

9.3.4 Diffusion along and across the magnetic field

In the two previous sections we discussed diffusion in one dimension. In two dimensions, there are two diffusion coefficients, one for each direction (plus cross terms that we can ignore for now). Consider the motion of particles in a turbulent magnetic field whose average points along the z direction. Then, for example, in the x–z plane, the diffusion equation (neglecting the advection term discussed above and cross terms) is given by

$$\frac{\partial f}{\partial t} = \frac{\partial}{\partial x}\left(\kappa_{\perp} \frac{\partial f}{\partial x}\right) + \frac{\partial}{\partial z}\left(\kappa_{\parallel} \frac{\partial f}{\partial z}\right), \tag{9.14}$$

where $\kappa_{\perp}$ and $\kappa_{\parallel}$ are the diffusion coefficients across the magnetic field and along it, respectively.

Because the time τ_s it takes for a charged particle in the heliosphere to scatter is generally much longer than the time it takes to gyrate about a magnetic field (i.e. $\Omega\tau_s \gg 1$), particles tend to move much more closely along the magnetic field than across it. As such, $\kappa_{\perp}$ is usually assumed to be much smaller than $\kappa_{\parallel}$. For this reason, many analyses simply neglect perpendicular transport. However, it is important to note that in many astrophysical plasmas of interest, perpendicular transport is the most important. A good example of this is the transport of galactic

† This assumes that the magnetic field is stationary in this frame of reference, which is not strictly true since there may be waves present with a variety of phase and group velocities. However, we are mostly concerned with the transport of energetic particles that have speeds that are much greater than the wave speeds (i.e. $v \gg v_A$, where v_A is the Alfvén speed).

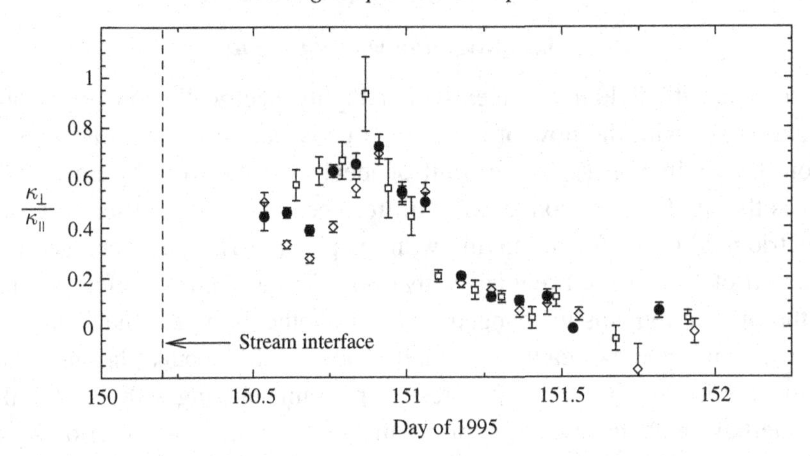

Fig. 9.6. The inferred value of the ratio of perpendicular to parallel diffusion coefficients for energetic ions, based on the observed particle streaming direction and magnetic field, during the passage of a corotating interaction region as seen by the Wind spacecraft at 1 AU. (Adapted from Dwyer *et al.*, 1997.)

cosmic rays to the Earth (see also Vol. III). The interplanetary magnetic field at large distances from the Sun is primarily perpendicular to the radial direction and is highly wound up by solar rotation because of the nature of the Parker spiral magnetic field. Therefore, in order for particles to reach the inner solar system, they must cross the magnetic field. If there were no cross-field transport, then the particles would have to move extremely long distances along the wound-up field, which is not reasonable because this would require extremely long mean free paths and would lead to very large pitch-angle anisotropies, which are not observed.

In addition, there is evidence that perpendicular diffusion may be much larger than previously thought. Shown in Fig. 9.6 is the result from an analysis of $\sim$400 keV helium ions associated with CIRs (Dwyer *et al.*, 1997). The figure shows the ratio $\kappa_\perp/\kappa_\parallel$ determined from the directions of the average magnetic field, and the directions of the gradient in particle intensity and anisotropy of low-energy helium ions observed by the Wind spacecraft during the passage of a CIR. The relatively large values of $\kappa_\perp/\kappa_\parallel$ imply that cross-field diffusion is very important.

The motion of particle across a magnetic field occurs in two ways: (1) the actual transfer of particles from one magnetic field line to the next resulting from scattering, or across the field arising from drifts, and (2) the motion of particles along magnetic lines of force that themselves meander in space in the direction(s) normal to the mean magnetic field. The contribution to the perpendicular diffusion from field-line random walk was first discussed by Jokipii (1966) and Jokipii and Parker (1969), and has received considerable attention in the literature (e.g. Forman *et al.*, 1974; Forman, 1977; Bieber and Matthaeus, 1997). Despite this, numerical simulations (Giacalone and Jokipii, 1999) yield a result that had not been explained

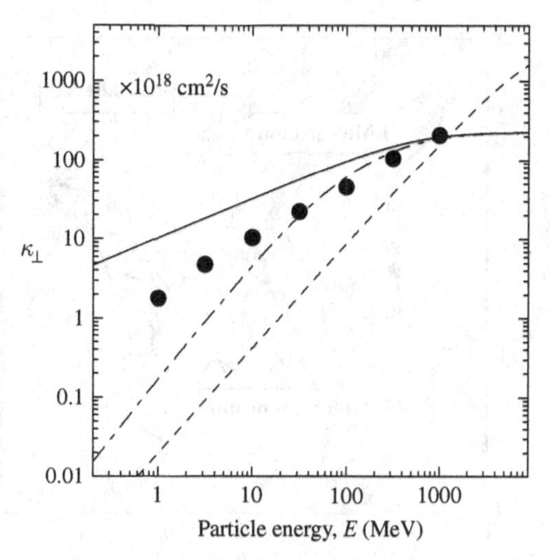

Fig. 9.7. The cross-field diffusion coefficient based on three different analytical approximations (curves) and numerical simulations (filled-in circle symbols).

theoretically before. This is shown in Fig. 9.7. The filled-in circles are the results of a numerical simulation for energetic ions moving in the magnetic fields that are typical of those seen at Earth's orbit and the curves are the results of various approximations for perpendicular transport (see Giacalone and Jokipii, 1999, for details). More recently, analytical work has emerged that agrees with the numerical simulations (Matthaeus *et al.*, 2003). This analytical work makes a different assumption about the way charged-particle trajectories decouple from individual magnetic lines of force than the quasi-linear approximation. In particular, it recognizes the important fact that when a particle scatters onto a new field line, the new field line is similar to the previous one. In the quasi-linear approximation, in contrast, when a particle scatters onto a new field line, the new field line is completely uncorrelated from the one that the particle was on previously.

An example of perpendicular transport is illustrated in Fig. 9.8. Two different particle trajectories, each with different energies and gyroradii, are shown moving in an irregular magnetic field that is representative of the heliospheric magnetic field at about 1 AU. The direction of the average magnetic field is along the vertical axis. Both types of cross-field transport discussed above can be seen in this figure.

This random walk of magnetic lines of force itself is an interesting problem receiving much attention (e.g. Matthaeus *et al.*, 1995; Gray *et al.*, 1996; Barghouty and Jokipii, 1996). Because of the fluctuations of the magnetic field, the field direction deviates from the average direction randomly leading to a random walk, or braiding of field lines. The relevant diffusion coefficients for the field lines (which

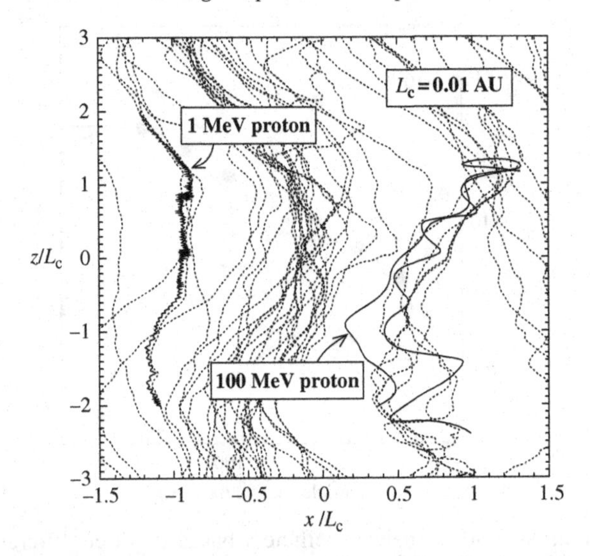

Fig. 9.8. Individual charged particles with different energies (as indicated) moving in an irregular magnetic field (grey lines). (From Giacalone and Jokipii, 1999.)

can be thought of, in some sense, in a similar manner as particle orbits) can be expressed in terms of statistics of the magnetic field, e.g. its power spectrum (see Vol. I, Chapter 7).

9.3.5 Compound diffusion, sub-diffusion, and super-diffusion

Here we wish to point out that there are terms in the literature with the names given in the subsection heading that contain the word "diffusion" but are not actually diffusive processes. Fundamentally, the diffusion coefficient that we have discussed can be related to the mean-square displacement of particles relative to some reference point. For example, consider particles released at the origin in a one-dimensional system. Particles scatter back and forth across $x = 0$ and their distribution is described by the diffusion equation given above. The mean-square displacement $\langle x^2 \rangle$ increases *linearly* with time for a strictly random, or diffusive process, in the limit that $t \to \infty$. That is, if we write

$$\langle x^2 \rangle \propto t^{\alpha}, \tag{9.15}$$

then, for $\alpha = 1$ the process is diffusion. If $\alpha \neq 1$, the process is *not* diffusive. The term sub-diffusion refers to the case $\alpha < 1$ and super-diffusion refers to the case $\alpha > 1$.

Physical models that give rise to either sub- or super-diffusion are particularly hard to obtain. However, one interesting example is that of particles moving exactly along magnetic field lines that themselves are meandering in space diffusively. This is the case of so-called compound diffusion (Kota and Jokipii, 2000). Consider a magnetic field whose mean points in the z direction, but diffusively meanders in the x direction due to fluctuations. Then, because the field lines are diffusive, $\langle x^2 \rangle \propto z$. Now suppose that a collection of particles are restricted to follow exactly along magnetic lines of force, but scatter diffusively back and forth along them (i.e. like the particle in Fig. 9.4). Then, the particles follow the behavior $\langle z^2 \rangle \propto t$. Simple substitution gives

$$\langle x^2_{\text{particles}} \rangle \propto \sqrt{t}. \tag{9.16}$$

This is not diffusive behavior. Thus, in order to get purely diffusive behavior, particles must actually transfer from one field line to another, via either scattering or drifts. Moreover, as related to our discussion above about restriction of motion in fields containing at least one ignorable coordinate, cross-field transport is inherently a three-dimensional process.

9.3.6 Energy change

In addition to scattering and advection with the flow, the particle speed itself can change. Principally, this can happen in two ways: (1) by scattering within a spatially varying flow, or (2) by diffusing in energy space because of collisions with randomly moving scattering centers. The latter of these is called second-order Fermi acceleration, or stochastic acceleration. This is an interesting topic, but is not considered in our discussion here. We examine further the first case.

Consider a particle moving in a given direction in an inertial frame that then scatters. Energy is conserved in the local plasma frame, but in the inertial frame the particle either gains or loses energy depending on whether it is moving initially against or with the flow. Suppose that at one scattering, it initially moves against the flow, and gains energy in the inertial frame (this is a head-on collision). When it next scatters, it will be moving initially with the flow and will lose energy. If the flow is everywhere uniform, then the particle loses exactly the same amount of energy it gained in the previous scattering and there is no net energy gain. But, if the second scatter occurs at a different flow speed, there is a net change in the particle's energy. The term that describes this behavior is given by

$$\frac{p}{3} \nabla \cdot \mathbf{v} \frac{\partial f}{\partial p}. \tag{9.17}$$

Particles gain energy if this term is negative and lose energy if it is positive.

A particularly good example of this is particle acceleration at a shock. Consider the energy of a particle in the frame of reference moving with the shock. As a particle scatters in the flow behind the shock, it loses energy because the particle was initially moving with the flow. The particle then returns upstream where it scatters off the incoming upstream flow leading to a gain in energy. The energy lost by the downstream scattering event is smaller than the energy gained by the upstream scattering event because the upstream flow speed is larger than that downstream. Thus, there is a net energy gain, which leads to an acceleration of particles. Note that at a shock the flow goes from large to small (in the shock frame) so that the divergence is negative and Eq. (9.17) is negative, giving rise to acceleration.

It is also noteworthy that the energy change term is positive for the case of a constant radial solar wind speed. So, all charged particles *lose energy* in the adiabatically expanding solar wind!

9.3.7 Parker's transport equation

The resulting superposition of the terms that we have discussed above leads to the generally accepted cosmic-ray transport equation first derived by Parker (1965). It is given by

$$\frac{\partial f}{\partial t} = \frac{\partial}{\partial x_i}\left[\kappa_{ij}\frac{\partial f}{\partial x_j}\right] - v_i\frac{\partial f}{\partial x_i} + \frac{p}{3}\frac{\partial v_i}{\partial x_i}\left[\frac{\partial f}{\partial p}\right] + \text{Sources} - \text{Losses}. \quad (9.18)$$

Note that we have written the diffusion coefficient in its full tensor form. This will be discussed further in the next section. Note also that we use the index notation so that a vector is written $v_i = \mathbf{v} = v_x\hat{x} + v_y\hat{y} + v_z\hat{z}$.

The cosmic-ray equation is remarkably general. It has been used widely in most discussions of cosmic-ray transport and acceleration over more than three decades. It is a good approximation provided there is sufficient scattering to keep the pitch-angle distribution nearly isotropic,† and if the particles move substantially faster than the speed of both the background fluid and the characteristic speed of the MHD waves contained in the plasma.

9.3.8 Anisotropies

The Parker transport equation is only strictly valid in the case of quasi-isotropic pitch-angle distributions, which is to be expected for the diffusive limit in which particles undergo many scatterings over the characteristic time scale of the problem. However, it is important to write down the diffusive anisotropy, $\vec{\delta}$, for purposes of completeness. The transport equation is applicable for the case $|\vec{\delta}| \ll 1$. This may not be satisfied in some cases, particularly for low-energy ions, or

† This should not be confused with anisotropic diffusion resulting when $\kappa_\perp \neq \kappa_\parallel$.

regions where there exist strong magnetic field gradients. The diffusive anisotropy is given by

$$\vec{\delta} = \frac{3\vec{S}}{vf}, \tag{9.19}$$

where $\vec{S}$ is the cosmic-ray streaming flux given by

$$\vec{S} = S_i = -\kappa_{ij}\frac{\partial f}{\partial x_j}. \tag{9.20}$$

One example of the use of the cosmic-ray streaming flux was in the determination of the large-scale structure of the termination shock based on the observed anisotropy of anomalous cosmic rays (Jokipii and Giacalone, 2004; Jokipii *et al.*, 2004). It was also used to derive the ratio of $\kappa_\perp/\kappa_\parallel$ shown in Fig. 9.6 based on the observed streaming direction of energetic ions associated with corotating interaction regions (Dwyer *et al.*, 1997).

9.4 The diffusion tensor

All of the quantities in the transport equation, except for the diffusion tensor, are directly observed by spacecraft or can be accurately determined by using the hydromagnetic approximation. Consequently, determining transport coefficients poses a fundamental challenge in the modeling of cosmic rays.

In general, the diffusion tensor κ_{ij} is related to the magnetic field vector B_i, the diffusion coefficients parallel and perpendicular to the mean field, $\kappa_\parallel$ and $\kappa_\perp$, and the antisymmetric diffusion coefficient, κ_A, as

$$\kappa_{ij} = \kappa_\perp\delta_{ij} - \frac{(\kappa_\perp - \kappa_\parallel)B_iB_j}{B^2} + \epsilon_{ijk}\kappa_A\frac{B_k}{B}, \tag{9.21}$$

where δ_{ij} is the Kronecker delta function ($\delta_{ij} = 1$ if $i = j$ and $\delta_{ij} = -$ if $i \neq j$), and ϵ_{ijk} is the Levi-Civita symbol: $\epsilon_{ijk} = 1$, or -1 if (i, j, k) is an even or odd permutation of $(1, 2, 3)$, respectively, and $\epsilon_{ijk} = 0$ if any index is repeated. We have also introduced the antisymmetric diffusion coefficient κ_A. Note that the symmetric terms reflect the diffusion due to small-scale turbulent fluctuations; in contrast, the antisymmetric term contains the particle drifts discussed in Section 9.2.3 caused by the spatial variations of the large-scale magnetic field.

9.4.1 Quasi-linear theory

We have stated previously that the diffusion coefficients, $\kappa_\parallel$ and $\kappa_\perp$, are related to the magnetic field, in particular the power spectrum of its fluctuating component.

Here, we show the derivation of $\kappa_\parallel$. According to the quasi-linear approximation (Earl, 1974; Luhmann, 1976), the spatial transport coefficient along the mean magnetic field is related to the pitch-angle diffusion coefficient, $D_{\mu\mu}$, as follows

$$\kappa_\parallel(v) = \frac{v^2}{4}\int_0^1 \frac{(1-\mu^2)^2}{D_{\mu\mu}}\mathrm{d}\mu, \tag{9.22}$$

where μ is the cosine of the pitch angle. $D_{\mu\mu}$ is related to the power in the random magnetic field fluctuations by

$$D_{\mu\mu} = \frac{\pi}{4}\Omega_0(1-\mu^2)\frac{k_{\mathrm{res}}P(k_{\mathrm{res}})}{B_0^2}, \tag{9.23}$$

where $k_{\mathrm{res}} = |\Omega_0/v\mu|$ is the resonant wave number, Ω_0 is the cyclotron frequency in the background magnetic field, B_0 (Jokipii, 1966).

Equation (9.23) is strictly applicable only for the case of one-dimensional turbulence in which the wave vectors are aligned with the mean field.

The relationship between $D_{\mu\mu}$ and the magnetic power spectrum follows from the equations of motions from individual charged particles. The method of solution is to solve the Lorentz force (Eq. 9.1) for a magnetic field that is composed of a mean component and a fluctuating component. The usual approach is to use a perturbation analysis in which the zeroth-order solution (the one for which there is no fluctuating component) is substituted into the original equation and only terms that are first order in $\delta B/B$ (where δB is the magnitude of the fluctuating component and B the magnitude of the mean component) are retained. The resulting equations are readily solved analytically. One then generally uses these solutions to obtain Fokker–Plank coefficients. For example, for the case of pitch-angle diffusion, the relevant Fokker–Planck coefficient is $D_{\mu\mu} = \langle\mu^2\rangle/(2\Delta t)$, where the angle-bracket notation refers to an ensemble average (average over many experiments, or realizations of the fluctuating field), and Δt refers to a relatively short time interval. The result is Eq. (9.23). This analysis was first performed by Jokipii (1966).

It is important to note that the quasi-linear theory is applicable when the time over which the zeroth-order solution to the equations of motion is valid is long compared to the time it takes a particle to traverse one coherence scale of the magnetic field. Additionally, we also generally require $\delta B/B \ll 1$. These criteria are not always well satisfied in space-physics applications, yet the method is widely used.

The interplanetary magnetic-field power spectrum has the form shown in Fig. 9.9, which comes from the work by Jokipii and Coleman (1968). It is flat for low frequencies (large wavelengths) and becomes power law at the coherence scale. The slope of the spectrum above this break is approximately equal

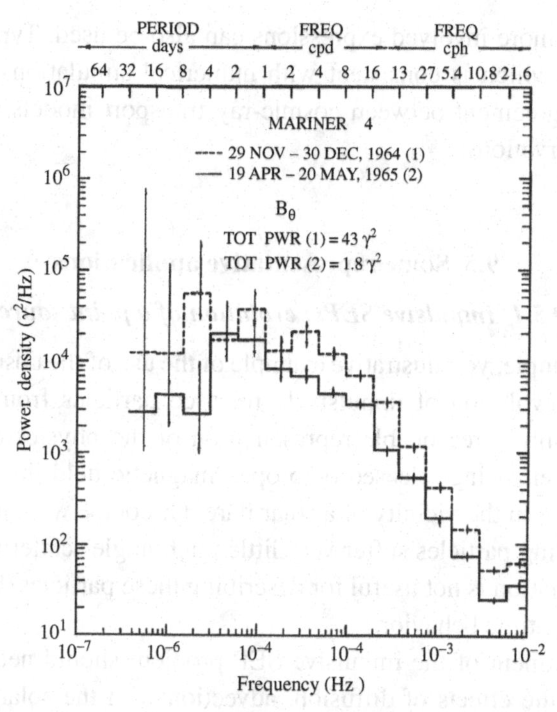

Fig. 9.9. The observed power spectrum of the latitudinal component of the interplanetary magnetic field. (From Jokipii and Coleman, 1968.)

to that predicted by Kolmogorov (1941). At very small wavelengths (high frequencies), the magnetic turbulence is dissipated at a scale called the dissipation scale. For a discussion of interplanetary magnetic-field turbulence, see Chapter 7 in Vol. I. A reasonable representation of the power in the magnetic fluctuations in the interplanetary magnetic field is

$$P(k) = \sigma^2 \frac{1}{1 + (kL_c)^{5/3}} \left(\int_0^\infty \frac{dk}{1 + (kL_c)^{5/3}} \right)^{-1}, \tag{9.24}$$

where L_c is the coherence length (usually taken to be about 0.01 AU based on spacecraft observations) and σ^2 is the turbulence variance (usually about $0.3B_0^2$). Note that this is normalized such that if the power spectrum is integrated from $k = 0$ to $k = \infty$, then the total variance is the result, as expected. Inserting this into Eqs. (9.21) and (9.22) we find

$$\kappa_\parallel(v) = \frac{3v^3}{20 L_c \Omega_0^2} \left(\frac{B_0}{\sigma} \right)^2 \csc\left(\frac{3\pi}{5} \right) \left[1 + \frac{72}{7} \left(\frac{\Omega_0 L_c}{v} \right)^{\frac{5}{3}} \right]. \tag{9.25}$$

Often modelers take $\kappa_\perp = \epsilon \kappa_\parallel$, where $\epsilon \ll 1$ is taken to be some constant. This has some justification from numerical simulations (Giacalone and Jokipii,

1999), although more involved expressions can also be used. Typically, one takes $\epsilon = 0.02$–0.05, which is consistent with numerical simulations and which also gives the best agreement between cosmic-ray transport models (both ACRs and GCRs) and observations.

9.5 Some representative applications

9.5.1 Impulsive SEPs: evolution of a point source

A particularly simple, yet illustrative example of the use of the cosmic-ray transport equation is the evolution of impulsively released particles from a point source. This is presumably a reasonable representation of the physics of SEP transport subsequent to their being released onto open magnetic field lines following their rapid acceleration in the vicinity of a solar flare. Of course we must recognize that the earliest arriving particles suffer very little pitch-angle scattering, and therefore the transport equation is not useful for describing these particles, but is adequate to describe the long-time behavior.

A proper treatment of the impulsive SEP problem should necessarily include, as a minimum, the effects of diffusion, advection with the solar wind, and adiabatic cooling. Spherical coordinates with the origin at the Sun would be a good choice. The resulting equation, even when simplified by making various assumptions about the choice of parameters can be quite difficult, or even impossible, to solve analytically. For our purposes here, which is simply for illustration and by no means is meant to be directly comparable to SEP observations, it suffices to consider a Cartesian geometry, a constant diffusion coefficient, and to neglect both advection with the flow and energy change. The result is simply Eq. (9.12), which is the one-dimensional diffusion equation. The solution for an impulsive injection of particles at $x = 0$ at time $t = 0$ is given by

$$f(x, t) = \frac{N_0}{\sqrt{4\pi\kappa t}} \exp\left(-\frac{x^2}{4\kappa t}\right), \tag{9.26}$$

where N_0 is the number of particles released.

Figure 9.10 shows a plot of the distribution of particles, given by Eq. (9.26), at the location $x = 1$ AU, as a function of time (in days). The diffusion coefficient was taken to be $\kappa = 2 \times 10^{21}$ cm^2/s, and $N_0 = 10^{14}$. If, for example, these are 10 MeV protons, then the corresponding mean free path would be about 0.1 AU. This profile has similarities to those seen at 1 AU following a flare or CME on the Sun, which illustrates the merit of the transport equation (of course, to do this problem correctly, one should include all of the transport effects). There has been a vast literature on the subject of modeling solar-energetic particles, many of which

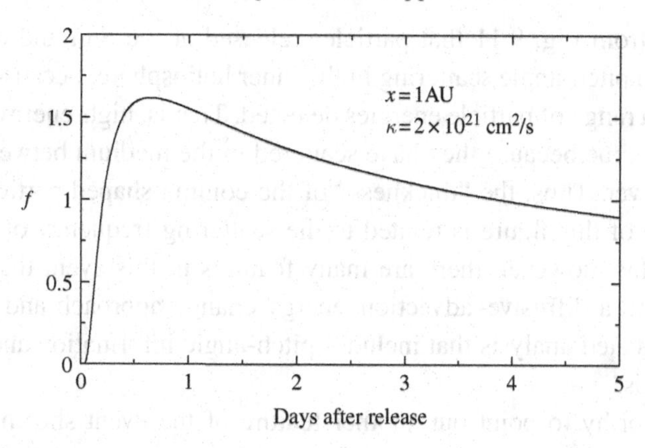

Fig. 9.10. Solution to the one-dimensional diffusion equation for a point-source release at a position 1 AU away from an observer: $f(1, t)$ from Eq. (9.25).

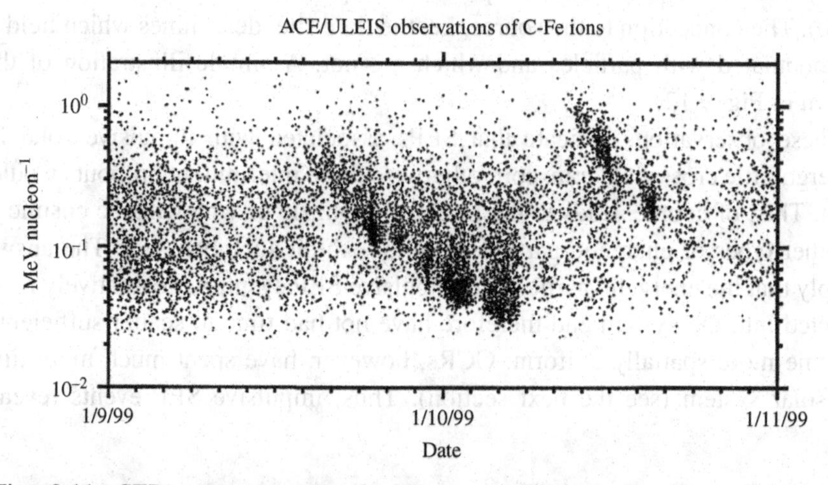

Fig. 9.11. SEP event, associated with an impulsive solar flare, seen by ACE/ULEIS. Each dot represents the detection of a particle by the detector. Two distinct events are shown. (Figure adapted from Mazur *et al.*, 2000. Reproduced by permission of the AAS.)

start with forms similar to the transport equation (the review by Reames, 1999 is a good starting point for a literature review of the subject).

An example of an impulsive-like SEP event observed at 1 AU by the ACE spacecraft (ULEIS instrument) is shown in Fig. 9.11. This plot is taken from the paper by Mazur *et al.* (2000), where each dot represents a detection by the instrument of an individual particle. Plotted is the particle kinetic energy versus time. The earliest arriving particles are the ones with the highest energy since they move with the highest speed. The slower ones arrive later. This velocity dispersion leads to the characteristic profile shown in the figure.

It is clear from Fig. 9.11 that particles released at the Sun and observed near Earth undergo pitch-angle scattering in the inner heliosphere, because at any given time there is a range of particle energies detected. That is, high-energy particles can arrive later in time because they have scattered in the medium between the source and the observer. Thus, the "thickness" of the comma-shaped particle event seen in the middle of this figure is related to the scattering frequency of the particles. Aside from this, however, there are many features in this event that are difficult to explain with a diffusive–advection–energy change approach and these require more sophisticated analysis that includes pitch-angle information and meandering magnetic fields.

It is noteworthy to point out another feature of the event shown in Fig. 9.11. There are intermittent dropouts in intensity during each of the two distinct events shown. These dropouts have been interpreted as resulting from the passage of alternately filled and empty "tubes" of particle flux past the spacecraft (Giacalone *et al.*, 2000). The connection to the source, i.e. the flare site, determines which field lines are populated with particles and which are not. A simple illustration of this is shown in Fig. 9.12.

These observations indicate that SEPs associated with impulsive solar flares undergo little cross-field transport, otherwise these intermittent dropouts would not exist. This, of course, leads to the interesting puzzle of why galactic cosmic rays, or other types of energetic particles, do not exhibit such behavior. The answer is simply that the energetic particles in impulsive SEP events were relatively recently injected into the system and therefore have not had time to scatter sufficiently to become more spatially uniform. GCRs, however, have spent much more time in the solar system (see the next section). Thus, impulsive SEP events reveal the

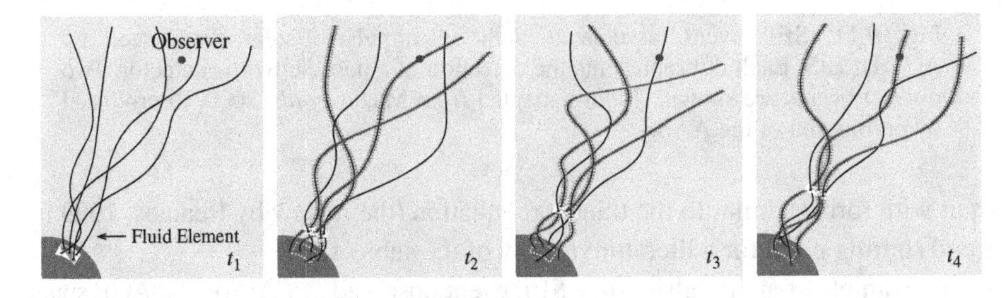

Fig. 9.12. An illustration of a possible interpretation of the intermittent intensity variations seen within the events shown in Fig. 9.11. The plots show five magnetic field lines, three of which are populated with particles at $t = 0$ (far left panel), and the other two are not. An observer is indicated towards the upper part of each plot. As the observer moves past various field lines that are advected with the solar wind flow, it sometimes sees energetic particles and sometimes not, depending on whether the field line it is presently seeing is connected to the source.

early time behavior of a collection of energetic charged particles moving in the heliospheric magnetic field.

9.5.2 Galactic cosmic-ray modulation

As we discussed in Section 9.1.1, GCRs are cosmic rays that pervade interstellar space and enter the heliosphere from the outside. The vast majority of them are swept out of the heliosphere before ever reaching Earth's orbit. The subject of GCR modulation has received much attention and is nicely reviewed by Potgieter (1998). However, for the purpose of a simple illustration of modulation, consider the steady-state Parker transport equation in one-dimensional spherical coordinates given by

$$\frac{1}{r^2}\frac{\partial}{\partial r}\left(r^2\kappa\frac{\partial f}{\partial r}\right) - v\frac{\partial f}{\partial r} + \frac{2vp}{3r}\frac{\partial f}{\partial p} = 0, \tag{9.27}$$

where v is the solar wind speed, which we take to be a constant (this simple illustration neglects the effect of the heliosheath and termination shock). Here we have taken the diffusion tensor to be symmetric and $\kappa_{rr} = \kappa$.

It is convenient to rewrite Eq. (9.27) in the following form:

$$\frac{1}{r^2}\frac{\partial}{\partial r}r^2\left(\kappa\frac{\partial f}{\partial r} - vf\right) + \frac{2v}{3rp^2}\frac{\partial}{\partial p}(p^3 f) = 0. \tag{9.28}$$

Generally this equation is not easy to solve, but if we assume that the second term on the left (describing the energy change of diffusing particles) is negligible, the resulting equation is readily solved to yield

$$f(r, p) = f(R, p)\exp\left(-\int_r^R \frac{v}{\kappa(r', p)}dr'\right). \tag{9.29}$$

Equation (9.29) gives an exponential decay of particles from the source ($r = R$) inward, into the solar system (where $r < R$). Moreover, it is reasonable to expect the diffusion coefficient to increase with momentum p so that higher-energy particles have a larger diffusion coefficient than lower-energy particles. Thus, higher-energy particles have a longer exponential-decay length, or diffusive skin depth, than do lower energy ones. Thus, they more easily reach the inner heliosphere than lower-energy cosmic rays. This leads to a turnover in the spectrum that is due to modulation. This is in qualitative agreement with the observed cosmic-ray spectrum at Earth as shown in Fig. 9.2.

9.5.3 The lifetime of GCRs in the heliosphere

Because the cosmic-ray spectrum decreases with increasing energy, most GCRs are of very low energy. Also, because of GCR modulation, low-energy particles are more easily swept out of the heliosphere. Therefore, the typical lifetime of a GCR in the solar system is very short. But how long do GCRs that reach Earth's orbit stay in the heliosphere? One way to address this question is through a simple dimensional analysis. If we assume that diffusion is the dominant process (of course advection with the solar wind is extremely important for the lowest energy particles, so this analysis should be considered to give an underestimate), then the characteristic time scale for diffusion in a sphere of radius R would be $\tau \sim \frac{R^2}{\kappa}$. It is reasonable to expect κ to depend inversely on the strength of the magnetic field based on our discussion in Section 9.4. That is, the weaker the field, the larger the diffusion coefficient (larger mean free path). The strength of the interplanetary magnetic field falls off with heliocentric distance for large distances from the Sun. Therefore, the appropriate diffusion coefficient to use in the above expression is that of the outer heliosphere, where it is the largest. A reasonable value turns out to be about $\kappa = 10^{24}\,\mathrm{cm}^2/\mathrm{s}$. Taking $R = 150\,\mathrm{AU}$, we find that the typical lifetime in the heliosphere of a GCR particle that reaches Earth orbit is of the order of a few months.

9.5.4 Cosmic-ray transport in the heliosphere

Shown in Fig. 9.13 is the daily count of neutrons produced by the impact of cosmic rays on the upper atmosphere, from ground-based neutron monitors. This is an indirect measure of the cosmic-ray flux in near-Earth orbit. The time–intensity profile shows a clear 11-year cycle that is coincident with the sunspot-number cycle. During periods of high solar activity, sunspot maximum, the cosmic-ray flux is low, and during periods of low solar activity, or solar minimum, the cosmic-ray flux is high. In addition to this, there is also a 22-year cycle present (the alternating

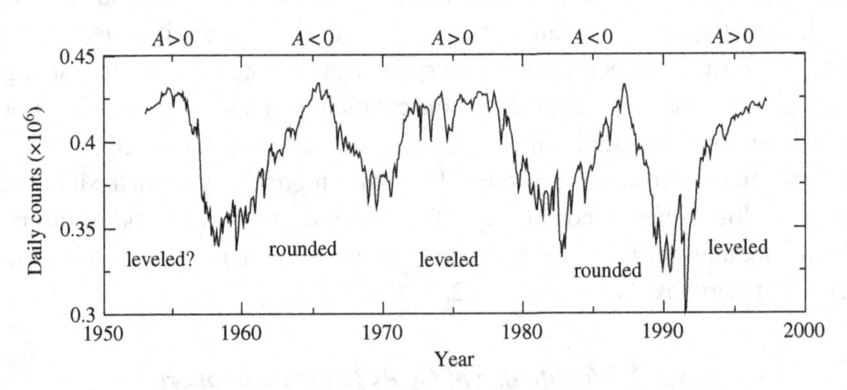

Fig. 9.13. Climax neutron monitor daily count rate of neutrons produced by the interaction of a primary cosmic ray with Earth's atmosphere. The meaning of A is defined in Fig. 9.14.

"leveled" vs. "rounded" cosmic-ray flux), which, as we discuss below, is related to the drift motions of cosmic rays.

The increased modulation during periods of solar maximum is related to a combination of effects related to the shedding of magnetic flux by the Sun at solar maximum. On the one hand, increased solar activity leads to more magnetic turbulence, which decreases the diffusion coefficient in the outer heliosphere leading to more modulation. On the other hand, and in addition to this, the merging of more numerous transient shocks and coronal mass ejections in the distant heliosphere creates magnetic barriers (so-called global merged interaction regions, or GMIRs; see Chapter 7) which also reduce the transport of cosmic rays into the inner heliosphere (leRoux and Potgieter, 1995). There is a lower level of magnetic turbulence and fewer magnetic barriers for cosmic rays to propagate through during solar minimum. This is a qualitative explanation for the 11-year cosmic-ray cycle and its relation to the sunspot-number cycle.

The 22-year cosmic-ray cycle seen in Fig. 9.13 is related to the 22-year solar magnetic polarity cycle (i.e. the polarity of the magnetic field during any given solar minimum is reversed from that in the previous minimum). The polarity of the Sun's magnetic field is important for the cosmic-ray drift that arises from the antisymmetric part of the diffusion tensor in Parker's transport equation. The role of drifts was first pointed out by Jokipii *et al.* (1977) and has been studied extensively (the interested reader should consult the articles by Jokipii and Thomas, 1981; Kota and Jokipii, 1983; and Potgieter and Moraal, 1985).

Including the drifts of cosmic rays has led to the widely accepted paradigm for cosmic-ray transport shown in Fig. 9.14. Drift motions for protons during two different solar polarity cycles are shown. During the period in which the solar magnetic field spirals outward in the north and inward in the south ($A > 0$, left panel) the GCR protons drift into the heliosphere from the polar regions of the

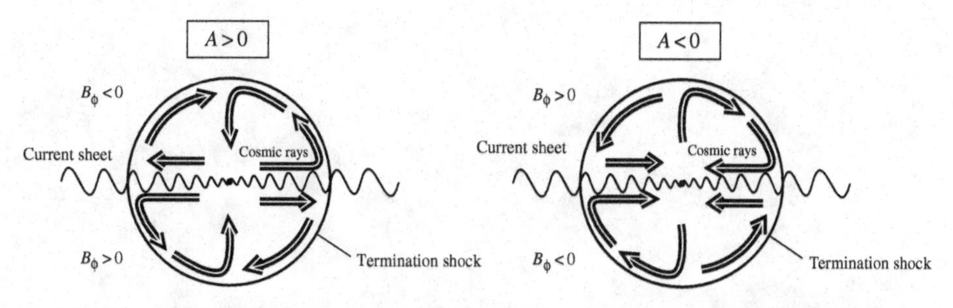

Fig. 9.14. Drift motion of cosmic rays in the heliosphere for two different solar magnetic-polarity cycles. The two polarities of the solar magnetic field are separated by the heliospheric current sheet. The value of $A > 0$ during the period in which the solar magnetic field is outward in the north and inward in the south. The termination of the solar wind is also shown.

heliosphere and outward along the heliospheric current sheet (which separates the two hemispheres and where the field reverses direction, hence the term "current sheet"). During the opposite polarity, in which the solar field is inward in the north and outward in the south ($A < 0$, right panel), GCR protons drift into the heliosphere along the current sheet. Note that, in addition to the drift along the current sheet, there is also a gradient-B drift along the termination shock resulting from the jump in the magnetic field strength across the shock.

The explanation for the alternating leveled and rounded cosmic ray intensity involves both the drift motions of the cosmic rays shown in Fig. 9.14 and the "waviness" of the heliospheric current sheet due to the offset of the solar magnetic axis and its rotation axis. When the "tilt" is large, the current sheet is very warped, whereas when it is small, the current sheet is much flatter (imagine the current sheet – as in Vol. I, Fig. 6.2 – forming above the rotating Sun with a tilted axial dipole, as in Vol. I, Fig. 8.1). The current sheet is generally known to be relatively flat during the center of the solar cycle minimum. So, during the cycle in which the cosmic rays come into the heliosphere along the current sheet, only when it is very flat will the full cosmic-ray flux be reached at Earth's orbit. Thus, during this phase, the cosmic-ray intensity will exhibit a rounded or "peaked" time–intensity profile. When the cosmic rays come in along the poles of the heliosphere, the full intensity is reached much sooner and remains at a high level throughout solar minimum, and hence, during this phase, the time–intensity profile is more level, or flat.

10

Energy conversion in planetary magnetospheres

VYTENIS VASYLIŪNAS

10.1 Introduction

Planetary magnetospheres, by their very nature, provide plenty of possibilities for the development of energy conversion processes. Fundamentally a planetary magnetosphere (see e.g. Vol. I, Chapter 10) is simply the interface between two distinct regions: on the outside, the solar wind; on the inside, the ionosphere, atmosphere, and surface of the planet. The quite different motions of matter within the two regions, together with the role of the magnetic field in mediating the interaction between them, lead (almost unavoidably, it seems) to configurations of changing energy; the changes occur on a variety of time scales, ranging from quasistatic to explosive.

In keeping with the general approach adopted in this series of textbooks, this chapter aims to present energy conversion in planetary magnetospheres in general terms as part of a sub-branch of physics, namely the discipline of magnetospheric physics (which in turn is a sub-branch of heliophysics). Many of the concepts and basic results, however, originate from specific observations at and near Earth; accordingly, the chapter begins (Section 10.2) with a phenomenological overview of geophysical processes related to space storms and radiation. The physical description of energy conversion processes is then developed (Sections 10.3, 10.4, 10.5) and applied to interpret the phenomenology of energy-conversion events, both at Earth (Section 10.6) and at other planets (Section 10.7). The chapter concludes (Section 10.8) with a sketch of a possibly universal process.

10.2 Overview of disturbances in Earth's space environment

Of the observed phenomena related to energy conversion processes in outer space, the polar aurora is the earliest known (with records and traces in history, mythology, literature, and the arts reaching back millennia; see Chapter 2,

Heliophysics: Space Storms and Radiation: Causes and Effects, eds. Carolus J. Schrijver and George L. Siscoe. Published by Cambridge University Press.

and e.g. Eather, 1980) and the easiest to observe, even without instruments. Next come disturbances of the Earth's magnetic field, detectable with relatively simple instruments available by the mid-nineteenth century. By the early twentieth century, the two phenomena were known to be connected, and the concept of *magnetic storm* was already current: geomagnetic disturbance of wide (global) extent on time scales of hours to days, unusually intense storms associated with occurence of aurora at unusually low latitudes, evidence of connection with solar activity. More localized auroral manifestations and intense geomagnetic disturbances at high latitudes, on time scales of minutes to hours, were studied under a variety of designations and synthesized much later (1960s) into the concept of *magnetospheric substorm*, with the help of *in situ* outer space observations which were becoming available and proved essential to establish the physical nature of the phenomenon. For brief historical accounts, see e.g. Chapman (1969; one of the key participants), Siscoe (1980), Egeland (1984), and Stern (1991).

The magnetic storm is defined nowadays (Gonzalez *et al.*, 1994) by the time variation of the geomagnetic Dst (disturbance storm time) index, illustrated schematically in Fig. 10.1. The Dst index (see e.g. Mayaud, 1980) is a measure of

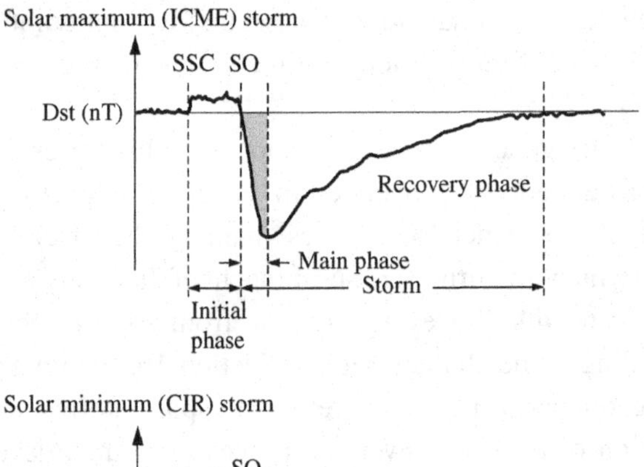

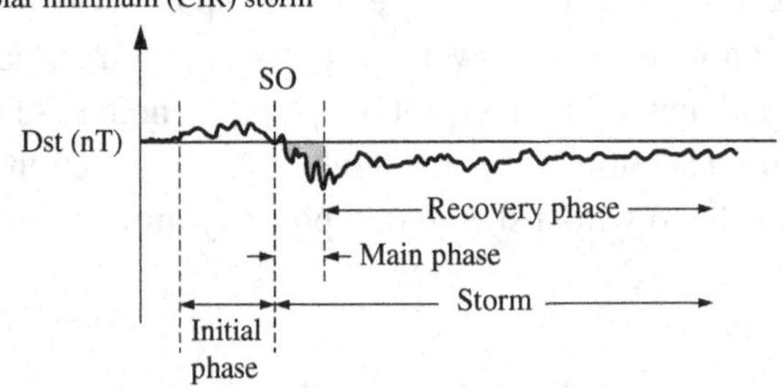

Fig. 10.1. Schematic time history of geomagnetic field variation for two characteristic magnetic storms. Time range: several days. Vertical variation range: ~100–200 nT. SSC: storm sudden commencement. SO: storm onset. The top panel shows the storm development in response to a characteristic interplanetary coronal mass ejection (ICME), and the bottom panel that for the passage of a corotating interaction region (CIR). (Figure adapted from Tsurutani *et al.*, 2006.)

a quasi-uniform magnetic disturbance field near the Earth, aligned with the dipole axis (northward for Dst > 0), such as would be produced by a ring of electric current (westward if Dst < 0) near the equatorial plane. A prolonged (hours to days) interval of negative Dst values constitutes a magnetic storm. The peak negative excursion is often taken as a measure of storm intensity: Dst −30 nT to −50 nT are weak storms, −50 nT to −100 nT moderate, and over −100 nT intense; storms over −300 nT occur at most a few times during a solar cycle (Earth's dipole field at the equator is about 31 000 nT, for comparison). The storm sudden commencement and the initial phase of positive Dst, which accompany many but not all storms, are no longer considered necessary ingredients of the storm concept.

As discussed in Section 10.6.2, the field depression quantified by Dst is the result of plasma pressure that inflates the dipole field. The essential phenomenon of the magnetic storm is thus the addition of a large amount of plasma energy to the dipolar field region of the magnetosphere. Furthermore, it is now well established that this energy addition results from a particular condition in the solar wind: "a sufficiently intense and long-lasting interplanetary convection electric field" (Gonzalez *et al.*, 1994), meaning $-\mathrm{v} \times \mathrm{B}/c$, for the IMF's southward component.

In contrast to the magnetic storm, there is much less unanimity on what defines a magnetospheric substorm (Rostoker *et al.*, 1980, 1987). Probably the most spectacular phenomenon, and the one most widely used as a unifying concept, is the auroral substorm, summarized in the classic figure of Akasofu (1964) reproduced here in Fig. 10.2, which illustrates schematically, by a time sequence of polar views of Earth, the development of the auroral forms (light-emitting regions) during what is called the *expansion phase* of the substorm: beginning with an initial brightening at the lowest latitudes near midnight (*onset*), the aurora intensifies greatly, becomes very complex in spatial structure (*auroral breakup*) and expands, predominantly westward and poleward but also eastward, eventually subsiding in a *recovery phase*. This auroral development is accompanied by strong geomagnetic disturbances (commonly reaching ~1500 nT and more), with a spatial distribution almost as complex as that of the aurora but describable roughly as equivalent to a current above the Earth (*auroral electrojet*) that is westward near and before midnight and eastward after midnight. Note: although the development shown in Fig. 10.2 is in the Northern Hemisphere only, essentially the same sequence also occurs simultaneously in the Southern Hemisphere, at the (more or less) magnetically conjugate locations (the resemblance to a two-ribbon solar flare, with ribbons of opposite magnetic polarity, has been repeatedly remarked upon).

Within the magnetosphere, the substorm expansion phase is marked by (1) greatly enhanced intensities and energies of charged particles, (2) changes of the magnetic field in the nightside magnetosphere and magnetotail, the initially tail-like field becoming more dipolar (*dipolarization*), and (3) fast ($\sim v_{\mathrm{A}}$) bulk flows

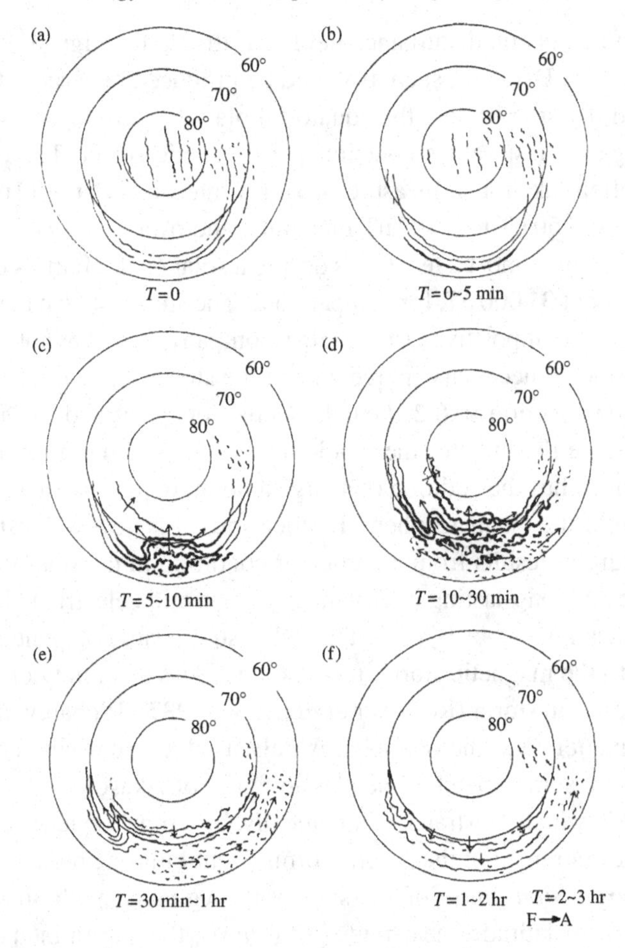

Fig. 10.2. Schematic diagram of an auroral substorm. View from above the North Pole, circles of constant geomagnetic latitude, Sun toward the top (Akasofu, 1964).

of plasma in the magnetotail, predominantly away from Earth at larger distances. This is the merest sketch of substorm phenomenology; for more detailed accounts, see e.g. Akasofu (1977), Kennel (1995) and Syrjäsuo and Donovan (2007).

Not shown in Fig. 10.2 is the substorm *growth phase* which, was not identified until some years after 1964: a time interval (~0.5−1 h) preceding the substorm onset, during which magnetospheric convection (see Vol. I, Section 10.4.3) observed in the ionosphere is enhanced, the amount of open magnetic flux in the magnetosphere increases, and quiet-time auroral forms move equatorward (to reach their locations shown in panel A of Fig. 10.2). Generally, the beginning of the growth phase is associated with a southward turning (or an enhancement of a pre-existing southward component) of the interplanetary magnetic field B_{sw}.

What changes, if any, of B_{sw} or other solar wind parameters are associated with the substorm onset and expansion phase is a still unsettled controversy; the two extreme positions are that the onset (1) is triggered by a northward turning of B_{sw} or (2) is a purely internal development of magnetospheric dynamics.

Because the presence of a southward component of B_{sw} (opposite to the dipole field in the Earth's equatorial plane) appears to be a prerequisite for the occurrence of both storms and substorms, the question may be raised: do magnetic storms and magnetospheric substorms constitute two physically distinct phenomena, or are they merely different-time-scale manifestations of a single phenomenon? Aside from matters of time scale and sequence, one essential conceptual difference is that the defining signature of a magnetic storm represents an enhanced *storage* of plasma energy, while that of a magnetospheric substorm represents in essence (independent of arguments about what it is in detail) an enhanced *dissipation* of energy.

In summary, geomagnetic and auroral phenomena involve particle energy, stored in the magnetosphere (e.g. to inflate the magnetic field) or transferred to the atmosphere (e.g. to excite the aurora); there are related changes of magnetic field configuration, and an evidently significant role is played by the component of the interplanetary magnetic field that can reconnect with the Earth's dipole field. A physical description of energy conversion in a general heliophysical context must also include other magnetospheres (see e.g. Vol. I, Chapters 10 and 13) in which the rotation of the planet may be more important than the solar wind.

10.3 Fundamentals of energy storage, transfer, and loss

10.3.1 Forms of energy

Throughout this chapter, I take a fundamental physical approach, treating energy as a *field* quantity, localizable to any point $(\mathbf{r}, t)$ of space and time (in contrast to an engineering approach, with energy assigned to a particular device, e.g. flywheel, capacitor, or inductor). For each form of energy, one has an *energy density* $U(\mathbf{r}, t)$ and an *energy flux density* $\mathbf{S}(\mathbf{r}, t)$, which satisfy the conservation equation

$$\frac{\partial U}{\partial t} + \nabla \cdot \mathbf{S} = \text{conversion rate}, \tag{10.1}$$

where the right-hand side represents the rate per unit volume of conversion of energy into or out of the particular form.

Three forms of energy are of direct importance for heliophysics: kinetic energy of motion, electromagnetic energy, and gravitational energy. The latter two are the energies of the two long-range fields which (as discussed in Vol. I, Chapter 1) act to organize matter in the cosmos. Nuclear energy (associated with the short-range fundamental forces) is of course the ultimate source of energy that powers the luminosity of the Sun and other stars, but its direct presence is confined to deep stellar interiors (fusion reactions) and, to a minor extent, planetary interiors (radioactivity); elsewhere it only appears in any significant amounts after conversion to other forms.

Kinetic energy of motion includes both energy of bulk flow and energy of thermal motions; the total, including both, is conveniently referred to as *mechanical* energy, for which the conservation equation (10.1) takes the form

$$\frac{\partial}{\partial t} U_{\text{mech}} + \nabla \cdot \left[\mathbf{v} U_{\text{mech}} + \mathsf{P} \cdot \mathbf{v} + \mathbf{q}\right] = \mathbf{E} \cdot \mathbf{J} + \rho \mathbf{v} \cdot \mathbf{g}, \tag{10.2}$$

$$U_{\text{mech}} \equiv \tfrac{1}{2}\rho v^2 + \epsilon, \qquad \epsilon = \text{Trace}\,(\mathsf{P})\,.$$

For electromagnetic energy, the conservation equation is given by Poynting's theorem

$$\frac{\partial}{\partial t}\frac{1}{8\pi}\left[B^2 + E^2\right] + \nabla \cdot \left[\frac{c}{4\pi}\mathbf{E} \times \mathbf{B}\right] = -\mathbf{E} \cdot \mathbf{J}. \tag{10.3}$$

For gravitational energy, an approximate expression adequate for most purposes of heliophysics and magnetospheric physics (Siscoe, 1983) is

$$\frac{\partial}{\partial t}\left[\rho \Phi_G\right] + \nabla \cdot \left[\rho \mathbf{v} \Phi_G\right] = -\rho \mathbf{v} \cdot \mathbf{g}. \tag{10.4}$$

(In the above equations, ρ is the mass density, $\mathbf{v}$ the bulk flow velocity, P the pressure tensor, $\mathbf{q}$ the heat flux vector, $\mathbf{B}$ the magnetic and $\mathbf{E}$ the electric field, $\mathbf{J}$ the electric current density, Φ_G the gravitational potential, and $\mathbf{g} = -\nabla\Phi_G$ the gravitational acceleration.)

The conversion rates between different forms of energy are given by

$$\begin{array}{ll} \mathbf{E} \cdot \mathbf{J} > 0 & \text{electromagnetic} \longrightarrow \text{mechanical} \\ \mathbf{E} \cdot \mathbf{J} < 0 & \text{mechanical} \longrightarrow \text{electromagnetic} \end{array} \tag{10.5}$$

and

$$\begin{array}{ll} \rho \mathbf{v} \cdot \mathbf{g} > 0 & \text{gravitational} \longrightarrow \text{mechanical} \\ \rho \mathbf{v} \cdot \mathbf{g} < 0 & \text{mechanical} \longrightarrow \text{gravitational;} \end{array} \tag{10.6}$$

there is no direct conversion between electromagnetic and gravitational energy (at least as long as general relativistic effects are neglected). If all the energy equations (10.2), (10.3), and (10.4) are added together, the conversion terms on the right-hand sides add to zero, implying conservation of total energy:

$$\frac{\partial}{\partial t} U_{\text{total}} + \nabla \cdot \mathbf{S}_{\text{total}} = 0. \tag{10.7}$$

Note that the conversion rates are *not* independent of frame of reference. All three quantities – energy density, energy flux density, and energy conversion rate – vary with choice of frame of reference, in such a way that the *form* of the energy equation remains invariant. Sometimes a profound significance is claimed for the sign of $\mathbf{E} \cdot \mathbf{J}$, regions with $\mathbf{E} \cdot \mathbf{J} < 0$ or > 0 being identified as "dynamo" or "load", respectively; since physics is frame-independent, this distinction cannot be fundamental.

10.3.2 Sources of energy for magnetospheres

Strictly speaking, there can be no source of energy as such: according to Eq. (10.7), energy can neither be created nor be destroyed but can only be converted from one form to another or transported from one region to another. For a region bounded in space such as a planetary magnetosphere, however, the term *energy source* is often applied to energy transported into the region across the boundary. The external source (in this sense) of energy for a planet and its associated system is the Sun, which supplies energy in two forms: electromagnetic radiation and the solar wind. The power carried by the electromagnetic radiation (solar luminosity) is observed to exceed that carried by the solar wind by a factor $\sim 10^6$; with $v_{sw}/c \sim 10^{-3}$, this implies that the rate at which the Sun is losing mass through relativistic energy-equivalent mass removal by the solar radiation is comparable to the mass outflow by the solar wind (Axford, 1985).

The solar radiation is the dominant energy source for the planet, the atmosphere, and part of the ionosphere. For the magnetosphere and the upper regions of the ionosphere, on the other hand, the solar wind is the only significant external source of energy available; solar radiation does not interact at all with these regions, where the density of matter is sufficiently low to make the mean free path for interaction with photons vastly larger than the size (column depth) of the system. (For the same reason, dynamics of the solar wind and the interplanetary magnetic field can be treated without reference to the omnipresent solar visible radiation: at 1 AU, for instance, one discusses magnetic fields typically of order ~ 10 nT and electric fields ~ 4 mV m^{-1}, while ignoring magnetic fields $\sim 10^3$ nT and electric fields ~ 30 V m^{-1} that are simultaneously present – albeit oscillating at $\sim 10^{15}$ Hz.)

When considering the solar wind as the energy source, only the kinetic energy of plasma bulk flow is of importance; the thermal and magnetic energies of the solar wind can be neglected, for a reason somewhat more subtle than might appear at first. They are small compared to the kinetic energy of the bulk flow, but not necessarily small compared to energies dissipated in the magnetosphere; the reason they are not important is that at the bow shock they are overwhelmed by additional thermal and magnetic energies extracted from the flow. Furthermore, to transfer magnetic energy across the magnetopause requires a normal component of the Poynting vector, hence a tangential component of the electric field, which interacts with the magnetopause current to extract more mechanical energy from the plasma; thus the Poynting vector just inside the magnetopause is in general completely different from the Poynting vector just outside (and also from the Poynting vector in the solar wind). The interplanetary magnetic field does exert a dominant influence on energy conversion processes in a planetary magnetosphere, but primarily by control of magnetic reconnection processes and open field lines – not by entry of the solar-wind Poynting flux into the magnetosphere.

An interior source of energy available for a planetary magnetosphere is planetary rotation (other sources of energy internal to the planet, e.g. heating by radioactivity or by slow contraction, have in general no direct interaction with the magnetosphere).

10.3.3 Energy loss and dissipation processes

Analogously to "energy source", the term *energy loss* (or *sink*) is often used to denote a process in which energy is transported out of the region under consideration, or else transformed into a form that allows it to escape from the system with no further interaction. A related concept is that of *energy dissipation*, a process in which energy is transformed into heat in the thermodynamic sense, with increase of entropy (for a detailed discussion in relation to the energy and momentum equations, see e.g. Vasyliūnas and Song, 2005; the dissipation rate so defined, unlike the energy conversion rate, is independent of frame of reference).

The following are among the principal loss and dissipation processes in planetary magnetospheres, energy being lost primarily to the atmosphere in (1) and (2) and being removed outside the system (to "infinity") in (3) and (4):

(1) **Collisional and Joule heating in the ionosphere** If the bulk flow of plasma differs from the bulk flow of the neutral atmosphere (usually as a consequence of magnetospheric dynamics), there is energy dissipation given by $\mathbf{E}^* \cdot \mathbf{J}$, where $\mathbf{E}^*$ is the electric field in the frame of reference of the neutral atmosphere. This is commonly referred to as "ionospheric Joule heating", but Vasyliūnas and Song (2005) have shown that in fact it is primarily frictional heating by collisions between plasma and neutral particles; Joule heating in the true physical sense ($\mathbf{E}' \cdot \mathbf{J}$, where $\mathbf{E}'$ is the electric field in the frame of reference of the plasma) contributing only a small fraction of the total. The energy is removed from the magnetic field and converted (via kinetic energy of relative bulk flow as an intermediary) to heat (thermal energy), with the heating rate per unit volume partitioned approximately equally between plasma and neutrals.

(2) **Charged-particle precipitation** Energetic charged particles that enter the atmosphere from above are usually said to be *precipitating*. They penetrate into the atmosphere to a depth that increases with increasing energy, until their energy is lost, going partly into heating the atmosphere and partly into ionization or other interactions.

One source of precipitating particles is simple loss from the radiation belts (see Chapter 11) or from the ring current and plasma sheet regions; the energy deposited in the atmosphere is taken from the mechanical (thermal) energy of the respective magnetospheric particle populations. In addition to these particles that precipitate merely because their velocity vectors are oriented in the appropriate direction, there are other sources of precipitating charged particles, in which the energy and the intensity of the particles have been enhanced by an acceleration process. In particular, the auroral phenomena that occur in nearly all of the planetary magnetospheres observed to date are generally interpreted as resulting from some special acceleration process that supplies the required intensities of precipitating charged particles. A widely accepted model, developed from extensive studies at Earth and applied to aurora at Jupiter and at Saturn, ascribes auroral acceleration to Birkeland (magnetic-field-aligned) electric currents accompanied by electric fields parallel to the magnetic field; the rate of energy supply to the precipitating particles is $E_{\|} J_{\|}$, hence the added energy is taken out of the magnetic field (in this model, aurora occurs only when the Birkeland current is directed upward, corresponding to electron motion downward). Auroral acceleration has also been associated with intense Alfvénic turbulence (which contains fluctuating Birkeland currents). For detailed reviews, see e.g. Paschmann *et al.* (2003) and references therein.

(3) **Emission of electromagnetic radiation** A variety of processes in planetary magnetospheres produce electromagnetic radiation of various types: atomic and molecular line emissions (from the aurora and from magnetospheric interactions with plasma and neutral tori), radio waves (wideband and narrowband), a veritable zoo of plasma waves, and even X-rays (bremsstrahlung from precipitating electrons and, possibly, nuclear line emissions excited by very energetic precipitating particles). Some aspects are discussed in Chapter 4; the emissions are of course of great interest for remote sensing of the associated processes. As far as the energetics of planetary magnetospheres are concerned, however, the amount of energy involved is negligibly small for most emissions, with only a few exceptions (UV radiation from the Io torus at Jupiter).

(4) **Energetic neutral particle escape** Neutral particles that remain within a magnetosphere must be gravitationally bound to the planet; plasma particles within the magnetosphere, on the other hand, typically have speeds that exceed (often by a large factor) the gravitational escape speed – plasma is held within the

magnetosphere by the magnetic field, not by gravity (the magnetic field itself, however, must be anchored to the planet by its gravity, as discussed in Vol. I, Chapter 1). Charge-exchange collisions between ions and neutrals, in which the outgoing neutral has the velocity of the incoming ion and vice versa, thus produce fast neutrals that escape from the system immediately, with their (newly acquired) kinetic energy. This process represents a loss (generally by quite significant amounts) both of neutral particles and of energy from the magnetosphere.

(5) **Dissipation processes in the magnetosphere** In regions of the magnetosphere with major departures from the MHD approximation (particularly where magnetic reconnection is occurring) dissipative processes such as Joule heating associated with effective resistivity may be significant. The primary effect is not energy loss but enhancement of conversion from magnetic to thermal energy.

10.3.4 Reservoirs of energy

The field approach to energy implies that energy may be regarded as *stored* in space, the energy density of the various forms being given by the terms that are time-differentiated in the energy equations (10.2), (10.3), and (10.4). The primary reservoir of stored mechanical energy in a planetary magnetosphere is the thermal energy of its various plasma structures, especially the *plasma sheet* of the magnetotail or magnetodisk, the *ring current*, and the plasma and neutral *tori* associated with the planet's moons (see e.g. the description of the structures in Vol I, Section 10.5.3); the kinetic energy of bulk flow of magnetospheric plasma also plays a role, particularly for plasma tori and in the case of rapid changes discussed in Section 10.5.

The primary reservoir of stored electromagnetic energy of importance for a planetary magnetosphere is the energy of the magnetic field; except for high-frequency radiation, which does not interact with the magnetosphere, the energy in the electric field is negligible in comparison to that in the magnetic field. Because the energy of the planetary dipole field itself does not change (except on time scales of the secular variation, $\sim 10^2 - 10^3$ years for Earth) and thus has no effect on the energetics of the magnetosphere, a convenient measure of stored electromagnetic energy is the energy of the total magnetic field minus the (unchanging) energy of the dipole field:

$$\frac{1}{8\pi} \int \left[B^2 - \left(B_{\text{dipole}}\right)^2 \right] \mathrm{d}V.$$

The stored gravitational energy can be changed only by a net radial displacement of matter; any such effects in the magnetosphere are for the most part negligible in comparison to changes of mechanical or magnetic energy.

10.4 Energy budget of magnetospheres

The topic of this chapter may now be formulated as follows: the primary sources of energy for a planetary magnetosphere being the kinetic energy of bulk flow, both exterior (solar wind flow) and interior (planetary rotation) to the magnetospheric volume, by what process and in what form does the energy enter the magnetosphere, what are its flow paths and conversions within the magnetosphere, what are its ultimate sinks, and what determines the time history of these developments? In this section, I first consider these questions without reference to explicit time variations, with particular attention to the role of stress balance and magnetic flux transfer (some of the issues are briefly discussed also in Vol. I, Chapter 11), leading to the construction (Section 10.4.3) of a schematic diagram for the magnetospheric global energy budget in a quasi-steady or time-averaged context. Then I consider in Section 10.5 the time-varying energy conversion processes, many of which can be described as consequences of time offsets or delays in the interactions corresponding to particular branches of the average energy budget diagram.

10.4.1 Extracting energy from bulk flow

Bulk flow of a medium carries not only kinetic energy but also linear momentum; extracting kinetic energy from the flow necessarily means also extracting linear momentum, which requires a force to be applied to the medium. Similarly, rotation of a body carries angular momentum; extracting kinetic energy from the rotation necessarily means also extracting angular momentum, which requires a torque to be applied to the body.

(1) **Relation between global energy input rate and force/torque** The net rate of energy extraction (power) $\mathcal{P}_{sw}$ from solar wind flow is equal to the difference of the solar wind kinetic energy flux across two surfaces perpendicular to the Sun–planet line, surface 1 ahead of the bow shock and surface 2 far downstream of the entire interaction,

$$\begin{aligned}\mathcal{P}_{sw} &= \frac{1}{2}\int_1 \rho v^3\,\mathrm{d}A - \frac{1}{2}\int_2 \rho v^3\,\mathrm{d}A \\ &= \frac{1}{2}\int \rho v(\overline{v}_1^2 - \overline{v}_2^2)\,\mathrm{d}A \\ &= S_{\mathrm{ft}}\,\overline{v}\Delta v \end{aligned} \tag{10.8}$$

(subscripts sw on ρ and v have been omitted, for simplicity), and the total force F is similarly equal to the difference of the linear momentum flux,

$$F = \int_1 \rho v^2\,\mathrm{d}A - \int_2 \rho v^2\,\mathrm{d}A = S_{\mathrm{ft}}\Delta v, \tag{10.9}$$

where $\Delta v \equiv \overline{v}_1 - \overline{v}_2$, $\overline{v} \equiv (\overline{v}_1 + \overline{v}_2)/2$ (bars indicate suitable averages) and

$$S_{\rm ft} = \int_1 \rho v \, \mathrm{d}A \simeq \int_2 \rho v \, \mathrm{d}A \qquad (10.10)$$

is the amount of mass per unit time flowing through the region of interaction between the solar wind and the magnetosphere, to be distinguished from $S_{\rm sw}$, the mass input rate from the solar wind into the magnetosphere discussed in Vol. I, Section 10.6.2. (Note: magnetic and thermal contributions to solar wind energy and momentum flux have been neglected as small in comparison to those of the bulk flow.) Combining Eqs. (10.8) and (10.9) yields a relation between the power and the force (in the direction of solar wind flow),

$$\mathcal{P}_{\rm sw} = F\overline{v}, \qquad (10.11)$$

which was used first by Siscoe (1966) and Siscoe and Cummings (1969) to estimate the energy input into the terrestrial magnetosphere, under the assumption that the relevant force F is the tangential (magnetotail) force acting primarily on the nightside, $F_{\rm MT}$ (see detailed discussion of forces in Vol. I, Section 10.3.2). (Note: if F is equated to the pressure force $F_{\rm MP}$ on the entire magnetopause, it can be shown that the associated $\mathcal{P}$ does not go into the magnetosphere but represents the power expended in irreversible heating at the bow shock (see also Section 10.4.3).)

Calculating the power extracted from planetary rotation is somewhat simpler. The angular momentum of the rotating planet is $\mathcal{I}\Omega_0$ and the kinetic energy of rotation is $\frac{1}{2}\mathcal{I}\Omega_0{}^2$, where $\mathcal{I}$ is the moment of inertia and Ω_0 the angular frequency of rotation (the subscript 0 designates the rotation frequency of the planet, as distinct from, e.g., the atmosphere or the magnetosphere). With $\mathcal{T}$ the torque on the planet (component along the rotation axis),

$$\mathcal{P}_{\rm rot} = \frac{\mathrm{d}}{\mathrm{d}t}\left(\tfrac{1}{2}\mathcal{I}\Omega_0{}^2\right) = \Omega_0 \frac{\mathrm{d}}{\mathrm{d}t}(\mathcal{I}\Omega_0) = \mathcal{T}\Omega_0, \qquad (10.12)$$

a relation between the power and the torque, completely analogous to Eq. (10.11). (In principle, Ω_0 decreases with time as the result of the torque, but in practice the rate of decrease is completely negligible. The time for the present magnetospheric torque to reduce appreciably the planet's rate of rotation is several orders of magnitude longer than the Hubble time, both at Jupiter and at Earth; for the latter, this implies that the magnetospheric torque is much smaller than the lunar tidal torque.)

(2) **Implications for linear/angular momentum** The linear or angular momentum that is extracted together with the kinetic energy is a conserved quantity; it cannot simply disappear, and its further transport must be accounted for.

What happens to the linear momentum extracted from the solar wind flow is well understood: it is transferred to and exerts an added force on the massive planet (Siscoe, 1966; Siscoe and Siebert, 2006; Vasyliūnas, 2007; see discussion in Vol. I, Section 10.3.2). The angular momentum extracted from the rotation of the planet, on the other hand, can only be removed to "infinity", and identifying the mechanism by which it is transported away is indispensable for understanding the interaction. There are several possibilities:

(a) In magnetospheres with a significant interior source S of plasma (from moons or planetary rings), angular momentum can be advected by the outward transport of mass. For the simple example of plasma corotating rigidly out to a distance R_H and coasting freely beyond R_H (an approximation to the partial-corotation model discussed in Vol. I, Section 10.4.4), angular momentum is transported outward at the rate $S{R_H}^2\Omega_0$, hence from Eq. (10.12) the extracted power is

$$\mathcal{P}_{rot} \simeq S\,{\Omega_0}^2\,{R_H}^2, \tag{10.13}$$

one half of which goes into the kinetic energy of bulk flow of the outflowing plasma (in this model), and the remainder is available for powering other magnetospheric processes (proposed for the magnetosphere of Jupiter by Dessler, 1980, and by Eviatar and Siscoe, 1980).

(b) If the solar wind exerts a tangential force on the magnetosphere, it will also exert a torque whenever the distribution of the force is not symmetric about the plane containing the solar wind velocity and the planetary rotation axis. The torque may be estimated as $\mathcal{T} \sim R_{MP}\Delta F$, where R_{MP} is the distance to the dayside magnetopause and ΔF is the difference between the force on the dawn and on the dusk side; this gives the ratio of power from rotation to power from solar wind flow as

$$\mathcal{P}_{rot}/\mathcal{P}_{sw} \sim (\Delta F/F)\,(\Omega_0 R_{MP}/v_{sw})\,. \tag{10.14}$$

In a slowly rotating magnetosphere such as Earth, $\Omega_0 R_{MP}/v_{sw} \equiv \epsilon \ll 1$ and one also expects $\Delta F/F$ to scale as $\sim\epsilon$; hence the power extracted from rotation by the solar wind torque is negligible.

(c) In a rapidly rotating open magnetosphere, on the other hand, magnetic field lines that extend from the planet into the solar wind may become twisted (by a process analogous to the formation of the Parker spiral in the solar wind), creating a Maxwell stress that transports angular momentum outward into the solar wind. This mechanism of extracting energy from planetary rotation was proposed by Isbell *et al.* (1984) for Jupiter (where it is now considered not important in comparison to mass outflow) and by Hill *et al.* (1983) for Uranus.

(d) If the magnetic moment of the planet is tilted relative to the rotation axis, electromagnetic waves that carry away angular momentum may be generated by the rotation. This is generally believed to be the primary mechanism for energy loss from pulsars but is negligible for systems that are very small in comparison to

c/Ω_0, the radius of the speed-of-light cylinder (which is the case for all planets in our solar system and their magnetospheres).

10.4.2 Role of magnetic flux transport

To extract kinetic energy from bulk flow, whether exterior (solar wind) or interior (planetary rotation), and inject it into the magnetosphere, the first step is to slow down the flow by the action of magnetic force at the interface. For the solar wind, this is sketched in Fig. 10.3a, which should be looked at in the context of a more complete representation of the open magnetosphere (e.g. Fig. 10.3 or Fig. 13.4 in Vol. I). As the plasma flows through the current layer implied by the sharp turn of the magnetic field, it is slowed down by the $\mathbf{J} \times \mathbf{B}$ force, by an amount Δv readily estimated as the Alfvén speed based on the internal field B_{T} and the external density ρ,

$$\Delta v \simeq B_{\mathrm{T}} / (4\pi\rho)^{1/2}, \tag{10.15}$$

and the (initially mechanical) energy flux density incident on the outside continues into the inside of the magnetotail as an electromagnetic energy flux density (Poynting vector). The interface is here idealized as a thin magnetopause, but in reality it must have appreciable thickness so that the amount of plasma S_{ft} flowing through the interaction region carries sufficient energy to account for the energy input into the magnetotail. The energy input rate from Eq. (10.11) with the force equal to F_{MT} given by Eq. (10.7) in Vol. I is

$$\mathcal{P}_{\mathrm{sw}} \simeq \left(B_{\mathrm{T}}^2/8\pi\right) A_{\mathrm{T}}\, v, \tag{10.16}$$

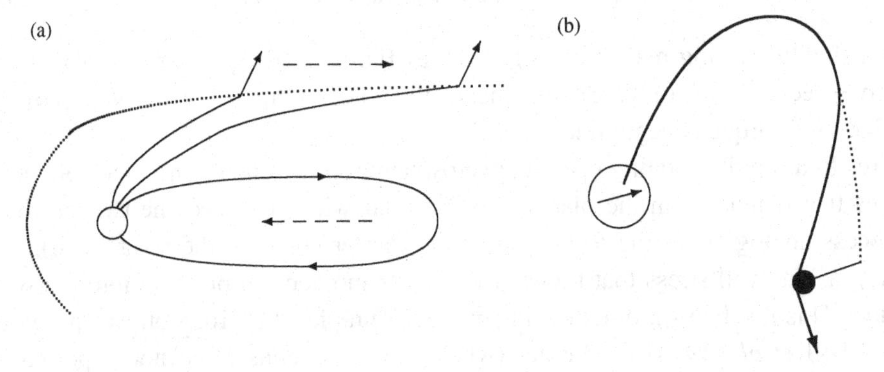

Fig. 10.3. (a) Deformation of magnetotail field by external plasma flow. Solid lines: magnetic field lines. Dashed arrows: plasma flow direction. Dotted line: magnetopause. (b) Deformation of planetary magnetic field by torque from magnetospheric plasma element (black sphere). Solid line: actual magnetic field line. Dashed line: undistorted magnetic field line. Arrow on planet's surface: direction of rotational motion.

which combined with Eqs. (10.8) and (10.15) gives

$$S_{\mathrm{ft}} \simeq \rho\, A_{\mathrm{T}}\, v \left[B_{\mathrm{T}} / \left(16\pi\rho v^2\right)^{1/2} \right]; \tag{10.17}$$

at Earth. This implies that mass flow through the interaction region must be a significant fraction ($\sim 1/4$) of solar wind flux through an area equal to the cross section of the magnetotail (for a more detailed discussion, see e.g. Vasyliūnas, 1987).

A qualitative but more physical way of looking at the interaction is to note that the flow of the solar wind plasma (massive in comparison to plasma in the magnetotail) is carrying the open magnetic field lines with it, while at the same time the feet of these field lines remain anchored to the planet (although free to move laterally, cf. Vol. I, Section 10.4.1); the length of a magnetic flux tube is thus increasing, but its cross-sectional area remains nearly constant (the field magnitude is fixed by the external pressure), hence the volume and with it the magnetic energy content is increasing. (For the plasma, the process is approximately a free expansion; hence the plasma energy content does not change much and is small in any case.) From this point of view, the energy input is closely related to the transport rate of open magnetic flux, from reconnection in the dayside to the magnetotail in the nightside; the question of the amount of energy involved is connected to the fundamental question of the length the magnetotail – how far can an open field line be stretched before it must reconnect and flow back as a closed field line?

Also sketched in Fig. 10.3a is a closed field line, flowing toward the planet and carrying the return magnetic flux. The volume of the flux tube is decreasing, and the plasma energy (greatly enhanced already by the reconnection process from the open to the closed field) is being increased by adiabatic compression. This can be shown to be a conversion of energy from magnetic to mechanical and is further discussed in Section 10.4.3 (energization by adiabatic compression is equivalent to energization by particle drift along the electric field; Hines, 1963).

For planetary rotation, the conversion of kinetic into magnetic energy is sketched in Fig. 10.3b. The prerequisite is a mechanical torque (directed against the rotation) in the magnetosphere; most easily visualized is simple inertia of a plasma element, which holds back the equatorial segment of a magnetic field line, while the feet of the field line at the planet continue to corotate, thus creating an azimuthal magnetic field and increasing the magnetic energy. If the plasma element were to remain at a fixed radial distance indefinitely, it would ultimately be brought up to full corotation and the azimuthal field would disappear; the outward transport process (see Vol. I, Sections 10.5.2, 13.2.1, and 13.3.4), however, removes the plasma in a finite time. The energy input rate thus depends on the rate of mass outflow, which in turn is coupled to circulation of magnetic flux (Vol. I, Chapter 10 and references therein).

Note that in both cases the kinetic energy is first converted into magnetic energy. The energy extracted from planetary rotation can be transported outward, at altitudes just above the top of the ionosphere, *only* by the Poynting vector – with the low density of matter in this region, any mechanical energy flux density is simply too small. That the energy input from the solar wind enters the magnetosphere predominantly in magnetic form is confirmed, at Earth, by the observation that the energy input rate is an order of magnitude larger than the mass input rate multiplied by $\frac{1}{2}v_{\mathrm{sw}}^2$ (Hill, 1979).

10.4.3 Energy budget diagram

A schematic diagram for the principal energy flow and transformation processes in a planetary magnetosphere–ionosphere system interacting with the solar wind is shown in Fig. 10.4. This is a simplified synthesis of more detailed energy flow charts derived for two extreme cases, solar wind energy source (Earth) and planetary-rotation energy source with planetary-moon mass source (Jupiter); for the more complicated case of Saturn, where these two sources are of comparable importance, detailed studies of the magnetosphere have only recently become possible (see Section 10.7.1). Here, I concentrate on energy aspects only; for a

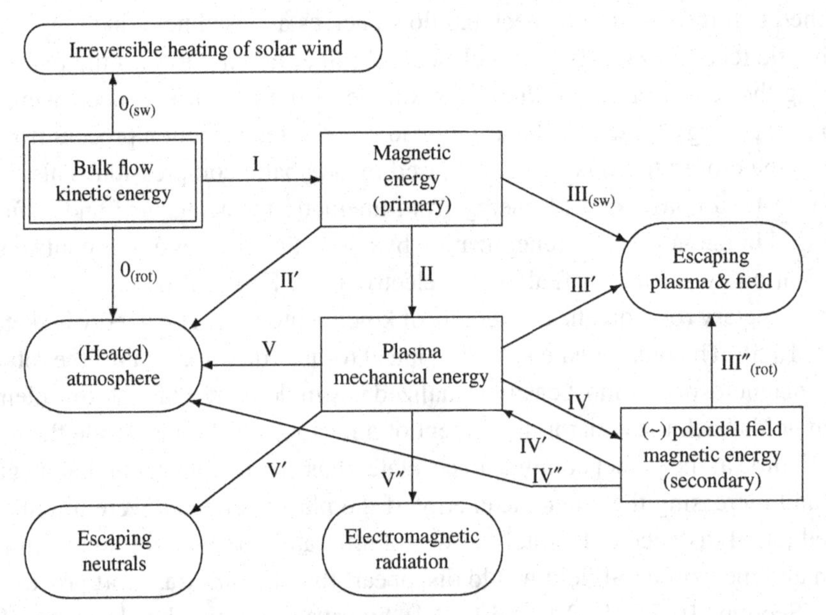

Fig. 10.4. (Simplified) general energy flow chart for planetary magnetospheres and ionospheres. Rectangular boxes: energy reservoirs. Rounded boxes: energy sinks. Lines: energy flow/conversion processes. (Note: only the energy-flow paths are shown, not the mass-flow paths.)

more general discussion, including plasma sources and transport, see Chapters 10, 11, and 13 in Vol. I.

In Fig. 10.4, the primary source of energy – bulk-flow kinetic energy, either of the solar wind or of planetary rotation – is shown by the double-lined box. Energy reservoirs are shown by rectangular boxes (with form of energy identified). Energy sinks, where energy is dissipated or leaves the system, are shown by rounded boxes (for all forms). Plasma mechanical energy is shown as a single reservoir, lumping together the various plasma regions and their thermal and bulk-flow energies. Magnetic energy is shown as two reservoirs: one coupled directly to the source (magnetotail, for the solar wind; wrapped-up field, for rotation), the other from inflation of the dipole field by plasma within the magnetosphere. Energy conversion processes are shown by connecting lines, with arrows indicating the direction of energy flow. The labels on the lines are keyed to the discussion of the corresponding process in the text; subscript (sw) or (rot) – process important only for dominant solar-wind or dominant rotation source, respectively; no subscript – important for both. Processes of minor importance on the scale of the entire magnetosphere (e.g. direct particle precipitation into the atmosphere from the magnetosheath or from the solar wind) have been left out.

(1) **Process I** is the initial conversion of bulk-flow kinetic energy into magnetic energy, described in Section 10.4.2. With the solar wind source, the magnetic energy is stored predominantly in the magnetotail. With the planetary-rotation source, the magnetic energy is stored predominantly in the azimuthal magnetic field, which is in the direction of lagging behind rotation of the planet (like the Parker spiral relative to rotation of the Sun).

(2) **Process II** is conversion of magnetic energy into mechanical energy that is then stored in magnetospheric plasma. It includes formation and energization of the plasma sheet (by magnetic reconnection and adiabatic compression, see Section 10.4.2) and energization of the plasma in the ring current region (predominantly by adiabatic compression during inward transport; although energy is also removed of course by adiabatic expansion during outward transport, the net effect is energy addition as long as there is a net inflow of plasma, to increase the ring current or maintain it against losses). Additionally, in magnetospheres with a significant interior source of plasma from a moon (Io at Jupiter, Enceladus at Saturn) there is the *pick-up* process: ionization of slow-moving (Keplerian) neutrals in the presence of flowing (nearly corotating) plasma, which imparts both flow and thermal energy to the ions.

(3) **Process II′** is also conversion of magnetic energy into mechanical energy which, however, goes directly into the ionosphere and the aurora (these are not shown in Fig. 10.4 – as far as the energy budget is concerned, they are simply

intermediaries in the process by which magnetic energy is converted to heat of the atmosphere). The process, closely associated with Birkeland currents, consists primarily of collisional and Joule heating of the ionosphere as well as auroral acceleration and precipitation (Section 10.3.3).

(4) **Processes III, III′, III″** represent the loss of energy (mechanical and magnetic) by outflow down the distant magnetotail; they all involve magnetic reconnection, since the magnetic field lines must sooner or later become disconnected from the planetary dipole. In a rotation-dominated magnetosphere, processes III′ and III″ are related to the formation of the planetary/magnetospheric wind (see Vol. I, Sections 10.4.4 and 13.2.4).

(5) **Processes IV, IV′** describe energy conversions between plasma in the inner regions of the magnetosphere (ring current, plasma torus) and the nearly dipolar but perturbed magnetic field. Process IV represents the deformation of the magnetic field by plasma pressure in the ring current region as well as by corotational stresses. Process IV′, in turn, represents energization of the plasma by adiabatic compression during inward transport related to the distorted poloidal magnetic field (rather than to the magnetotail field, as in process II; the distinction between the two is not always clear-cut).

(6) **Process IV″** is conversion of magnetic energy into heating of the atmosphere, via the ionosphere and the aurora, analogous to process II′ but taking energy from the distorted magnetic field in the outer magnetosphere rather than from the magnetotail or the wrapped-up azimuthal field. The related auroral processes may be important for substorm onset (Section 10.6).

(7) **Energy sinks**: loss of magnetic and mechanical energy by outflow down the magnetotail (processes III, III′, III″) and loss of magnetic energy by ionospheric and auroral processes into the atmosphere (processes II′, IV″) have already been discussed. In addition, mechanical energy of plasma in the magnetosphere is lost to the atmosphere by particle precipitation (**process V**); it is lost to "infinity" by escape of fast neutrals from charge exchange (**process V′**) and by electromagnetic radiation (**process V″**). In the case of the Io torus at Jupiter, radiation (mainly UV) produced by atomic/molecular collision and excitation processes carries an amount of energy that is significant (possibly even dominant) for the magnetospheric energy budget (Thomas *et al.*, 2004, and references therein). Electromagnetic radiation from the atmosphere (including auroral emissions) is not shown in Fig. 10.4, the power involved being in general negligible on the scale of the magnetospheric energy budget; discussions of energy supply in the aurora usually refer to energy in the precipitating particles that excite the auroral emissions, not energy in the emissions themselves.

(8) **Processes 0** extract energy from bulk flow but do not put it into the magnetosphere. For the solar wind source, the net energy extracted from bulk flow as the solar wind is slowed down and forced to go around the magnetospheric obstacle represents the power expended in irreversible heating of solar wind plasma at the bow shock: when the plasma that has been compressed and heated at the bow shock expands far downstream to its initial (ambient solar wind) pressure, it has a higher temperature (because of the increase of the entropy) and hence (by Bernoulli's law) a slower velocity. For the planetary-rotation source, the torque that extracts energy from rotation is transmitted by magnetic stresses, which can act only as far as the bottom of the ionosphere; farther down, between the ionosphere and the planet, the torque must be transmitted purely by stresses in the neutral medium – effective viscous stresses from velocity shear, accompanied in general by energy dissipation and thus heating of the neutral atmosphere. In both cases, the power involved may exceed by an order of magnitude the entire energy input into the magnetosphere. This is most obvious for Earth, where (see Section 10.4.1) F_{MP} exceeds F_{MT} typically by a factor $\sim$10; for Jupiter, this is suggested by the inference that, when ionospheric plasma slips relative to corotation, the neutral atmosphere very nearly moves together with the plasma (Huang and Hill, 1989; Pontius, 1995), implying a large velocity shear below the ionosphere. Processes 0 may thus constitute the largest energy dissipation processes in the entire interaction of the solar wind with a magnetized planet.

10.4.4 Overview of rates and constraints

It is of interest to consider what can be said about the energy conversion rates for the various paths of Fig. 10.4 in the two extreme cases, solar wind energy source (Earth) and planetary-rotation energy source with planetary-moon mass source (Jupiter), with particular attention to the constraining processes; these play an essential role in the origin of time-varying energy releases discussed in Sections 10.5, 10.6, and 10.7.

(1) **Solar-wind-dominated magnetosphere** The total power $\mathcal{P}_{\mathrm{total}}$ supplied by the solar wind energy source can be considered to be a known quantity, fixed by the solar wind parameters and the size of the magnetosphere; in order of magnitude it is equal to the flux of solar wind kinetic energy through an area equal to the cross section of the magnetotail, $\frac{1}{2}\rho_{\mathrm{sw}}(v_{\mathrm{sw}})^3 A_{\mathrm{T}}$. The power in paths 0_{sw} and I is fixed by force balance considerations: $\mathcal{P}_{0_{(\mathrm{sw})}}$ and $\mathcal{P}_{\mathrm{I}}$ are obtained from Eq. (10.11) with F set equal to F_{MP} and F_{MT}, respectively. This gives $\mathcal{P}_{0_{(\mathrm{sw})}}$ nearly equal to (but of necessity slightly less than) $\mathcal{P}_{\mathrm{total}}$, as discussed in Section 10.4.3, and $\mathcal{P}_{\mathrm{I}}$ equal to $\mathcal{P}_{\mathrm{sw}}$ of Eq. (10.16), smaller than $\mathcal{P}_{\mathrm{total}}$ by an order of magnitude and determined to

a large extent by the amount of open magnetic flux $\sim B_T A_T$, in agreement with the discussion of Section 10.4.2.

For the remaining paths, there are no obvious general estimates of the expected power. There have been numerous empirical estimates, however, of the power in paths II and II′, along with a search for its dependence on solar wind parameters (e.g. Weiss *et al.*, 1992; Koskinen and Tanskanen, 2002, and references therein). The ratio $\mathcal{P}_{II}/\mathcal{P}_{II'}$ of dissipated to stored energy is uncertain (estimates range from $\sim$0.1 to >0.5), and most studies concentrate on the sum $\mathcal{P}_{II} + \mathcal{P}_{II'}$, which is found to vary with the rate of open magnetic flux transport, similarly to $\mathcal{P}_I$ on the average. Several empirical formulas for the dependence on solar wind parameters have been proposed (Burton *et al.*, 1975; Perrault and Akasofu, 1978, and others; review by Gonzalez, 1990); the differences are not very significant in view of the uncertainties. The magnitude of $\mathcal{P}_{II} + \mathcal{P}_{II'}$, however, is in general nearly an order of magnitude smaller than $\mathcal{P}_I$ estimated from Eq. (10.16), for comparable solar wind conditions. This implies that, at least on the average, a large part of the power $\mathcal{P}_I$ supplied to the magnetosphere escapes down the magnetotail, via paths $III_{(sw)}$ and III′, and only a fraction enters near-Earth space – the space weather effects discussed in Chapter 2 are produced by something like a small percentage of the total power in the solar wind interaction with the Earth system.

(2) **Rotation-dominated magnetosphere with internal mass source** In this case, the total power supplied by the rotational energy source *cannot* be considered a quantity known a priori: it is determined by the applied torque, which depends in detail on the dynamics of the magnetosphere (in contrast to the solar wind case, where the mere deflection of the solar wind specifies the dominant force). What can be considered as known is the internal mass source of the magnetosphere: the total rate S of *mass* (not energy) flow associated with path III′. The requirement of outward transport of mass S determines, among other magnetospheric parameters, the torque and thence, by Eq. (10.12), the total power $\mathcal{P}_{rot}$ extracted from planetary rotation. An example is provided by the simple model of Eq. (10.13) in which R_H, given by Eq. (10.23) in Vol. I, itself depends on S and other magnetospheric and ionospheric parameters. Note that the power along path $\mathcal{P}_{0(rot)}$, direct heating of the atmosphere, is contained in the total, leaving only the difference $\mathcal{P}_{rot} - \mathcal{P}_{0(rot)}$ as the power supplied to the magnetosphere. Because plasma flow is coupled to magnetic flux transport, maintaining the given outflow S imposes self-consistency constraints on other energy flow paths besides III′.

At Jupiter, the average loss rates of energy by radiation (path V″) and by escape of neutral particles (path V′) have been empirically determined and the associated collisional/radiative processes extensively modeled (Thomas *et al.*, 2004; Vol. I, Chapter 10 and references therein). The energy loss in precipitating particles that produce the observed aurora has also been empirically estimated (Clarke *et al.*, 2004, and references therein); the main auroral oval is generally attributed to Birkeland currents of partially corotating plasma (Cowley and Bunce, 2001; Hill, 2001) and is thus part of path II′. Little can as yet be said about power in paths I and II, other than inferences from summing the empirical loss rates.

10.5 What leads to explosive energy releases?

The discussion so far has ignored time variations and has proceeded on the tacit assumptions that all the energy supply, conversion, and dissipation processes are more or less in balance. There is no general requirement for this to be the case, and in fact often it is not the case, as evidenced by the occurrence of rapid or even explosive processes (e.g. substorm onset at Earth). Energy balance presupposes a more general equilibrium of the entire system; as the system evolves in response to, for instance, the changing external boundary, the various terms initially in balance may change differently, so that the system no longer is in equilibrium but varies in time (possibly much faster than the variation of the boundary conditions).

The prototypical example is kinetic energy from the solar wind being converted into magnetic energy of the magnetotail at an increased rate due to enhanced dayside reconnection (in response to changed solar wind conditions), but the rate of removal by conversion of magnetic energy into mechanical energy of magnetospheric plasma plus escape down the magnetotail not being equally enhanced (for reasons that need to be identified). In this case, the magnetic energy reservoir increases with time and reaches a point at which (again, for reasons that need to be identified) the magnetic energy content can no longer be maintained but must be converted to other forms.

10.5.1 Magnetic topological changes

As noted in Section 10.4.2, magnetic flux transport and the increase of magnetic energy by stretching the field play an important role in supplying energy to the magnetosphere. Non-equilibrium configurations of the magnetotail that change the magnetic topology and allow different paths of flux transport are therefore of particular interest. (For a discussion of magnetic topology, see e.g. Vol. I, Chapter 4.)

A simple sketch of a model widely invoked to interpret magnetospheric substorms at Earth is shown in Fig. 10.5 (Vasyliūnas, 1976), which displays a time sequence of magnetospheric configurations. Each panel shows the magnetic field line configuration in the noon–midnight meridian plane (left) as well as the configuration of magnetic singular X- and O-lines in the equatorial plane (right)

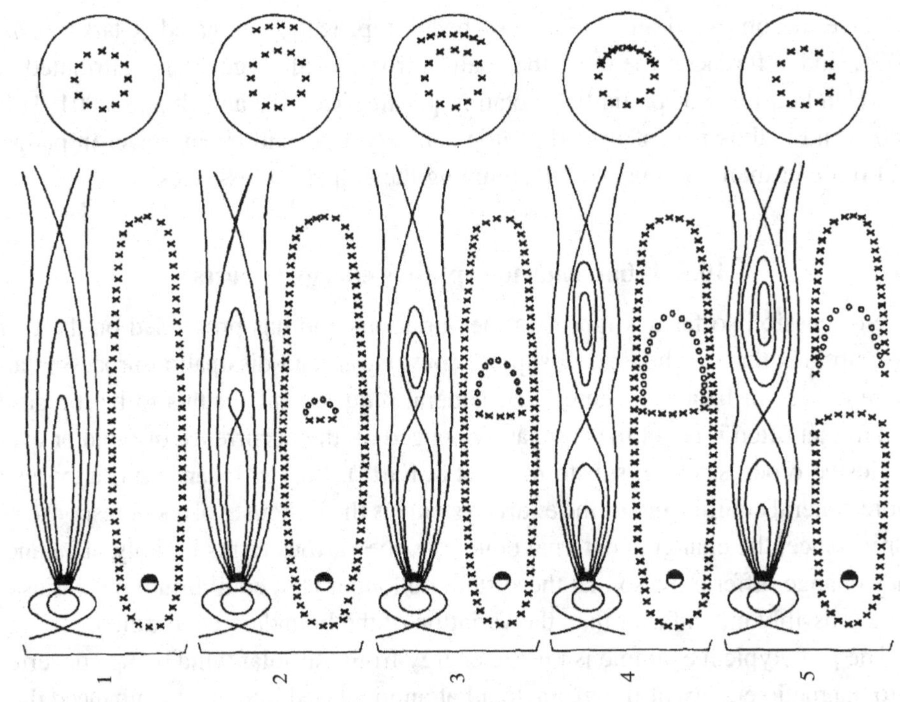

Fig. 10.5. Possible changes of the magnetic field topology in the magnetotail of a solar-wind-dominated magnetosphere. The diagram (from Vasyliūnas, 1976) is shown rotated to facilitate comparisons with diagrams of filament eruptions in e.g. Chapter 6: the solar wind here blows from bottom to top, rather than from left to right as in the original and in the analogous figures of Vol. I, Chapter 10. Each panel in the sequence shows a side view of the magnetic field (left), the outline of the X-lines seen from above the north pole (right), and a top-down view of the mapping of the reconnection region onto the Earth (top).

and projected to the ionosphere (top); the equatorial projection, absent in many later versions (e.g. Hones, 1977), is essential for describing the three-dimensional structure of the magnetic field. Panel 1 is the simplest topology of the open magnetosphere (cf., e.g., Fig. 10.3 in Vol. I). In panel 2, a small volume usually called a *plasmoid* appears deep within the closed-field-line region, bounded on the earthward side by a newly formed *near-Earth X-line* (NEXL) and threaded by magnetic field lines that encircle the attached O-line; ideally, the field lines are confined within the plasmoid and connect neither to the Earth nor to the solar wind (what the real topology is, however, is still uncertain). For the ideal topology, the plasmoid can be visualized in three dimensions as shaped roughly like a banana, oriented approximately dawn to dusk and tapering to zero thickness at both ends, with the X-line on its surface and the O-line running through the middle of its volume. The plasmoid grows (panel 3) by magnetic reconnection until it touches the separatrix of the open field lines (panel 4, *onset of lobe reconnection*); afterwards (panel 5), the plasmoid is on interplanetary field lines and is carried away (presumably) by the solar wind.

A model of topological changes for a rotation-dominated magnetosphere has been developed (Vasyliūnas, 1983) and likewise widely invoked to interpret events at Jupiter and Saturn believed analogous to magnetospheric substorms at Earth. Shown in Fig. 10.7 in Vol. I, it is in essence a direct adaptation of Fig. 10.5, different only in three respects: (1) the time sequence has been translated into an azimuthal-angle sequence, (2) field lines are stretched by the outflow of plasma from an internal magnetospheric source (planetary/magnetospheric wind) rather than by the flow of the solar wind past the magnetosphere, (3) there are no counterparts to panels 4 and 5, since field lines connected to the solar wind are not considered.

Numerous examples of magnetic topological changes, some similar to those of Fig. 10.5, others more complicated (possible in the absence of a strongly constraining planetary dipole field) have been discussed in relation to solar flares and coronal mass ejections (see Chapter 6).

10.5.2 Role of instabilities

Instabilities have attracted much attention as a possible way of inducing rapid change from equilibrium to non-equilibrium configurations – an alternative to straightforward evolution to non-equilibrium as the result of changing boundary conditions. (Actually, the two possibilities are related: if a system evolves from equilibrium to non-equilibrium, the configuration at the transition point is one of unstable equilibrium.) Specific types of instabilities have been invoked to interpret particular aspects of rapid energy conversion processes in planetary magnetospheres, especially at Earth.

(1) **Tearing-mode instabilities** "Tearing mode" is a generic term for instabilities that result in the reconnection of initially oppositely directed magnetic fields. They are obvious candidates for initiating topological changes of the magnetotail (in particular, those envisaged in Fig. 10.5), as proposed by Schindler (1974) and others; see e.g. Wang and Bhattacharjee (1993); Chapter 10 of Schindler (2007), and references therein.

(2) **Current-driven instabilities** The concept that a sufficiently intense electric current may bring about its own breakdown, by creating conditions that impede current flow, was first suggested by Alfvén and Carlquist (1967) as a model for solar flares. Under the name "current disruption" it has been widely discussed as a model for substorm onset and expansion. Various instabilities that develop when

the current density exceeds some threshold value have been proposed; see e.g. Lui (1996, 2004), and references therein.

(3) **Interchange and ballooning instabilities** Interchange instabilities that do not appreciably change the magnetic field are thought to be essential for plasma transport in rotation-dominated magnetospheres (see Vol. I, Sections 10.5.2 and 13.2.1). Ballooning instabilities can be viewed roughly as interchange that does change the magnetic field. As a model for substorms, they have been invoked particularly at the transition between the dipole field and the magnetotail, in several variants; see e.g. Hurricane *et al.* (1998), Samson (1998), Cheng (2004), and references therein.

10.6 Applications: Earth

Fundamentally, time-varying energy conversion events in the magnetosphere are produced when the various energy-flow paths in Fig. 10.4 are not in balance. The task is to understand which paths are out of balance, on what time scales, and for what physical reasons. The fact that the incident solar wind is itself always varying on many different time scales ensures the occurrence of a whole spectrum of time-varying magnetospheric phenomena, but it also makes it difficult to determine the extent to which they are governed either by internal dynamics of the magnetosphere or by changing solar wind conditions.

10.6.1 Magnetospheric substorms

The phenomenological description of the magnetospheric substorm (sketched briefly in Section 10.2) leads to a physical description that can be summarized (equally briefly) as a two-stage process. Stage 1 (growth phase): as a consequence of a southward interplanetary magnetic field, the configuration of the magnetosphere changes, its magnetic field becoming highly stretched (increased magnetic flux in the magnetotail, reduced flux in the nightside equatorial region). Stage 2 (expansion phase, initiated by the onset): the magnetic field changes to more nearly dipolar (increased flux on the nightside), and there is enhanced energy input and dissipation to the inner magnetosphere and the ionosphere/atmosphere; the process occurs on dynamical time scales (comparable to or shorter than wave travel times) and is accompanied (most probably) by changes of magnetic topology.

In terms of energy flow paths of Fig. 10.4: during stage 1, $\mathcal{P}_{\mathrm{I}}$ (power in path I) is enhanced and is appreciably larger than the sum $\mathcal{P}_{\mathrm{II}} + \mathcal{P}_{\mathrm{II'}} + \mathcal{P}_{\mathrm{III(sw)}}$. During stage 2, $\mathcal{P}_{\mathrm{II}}$ and particularly $\mathcal{P}_{\mathrm{II'}}$ are enhanced; $\mathcal{P}_{\mathrm{III(sw)}}$ and $\mathcal{P}_{\mathrm{III'}}$ presumably are enhanced in connection with topological changes exemplified by Fig. 10.5.

The substorm growth phase is in essence the increase of open magnetic flux in the magnetosphere, which occurs for a two-fold reason. First, the flux addition rate at the dayside reconnection region increases as the solar wind transports more magnetic flux, of the sense opposite to the terrestrial dipole flux, toward the magnetosphere; the reasons for this are assumed to lie in the physics of magnetic reconnection (e.g. Vol. I, Chapter 5). Second, the flux return rate at the nightside reconnection region does *not* increase to match the addition rate; the reasons for this are not at all well understood. One obvious possibility is to assume that the nightside reconnection rate is controlled by local solar wind conditions, just like the dayside rate, so that any increase is delayed by the solar wind flow time to reach the distant X-line of the open magnetosphere (e.g. Fig. 10.3 in Vol. I), but this is unlikely for at least two reasons: (a) in most models, the distant X-line is located well within the magnetotail, not in direct contact with solar wind plasma; (b) any effect of enhanced dayside reconnection can be communicated to the magnetotail by wave propagation within the magnetosphere much faster than by advection in the solar wind. Another possibility is related to stress balance in the magnetotail: the earthward-directed magnetic tension force is opposed by a tailward-directed total (plasma plus field) pressure gradient force, which may impede the earthward flow of plasma and hence the return of magnetic flux. Within the magnetosphere, the net effect of the substorm growth phase is to remove magnetic flux from the nightside magnetosphere by flow toward the dayside reconnection region and to add magnetic flux to the magnetotail (enhanced stretching of magnetotail field lines).

The substorm expansion phase does return the magnetic flux, rapidly and spectacularly, from the magnetotail to the nightside magnetosphere (dipolarization of a previously stretched tail-like field); given that plasma in the magnetotail beyond a distance typically ~15–20 Earth radii is observed to flow away from Earth, the process must almost unavoidably proceed by topological changes of the type sketched in Fig. 10.5. The energy input into plasma, energetic charged particles, and the aurora can be largely accounted for by adiabatic compression and Birkeland current effects. What remains highly controversial is how does the process start and why is it so sudden and catastrophic. Two distinct views have been in contention for decades. One (commonly, albeit inaccurately, called "current disruption model" or sometimes "inside-out scenario") postulates that the substorm onset begins deep within the magnetosphere, at or near the interface between the tail-like and the dipolar magnetic fields, most likely as a result of one or more of the current-driven or ballooning instabilities mentioned in Section 10.5.2; topological changes of the magnetotail are regarded as consequences of the onset. The other ("NEXL model" or "outside-in scenario") postulates the topological sequence of Fig. 10.5 (or some equivalent) as the essential phenomenon and regards the inner-magnetosphere and auroral effects as consequences; the onset itself is identified either with the appearance of the plasmoid (panel 2) or, less commonly, with the onset of lobe reconnection (panel 4). For references and discussion of physical issues distinguishing the models, see e.g. Vasyliūnas (1998).

A further complication is the question of external versus internal influences. That the growth phase is initiated by changing solar wind conditions is the consensus view. The onset and expansion phase, on the other hand, are regarded by the majority as basically the result of internal dynamical processes, although subject to solar wind influences (e.g. if the system is evolving toward instability, it may be pushed over the threshold by a change in the solar wind). A substantial minority, however, considers the substorm onset intrisically as triggered by a solar wind change (typically toward a more northward interplanetary magnetic field).

10.6.2 Magnetic storms

Our understanding of magnetic storms has been decisively influenced by a remarkable theoretical result, the Dessler–Parker–Sckopke theorem, which relates the external magnetic field at the location of a dipole to properties of the plasma trapped in the field of the dipole. First derived by Dessler and Parker (1959) for special pitch-angle distributions and extended to any distribution by Sckopke (1966), the theorem states that $\mathbf{b}(0)$, the magnetic disturbance field of external origin at the location of a dipole of moment $\boldsymbol{\mu}$, satisfies

$$\boldsymbol{\mu} \cdot \mathbf{b}(0) = 2U_K, \tag{10.18}$$

where U_K is the total kinetic energy content of plasma in the magnetosphere. What is remarkable is that the right-hand side does not depend on the spatial distribution, the partition between bulk-flow and thermal energy, or any properties of the energy spectrum.

Originally derived by Biot–Savart integration of axially symmetric drift currents, the theorem was subsequently derived from a virial-theorem argument and thereby considerably generalized, with the addition of a few terms on the right-hand side (which, however, are mostly ignored in practice except for a negative contribution from solar wind dynamic pressure, $\rho_{sw} v_{sw}^2$); see Carovillano and Siscoe (1973), Vasyliūnas (2006), and references therein.

Although $\mathbf{b}(0)$ nominally is evaluated at the center of the Earth, it is also equal to the (vector) average of $\mathbf{b}(\mathbf{r})$ over the surface of the globe (by a theorem for solutions of Laplace's equation, satisfied within the globe by each Cartesian component). The Dst index is the average, over a low-latitude strip of the globe, of the disturbance field component aligned with the dipole; after some corrections (chiefly removing the contribution from induced earth currents), −Dst may be considered a reasonable proxy for the left-hand side of Eq. (10.18), as long as Dst < 0. The Dessler–Parker–Sckopke theorem then provides a method of inferring the plasma energy content – the energy contained in the box "plasma mechanical energy" in Fig. 10.4 – simply from the value of the Dst index. (The empirical estimates of $\mathcal{P}_{\mathrm{II}}$ mentioned in Section 10.2 were obtained largely by this method from observed time variations of Dst.) Direct *in situ* observations have established that the greater part of the energy resides in what is called the ring current region (see Vol. I, Section 10.5.3).

Geomagnetic storms, particularly the intense ones, are characterized by unusually large amounts of energy stored as mechanical energy of plasma in the ring current region, in comparison to other storage regions. This implies that during the development of an intense storm the power in path II is unusually large, on the average. Whether this enhanced conversion rate from magnetic energy into mechanical energy of ring current plasma results from a different interaction process or simply from a different time sequence of solar wind parameters is an unresolved question. More specifically, can the energy for storms be suplied by a sequence of substorms (perhaps unusually frequent and/or unusually intense), or is some other process required? A related question is that of *geoeffectiveness*: when interplanetary structures such as CMEs (see e.g. Chapter 6) impinge on the Earth, under what conditions do they produce intense magnetic storms? (prolonged southward B_{sw} is one that is well established). For discussion and references, see e.g. Tsurutani *et al.* (1997), Kamide *et al.* (1998), and Song *et al.* (2001).

10.7 Applications: other planets

10.7.1 Survey of processes

Our knowledge of energy conversion processes in the magnetosphere of planets other than Earth is strongly conditioned by available observations. For the most part, these are measurements of magnetic fields and charged particles by instruments on spacecraft on flyby trajectories and, more recently, in orbit around the planet. (Remote sensing, e.g. of the aurora, is mostly limited to special campaigns; sufficient observations have been accumulated to establish a reasonable picture of the general morphology of aurora at Jupiter and Saturn, but detailed studies of the time-varying aspects, with the use of concurrent *in situ* observations, are just beginning.) Energy conversion events observed at other planets so far have been generally classified as analogous or at least similar to terrestrial magnetospheric substorms, on the basis of features in the data that resemble what is observed at Earth.

Substorm-like events were first described at Mercury (Siscoe *et al.*, 1975); they occurred during the first Mariner 10 flyby in 1974 and were identified on the basis of observed energetic electron and magnetic field changes that were similar in almost every respect to substorm-related changes in the Earth's magnetotail, except for a much shorter time scale (~20 times faster at Mercury than at Earth). This is in agreement with the supposition that the magnetosphere of Mercury is essentially just a scaled-down version of that of Earth (Ogilvie *et al.*, 1977). (Note: results from the Messenger spacecraft currently in orbit around Mercury are just beginning to be available and have not been taken into account in writing this chapter.)

The magnetosphere of Uranus has been investigated only once, during the flyby of Voyager 2 in 1986. Observed temporal variations of plasma (McNutt *et al.*, 1987), energetic particles (Mauk *et al.*, 1987), and magnetic field (Behannon *et al.*, 1987) were interpreted (on the basis of similarity to observations at Earth) as suggestive of substorm-like events.

By contrast, at Jupiter the extensive data set, from six flybys and above all from the Galileo orbiter, has made it possible to establish unambiguously the existence of characteristic energy conversion events and to determine their main features. These include: magnetic field change, first stretched or more tail-like, followed by relaxed or more nearly dipolar; enhanced plasma flow along approximately the radial direction, alternating between toward and away from the planet; increase of energetic particle intensities, interpreted as heating of the plasma. The duration of an event is typically one to a few hours; there is some indication of a possible recurrence tendency at an interval of a few days. The most common interpretation is that these are rotational counterparts of the terrestrial substorm, involving topological changes similar to those of Fig. 10.5 but driven by the rotational stresses of outflowing plasma rather than by the solar wind drag on open field lines, hence described by some variant (possibly time-dependent or small-scale) of Fig. 10.7 in Vol. I. For references and more detailed description see e.g. Krupp *et al.* (2004); I discuss the physics of the energy conversion briefly in Section 10.7.2.

At Saturn, following two flybys, the accumulation of data by the Cassini orbiter is still in progress. Substorm-like events quite similar to those at Jupiter and interpreted by the same basic concepts have been reported (Jackman *et al.*, 2007; Hill *et al.*, 2008).

10.7.2 Analogs of magnetospheric substorms in strongly rotating magnetospheres

Because the observations at Jupiter and Saturn suggest that the substorm-like events may represent a two-stage process, we may ask how this can be accounted for by imbalances of paths in the energy flow diagram, Fig. 10.4. In a rotation-dominated magnetosphere with internal mass source, the rate of mass flow S along path III′ may be considered as given (Section 10.4.4). Plasma outflow carries magnetic flux with it and would (in the absence of flux return) increase the energy in the magnetic field by stretching the field lines; hence the outward transport magnetic flux may be associated with path IV and the return flux with path IV′. An explosive energy release can now occur in a way that closely parallels the two stages of the terrestrial magnetospheric substorm as described in Section 10.6.1: first, magnetic flux is transported outward, but the return flux is impeded, for a reason to be identified (possibly by the adverse pressure gradient of a stretched-out field, as discussed for Earth); second, a fast return of the accumulated flux is initiated by some process, to be identified (possibly an instability of some type).

10.8 Concluding remarks

Magnetospheric substorms at Earth, analogous events at Jupiter and Saturn, and solar flares and other events discussed in Chapter 6, most of which are interpreted as explosive releases of energy stored in the magnetic field, may perhaps be viewed as manifestations of an underlying universal process, which I summarize tentatively as follows:

(1) The process occurs in two steps: first, mechanical stresses deform the magnetic field (on the Sun, the emergence of new flux – as flux ropes – from below the surface, associated, of course, with a plasma flow, often plays a part in this active-region environment) into a configuration of increased energy; second, the magnetic configuration becomes unsustainable and changes quickly, releasing the energy. Both steps are in general associated with magnetic topological changes.

(2) In most cases, the mechanical stress is related to plasma flow, which transports magnetic flux and, with field lines attached to a massive body, increases the magnetic energy.

(3) Why the magnetic configuration becomes unsustainable and what causes the quick change remain highly disputed questions; many possibilities can be imagined, and there may not be a universal answer.

(4) A potentially universal aspect is magnetic flux return: inability to return the flux smoothly seems to play a role (for Earth at least).

[illegible] flux with [illegible] and [illegible] the [illegible] [illegible] in the magnetic field [illegible] the [illegible] [illegible] magnetic flux may be [illegible] with [illegible] and the return flux with [illegible] [illegible] [illegible] [illegible] [illegible] [illegible] [illegible] leaves of the [illegible] [illegible] as described in Section 10.[illegible] [illegible] [illegible] [illegible] [illegible] [illegible] [illegible] [illegible] reasons [illegible] [illegible] [illegible] [illegible] [illegible] [illegible] [illegible] to be identified [illegible] [illegible] of [illegible]

10.7 [illegible] remarks

[illegible] [illegible] [illegible] [illegible] [illegible] [illegible] [illegible] and [illegible] [illegible] [illegible] of which are [illegible] [illegible] [illegible] [illegible] [illegible] may [illegible] [illegible] [illegible] [illegible] [illegible] [illegible] [illegible] [illegible] [illegible]

(1) The [illegible] [illegible] [illegible] [illegible] [illegible] [illegible] [illegible] [illegible] new flux [illegible]

(2) In most cases the mechanism [illegible] [illegible] magnetic flux, which [illegible] [illegible] magnetic flux and [illegible] [illegible] [illegible] [illegible] [illegible]

[illegible] the magnetic [illegible] [illegible] [illegible] [illegible] and [illegible] [illegible] [illegible] [illegible] [illegible] [illegible] [illegible] [illegible] [illegible] [illegible] [illegible] [illegible] [illegible] [illegible]

(3) [illegible] [illegible] [illegible] [illegible] [illegible] [illegible] in the [illegible] [illegible] [illegible] role [illegible] [illegible]

11

Energization of trapped particles

JANET GREEN

11.1 Heliophysical particles: universal processes and problems

At the time the very first satellites were launched half a century ago, the space environment was portrayed by mainstream media as the science fiction home where Flash Gordon fought evil aliens. The very real threat of Earth's radiation belts was not even imagined in either the fantasy or science worlds so no consideration was given to how the very energetic particles of the belts, traveling near the speed of light and capable of penetrating solid material, might affect instrumentation. Yet it was the diminished performance of the Geiger counter designed by James Van Allen (Van Allen *et al.*, 1958; see also Section 3.1), that led to the eventual discovery of the belts. Van Allen speculated that the unusually low flux measurements returned by his experiment were actually a sign that the instrument had saturated, overwhelmed by a previously unknown and very large population of energetic particles.

The conjecture was confirmed by the dozens of satellites launched to probe Earth's magnetosphere providing a qualitative depiction and understanding of the radiation belts. (For a list of satellites with radiation belt particle data visit the Virtual Radiation Belt Observatory on line.) The belts consist of protons and electrons trapped in Earth's magnetic field forming torus-shaped regions extending from ~1.5 to ~10 Earth radii (R_E) (see Fig. 11.1). The protons form only a single belt. The electrons form two belts separated at ~2.5 R_E by a minimum flux region known as the slot (Lyons and Thorne, 1973). The name "radiation belts" refers to only the most energetic particles, but no exact energy separates these from the lower energy plasma that fills the magnetosphere. These highest energy particles, typically above ~0.3 MeV, are usually separated out for study because their rapid orbits about Earth (less than ~10 min) and their fast gyration about magnetic field lines (~0.01 s) subject them to slightly different acceleration and loss processes than their lower energy counterparts. Electrons with such high energies

Heliophysics: Space Storms and Radiation: Causes and Effects, eds. Carolus J. Schrijver and George L. Siscoe. Published by Cambridge University Press.

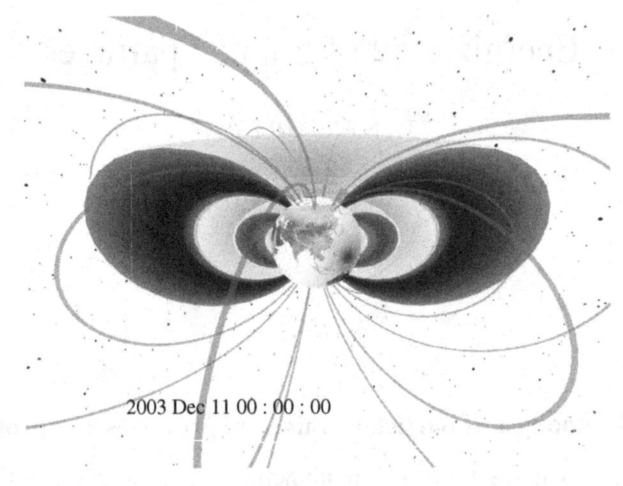

Fig. 11.1. Schematic depiction of Earth's electron radiation belts. (Courtesy of the NASA/Goddard Space Flight Center Scientific Visualization Studio.)

and small mass have velocities approaching the speed of light where relativistic effects become significant; therefore, the radiation belt electrons are often called relativistic electrons.

Earth's radiation belts have gained notoriety because of their harmful effects on satellites and humans working in space (see Chapters 13 and 14). The energetic particles can cause satellite surface charging, deep dielectric discharge, single event upsets (SEU), and solar power degradation. Surface charging, usually attributed to ~30 keV electrons, occurs when electrons collect on the satellite's exterior, thereby charging sections of the satellite to different potentials, which in turn leads to arcing. Deep dielectric discharge occurs when high-energy electrons (typically >1 MeV) penetrate through thick shielding and build up in dielectric material. If the charge increases faster than it dissipates, a sudden discharge may occur and permanently damage electronics. Single event upsets are normally attributed to >50 MeV protons that can penetrate through solid material and flip a bit in memory, sometimes sending electronics into an unwanted or unrecoverable state. Finally, these energetic particles can penetrate into solar panels causing the cover to turn brown and ultimately reducing the efficiency of the panel as less light gets through. Astronauts must also contend with the same intense radiation environment as the satellites. The International Space Station operates in an orbit that sometimes skims the low-altitude footprint of the radiation belts. Careful planning is required to avoid repairs or other activities that require the astronauts to leave the protective shell of the spacecraft when the radiation belt fluxes are high.

While these effects can be costly, many can be mitigated by understanding the radiation environment and its temporal variability. Satellite maneuvers and maintenance can be delayed until radiation flux levels decrease and astronauts can remain indoors until the storm has passed. Thus, research efforts have focused on understanding and predicting global radiation-belt particle flux variability. Research relating to Earth's radiation belts has flourished through NASA's Targeted Research and Technology grants intended to develop "understanding of Heliophysics science that may affect life and society". Yet even after these focused research campaigns and decades of investigation, some key science questions remain unsolved. Basic questions are still posed such as "What acceleration and loss processes contribute to the radiation belt variability?" Answers have been elusive, in part because the particles sail through different regions of the magnetosphere and intense wave areas with such speed that it is difficult to say which interaction caused significant change. Relativistic electrons orbit Earth in less than 10 minutes. A change in the particle flux or pitch angle distribution measured locally by one satellite may be caused by the particles' brief transit through a region of waves on the other side of the magnetosphere that may not be sampled by *in-situ* sensors at that time. Or perhaps, the observed changes are caused by many short transits through several different wave regions. Additionally complicating the analysis is the fact that a satellite at a fixed radial distance may observe flux variations simply because the global magnetic field configuration has changed such that the satellite is now measuring a different particle population. Finally, radiation belt particles are difficult to measure (see Chapter 3). Most instruments rely on solid-state technology. The data from these instruments are sometimes contaminated by penetrating particles other than those intended to be measured and saturation effects during the most intense and generally most interesting events. To address some of the still-debated basic science questions, NASA will launch the Radiation Belt Storm Probes that will carry some of the most sophisticated wave and particle instruments yet flown. The two probes, along with two GOES satellites and possibly a Japanese satellite known as ERG, will provide the best instrumentation set ever to measure the global radiation environment.

Looking even farther into the future, as NASA makes plans to go back to the Moon and onward to Mars, high-energy particle radiation outside of Earth's protective magnetic cocoon takes on new importance. Unlike Earth, the Moon and Mars do not have internally generated magnetic fields that can contain trapped radiation belts, but they do sit directly in the stream of energetic particles that are sometimes released from the Sun. The absence of a magnetic field means astronauts on the surface would not be shielded from the particle radiation streaming from the Sun. Some estimates indicate that the radiation at Mars during the Halloween 2003 storm would have been fatal to unprotected astronauts. Thus, there is now renewed interest in understanding acceleration processes at the Sun. Still other planets visited by NASA probes, such as Jupiter and Saturn, have internal magnetic fields and radiation environments that are both similar to Earth's and yet strangely unique. Compared to Earth, Jupiter's magnetic field and its surrounding magnetosphere are massive. The magnetic field strength at Jupiter's surface is 10 times that of Earth, the magnetopause stand-off distance is 50–100 R_J ($R_J = 71\,492$ km) or nearly 100 times that at Earth. At Saturn, the surface magnetic field is comparable to Earth's, the magnetopause distance is 30–40 R_s, roughly 10 times the distance at Earth (see figures and tables in Chapter 13 in Vol. I). Yet even in these strange environments, the debate about dominant acceleration and loss mechanisms is familiar. Many of the same mechanisms are believed to operate at these planets as well. In fact, radiation belt models developed initially for Earth have been used with great success for both Saturn and Jupiter by making allowances for such things as plasma injections from Jupiter's volcanic moon, Io, and charge exchange with Saturn's rings.

In Section 11.2 we begin our examination of particle radiation with an explanation of how these very energetic particles move. Section 11.3 gives a general description of particle radiation in the different planetary environments. Section 11.4 describes the acceleration processes that give these particles their high energies, and Section 11.5 describes how the particles are eventually lost.

11.2 Particle motion

Understanding planetary particle radiation physics requires a basic understanding of how particles move. We begin by describing the motion of a single particle in very simple field geometries. However, the complex magnetic and electric fields of the heliosphere very quickly complicate that motion. Since individual particles cannot be tagged and tracked we will gradually simplify the description of motion to illustrate how collections of particles move and define observables that are useful for analyzing acceleration and loss processes.

11.2.1 Single particle motion

The motion of every individual charged particle in the heliosphere can be described by the Lorentz force equation, Eq. (9.1). That equation shows that a charged particle in an electric field and no magnetic field increases its velocity following the direction of the electric field. Now considering only a magnetic field, the particle gyrates in a circle perpendicular to the magnetic field direction, as described by Eq. (9.2), which shows that the motion is constant along the field.

A very important aspect of the Lorentz equation when discussing particle acceleration is that the electric field may change the energy of the particle but the magnetic field does not. This relation is shown by taking the dot product of the Lorentz equation with $\mathbf{v}$ giving (expressed in cgs units)

$$\mathbf{F} \cdot \mathbf{v} = q(\mathbf{v} \cdot \mathbf{E}) + \mathbf{v} \cdot (\mathbf{v} \times \mathbf{B}), \tag{11.1}$$

or

$$\frac{\mathrm{d}W}{\mathrm{d}t} = q\mathbf{v} \cdot \mathbf{E}, \tag{11.2}$$

where W is the kinetic energy.

These simplified geometries are useful for textbook illustrations but in realistic situations magnetic and electric fields rarely occur in separate and uniform configurations. Even taking a very basic view of Earth's magnetosphere as a perfect dipole magnetic field with no electric fields greatly complicates the motion. In this case, the motion separates into three oscillatory types occurring at increasingly slower time scales. On the fastest time scale, a particle gyrates around the field line as described above.

The second oscillatory type motion in the dipole relates to the particle's velocity parallel to the magnetic field. As the particle follows the field line towards the poles, it moves into a gradient because the magnetic dipole field increases near Earth's surface. The effect of this gradient is to convert the parallel motion of the particle into perpendicular motion as shown schematically in Fig. 11.2. As the particle's trajectory moves towards the pole, the gradient effectively creates a Lorentz force opposite to the parallel motion. Eventually, the parallel velocity will go to zero and then reverse direction, now moving away from Earth causing the particle to bounce between the southern and northern poles. The point at which the parallel velocity goes to zero is called the mirror point and the oscillation between the two poles is referred to as the bounce motion.

In addition to the gyromotion and bounce motion, the particle will circle around Earth in an oscillatory manner known as drift motion. The azimuthal drift is caused by the radial gradient of the dipole field. Intuitively, this drift can be attributed to the changing gyroradius in different magnetic field strengths. In the stronger

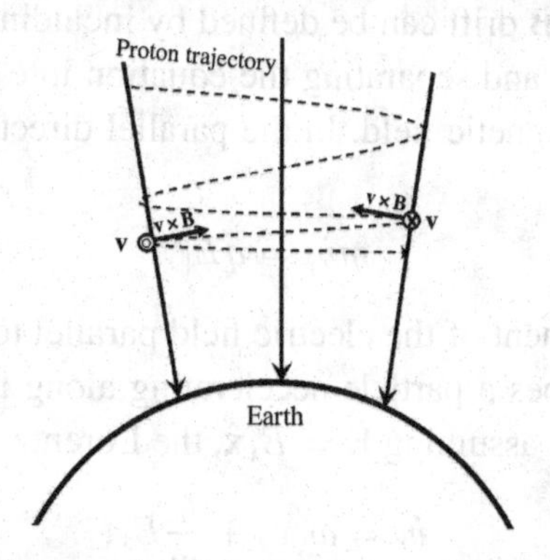

Fig. 11.2. Schematic diagram showing the Lorentz force as a particle moves into the magnetic field gradient at Earth's poles.

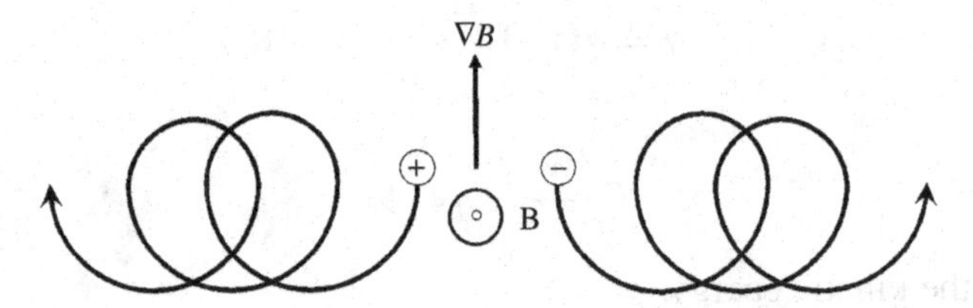

Fig. 11.3. Schematic diagram for the gradient-B drift.

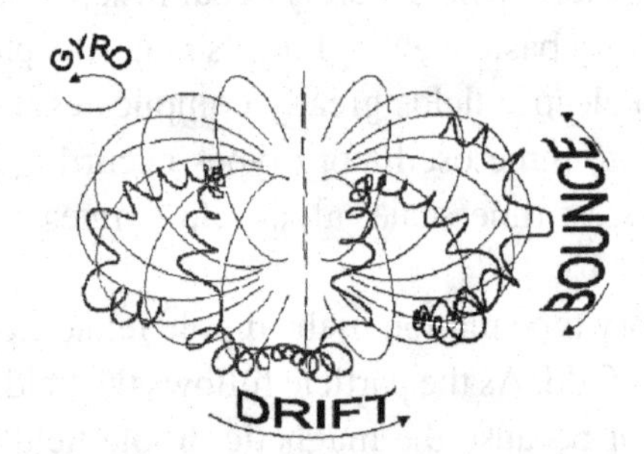

Fig. 11.4. Schematic diagram of particle motion in a dipole magnetic field.

magnetic field the gyroradius will decrease and in the weaker field the gyroradius will increase creating the orbit shown in the schematic of Fig. 11.3. As protons and electrons gyrate in opposite directions, they also drift in opposite directions.

Putting all three together gives the full motion as visualized in Fig. 11.4.

11.2.2 Guiding center motion

Often, particle motion can be described by separating it into a drift velocity with gyromotion superimposed as in the examples here.

E × B drift: The **E** × **B** drift can be defined by including a uniform electric field in the Lorentz equation and separating the equation into components parallel and perpendicular to the magnetic field. In the parallel direction the Lorentz equation becomes

$$m\dot{v}_{\parallel} = qE_{\parallel}, \tag{11.3}$$

where $E_{\parallel}$ is the component of the electric field parallel to the magnetic field. This equation simply describes a particle accelerating along the magnetic field. In the perpendicular direction, assuming $\mathbf{E} = E_x\hat{\mathbf{x}}$, the Lorentz equation becomes

$$\dot{v}_x = \omega_g v_y + \frac{q}{m}E_x, \tag{11.4}$$

$$\dot{v}_y = -\omega_g v_x. \tag{11.5}$$

Taking the second derivative of the velocity gives

$$\ddot{v}_x = -\omega_g^2 v_x, \tag{11.6}$$

$$\ddot{v}_y = -\omega_g^2 \left(v_y + \frac{E_x}{B} \right). \tag{11.7}$$

These equations describe gyration superimposed on an additional E/B drift in the $\mathbf{E} \times \mathbf{B}$ direction.

General force drift: Using $\mathbf{F} = q\mathbf{E}$ to find $\mathbf{E}$ and substituting into the $\mathbf{E} \times \mathbf{B}$ drift equation creates a general force equation,

$$v_F = \frac{1}{\omega_g} \left(\frac{\mathbf{F}}{m} \times \frac{\mathbf{B}}{B} \right). \tag{11.8}$$

This equation can be used to define the drift velocity caused by any general force. Other types of drift include curvature drift caused by a centrifugal force related to the curvature of the dipole field lines, polarization drift that results from a slowly varying electric field, and a gravitational drift. For a more thorough discussion of the various drifts, see Baumjohann and Treumann (1996).

11.2.3 Collective motion described by particle invariants

The Lorentz equation and drift velocity derivations provide a feel for how single particles move throughout the magnetosphere, but to analyze satellite measurements requires a more generalized view of particle motion because detectors do not measure the position and velocity of every particle in space to be propagated forward in time using the Lorentz equation. To this end, it is instructive to describe particle motion using aspects of the motion that are conserved when time variations of the magnetic field are slow. For charged particles in the magnetosphere, there are three such invariants associated with the gyro, bounce, and drift motion (described by Eq. (9.6)). Assuming that the invariants are conserved confines the particle location to within a shell about Earth.

First invariant The first invariant is associated with the gyromotion of the particle about the field line and is given by

$$\mu = \frac{p_\perp^2}{2m_0 B}. \tag{11.9}$$

Here $p_\perp$ is the relativistic momentum in the direction perpendicular to the magnetic field, m_0 is the rest mass of the electron, and B is the field strength.

Second invariant The second invariant corresponds to the bounce motion of a particle along a field line and is given by

$$J = \oint p_{\parallel} \, \mathrm{d}s, \tag{11.10}$$

where $p_{\parallel}$ is the particle momentum parallel to the magnetic field and ds is the distance a particle travels along the field line. It is convenient to rewrite the second invariant in terms of only the magnetic field geometry by the following manipulation. If no parallel forces act on a particle then momentum is conserved along a bounce path and $J = 2pI$ where p is momentum and

$$I = \int_{s_\mathrm{m}}^{s'_\mathrm{m}} \left(1 - \frac{B(s)}{B_\mathrm{m}}\right)^{1/2} \mathrm{d}s. \tag{11.11}$$

Here s_m is the distance of the particle mirror point, $B(s)$ is the field strength at point s, and B_m is the mirror point magnetic field strength. If the first invariant is conserved then K, as defined below, is also conserved:

$$K = \frac{J}{2\sqrt{2m_0\mu}} = I\sqrt{B_\mathrm{m}} = \int_{s_\mathrm{m}}^{s'_\mathrm{m}} (B_\mathrm{m} - B(s))^{1/2} \, \mathrm{d}s, \tag{11.12}$$

where m_0 is the rest mass of an electron. Throughout this chapter we will refer to K when speaking of the second invariant.

Third invariant The third and final invariant corresponds to the drift motion of a particle about the Earth and is given by

$$\Phi = \oint A_\Phi \, \mathrm{d}l. \tag{11.13}$$

In this equation A_Φ is the magnetic vector potential and dl is the curve along which lies the guiding center drift shell of the electron. Using Stokes' theorem the third invariant can be written as

$$\Phi = \int (\nabla \times \mathbf{A}) \, \mathrm{d}S = \int \mathbf{B} \, \mathrm{d}S, \tag{11.14}$$

where $\mathbf{B}$ is the magnetic field and dS is area. Therefore, conservation of this invariant requires that an electron gyration always encloses the same amount of magnetic flux as it drifts about the Earth. In a dipole field this is equivalent to saying that the electron remains at fixed radial distance. The Roederer L parameter, commonly written as L^*, is another useful form of the third invariant:

$$L^* = \frac{2\pi \mathbf{M}}{\Phi R_\mathrm{E}}, \tag{11.15}$$

where **M** is the magnetic moment of the Earth's dipole field. The L^* parameter is the radial distance to the equatorial location where an electron would be found if all external magnetic fields were slowly turned off leaving only the internal dipole field.

11.2.4 Phase space density and Liouville's theorem

Two more concepts are needed to finally interpret particle measurements from satellites: phase space density and Liouville's theorem. Phase space density, $f(\mathbf{x}, \mathbf{p})$, is the distribution or density of particles in position and momentum space (cf. Sections 3.2 and 9.3.1). Our interest in working with phase space density is that it can be used to understand how collections of particles move rather than individual particles. More specifically, Liouville's theorem states that as the system evolves or moves along a trajectory in phase space the density must remain constant. The proof of this theorem is illustrated intuitively by considering a volume of phase space. As the particles in the volume are subjected to forces, their position and momentum will change but the trajectories of particles in phase space can never cross. Trajectories crossing would imply the physical impossibility that two particles with the same position and momentum subjected to the same forces go in different directions. Thus, the particles act as an incompressible fluid. As they move, the volume can change shape but the density remains the same.

At first glance, Liouville's theorem seems to be an esoteric statement, but in fact its application is quite powerful. The particle flux (number of particles per $\mathrm{cm}^2\,\mathrm{s}\,\mathrm{str}\,\mathrm{keV}$) measured by a particle detector on a satellite, $J(E, \alpha, \varphi, \mathbf{x})$ where E is the energy, α is the pitch angle, φ is the gyrophase, and $\mathbf{x}$ is the position, can be directly related to the phase space density through the relation $J(E, \alpha, \varphi, \mathbf{x}) = f(\mathbf{x}, \mathbf{p})/p^2$ (Baumjohann and Treumann, 1996). Liouville's thereom states that the phase space density does not change as the particles move along a trajectory. We also know that, if time variations of the magnetic field are slow, a particle's trajectory must move along a contour of constant adiabatic invariants. Putting these two concepts together means that $f(\mu, J, L^*, \varphi_1, \varphi_2, \varphi_3)$ wherever it is measured must remain constant. (Here $\varphi_{1,2,3}$ are phase angles associated with each invariant. For simplicity, it is generally assumed that the phase space density does not vary with the phase angles.) Any change of phase space density implies that one of the invariants is broken. In fact, acceleration mechanisms always violate an invariant. Thus, an increase in phase space density expressed as a function of the adiabatic invariants is a sign that acceleration has occurred. Flux measurements, in contrast, can change simply because the magnetic field topology has changed, making these data very difficult to interpret.

11.3 General characteristics of heliospheric particle radiation

11.3.1 Earth's radiation belts

Earth's radiation belts consist of protons and electrons that fill a torus-shaped region extending out to approximately 10 R_E. Typically, particles with energies greater than a few hundred keV are considered part of the radiation belts although there is no exact energy that divides the high and low energy particle spectrum. The populations are separated for study because their trajectories in the magnetosphere and temporal variations differ. The gradient drift discussed in Section 11.2 is directly proportional to a particle's perpendicular energy. Therefore, high-energy particle motion is dominated by the gradient drift while low-energy particle motion is dominated by the $\mathbf{E} \times \mathbf{B}$ drift. The two different drifts produce very different trajectories. The high-energy particles tend to follow closed drift paths about Earth while the lower-energy particles follow paths that flow Earthward from the magnetotail and directly out of the dayside magnetosphere.

The radiation belts have been measured and described by *in-situ* measurements from numerous satellites now for decades. Figure 11.5 shows how the flux of radiation belt electrons varies over 10 years and nearly a complete solar cycle as measured by the SAMPEX satellite. The SAMPEX satellite was in a low-altitude, highly inclined polar orbit. Such an orbit allows the satellite to sample the entire belt four times per orbit. The plot shows the logarithm of the average flux in 1 day and 0.25L bins. The inner belt is apparent inside of $L = 2$ and is fairly stable. The outer belt outside of $L = 2$ shows tremendous variability. Both the radial extent and the intensity vary on short time scales of several hours. Although not obviously apparent from the figure, more detailed analysis shows that the outer belt electron flux varies as a function of the year and the phase of the solar cycle (Baker *et al.*, 2001; Li *et al.*, 2001). The fluxes are most intense and move to lower L shells near the equinoxes and near the declining phase of the solar cycle when high-speed wind streams from the Sun are most prevalent. Sometimes changes are correlated with geomagnetic activity as described by the Dst index (see Section 10.2). Typically, during a geomagnetic storm, the flux measured at a fixed location will decrease during the storm main phase as the ring current intensifies. Often, but not always, the flux will increase during the recovery phase. One puzzling aspect of the electron radiation belts is that not all geomagnetic storms produce a flux increase even though the storm indicates enhanced energy input from the solar wind (Reeves *et al.*, 2003). Nearly 20% of storms actually produce a decrease, making it essential to understand both loss and acceleration as discussed in the rest of this chapter.

The proton radiation belt also varies as seen in Fig. 11.6 but in a very different way from the electron belts. Figure 11.6 shows proton flux (number per cm^2 s str) in the energy band from 2.5 to 6.9 MeV as measured by Space Environment Monitor (SEM-2) on the National Oceanic and Atmospheric Administration NOAA-15

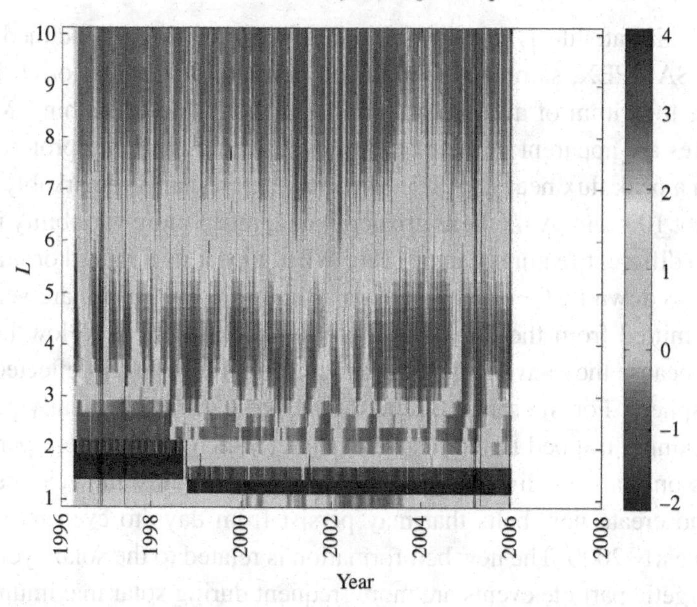

Fig. 11.5. Radiation belt electron flux ($^{10}\log(\text{counts/s})$) as measured by the Proton Electron Telescope (PET) Elo channel that measures electrons with energies > 1.5 MeV on the SAMPEX satellite. The data are averaged in 0.25 *L* and 1 day bins. See also Plate 7 in the color-plate section.

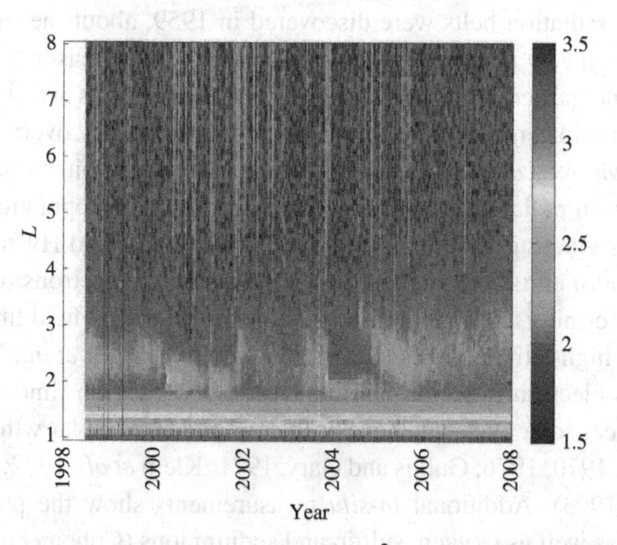

Fig. 11.6. Radiation belt proton flux (number per cm^2 s str on a logarithmic scale) from the SEM-2 instrument that measures protons with energies between 2.5 and 6.9 MeV on the NOAA-15 satellite. The data are averaged in 0.2 *L* and 1 day bins. See also Plate 8 in the color-plate section.

satellite. This satellite is in a low-altitude (~850 km) highly inclined polar orbit and, like SAMPEX, samples the radiation belts four times per orbit. Figure 11.6 shows the logarithm of average proton flux in 1 day and 0.2 L bins. Many different features are apparent in the plot. At low L values the inner proton belt can be seen with a peak flux near $L \sim 1.5$. The inner belt remains remarkably stable even over nearly 10 years. At L shells greater than 2, much more variability is observed. Here, two different features can be seen. What appear as long red or orange stripes from $L = 8$ down to $L \sim 2$ are not data artifacts. These stripes are very energetic protons emitted from the Sun. These protons can penetrate to low latitudes and small L because they have large gyroradii and are not strongly deflected by Earth's magnetosphere. For the same reason, their large gyroradii typically prevent them from remaining trapped in the magnetosphere, which is why these particle events appear as only a short-lived stripe. However, occasionally they can be trapped at low L and create new belts that may persist from days to even months as seen clearly in early 2000. The new belt formation is related to the solar cycle in that the solar energetic particle events are more frequent during solar maximum and rarely occur near solar minimum.

11.3.2 Jupiter's radiation belts

Jupiter's radiation belts were discovered in 1959, about the same time as Earth's but through very different means. No satellite would measure *in-situ* particles until 1973. The earlier discovery of Jupiter's radiation belts was based on decimetric radio emissions measured at Earth. Shortly after the discovery of Earth's radiation belts it was realized that these radio emissions from Jupiter could be interpreted as synchrotron radiation (see Chapter 4) from a similar population of very energetic electrons gyrating about magnetic field lines (Drake and Hvatum, 1959). Imaging of the radio emissions depicts two populations of electrons: one confined to the equator forming a very thin disk and another following field lines to high latitudes forming high-latitude lobes (Levin *et al.*, 2001; Garrett *et al.*, 2005). Variations in Jupiter's electron radiation belt have been observed on time scales of months to the 11-year solar cycle and have been correlated with solar wind dynamic pressure (Gerard, 1970, 1976; Gulkis and Gary, 1971; Klein *et al.*, 1972; Bolton *et al.*, 1989; Kaiser, 1993). Additional *in-situ* measurements show the presence of >1 MeV protons as well as oxygen, sulfur, and sodium ions (Cohen *et al.*, 2001).

11.3.3 Saturn's radiation belts

Unlike Jupiter, there was no prior indication from radio emissions that Saturn was surrounded by radiation belts. Saturn's belts were first confirmed in 1979 by *in-situ* particle measurements from the Pioneer 11 flyby (McKibben and Simpson, 1980; Simpson *et al.*, 1980; Van Allen *et al.*, 1980). The observations showed energetic protons and electrons surrounding the planet into the outer edge of ring A at 2.3 R_s with almost undetectable flux levels inside of the ring boundary (Chenette *et al.*, 1980). The almost complete absorption of particles by ring A explained the lack of synchrotron radiation and radio emissions. Comparison between the inbound and outbound passes of Pioneer 11 as well as *in-situ* measurements from Voyager 1 and 2 show some temporal variability that may be related to solar wind changes (McDonald *et al.*, 1980; Simpson *et al.*, 1980; Krimigis *et al.*, 1983). However, with only a few flybys any temporal or spatial differences are difficult to discern. Longer term data from the Cassini satellite that went into orbit in 2004 may yet yield some insight to the radiation belt variability.

11.4 Radiation belt acceleration mechanisms

Although the mechanisms proposed to explain the acceleration of radiation-belt particles are based on universal physical principles, they do not affect all particles in all planetary environments in the exactly the same manner. Protons and electrons are subject to different acceleration processes because the protons are more massive and thus have much slower gyrofrequencies and larger gyroradii. At each planet, the mechanisms operate in distinct fashions because of differences in particle sources and the magnetic field strength and topology. To accommodate and illustrate these differences, we first discuss the acceleration of relativistic electrons at Earth, then discuss proton acceleration at Earth, and lastly highlight these processes at Jupiter and Saturn.

11.4.1 Relativistic electron acceleration mechanisms

Many acceleration mechanisms have been proposed to explain electron radiation belt flux increases at Earth but their exact contributions are still debated. Proposed acceleration mechanisms are often separated into two categories: internal (or local) source acceleration and external source acceleration. External source acceleration mechanisms are so named because they move electrons from outside geosynchronous orbit (6.6 R_E) to the inner magnetosphere, accelerating electrons through the transport process. They operate over large spatial and time scales that violate the particles' third adiabatic invariant. Internal source acceleration mechanisms, on the other hand, locally accelerate electrons in the inner magnetosphere inside of 6.6 R_E. They operate on fast time scales and small spatial scales and violate all three adiabatic invariants. The most prominent of the proposed mechanisms

Table 11.1. *Particle acceleration mechanisms*

External source acceleration	Internal source acceleration
Substorm acceleration	Chorus-EMIC diffusive wave acceleration
Shock acceleration	Non-linear whistler wave acceleration
Radial diffusion	

in each category are listed in Table 11.1. For a more complete list see the review by Friedel *et al.* (2002).

11.4.1.1 External acceleration

The manner in which external mechanisms accelerate particles can be illustrated starting with the assumption that the first adiabatic invariant is conserved. These mechanisms move electrons radially inward where the magnetic field is stronger. Since μ is conserved during the transport process, the increase in field strength requires that the particles' perpendicular energy also increase. The total energy gain is directly related to the amount of radial transport. The relationship between transport and acceleration is easy to describe using the conservation of the first adiabatic invariant but the explanation hides the complex physics of the acceleration. Ultimately, it is an electric field that transports and accelerates the electrons because the magnetic field cannot change the particle energy (see Section 11.2). What separates the acceleration mechanisms is the exact form and time scale of that electric field.

The electric field in both shock-induced acceleration and substorm-induced acceleration is a large-scale inductive electric field that sweeps through the magnetosphere as the global magnetic field changes. The shock-induced electric field is caused by the compression of the magnetosphere as shocked solar wind passes Earth. Using a theoretical description of such an electric field and a particle transport code, Li *et al.* (1993) were able to reproduce observed electron flux signatures during the passage of one extremely large shock in March of 1991. Kress *et al.* (2007) showed similar acceleration due to shock-induced electric field for the October 29, 2003 event. However, such large sudden events are rare. A parametric study by Gannon *et al.* (2005) of shock-induced acceleration suggests that smaller more pervasive compressions do not contribute significantly to electron radiation-belt flux increases. Thus, shock acceleration is usually only discussed for specific events and not the very common flux increases that occur with most geomagnetic storms.

The substorm electric field is produced when the stretched magnetotail is pinched off near $10\,R_E$ and the remaining plasma is hurled Earthward resulting in a more dipolar magnetic field configuration (Birn *et al.*, 1998). Li *et al.* (1998) were able to model the injection of low-energy (<315 keV) electrons into geosynchronous orbit and reproduce observed flux features using transport due to inductive substorm electric fields, but the study did not consider higher-energy radiation-belt electrons. Investigating the effect of substorms on higher-energy electrons, Ingraham *et al.* (2001) concluded that some features of >1 MeV electron flux observed during one interval where the radiation belt electron flux was rising were consistent with repeated injections from multiple substorms. However, Kim *et al.* (2000) used magnetic and electric fields from a substorm MHD simulation to show that electrons could not be transported inside of 10 R_E. The results suggest that substorms may contribute to a seed population of electrons at large radial distance but some other mechanism, such as radial diffusion, is necessary to bring the electrons into the inner magnetosphere. Hence, much of the acceleration debate focused on radial diffusion.

In the case of radial diffusion, the electric field is that of ultra-low-frequency (ULF; 300 Hz to 3 kHz) waves that continuously agitate the magnetosphere. The details of this interaction have taken decades to develop and contributions from many researchers. The basic premise of the mechanism is that electric fluctuations induce small random perturbations of the electrons' positions, causing them to diffuse radially throughout the magnetosphere. The process is similar to diffusion in a gas only in this case the random walk motion of the particles is caused by electric fields instead of collisions. The random walk can be described by the Fokker–Planck equation:

$$\frac{\partial f}{\partial t} = -\frac{\partial}{\partial r} D_1(L) f + \frac{\partial^2}{\partial r^2} D_2(L) f, \tag{11.16}$$

$$D_1 = \langle L - L_0 \rangle / \tau, \quad D_2 = \langle (L - L_0)^2 \rangle / \tau. \tag{11.17}$$

In this equation $f(L, \mu, K, t)$ is the phase space distribution of electrons, L is the radial coordinate, and L_0 is initial radial position. The equation relates the changing distribution of electrons at radial position L to the distance that an electron moves, $(L - L_0)$, and at first seems too simple to instigate decades of discussion. The challenge that has tried many researchers is to define the forces that perturb the radial position and from those forces define diffusion coefficients D_1 and D_2 so that the radial diffusion equation can be used to predict the radiation belt evolution.

In fact, defining the diffusion coefficients was challenging enough that radial diffusion was initially not considered a viable acceleration mechanism. Kellogg (1959) was one of the first to propose radial diffusion as an acceleration mechanism but he focused on inductive electric fields produced from the changing magnetic field topology during storms and derived acceleration time scales of 30 years that

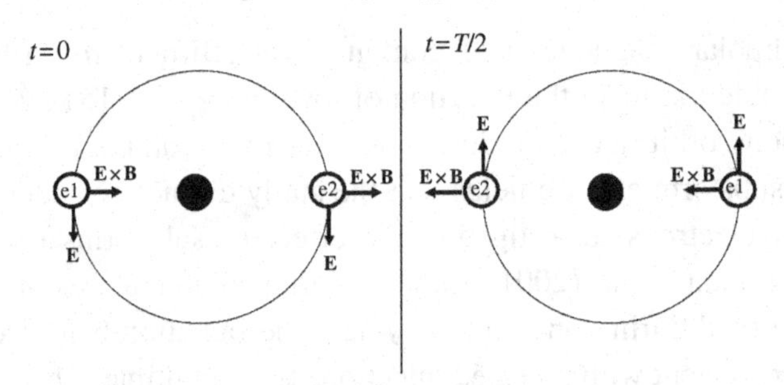

Fig. 11.7. Schematic diagram of an electron in drift resonance with a ULF wave. The left panel shows two electrons labeled e1 and e2, the direction of the wave electric field, and the direction of the particles' $\mathbf{E} \times \mathbf{B}$ drift at time $t = 0$. The right panel shows the same properties half a wave period and electron drift period later.

were much too long to explain observations. Parker (1960) as well as Davis and Chang (1962) used sudden compressions of the magnetosphere to induce diffusion but again found long time scales of a few years. These authors wrongly assumed that fluctuating fields like those of a ULF wave would average over time to zero and cause no net radial motion.

Falthammer (1965) finally recognized that time-varying fields fluctuating specifically at the same frequency as an electron drifting about Earth could cause rapid acceleration through a "drift resonance". Figure 11.7 gives a pictorial explanation of an electron drift resonance. First assume a time-varying azimuthal electric field of the form $\mathbf{E} = \mathrm{d}E \sin(m\phi - \omega t)\hat{\phi}$. Figure 11.7 shows the electric field of a wave with $m = 1$ at time $t = 0$ in the left panel and to the right is the wave electric field half a period later. The electron drifting around Earth labeled e1 gains energy from this time-varying electric field in the following way. The e1 electron, at time $t = 0$, experiences an electric field and an $\mathbf{E} \times \mathbf{B}$ drift that moves it radially inward. Half a period later (right panel) the e1 electron experiences an oppositely directed electric field since the electron drift frequency and wave frequency are the same and once again it is pushed radially inward. Ultimately the electron experiences an azimuthal electric field that continuously moves it inward causing the electron to gain energy. However, the electron e2, which began at time $t = 0$ on the opposite side, would have seen an electric field that pushed it radially outward in the same manner. Thus, the drift resonance causes electrons to diffuse radially inward and outward and decelerates as well as accelerates electrons.

Falthammer (1965) hypothesized that this type of perfectly sinusoidal wave might exist in controlled cyclotron accelerators but not in Earth's magnetosphere. However, even a randomly varying electric field with power at the drift frequency of the electron can cause radial motion in a diffusive manner. The most probable

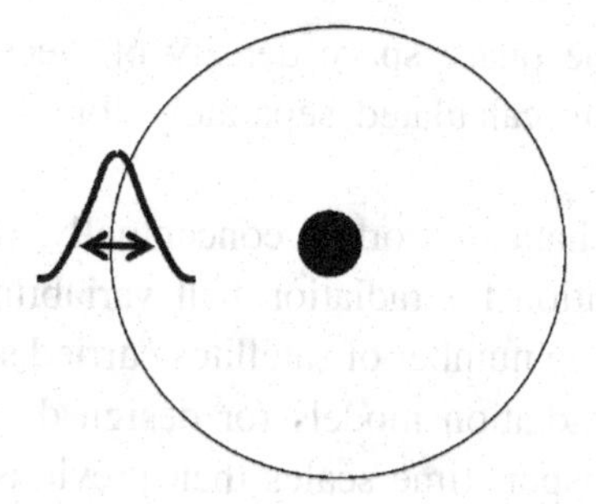

Fig. 11.8. Schematic diagram showing how a distribution of electrons spreads in L, where the black circle represents Earth and the light circle represents the drift path about Earth. Electrons spread uniformly towards and away from Earth.

effect of a randomly varying electric field is that an electron will sometimes experience an electric field pushing it inward and sometimes one pushing it outward. The result on average leads to no net radial motion. But in the random field, as in any random walk problem, there exists some probability that an electron will sample just the right electric field perturbations that move it a net distance inward. Likewise, some small probability exists that the electron will experience an electric field that moves it a net distance outward. If electrons are acted on by the random field, a few will move radially inward and a few will move radially outward while the bulk of the distribution remains stationary as demonstrated by the schematic of Fig. 11.8.

The net energy gain of the distribution of electrons in the random field depends on how many electrons gain energy compared with how many lose energy. The net energy gain is determined by the phase space density as a function of L. If electrons are uniformly distributed in L then the same number of electrons move inward and gain energy as those that move outward and lose energy with no net energy gain. If the slope of f versus L is positive, more particles move inward and gain energy than particles move outward and lose energy and the distribution of electrons gains energy. If the slope of f versus L is negative then the opposite occurs. More particles move radially outward and the result is a net loss of electron energy. The same theory applied here to the specific example of time-varying electric fields also applies to time-varying magnetic fields but the details are omitted here (see Schulz and Lanzerotti, 1974).

Falthammer (1966) showed a relationship between the diffusion coefficients, D_1 and D_2, of the Fokker–Plank equation and determined the final form of the radial diffusion equation given below:

$$\frac{\partial f(L, \mu, K, t)}{\partial t} = L^2 \frac{\partial}{\partial L}\left(\frac{D}{L^2}\frac{\partial}{\partial L}\left[f(L, u, K, t)\right]\right). \tag{11.18}$$

Here $f(L, \mu, K, t)$ is the phase space density of electrons and D is the diffusion coefficient, which is calculated separately for electric and magnetic field perturbations.

Satisfied that the mechanism worked conceptually, radial diffusion became a widely accepted explanation for radiation belt variability. Later, the theory was challenged as an increasing number of satellites carried sophisticated technologies that demanded precise radiation models for design development. New observations revealed faster transport time scales then previously expected or predicted by radial diffusion. To meet this challenge the theory was revisited and elaborated to include higher order resonances caused by electron drift motion in more realistic non-dipolar fields that increase diffusion (Elkington *et al.*, 1999, 2003; Perry *et al.*, 2005; Fei *et al.*, 2006). However, doubt about the ability of radial diffusion to fully explain observations led to the development of new competing ideas regarding electron acceleration, including the internal source acceleration mechanisms.

11.4.1.2 Internal source acceleration mechanisms

The internal source acceleration mechanisms discussed here accelerate electrons through interaction with the electric field of a VLF (3 kHz to 30 kHz) wave. The interaction is similar to the ULF wave resonance, but in this case the resonance occurs between the wave electric field and the gyration of the particle about the magnetic field instead of the drift about Earth. The EMIC (electromagnetic ion cyclotron)-chorus wave mechanism assumes the interaction with the wave can be described as a random walk diffusive process very similar to radial diffusion. This assumption is only valid when wave amplitudes are small. The non-linear whistler wave acceleration mechanisms describe how electrons interact with a monochromatic set of large-amplitude waves when diffusion is no longer valid.

The resonance between an electron and a VLF wave can be illustrated by considering a VLF wave propagating at an angle θ from the direction of the magnetic field with magnetic and electric field perturbations perpendicular to the direction of propagation. The electron gyrating about the magnetic field will experience a constant electric field from the wave when the gyrofrequency of the electron equals the Doppler-shifted frequency of the wave such that the following equation is obeyed (Brice, 1964):

$$\omega - k\cos\theta v_{\parallel} = \frac{\Omega_g}{\gamma}, \tag{11.19}$$

$$\gamma = \left(1 - \frac{v^2}{c^2}\right)^{1/2}, \quad \Omega_g = \frac{qB}{m_0}. \tag{11.20}$$

The parameters are defined as follows: γ is the relativistic correction, v is the electron velocity, c is the speed of light, θ is the angle between the propagation direction of the wave and the magnetic field direction, $v_{\parallel}$ is the velocity of the particle parallel to **B**, ω is the frequency of the wave, Ω_g is the gyrofrequency of the electron, q is the charge, B is the magnetic field magnitude and m_0 is the rest mass of the electron. The equation, called the resonance condition, shows that only electrons with a specific velocity will be accelerated or decelerated by a wave of a given frequency. In the non-relativistic limit, $\gamma = 1$; therefore the resonance condition specifies only the parallel velocity of electrons affected by the waves and particles with arbitrary perpendicular velocity are resonant. In the relativistic case, the resonance condition applies for parallel and perpendicular velocity with a known functional relationship.

In contrast to the ULF wave resonance, the VLF wave resonance will affect both the electrons' energy and pitch angle. To understand how the wave affects the energy and pitch angle of an electron, we start with the fact that, in the reference frame moving with the wave, the energy of the particle does not change. This assumption is written in the wave reference frame as

$$v'_{\perp}\mathrm{d}v'_{\perp} + v'_{\parallel}\mathrm{d}v'_{\parallel} = 0, \tag{11.21}$$

and is the same for non-relativistic and relativistic conditions. Here $v'_{\perp}$ is the particle velocity perpendicular to the magnetic field and $v'_{\parallel}$ is the velocity parallel to the magnetic field in the wave reference frame. Using a Lorentz transformation to translate the equation from the reference frame of the wave to our inertial reference frame gives:

$$\text{Non-relativistic:} \quad v_{\perp}\mathrm{d}v_{\perp} + (v_{\parallel} - u_{\mathrm{ph}})\mathrm{d}v_{\parallel} = 0, \tag{11.22}$$

$$\text{Relativistic:} \quad \left(1 - \frac{u_{\mathrm{ph}}v_{\parallel}}{c^2}\right)v_{\perp}\mathrm{d}v_{\perp} + \left(v_{\parallel} - u_{\mathrm{ph}} + \frac{U_{\mathrm{ph}}v_{\perp}^2}{c^2}\right)\mathrm{d}v_{\parallel} = 0, \tag{11.23}$$

where v_{ph} is the phase velocity of the wave.

Integrating the equations over $v_{\perp}$ and $v_{\parallel}$ gives a function that constrains how $v_{\perp}$ and $v_{\parallel}$ of the electron change throughout the interaction with a wave. The equation plotted as a function of $v_{\perp}$ and $v_{\parallel}$ is called the pitch angle diffusion curve.

Gendrin (1981) gives an intuitive method for determining the net energy gain or loss as a distribution of particles interacts with a wave. Our discussion here follows that method. Figure 11.9a, shows a contour of constant phase space density for an isotropic electron pitch angle distribution (grey trace) and a diffusion curve slightly peaked at 90° (black trace). Electrons that satisfy the resonance condition diffuse in both directions along the diffusion curve (black arrows). A diffusion imbalance occurs when there is a phase space density gradient (grey arrow). In the example shown, the gradient causes more electrons to diffuse towards 90° along the negative gradient of phase space density and to higher energy (black arrow).

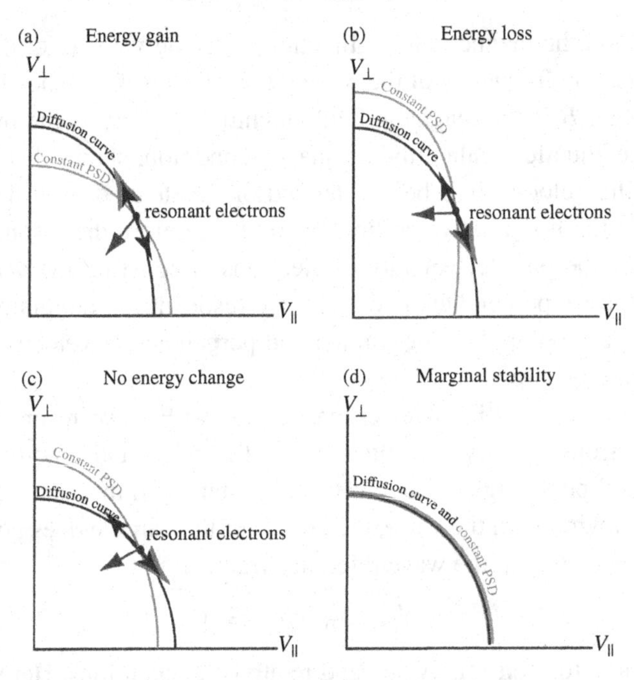

Fig. 11.9. Schematic diagram showing how a distribution diffuses in pitch angle and energy while interacting with a VLF wave.

Thus, the distribution of electrons gains net energy. Figure 11.9b shows the same diffusion curve with an electron distribution initially peaked at 90°. In this example, more electrons diffuse towards small pitch angle and lower energy. Part (c) shows a diffusion curve that lies on a circle of constant energy with an initial electron distribution peaked at 90°. In this example, more electrons diffuse along the curve to smaller pitch angles but the energy remains constant. Diffusion occurs until the interaction reaches marginal stability and the contour of constant electron phase space density lies along the diffusion curve (Fig. 11.9d).

The EMIC-chorus wave mechanism proposes that electrons interact with both whistler chorus and EMIC waves as the electrons drift about Earth in such a fortuitous way that the distribution is steadily pushed to higher energy. In this model, EMIC waves at dusk interact with electrons to produce an isotropic pitch angle distribution. The electrons continue their drift to the dawn side of the magnetosphere where chorus waves are predominantly found. The diffusion curves for chorus waves are such that an isotropic distribution will diffuse towards higher energy and larger pitch angles as in Fig. 11.9a. The energized electrons now peaked near 90 degrees continue around to the dusk side of the magnetosphere where the EMIC waves are found. The EMIC waves interact with the electrons to again produce an isotropic pitch angle distribution but with no energy loss as in Fig. 11.9c. This isotropic distribution is now primed to interact with the chorus waves once again and gain energy. Since the electrons traverse the magnetosphere in less than 10 minutes, the mechanism can effectively increase the energy over periods of days.

The non-linear whistler wave mechanisms assume that electrons are energized through a resonant interaction with whistler waves. However, the previous diffusion model requires that wave amplitudes are small in order for diffusion to be an adequate approximation. If this is not the case, the interaction must be described in a more detailed manner. The mechanism has been described theoretically by Albert (2003) and also Summers and Omura (2007), who concluded that under the right conditions a 100 keV electron could be accelerated to MeV energies within minutes. These mechanisms have yet to be compared in detail with observations or included in any kind of global model of electron flux. However, new measurements of whistler waves suggest that the small amplitude assumption is very often invalid, making non-linear modeling an active area of interest (Cattell *et al.*, 2008).

11.4.2 Differentiating external and local source acceleration

The two types of mechanisms make predictions that can be used to test whether they are significant processes in the radiation belts. The simplest expectation is that relativistic electron flux enhancements will correlate with either VLF or ULF wave activity. Unfortunately, this prediction does not distinguish between the mechanisms because both types of waves correlate well with electron flux enhancements (Rostoker *et al.*, 1998; Mathie and Mann, 2000; Meredith *et al.* 2002; O'Brien *et al*, 2003; Smith *et al.*, 2004). The common correlations are not surprising because flux enhancements tend to occur during geomagnetically active periods when waves are also enhanced.

Alternatively, the mechanisms can be differentiated by following how the gradients of electron phase space density (PSD) given for fixed adiabatic invariants evolve. The radial diffusion mechanism predicts that electrons will diffuse radially from areas of high to low PSD. Thus, to produce an increase in the inner magnetosphere requires an initial gradient with higher PSD at large L. Diffusion acts to reduce gradients, which implies that the PSD in the inner magnetosphere cannot increase above the level of the outer source region (Walt, 1994; Green and Kivelson, 2004). Local acceleration mechanisms, on the other hand, move particles from one energy and pitch angle to a new energy and pitch angle or from one μ and K to a new μ and K. These mechanisms increase the PSD wherever the waves are present, producing PSD peaks in the inner magnetosphere that increase with time.

Using PSD gradients to differentiate the mechanisms is not always straightforward. Processes other than acceleration, such as losses, may also affect gradients. For example, radial diffusion may increase the PSD followed by losses at large

L that carve away a peak. Alternatively, the outer source of electrons from the plasma sheet may decrease causing outward radial diffusion to ensue in an attempt to diminish the sharp negative gradient at the outer boundary (Selesnick and Blake, 2000; Shprits *et al.*, 2006). These types of peaks may be confused with those caused by local acceleration unless the time evolution is tracked consistently. Also an important consideration is that equatorial and non-equatorial particles may evolve differently. For example, electrons may be transported inward by radial diffusion at the equator then scattered locally at constant energy by EMIC waves to off-equatorial regions, producing PSD peaks in the off-equatorial electron population unrelated to local acceleration. Thus, sampling of both equatorial and off-equatorial particles is required to fully evaluate acceleration mechanisms. Additionally, the transformation from flux to PSD is fraught with errors. As a result, many studies comparing gradients to predictions have come to different conclusions (for an explanation of the types of errors and a review of the many different studies see Green, 2006). The most comprehensive study to date, which went to great lengths to reduce PSD errors, found PSD peaks indicative of internal acceleration mechanisms (Chen *et al.*, 2007). The results do not imply that radial diffusion is not relevant. Instead, they indicate that radiation belt models in the inner magnetosphere must include both types of mechanisms.

11.4.3 Proton radiation belt acceleration

The structure and temporal variability of the proton radiation belt is strikingly different from its electron counterpart; yet, some of the same mechanisms are proposed to explain the acceleration of these particles. The protons normally form only one belt with fluxes that peak near $L = 1.5$ and they tend to be more stable. However, during highly geomagnetically active periods, such as brought about by the passage of a large shock and sometimes an accompanying solar energetic particle (SEP) event, fast and dramatic changes occur. Often these changes mean a complete reconfiguration where entirely new, sometimes transient proton belts are formed that may last days to years.

Simulations of proton motion in both analytical and MHD magnetic field models suggest that the new proton belts are formed when protons are transported radially inward by large induced electric fields that arise as a large shock passes the Earth (Hudson *et al.*, 1995c, 1997, 2004; Kress *et al.*, 2005). The mechanism is almost the same as proposed for some electron radiation belt acceleration events at Earth, except that forming a new proton belt requires an additional source of protons from the solar wind. Often large shocks are accompanied by very high fluxes of protons that are released from the Sun and further accelerated by the shock. Normally, Earth's magnetic field acts as a protective bubble that only allows these solar protons to enter over the polar caps where they are absorbed into the atmosphere. However, as the shock passes Earth, the magnetic field is distorted such that the accompanying protons can gain access to the inner magnetosphere and become trapped in the field. Once trapped, they are swept up by the induced field and pushed to small radial distances and higher energies to form a new belt.

11.4.4 Radiation belt acceleration at Jupiter and Saturn

It may seem surprising that even in the very different space environments of Jupiter and Saturn some of the same acceleration mechanisms are invoked, but the connection is less astounding when the paucity of observations at these two planets is considered. Many theories of acceleration were first derived for Earth, where observations are comparably plentiful, and then modified to account for differences in the magnetic field geometries and source populations of the outer planets. As a result, the familiar debate about whether the interaction with ULF or VLF waves dominates acceleration also applies to these planets.

Radial diffusion has been used to explain the acceleration of radiation belt particles at both Jupiter and Saturn using the Salambo code, which was initially developed for Earth (Santos-Costa, 2003; Sicard and Bourdarie, 2004), although the mechanism's ability to fully explain steep particle spectra has been questioned (de Pater and Goertz, 1990). Internal acceleration mechanisms have been adapted to the conditions of Jupiter and Saturn. Theoretical calculations of both the chorus-wave induced pitch angle and energy diffusion (Horne *et al.*, 2008) and non-linear whistler acceleration (Summers and Omura, 2007) have shown them to be feasible at these outer planets. While radial diffusion may be able to produce observed flux levels, these two studies suggest that it may not be the only possible explanation. More detailed comparisons to phase space density gradients and pitch angle distributions are needed to confirm the theories.

11.5 Radiation belt particle losses

No discussion of radiation belts would be complete without mentioning losses, which often play an equal role in determining how flux levels evolve. As in the previous section, we will first discuss electron radiation belt losses at Earth, followed by a discussion of proton losses at Earth, and finally losses at the outer planets.

11.5.1 Electron radiation belt losses at Earth

Often electron radiation belt discussions at Earth dwell on the acceleration processes that move electrons to very high energies because it is these high energies that are so detrimental to people and electronics in space. However, a survey of electron radiation belt changes during storms by Reeves *et al.* (2003) indicates that losses play an equally important role in determining whether fluxes reach threatening levels. The survey found that only 53% of storms cause radiation belt flux levels to increase even though these storms signify increased energy input to the magnetosphere. In 19% of storms the flux actually decreased and in 28% the flux did not change. The variable response to energy input suggests that loss and acceleration rates are often comparable and ultimately compete to determine final flux levels.

Before delving into an explanation of mechanisms proposed to explain these losses we will briefly discuss the notion of an adiabatic flux decrease versus "actual" loss. An adiabatic decrease describes one associated with a change in the magnetic field topology. If the magnetic field changes slowly, electrons will move such that all three adiabatic invariants are conserved, causing the flux to decrease temporarily. The loss is only an apparent loss because the adiabatic decrease is reversible. If no other loss or violation of the invariants occurs, the fluxes will return to their previous levels when the field relaxes to its original configuration. An often cited example of an adiabatic flux decrease is the "Dst effect" (Kim and Chan, 1997; Kim *et al.*, 2002). The "Dst effect" describes the motion and apparent loss of relativistic electrons that occurs during the main phase of storms. As the Dst index decreases during the main phase, a relativistic electron must move outward to enclose the same amount of flux within its drift orbit and conserve its third adiabatic invariant. A spacecraft sitting at a fixed radial distance now samples the population of electrons that once resided at lower L as shown in the schematic in Fig. 11.10. As the electrons move outward they move to a lower magnetic field region. If the field changes slowly, they must also conserve their first adiabatic invariant so their perpendicular energy must decrease. So, in fact, the satellite at

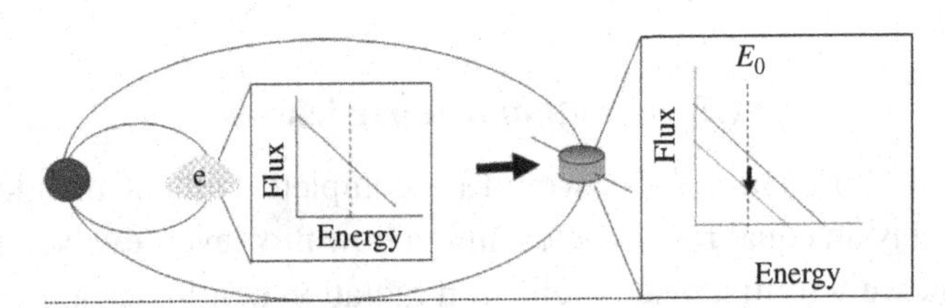

Fig. 11.10. Schematic diagram showing the adiabatic motion and flux decrease observed by a satellite caused by the "Dst effect". The dark circle represents Earth. The left hand box represents the spectrum of electron flux versus energy at a position initially Earthward of the satellite. The right hand box shows how that spectrum appears after the electrons move outward to the position of the satellite. The entire spectrum shifts to lower energy generally resulting in a measured flux decrease at constant energy.

fixed radial distance now measures the population of electrons that once resided at lower L with an energy spectrum shifted to lower energy. Figure 11.10 shows the initial energy spectrum measured by the satellite and the shifted energy spectrum. Since the satellite measures a fixed energy, the shifting spectrum generally causes a measured flux decrease. While some portion of the observed flux decrease during the main phase of storms is likely due to the "Dst effect", some also appears to be true loss (Kim and Chan, 1997). Electron flux decreases are also observed during relatively quiet times not associated with large storms (Onsager *et al.*, 2002; Green *et al.*, 2004). These depletions occur when the magnetic field changes to a more stretched configuration. However, these also cannot be completely explained by adiabatic motion because the electron flux often remains low for many days even after the field returns to its nominal dipolar.

The mechanisms that have been proposed to explain the loss of relativistic electrons are: drift out of the magnetopause boundary, outward radial diffusion, and scattering into the atmosphere. Scattering can be caused by interactions with a thin current sheet, EMIC waves or chorus waves.

Loss of electrons through the magnetopause boundary occurs when the drift paths of electrons are altered as the magnetic field changes from a quiet time configuration to more disturbed conditions. During quiet times, the drift motion of an electron starting in the magnetotail is dominated by an electric potential field directed from dawn to dusk that moves electrons Earthward. As the electrons get closer to Earth, the magnetic radial gradient causes a westward drift. Some of these drift trajectories will cross Earth's magnetopause and the electron will be swept away by the solar wind. Closer to Earth, the trajectories of the electrons will be dominated by the gradient drift. Undisturbed, electrons in this near Earth region will simply drift about continuously on closed almost circular paths. Ukhorskiy *et al.* (2006) calculated drift paths defined using the TS05 magnetic field for quiet and disturbed conditions. During quiet conditions the model shows electrons inside of $\sim 7\ R_E$ on closed drift orbits. However, the drift paths calculated for a moderately large storm using the TS05 model indicate that electrons inside of $\sim 5\ R_E$ will now encounter the magnetopause. The results suggest that during geomagnetic storms most of the outer electron radiation belts are emptied into the solar wind and replaced by an entirely new belt of accelerated electrons. While this suggestion seems plausible it has not been explicitly verified with observations. Green *et al.* (2004) concluded that loss to the magnetopause was not an adequate explanation for electron flux depletions observed during more quiet conditions because the flux of energetic protons on similar drift paths did not decrease.

Radial diffusion, which was discussed in detail as an acceleration mechanism, has also been proposed as a loss process. Radial diffusion acts to reduce gradients by pushing particles from high phase space density to low phase space density. The outermost closed drift orbit of the radiation belts represents a very steep gradient where the phase space density goes to zero. If ULF waves are present, then radial diffusion will push particles outward to the magnetopause. Shprits *et al.* (2006) showed qualitatively that radial diffusion with an outer boundary that experiences significant decreases can cause losses down to low L values.

Losses into the atmosphere occur when some mechanism scatters electrons to smaller pitch angles causing them to travel farther down the field line and collide with the neutral atmosphere. The current sheet, which forms in the magnetotail as the lobes are stretched and forced together by solar wind dynamics, is an effective scattering region. Scattering occurs when the magnetotail becomes stretched to the point that an electron bouncing along a field line can no longer make it around the kinked field without violating its first invariant. Traversing the kink changes the particle's pitch angle. Under certain conditions the pitch angle changes can be described as a diffusive process. Many authors have calculated diffusion coefficients and lifetimes for this type of interaction using different magnetic field models (Young *et al.*, 2008, and references therein) but the significance of this loss contribution has yet to be verified.

Chorus and EMIC waves were invoked in Section 11.3 as a means to accelerate electrons. However, these waves may also cause rapid loss into the atmosphere. Whether or not the waves produce net acceleration or loss depends on the initial gradients of the electron distribution as a function of pitch angle. Assuming the appropriate distribution exists, EMIC waves are expected to cause losses on the time scales of several hours to a day (Summers and Thorne, 2003; Albert, 2003). Whistler chorus may cause losses on time scales of one day, but these estimates are sensitive to parameters such as the cold plasma density. Loss rates may increase to time scales less than a day during storm main phase when the plasma density is expected to vary (Summers and Thorne, 2003).

Quantifying loss into the atmosphere and identifying its cause has been challenging because features in the observations vary with energy and the terminology describing these feature has not been consistent (for a thorough review of mechanisms that cause loss to the atmosphere and observations see Millan and Thorne, 2007). However, a more coherent picture is starting to emerge. Two types of electron precipitation into the atmosphere have been identified in low-altitude observations: microbursts and bands. Microbursts are short enhanced bursts of electrons that last approximately tens of seconds and extend up to MeV energies. The microbursts are clearly related to whistler chorus because they are co-located with the wave measurements on the dusk side of the magnetosphere and have structure consistent with the bursty nature of the chorus waves. In contrast, the band precipitation typically spans a wider spatial region, is observed from dusk through midnight, and may last for hours. The connection between the band precipitation and loss mechanisms is still unclear. Some speculate that the bands at dusk may be caused by EMIC waves while those in the midnight region are caused by current sheet scattering.

11.5.2 Proton radiation belt losses at Earth

The proton losses from the radiation belts have not been analyzed in the same detail as the dramatic formation of new belts. New belts last from days to years. Mechanisms proposed to explain the disappearance of these belts include scattering caused by the kinked field (Anderson *et al.*, 1997; Young, 2008), and interaction with EMIC waves (Hudson *et al.*, 1998a). Lorentzen *et al.* (2002) analyzed a large variety of data during a 2-year period and found no consistent description, so that "many questions remain to be answered, and it may be that more than one mechanism plays a role in each event."

11.5.3 Radiation belt losses at Jupiter and Saturn

When discussing radiation belts of the outer planets one new loss mechanism must be included that is not relevant at Earth: absorption by moons and rings. The moons and rings effectively sweep up the particles along their orbit. This type of loss may at first not appear significant. However, it can effectively create a barrier that prevents particle fluxes from building up to high levels inward of the orbit. For example, if radial diffusion is the dominant acceleration mechanism, the radiation belts can exist inward of the orbit only if the diffusion time scales are fast enough for particles to be pushed past the orbit before being swept clean again. For discussions of this type of loss, see Sicard and Boudarie (2004), de Pater *et al.* (1997), and Mead and Hess (1973).

12

Flares, coronal mass ejections, and atmospheric responses

TIMOTHY FULLER-ROWELL AND STANLEY C. SOLOMON

12.1 Introduction

The tenuous, partially ionized plasma in planetary upper atmospheres is vulnerable to explosive and dynamic events from both the Sun and the lower atmosphere. The power of the Sun is continuously bombarding the atmospheres of planets with photons, energetic particles, and plasma. Some of the most dramatic solar events are the sudden release of electromagnetic energy during solar flares, and plasma from interplanetary coronal mass ejections (ICME). The intense solar radiation from a flare is the first to impact a planetary system, shortly followed by the arrival of relativistic energetic particles. Some time later, hours to days depending on the planet's distance from the Sun, the bulk of the plasma arrives to interact with, in some cases, the planetary magnetosphere; energy is then channeled into the upper atmospheres and ionospheres. The upper atmospheres are subjected to dramatic changes in external forcing by these types of events, by as much as a factor of two in total energy deposited, by an order of magnitude for individual processes, and by several orders of magnitude in some wavelength bands.

The upper atmospheres of planets are also being pushed and jostled by energy and momentum propagating upward from the dynamic chaotic lower atmospheres. The total solar irradiance driving the lower atmospheres is invariant except for the fraction of one percent changes observed over a solar cycle. Estimates have been made of the impact of longer-term changes in solar radiative output on Earth's climate, an area that is explored further in Vol. III. On shorter time scales, it has been known for a long time that waves propagate from the lower to the upper atmosphere. Unlike the upper atmosphere, lower atmosphere variability is not due to changes in forcing, but because the weather systems are naturally chaotic. On the large scale, the day-to-day changes in planetary waves, tides, and gravity waves impose variability on upper atmosphere winds, composition, the ionosphere, and electrodynamics. This aspect is also expanded upon in Vol. III. On the small scale,

Heliophysics: Space Storms and Radiation: Causes and Effects, eds. Carolus J. Schrijver and George L. Siscoe. Published by Cambridge University Press.

when a thunderstorm explodes with a bolt of lightning for instance, electromagnetic waves propagate throughout geospace, and at the same time, gravity waves are launched that can penetrate into the upper atmosphere. The nature of these forcing mechanisms from the lower atmosphere, including the range of waves and electromagnetic processes, and their impact on the upper atmosphere and ionosphere, is a relatively new field within heliophysics.

Explosive events can also occur internally in the upper atmosphere, in the same chaotic way that terrestrial weather systems form and convective thunderstorms emerge. These internal "instabilities," like the weather, are not forced directly but can grow and bloom as spectacular displays of ionospheric irregularities. The physical processes were touched upon in Chapter 12 of Vol. I. One of the favored mechanisms for equatorial irregularities, or spread-F, as it is also referred to, is the generalized Rayleigh–Taylor (R-T) instability (see Kelley, 1989; Basu *et al.*, 1996, 2001). However, the growth rate is relatively slow for the instability to form and requires excessive forcing by gravity wave seeding. Recently, the "collisional shear instability" has come to light as a potentially important mechanism in the initiation or seeding of the subsequent R-T instability (Hysell *et al.*, 2005). This collisional shear instability is thought to have a faster growth rate than R-T, and also induces structure at wavelengths closer to that observed in nature. Several other plasma processes occur at other latitudes, such as the Perkins, gradient-drift, and two-stream instabilities (see Perkins, 1973; Zhou and Mathews, 2006; and reviews by Kelley, 1989, and by Fejer and Kelley, 1980, for more information).

This chapter focuses on the atmospheric response to the explosive events initiated on the Sun, rather than those from the lower atmosphere or generated internally. Although the electromagnetic radiation is the first to arrive after a solar flare, the impact is relatively modest compared to the response to a typical ICME. We will therefore start with the dynamic response of the upper atmosphere to an ICME striking a planet's magnetosphere, such as Earth's, that drives the processes collectively referred to as a geomagnetic storm. The magnetosphere filters the ICME transients in the solar wind and interplanetary magnetic field, and modulates the electrodynamic and particle energy and momentum that are finally deposited into the upper atmosphere. The sequence of neutral temperature, density, winds, composition, plasma, and electrodynamic response will be traced through the history of an event.

This is followed by a description of the physical processes and the atmospheric response to intense solar flares, including photochemical and dynamical changes. Until fairly recently, solar flares were thought to produce substantial changes only in the X-ray part of the solar spectrum at wavelengths shorter than about 2 nm, and to impact mostly the lowest layers of the ionosphere, the D region below 100 km altitude, and the E region from about 100 to 130 km altitude. Improvements in solar observational capability and observations of extreme flare events during the declining phase of solar cycle 23 have revealed that flares can have a much broader signature in the solar ultraviolet spectrum and a more substantial impact on the upper atmosphere and ionosphere than previously thought.

First, a word about nomenclature. Heliophysics covers physical processes in media that range from almost completely ionized, as in stellar coronae, to virtually zero ionization fraction, as in planetary lower atmospheres. Historically, the planetary upper atmospheres have been considered as tenuous extensions of the lower atmospheres, and use the same basic equations as meteorology, except that the properties of a new minor species of charged particles, the ionosphere, has to be accommodated. It is also important to think of the upper atmosphere as an extension downward of the space plasma domain, and accommodate the terminology and equations generally used there. The upper atmosphere can therefore be thought of as a transition region between the dense, neutral atmospheric fluid and space plasmas. As was discussed in Vol. I, Chapter 12, the upper atmosphere fluid is more than 99% neutral, but even with the ionization fraction of less than 1% it is technically still a plasma, albeit weakly ionized. We use the term thermosphere to refer to the predominant neutral component, and the term ionosphere as the minor species of charged particles. Due to the mass of the thermosphere, it is not surprising that the ionosphere is affected by the dynamics and composition of the neutrals. Similarly, and maybe surprisingly, although only a relatively minor species in the partially ionized plasma of the upper atmosphere (see Fig. 12.13 in Vol. I for ionization fractions in the Earth's ionosphere compared to those in the solar atmosphere), the ionosphere has a significant, and sometimes dominant, impact on the neutral gas. The coupling and interaction between the neutral and ionized state is therefore crucial to a thorough understanding of either component.

12.2 ITM responses to geomagnetic storms

The term geomagnetic storm, in the context of aeronomy, i.e. the physics and chemistry of the upper atmosphere (the ionosphere, thermosphere, and mesosphere, or ITM), is somewhat of a misnomer, since the total magnetic field change during a storm is relatively modest (<1%) below 1000 km altitude. Part of the nomenclature evolved from early instrumentation, such as the ground-based magnetometer, which indicated that the horizontal component of Earth's magnetic field fluctuated by a small percentage during strong auroral displays (see Chapter 1). Other descriptions, such as ionospheric and auroral storms, are also used and are equally appropriate to describe these episodic events and their impact.

We begin this section by quantifying the energy sources during a typical geomagnetic storm, and contrast these with the quiet-time solar radiative power. The section also reviews our understanding of the response of the upper atmosphere to these types of injections of energy, and describes the range of physical processes that follow. Understanding the neutral dynamics, composition, temperature, mass density, plasma, and electrodynamic response of the upper atmosphere to geomagnetic storms is a significant challenge for heliophysics, and is particularly relevant due to the impact of space weather on operational systems.

12.2.1 High-latitude energy injection

On a typical quiet geomagnetic day, the total magnetospheric energy deposited in Earth's upper atmosphere is significantly less than that from solar radiation (Roble *et al.*, 1987). In this quiet case, the globally averaged upper thermospheric, or exospheric, temperature increases by 10 to 15% due to the solar wind/interplanetary magnetic field sources that are channeled through the magnetosphere into the upper atmosphere. The combined solar radiative power input from the two main energy channels EUV ($\lambda \leq 102.7$ nm) and FUV ($102.7 < \lambda < 200$ nm) are typically ~440 and ~1080 GW, respectively, for moderate solar activity (as measured by the solar radio flux, F10.7 ~150, compared to monthly averages from ~70 up to ~300 through a strong cycle). In contrast, the total for magnetospheric power sources during geomagnetic quiet times is typically 50 to 100 GW, including the electromagnetic and particle components. On a typical day, however, the magnetospheric sources contribute about half of the variability of the system. During a storm, this geomagnetic source can increase ten to twenty fold, and can overwhelm the solar radiation source, both as the source of variability and in terms of the total energy injection. Because of the nature of the interaction between the solar wind and magnetosphere, most of this energy is initially deposited towards the mid and high latitudes.

The magnitude of the magnetospheric energy source, we expect, can be surprisingly well quantified, not because we can estimate the magnitude of the energy source itself, but because we can measure the atmospheric response to a storm. The energy injection causes heating, a thermal expansion, and an increase in atmospheric density at low Earth orbiting (LEO) satellite altitudes. For instance, the accelerometer onboard the CHAMP satellite measures the neutral mass density at around 400 km (see for example Liu and Lühr, 2005). The increase in density is a good measure of the thermal expansion and hence the energy injection. The upper panel of Fig. 12.1 shows a comparison of CHAMP satellite orbit-averaged neutral mass density with the results of a numerical model for the first 15 days of January 2005 (Fedrizzi, private communication, 2009). The black and red lines are the orbit averages of density for CHAMP and the model, respectively. The model is able to follow the density response and recovery with quite high fidelity.

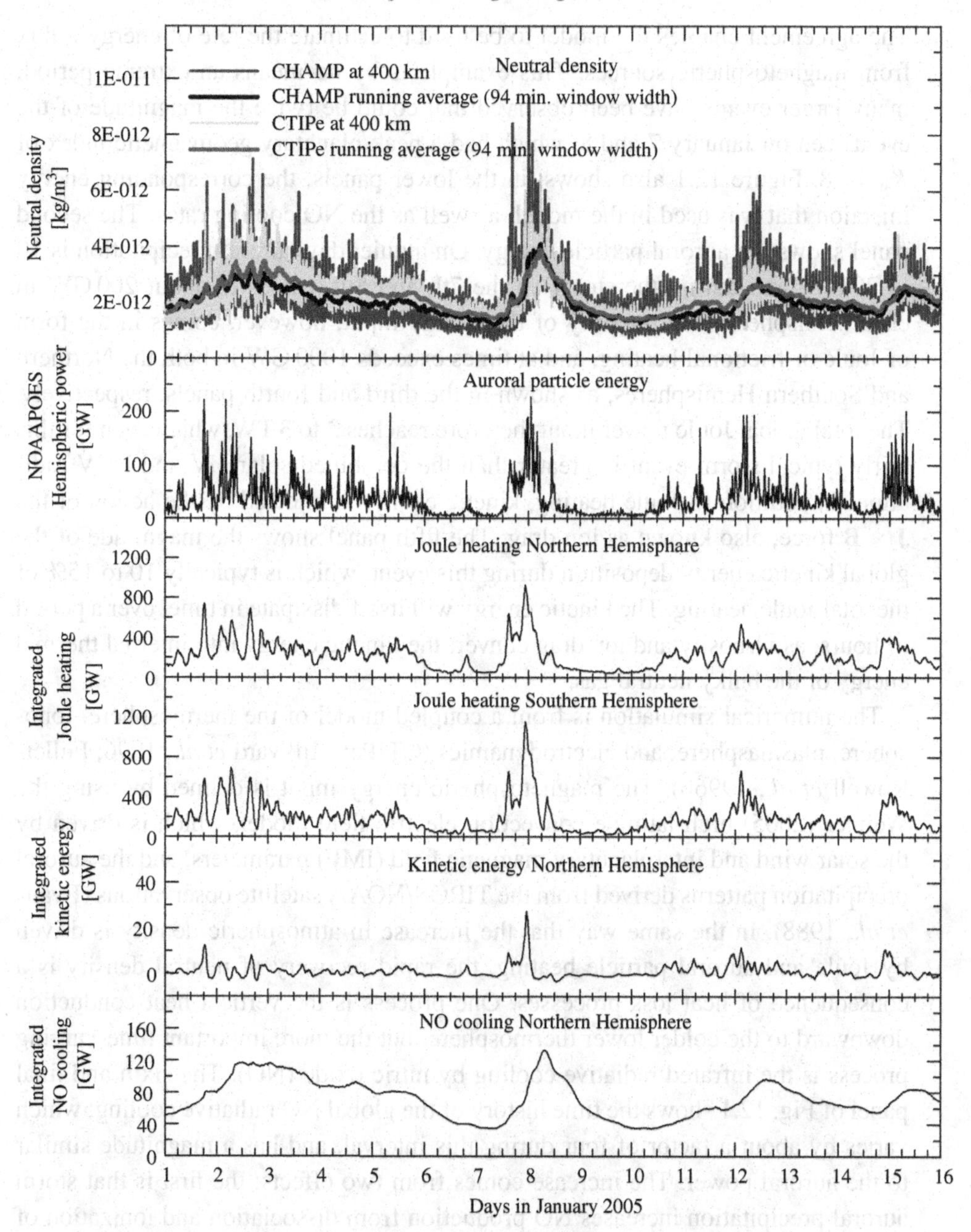

Fig. 12.1. Ionospheric properties during a geomagnetic storm. The upper panel shows a comparison of CHAMP neutral density measurements at 400 km altitude with a numerical simulation, for a stormy period in January 2005. The lower panels show, from top to bottom, estimates of auroral power, Joule heating in the Northern and Southern Hemispheres, kinetic energy deposition, and nitric oxide infrared cooling rates. (Courtesy of M. Fedrizzi.) See also Plate 9 in the color-plate section.

The agreement enables the model to be used to estimate the rate of energy influx from magnetospheric sources. This example is by no means an extreme period; many larger events have been observed that could be twice the magnitude of the event seen on January 7 and 8, which had a peak planetary geomagnetic index of $K_p = 8$. Figure 12.1 also shows, in the lower panels, the corresponding energy injection that was used in the model, as well as the NO cooling rates. The second panel shows the auroral particle energy. On a quiet day auroral precipitation is 10 to 20 GW, and during the storm on the 7th and 8th it rises to about 200 GW in each hemisphere. The majority of the energy input, however, comes in the form of Joule or frictional heating, and at times exceeds 1000 GW in both the Northern and Southern Hemispheres, as shown in the third and fourth panels, respectively. The total global Joule power input therefore reaches 2 to 3 TW, which even for this fairly typical storm event is greater than the combined solar UV and EUV radiation. In addition to Joule heating, kinetic energy is injected by the action of the $\mathbf{J} \times \mathbf{B}$ force, also known as ion drag. The fifth panel shows the magnitude of the global kinetic energy deposition during this event, which is typically 10 to 15% of the total Joule heating. The kinetic energy will itself dissipate in time, over a period of hours, as viscosity and ion drag convert the kinetic energy into internal thermal energy of the bulky neutral gas.

The numerical simulation is from a coupled model of the thermosphere, ionosphere, plasmasphere, and electrodynamics (CTIPe; Millward *et al.*, 1996; Fuller-Rowell *et al.*, 1996b). The magnetospheric energy input is defined by using the Weimer (2005) high-latitude convection electric field model, which is driven by the solar wind and interplanetary magnetic field (IMF) parameters, and the auroral precipitation patterns derived from the TIROS/NOAA satellite observations (Evans *et al.*, 1988). In the same way that the increase in atmospheric density is driven by Joule and auroral particle heating, the rapid recovery of neutral density is a consequence of heat loss processes. One process is the vertical heat conduction downward to the colder lower thermosphere, but the more important time-varying process is the infrared radiative cooling by nitric oxide (NO). The sixth and final panel of Fig. 12.1 shows the time history of the global NO radiative cooling, which varies by about a factor of four during this interval, and has a magnitude similar to the auroral power. The increase comes from two effects: the first is that storm auroral precipitation increases NO production from dissociation and ionization of molecular nitrogen; the second is that the O atom vibrational relaxation of excited NO is temperature dependent. The CTIPe simulation in Fig. 12.1 uses the time-dependent estimates of NO from the Marsh *et al.* (2004) empirical model based on SNOE satellite data. The empirical model is used in CTIPe rather than solving for minor species photochemistry self-consistently, which is discussed further in Section 12.3.

The agreement of the simulation with the observed time evolution of neutral density implies that the estimates of total energy dissipation and cooling rates are reasonably well simulated by the model. The sum of the Joule heating plus kinetic energy dissipation, or total electromagnetic energy $\mathbf{J} \cdot \mathbf{E}$, is also referred to as the Poynting flux (Thayer *et al.*, 1995). In most cases this flux is downward from the magnetosphere, the power being generated by the solar wind dynamo. During the recovery from a storm, the stored kinetic energy in the neutral winds that have been "wound-up" during the driven phase can be released. This "flywheel effect," as it is known, is elaborated upon in Section 12.2.4.

The spatial distribution, day/night differences in Joule and particle heating, and the difference between the hemispheres, are modulated by solar-produced conductivities. If the magnetosphere acts as a simple battery, or source of potential, then the current flow and dissipation through the ionosphere–thermosphere system is affected by conductivity produced by ionization from both auroral particle precipitation and solar radiation. The solar-produced conductivity is expected to lead to differences between the energy dissipation in the summer and winter hemispheres, and between the day and night sides. However, it is difficult to quantify this effect because of changes in other parameters, such as the angle of attack of Earth's dipole to the flowing solar wind, the degree to which the solar wind/magnetospheric dynamo acts more like a current source than a source of potential, and the response of the magnetosphere to the changes in ionospheric conductivity. Note in Fig. 12.1 that the southern summer Joule heating rates are slightly higher than in the winter hemisphere, although for this simulation there has been no self-consistent feedback with the magnetosphere.

12.2.2 Auroral heating and ionization

Auroral precipitation, as described above, contributes only a modest 10 to 20% of the energy influx during a geomagnetic storm. Yet it is the auroral displays that are the visual manifestation of geomagnetic storms, and that capture the imagination. Although a modest energy source, the auroral ionization provides the catalyst for the electromagnetic energy transfer. Auroral precipitation is most effective in increasing the E-region plasma densities, which produces an increase in the peak of the Pedersen conductivity profile around 125 km altitude (see Vol. 1, Section 12.5). A doubling of the plasma density doubles both the conductivity and Joule heating dissipation from the auroral currents, a contribution to the energy budget significantly larger than the particle energy itself. Figure 12.2 illustrates the statistical pattern of auroral energy influx generated from the TIROS/NOAA satellite observations (Evans *et al.*, 1988). Power flux observations accumulated during a single transit over the polar region (which requires about 25 minutes as the satellite

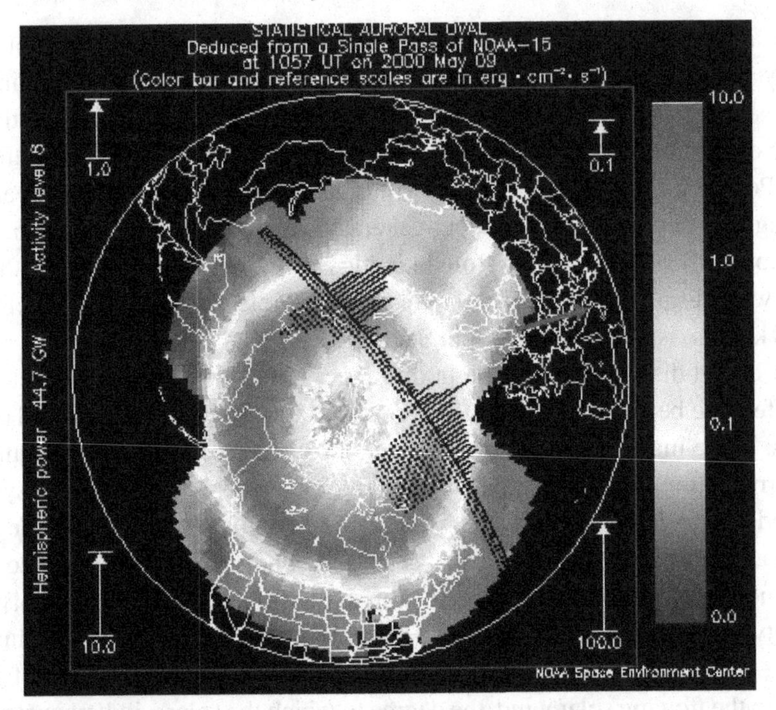

Fig. 12.2. Statistical pattern of auroral energy input derived from TIROS/NOAA satellite data during a single transit of the polar region. (From Evans *et al.*, 1988.) See also Plate 10 in the color-plate section.

moves along its orbit) are used to estimate the total power input by auroral particles to a single polar region. This estimate, which is corrected to take into account how the satellite passes over a statistical auroral oval, is a measure of the level of auroral activity, much as K_p or A_p are measures of magnetic activity. A particle power input of less than 10 GW to a single polar region, either the North or South, represents a low level of auroral activity. A power input of more than 100 GW represents a high level of auroral activity. The auroral power in this figure is 44.7 GW.

The characteristic energy of the auroral particles is typically 2 to 3 keV, which produces maximum ionization in the *E* region, around 125 km altitude. The ionospheric *E* region is close to ionization equilibrium, meaning that ions recombine rapidly so production is balanced by loss and there is little time for transport. Therefore, the increases and decreases in plasma density follow the spatial and temporal characteristics of the auroral particle source. In the upper thermosphere, around 300 km or so, the ionization rates also increase from auroral precipitation, but less dramatically than in the *E* region. At these higher altitudes, the ionospheric plasma is no longer in chemical equilibrium so transport can influence the distribution, producing tongues of ionization and plasma troughs. In the middle and upper thermosphere, the neutral composition can also change more drastically than below, which can change ion loss rates and at times dominate the response (see Fig. 12.3).

12.2.3 Storm dynamics at high latitudes

As discussed in Vol. I, Section 12.1, in contrast to the treatment of electric and magnetic fields in the magnetosphere, electric fields in the upper atmosphere are treated as a gradient of a potential, so that in the collision-rare upper thermosphere, $\mathbf{E} \times \mathbf{B}$ plasma drifts are synonymous with the electric field. Magnetospheric and ionospheric physicists use different terminologies. At high latitudes, ionospheric plasma responds directly to the strong magnetospherically imposed electric fields, which cause ion drifts of many hundreds, if not thousands, of meters per second. Volume I, Section 12.6, showed typical neutral winds, ion drifts, and plasma densities at mid to high latitudes during fairly average geomagnetic activity ($K_p \sim 3$) in response to a fairly typical two-cell pattern of magnetospheric convection. Figure 12.3 shows the response of the upper atmosphere in a similar format, but for the Southern Hemisphere, shortly after impact on the magnetosphere of an ICME. On the left, the vectors represent plasma drift velocity in the upper thermosphere, where the ion drift motion is close to the $\mathbf{E} \times \mathbf{B}/B^2$ because the collisions with the thermosphere are relatively infrequent. The ion drift exceeds 1300 m/s over wide areas. The color contours represent plasma density near the F-region peak, close to 300 km altitude. In contrast to the quiet case depicted in Vol. I, the high-latitude plasma densities are actually depleted in spite of the increase in auroral ionization. The cause of the depletion is the change in the neutral composition resulting from the Joule heating, which will be expanded upon later. Increased upwelling enhances the molecular oxygen and nitrogen neutral species, compared to atomic oxygen, so there is a significant increase in recombination rates, referred to as ionospheric loss rates.

On the right of Fig 12.3, the neutral wind and temperature response over the same region is shown. Peak winds in the quiet case in Vol. I were around 300 to 400 m/s. With the imposition of more intense magnetospheric convection, even with the infrequent collisions at this altitude, there is a sufficient momentum source to accelerate the medium to over 800 m/s after about four hours (see Killeen *et al.*, 1984, 1988, for observations of these winds). The dynamical properties of the vortices were described extensively in Vol. I, Chapter 12, and the same basic physical processes are operating in this more intense case. For this Southern Hemispheric view, the strong dusk sector vortex is anticlockwise, or cyclonic, due to the inertial resonance between the ion and neutral convection (Fuller-Rowell *et al.*, 1984; Fuller-Rowell, 1995). Overall, the increase in Joule heating raises the average temperature at high latitude by two to three hundred kelvin in four hours. The weaker anticyclonic dawn vortex, which does not resonate with the plasma motion,

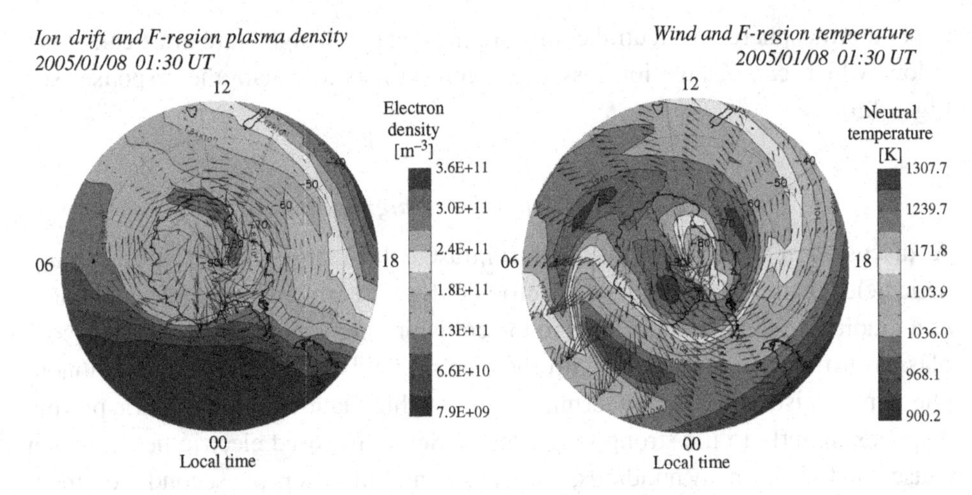

Fig. 12.3. Simulated response of the F-region plasma densities (left) and neutral winds and temperature (right) at the peak of the storm event at 1:30 UT on January 8, 2005, in the Southern Hemisphere. Both represent the response in the upper thermosphere and ionosphere at about 300 km altitude. Peak neutral winds are in excess of 800 m/s. (Courtesy of M. Fedrizzi.) See also Plate 11 in the color-plate section.

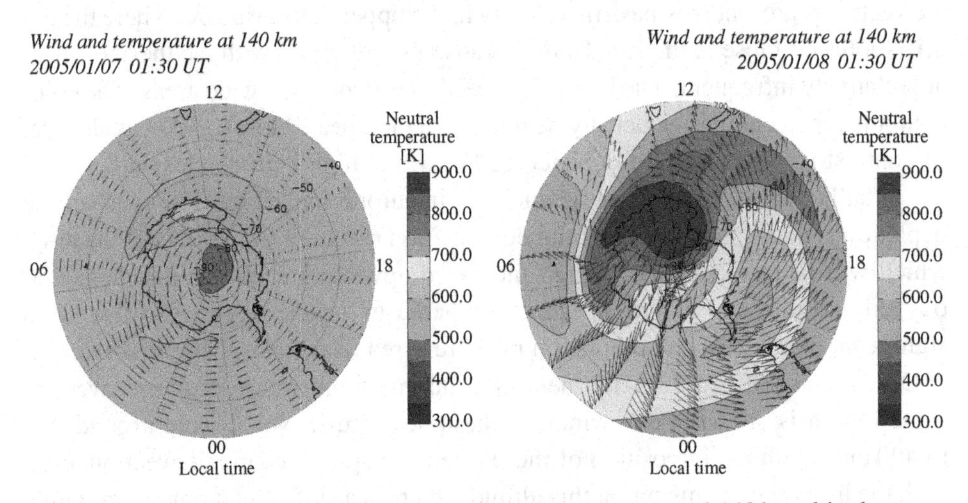

Fig. 12.4. Neutral winds in the lower thermosphere at around 140 km altitude at the peak of the storm at 1:30 UT on January 8, 2005, in the Southern Hemisphere (right), and at the same UT on the quiet day preceding the storm (left). Winds in the lower thermosphere increase dramatically in response to the storm, but peak magnitudes are about half those at 300 km. Lower thermosphere winds driven by the storm also tend to be slower to dissipate, sometimes acting as a "flywheel" driving Poynting flux upward from the thermosphere/ionosphere to the magnetosphere. (Courtesy of M. Fedrizzi.) See also Plate 12 in the color-plate section.

is always divergent and tends to form a cool, low-density region, superimposed on the overall increase from Joule heating. The dusk cell in the quiet case (Vol. I, Chapter 12) was not strong enough to be divergent; in this more intense case it is divergent, and so also tends to form a cold, low-density region at its core (Crowley *et al.*, 1996). The numerical results depicted in Fig. 12.3 are taken from the same CTIPe simulation as in Fig. 12.1, at the time of the peak of the storm event at 1:30 UT on January 8, 2005.

12.2.4 High-latitude storm dynamics in the lower thermosphere and the flywheel effect

The wind and temperature in the lower thermosphere for the Southern Hemisphere at 140 km altitude, at the same storm time as above, are shown on the right in Fig. 12.4, which shows a similar pattern as at 300 km, but reduced in magnitude. The panel on the left of Fig. 12.4 shows the wind and temperature on the quiet day preceding the event. At 140 km altitude, collisions between the neutrals and ions are more frequent, but the inertia of the more massive neutral atmosphere more than outweighs the increased collisional force. The peak winds at this altitude are about half those at 300 km, and the balance between the main forces of inertial, Coriolis, ion drag, and pressure gradients is different (see Kwak and Richmond, 2007). Forty kilometers lower in altitude, the impact of the ion motion would be barely discernable. The neutral temperature, however, has increased by three to four hundred kelvin, at least as much as at 300 km.

One hundred and forty kilometers is above the altitude where eddy diffusion tends to mix and damp the winds, and molecular diffusion is still slow in the dense lower thermosphere. The momentum in the neutral winds is therefore slow to dissipate; the winds remain elevated for several hours, at speeds exceeding the ion drifts driven by the imposed, but waning, electric field; this is the so-called "flywheel" effect. The charged particles recombine rapidly in this dense chemically controlled region as auroral precipitation returns to its quiet-time levels. The elevated winds and the background conductivity produce a dynamo action, which drives electromagnetic energy, or Poynting flux, upward from the lower thermosphere to the magnetosphere. The total power is fairly small compared with the downward flux during the solar wind driven phase; the impact on the magnetosphere is likely to be fairly modest.

12.2.5 Global thermosphere dynamics, temperature, and density response

The dynamic changes during a storm provide the conduit for many of the physical processes that ensue in the upper atmosphere. For instance, the increased meridional winds at mid latitudes push plasma parallel to the magnetic field to regions of different neutral composition. The global circulation carries molecular rich air from the lower thermosphere upward and equatorward, changing the ratio of atomic and molecular neutral species, and changing loss rates for the ionosphere. The physical processes are similar to those behind the interhemispheric seasonal circulation driving the global composition structure, as described in Vol. I, Section 12.3. The storm wind system also drives the "disturbance" dynamo, which through plasma transport modifies the strength and location of the equatorial ionization anomaly peaks. These processes are elaborated upon in Section 12.2.9.

During geomagnetic storms, the global dynamics of the upper atmosphere changes dramatically (Buonsanto, 1999; Fuller-Rowell *et al.*, 1994, 1997; Fejer *et al.*, 2002; Emmert *et al.*, 2001, 2002, 2004). The response is complex even during the simplest of events. The thermosphere, although thought of as a sluggish medium, can respond quite quickly (in tens of minutes) and can support high-speed, large-scale gravity waves that propagate globally, initiated by impulsive forcing at high latitudes (Richmond and Matsushita, 1975). The large-scale waves have typical wavelengths of 1000 km or more and phase propagation speeds ranging from 400 to 1000 m/s (Hunsucker, 1982; Shiokawa *et al.*, 2002). Gravity waves propagate at close to sound speeds, so waves launched by auroral heating can reach mid latitudes in an hour, and can reach the equator and penetrate into the opposite hemisphere within three hours. Waves launched from both hemispheres interact to form a quite complex wave train (Shiokawa *et al.*, 2002; Lu *et al.*, 2008), even for the simplest forcing time histories. Real events with complex time histories are more difficult to unravel.

Large-scale gravity waves provide the mechanism for transmitting changes in pressure gradients around the globe. A new global circulation can therefore be imposed on the same time scale as gravity-wave propagation; it does not rely on, nor require, the bulk physical transport of mass by the wind field, which is typically much slower at mid latitude, 100 to 200 m/s.

Figure 12.5 shows the change in neutral wind at mid and low latitudes at 250 km, three hours into a numerical simulation of a step-function increase in high-latitude forcing in the auroral oval (65°–75° geomagnetic latitude). The wind response is shown within 50° latitude of the geographic equator, to allow for a scale that clearly shows the mid- and low-latitude dynamic response. Whereas at auroral latitudes the peak neutral winds would be close to 1000 m/s, at mid and low latitudes the winds are much more modest, with 100 to 200 m/s wind surges above the background circulation. At this time, three hours into the simulation, the disturbance winds have reached the equator and are beginning to penetrate the other hemisphere and interact with the opposing wave front from the other pole. The arrival of the wave front at the geographic equator within three hours indicates a propagation speed of

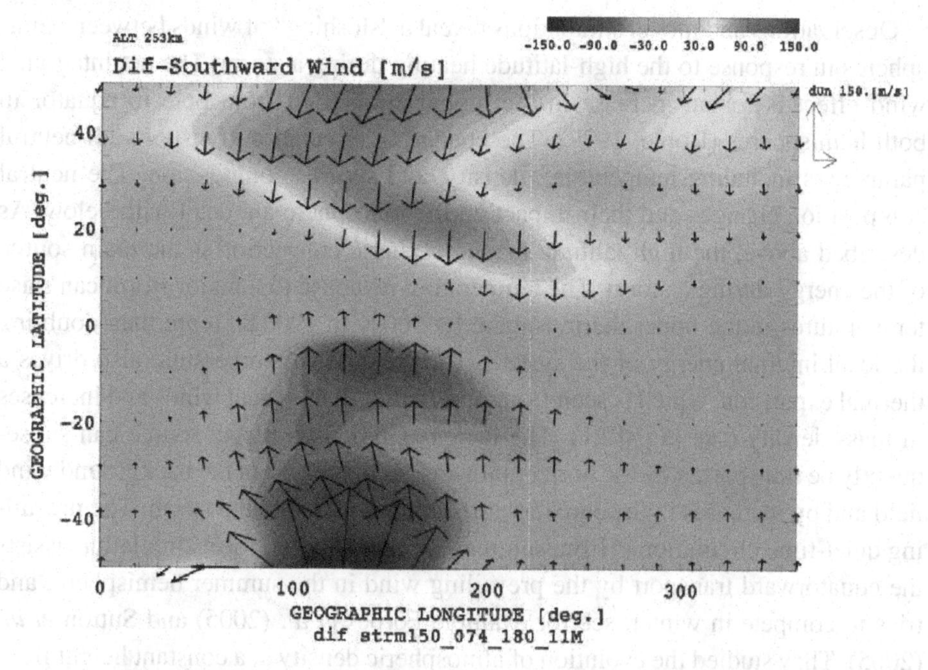

Fig. 12.5. Simulation of the response of the neutral winds at mid and low latitudes at 250 km altitude, shortly after a sudden increase in high-latitude Joule heating. The region within 50° of the geographic equator is shown at 15 UT, three hours after the increase in high-latitude magnetospheric forcing, equivalent to a $K_p \sim 7$. Wind surges of ~150 m/s are produced, mainly on the night side. See also Plate 13 in the color-plate section.

about 700 m/s, in this case. A vertical cut through the thermosphere would reveal a tilted wave front with the wave propagating more slowly at the lower altitudes (Richmond and Matsushita, 1975).

The dependence on longitude, or local time, is quite prominent with the strongest intensity of propagation in the 100° to 200° longitude sector, which for the Universal Time of the image (15 UT) is on the night side. The peak response appears to be more dependent on day or night difference, rather than the longitude sector of the magnetic pole. Stronger night-side wave propagation can be due to reduced ion drag (Fuller-Rowell *et al.*, 1994).

Two hours later in the numerical simulation, the wave surges penetrate the opposite hemisphere and drive poleward winds at mid latitude, at a time when the high-latitude forcing is still at its strongest. The complex wave train of equatorward and poleward winds during geomagnetic disturbances is a typical characteristic of neutral wind observations (Shiokawa *et al.*, 2002), and will have a signature in many of the neutral and plasma parameters.

Observations and model simulations reveal a "sloshing" of winds between hemispheres in response to the high-latitude heating during a storm. The net integrated wind effect is for an increase in the global circulation from pole to equator in both hemispheres (Roble, 1977). The change in circulation transports all neutral parameters including temperature, density, and species composition. The neutral composition changes and their impact on the ionosphere are dealt with below. As described above, the high-latitude magnetospheric convection is the main source of the energy during a storm. The temperature response to a major storm can raise temperatures in the upper thermosphere by 500 K to 1000 K, more than doubling the local internal energy of the system. The increase in temperature also drives a thermal expansion, which is seen from observations of vertical winds and increases in mass density (see Fig. 12.1). The increase in density at the source can subsequently be transported by the horizontal wind field, both from the background wind field and by increases in the equatorward circulation due to the storm. The prevailing quiet-time circulation is from summer to winter so the storm circulation assists the equatorward transport by the prevailing wind in the summer hemisphere, and tries to compete in winter, see for example Forbes *et al.* (2005) and Sutton *et al.* (2005). They studied the evolution of atmospheric density at a constant height from a fixed local time sector from the CHAMP satellite. The data clearly show the penetration of regions of increased neutral density from the summer polar regions, and the lack of such penetration in the winter hemisphere.

12.2.6 Neutral composition response

The change in the global circulation induces upwelling at high latitudes and transport of molecular-rich air (O_2, N_2) upward and equatorward from the mid and lower thermosphere (Rishbeth *et al.*, 1987; Burns *et al.*, 1991; Prölss, 1997). The circulation during prolonged storms can transport neutral composition to low latitudes, which has been observed by space-based composition measurements, as shown lates in Fig. 12.7, from the review by Crowley and Meier (2008). The same can happen during solstice for even quite modest storms due to the additional transport by the prevailing summer to winter circulation (Fuller-Rowell *et al.*, 1996a).

The region of increased mean molecular mass during a storm has been termed a composition "bulge," which is distinct from the background seasonal/latitudinal structure described in Vol. I, Section 12.3. The bulge can be transported by the background and storm-induced wind fields. The seasonal dependence in the transport of the composition bulge is depicted in Fig. 12.6. The left shows summer, the middle shows winter, and the right reflects the equinox case. Each case shows a snapshot from simulations at the three seasons of the storm-time change in mean molecular mass in the Northern Hemisphere from 10° latitude to the pole, on a

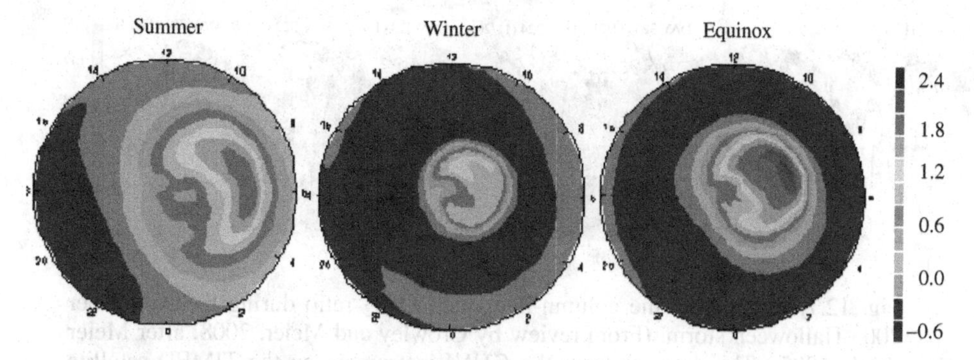

Fig. 12.6. Numerical simulations of the equatorward extent of the "composition bulge" for equivalent storms in the Northern Hemisphere for summer (left), winter (middle), and equinox (right). The seasonal circulation assists the transport to low latitudes in the summer hemisphere and inhibits the transport in winter. See also Plate 14 in the color-plate section.

pressure surface close to 300 km altitude, 6 hours after a 12-hour storm. The figure is from numerical simulations using CTIPe in response to a substantial storm-like increase in high-latitude Joule heating. The composition change is an integrated effect of upwelling over the storm period, which gradually decays as transport by molecular diffusion tries to restore equilibrium. The diffusion time scale, however, can be long, typically a day or two. For reference, a change in mean mass from 19 to 22 atomic mass units (amu) is equivalent to a change in the proportion of molecular nitrogen from 25% to 50%. In the summer (left), the bulge of increased mean molecular mass (which is also equivalent to a decrease in the height-integrated ratio of O/N_2 shown in Fig. 12.7) has been transported by the wind field to low latitudes. Durng the northern winter solstice (middle), the composition bulge has been constrained to high latitudes. The equinox (right) is the intermediate case. In summer, the storm-induced circulation augments the normal seasonal circulation from summer to winter, so the composition disturbance can very easily be transported to mid and low latitudes. In winter, the composition is constrained to high latitudes, because the storm circulation competes with the seasonal flow. Stronger storms can drive an equatorward circulation that can overpower the seasonal circulation, but it still has to compete with the opposing forcing of winds from the opposite hemisphere. The seasonal effect on meridional transport can also be modulated by hemispheric asymmetries in the strength of the magnetospheric sources themselves. This can arise either from asymmetries induced by the tilt of the magnetosphere with respect to the solar wind, or from different ionospheric conductivities in the polar region from solar illumination. Joule heating in the North and South polar regions is very likely to be different, as in Fig. 12.1, and to be further modulated by UT.

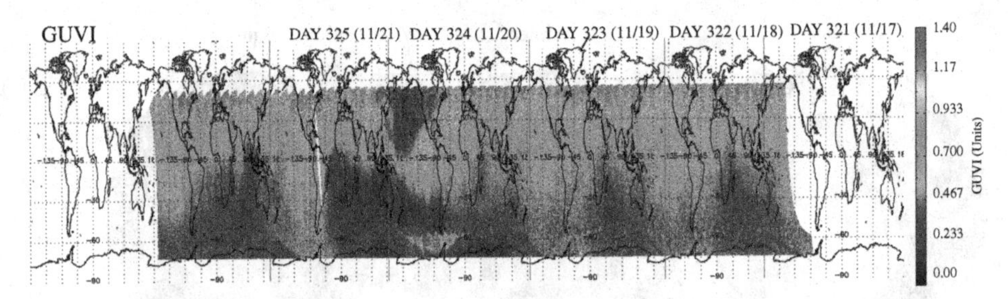

Fig. 12.7. Changes in the column-integrated O/N_2 ratio during the November 2003 Halloween storm. (From review by Crowley and Meier, 2008; after Meier *et al.*, 2005.) The data are from the GUVI instrument on the TIMED satellite (Paxton *et al.*, 1999). Five days of GUVI data are plotted as individual day-side orbits and assembled as a montage; time runs from right to left. The storm event on day 324 causes a decrease in the column-integrated O/N_2 in both hemispheres. The Southern Hemisphere depletion penetrates further equatorward as expected from the transport effect of the global seasonal circulation. See also Plate 15 in the color-plate section.

The composition-bulge scenario also predicts a diurnal modulation of the storm-time composition change. In the same way that the seasonal circulation transports the bulge, the normal background diurnal variation of the upper thermosphere wind at mid latitude can influence the bulge. On the day side, upper thermosphere winds are poleward, responding to the day-side, solar-generated increase in temperature and pressure, so limiting the equatorward transport. Similarly, on the night side, the diurnal equatorward winds assist the flow to mid and low latitudes. As the globe rotates the bulge rotates with it, and the transport is diurnally modulated by this alternating day-side poleward and night-side equatorward wind field.

12.2.7 Positive and negative ionospheric storms

It is well known that changes in neutral composition can impact the ionosphere by changing the ion loss rate. A decrease in the O/N_2 ratio can cause substantial decreases in plasma density (Strickland *et al.*, 2001), often referred to as a "negative phase" ionospheric storm (Prölss, 1997; Rodger *et al.*, 1989). The depleted F-region plasma in the polar region in Fig. 12.2, shown previously, was a manifestation of this effect.

In spite of the complexities in the observed response of the ionosphere to a geomagnetic storm, systematic features are apparent. One of the breakthroughs in understanding the storm-time ionosphere came from analysis of extensive ionospheric observations, and from interpretation of the data by physical models. Figure 12.8 shows that the storm-time response of the ionosphere reveals both

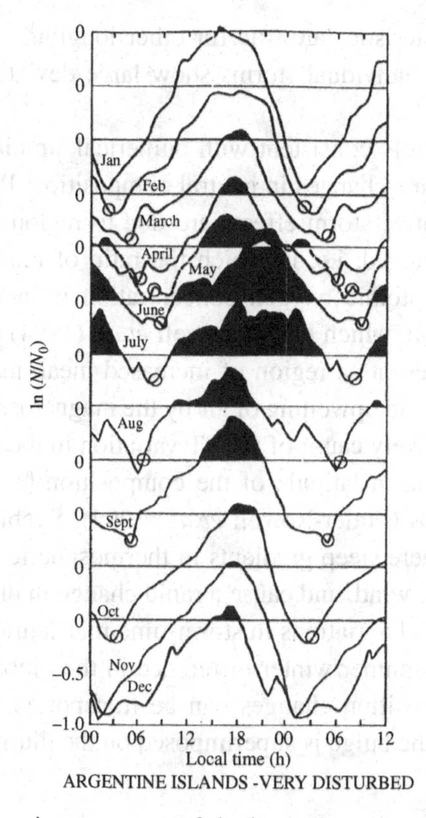

Fig. 12.8. The storm-time response of the ionosphere reveals both seasonal and local-time (LT) dependencies. The figure shows the diurnal variation of the natural logarithm of the ratio of the storm-to-quiet peak F-region plasma density, NmF_2, at Argentine Islands (65° S) for 1971–81. For reference, a decrease of 0.5 indicates a decrease in the plasma density by 40% (from Rodger *et al.*, 1989).

seasonal and local-time (LT) dependencies. The figure shows the diurnal variation of the natural logarithm of the ratio of the storm-to-quiet peak F-region plasma density, designated NmF_2. For reference, a decrease of 0.5 indicates a decrease in the plasma density by 40%. Rodger *et al.* (1989) demonstrated that, at this southern magnetic mid-latitude station (Argentine Islands), a consistent local time signature in the ratio of disturbed to quiet NmF_2 existed throughout the year, with a minimum in the morning hours around 06 LT and a maximum in the evening around 18 LT (see Fig. 12.8). The local time "AC" variation was superimposed on a "DC" shift of the mean level that varied with season, being most positive in winter (May–July) and most negative in summer (October–February). The data supported the widely held belief that "positive phases" of storms (increases in electron density) are more likely in winter mid-latitudes, and "negative phases" of storms (decreases in electron density) are more likely in summer. Field and Rishbeth (1997) showed that these same characteristics are true for other longitude sectors. Rodger *et al.* stressed the point that individual storms show large deviations from the average behavior.

The response is entirely consistent with numerical simulations and the discussion above on storm-time changes in neutral composition. Prölss (1997) reviewed the evidence that negative storm effects are due to regions in which the neutral gas composition is changed, i.e. in which the ratio of molecular gas concentration ($N_2 + O_2$) to the atomic oxygen concentration is increased. It was shown above that such a region, which Fuller-Rowell *et al.* (1994) called a "composition bulge" because it represents a region of increased mean mass, is originally produced through heating and upwelling of air by the magnetospheric energy inputs at auroral latitudes. The likely cause of the LT variation in the ionosphere, therefore, is simply the oscillation in latitude of the composition bulge in response to the diurnally varying winds (Fuller-Rowell *et al.*, 1994). Skoblin and Förster (1993) also showed a case where steep gradients in thermospheric composition could be advected by meridional wind, and cause a rapid change in plasma content.

Similarly, the seasonal variations in storm-time mid-latitude plasma density are a consequence of the summer/winter difference in the global circulation, and the degree to which composition changes can be transported to mid latitudes. The seasonal migration of the bulge is superimposed on the diurnal oscillation.

12.2.8 Structure in the plasma response

Not too many years ago, storm-time ionospheric changes were often characterized purely as "positive" and "negative" phases, as illustrated in Fig. 12.8. This characterization was appropriate for the interpretation of a limited number of point or local measurements from the few ionosonde stations scattered around the world and the still fewer incoherent scatter radar facilities. The conventional wisdom was that the negative phase at mid latitude was a consequence of neutral composition change (a basic concept that is still largely accepted). At the same time, the cause of the large-scale positive phase was still very much open to question, but was thought to be mainly due to winds or decreases in mean mass (a tenet now challenged). The wind effect is the concept whereby the equatorward winds at mid latitudes, in the presence of an inclined dipole-like magnetic field, tend to raise the plasma to regions where there are fewer heavy molecular neutral species, and hence experience reduced loss rates. The composition effect is the reverse of the argument for the negative phase: downwelling at mid or low latitudes, i.e. a closing of the global circulation, or Hadley cell, increases the O/N_2 ratio and drives a positive phase. This basic interpretation of observations is described in the reviews by Prölss (1997), Fuller-Rowell *et al.* (1997), and Buonsanto (1999).

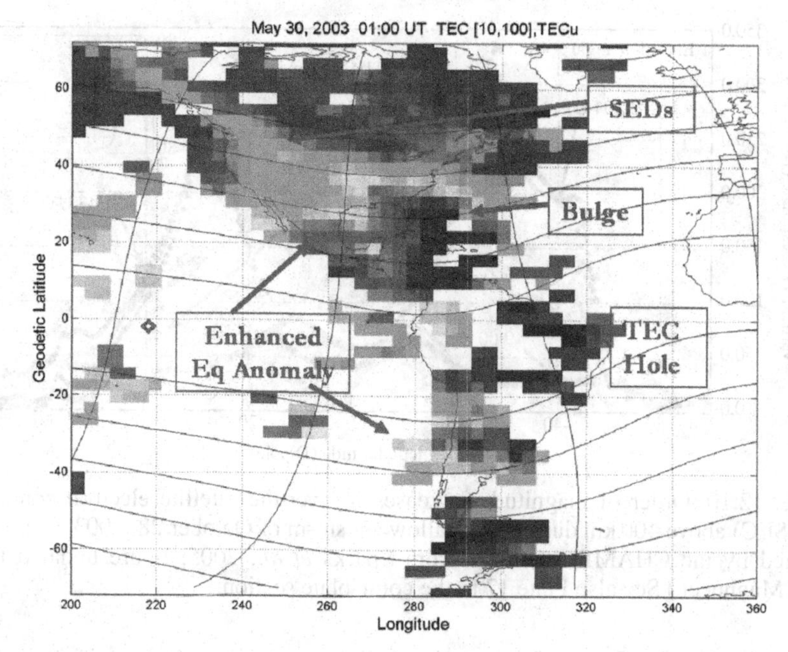

Fig. 12.9. Illustration of the large enhancement "bulge" in TEC at mid-latitudes during a geomagnetic storm, and showing the plume of plasma (storm-enhanced density, or SED) connecting the bulge to the high latitudes. (Courtesy of J. Foster.) See also Plate 16 in the color-plate section.

Three significant events have altered this perspective: (1) "mapping" of the plasma response to storms is now possible due to the explosion of the number of ground-based, dual-frequency Global Positioning System (GPS) observations of total electron content (TEC) (e.g. Foster *et al.*, 2002; Coster *et al.*, 2003); (2) the global mapping of neutral composition change from the GUVI instrument on the TIMED satellite (Paxton *et al.*, 1999); and (3) the spectacular images of the plasmasphere from the IMAGE spacecraft (Goldstein *et al.*, 2003). An example of the first is shown in Fig. 12.9, from Foster (private communicationi, 2009), based on the work of Foster *et al.* (2005) and Foster (2008), the second is shown in Fig. 12.7, and the third is shown later in Fig. 12.11. Figure 12.9 shows large increases in TEC at mid latitudes, including features described as TEC plasma "bulges" and "SEDs" (storm-enhanced densities). These features are in addition to the formation of the equatorial ionization anomaly (EIA), which is the most prominent ionospheric feature during quiet geomagnetic activity. The EIA is driven by plasma transport from the typical dayside eastward electric field at low latitudes, see discussion in Vol. I, Section 12.6. The new ground-based TEC "imaging" capability can follow changes in the EIA as the storm-time electrodynamics evolves, but can also reveal these new features, raising new science

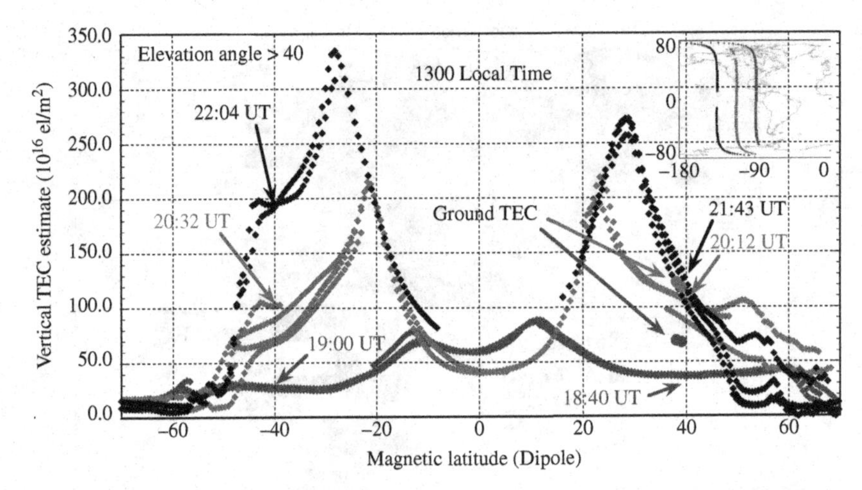

Fig. 12.10. Order of magnitude increases in over-the-satellite electron content (OSEC) above 400 km during the Halloween storm of October 28, 2003, as measured by the CHAMP satellite. (From Sparks *et al.*, 2005; figure updated by A. Mannucci.) See also Plate 17 in the color-plate section.

questions, and stimulating the introduction of a new vocabulary of ionospheric phenomena.

12.2.8.1 Plasma "bulges" at mid latitude

A dramatic increase in TEC during a storm was shown by Mannucci *et al.* (2005), see Fig. 12.10. They used upward-looking GPS data from the CHAMP satellite at ∼400 km altitude. The data show a huge increase in upward-looking over-the-satellite electron content (OSEC), which increases at mid latitudes from about 50 to over 300 TEC units (1 TEC unit equals 10^{16} electrons/m^2). At least two mechanisms have been proposed for the increase. The first, from Foster (private communication, 2009), suggests that the storm-time electrodynamics at low latitudes transports plasma from the equatorial ionization anomaly towards mid latitudes. The mechanism relies on the idea that strong polarization electric fields are established as the E region moves into darkness at either end of an ionospheric flux tube. The process is also strongly influenced by the distortion of the Earth's magnetic field in the American longitude sector.

The second mechanism suggests that the buildup of plasma at mid latitudes is simply a consequence of the expansion of the high-latitude magnetospheric convection (Heelis *et al.*, 2009). In the mid-latitude dusk sector, the expanded two-cell pattern of high-latitude electric potential would produce an electric field that is directed poleward and eastward. The plasma drift from the poleward field would tend to stagnate the plasma in local time, holding it in sunlight for longer and allowing plasma densities to build up from solar ionization. At the same time, because of the more inclined magnetic field at mid latitudes, the drift induced by the eastward electric field tends to raise the height of the ionosphere to regions of reduced neutral molecular species, and hence reduced ion recombination rates, which also tend to store plasma.

Lei *et al.* (2008) showed an increase of mid-latitude TEC during a storm from numerical simulations of a coupled magnetosphere–ionosphere–thermosphere model (CMIT). In their simulations, the increase appeared to arise from an increase in the equatorial eastward electric field followed by an increase in the strength of the equatorial ionization anomaly.

Whatever the mechanism, it is now clear that electrodynamics plays a much more important role in understanding the storm-time increases in plasma density at mid latitudes. Observing the nature of these more localized (in local time) regions of increase, and understanding the mechanisms responsible for their formation, was elusive before the "imaging" techniques became widely used.

12.2.8.2 *Storm-enhanced densities*

The spatially resolved feature in Fig. 12.9, now referred to as the storm-enhanced density (SED; Foster *et al.*, 2005), was originally called the "dusk effect" by Mendillo *et al.* (1970) when discovered in earlier observations. An SED is a plume of increased plasma density that appears to emanate from the mid-latitude plasma bulge. SEDs are likely to be the ionospheric counterpart of the plasmaspheric plumes seen by the IMAGE satellite shown in Fig. 12.11. Foster *et al.* (2002)

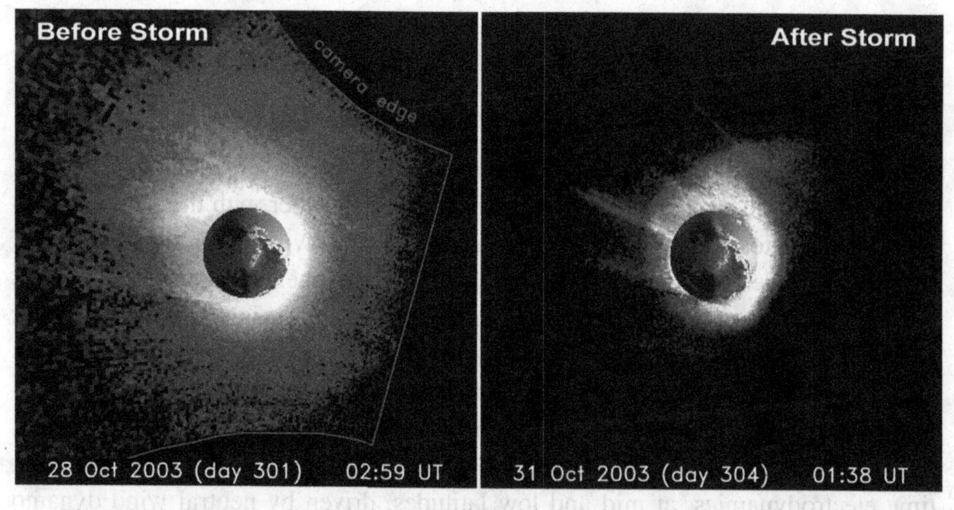

Fig. 12.11. Satellite observations of the erosion of the plasmasphere during a storm, from observations by the IMAGE satellite before and after the Halloween storm of October 28, 2003. (Courtesy of J. Goldstein.) The plasmaspheric tail, or plume, can be seen in the dusk sector during the storm event.

traced the expected field-line geometry between the two features, which indicates the two structures are the respective plasma signatures at either end of the flux tubes responding to the same electric field.

The SED feature is likely to be associated with plasma transport by a fairly narrow mid-latitude electric field structure, now referred to as a sub-auroral polarization stream (SAPS). SAPS result from a polarization of inner magnetosphere plasma in the dusk sector during storms (Sazykin *et al.*, 2005). This tends to form in regions of low E-region conductivity, which inhibits the closure of magnetosphere currents and leads to strong polarizing electric fields, often exceeding tens of millivolts per meter. The argument is that the fast convection from the SAPS picks up plasma from the bulge and transports it westward and poleward. This plasma can subsequently be carried onto the dayside and over the pole, in the traditional way that a tongue of ionization forms, as day-side plasma is picked up by the magnetospheric convection and transported into the polar cap.

The apparent dichotomy is that the SAPS requires low conductivity, but is in the vicinity of the high F-region plasma densities associated with the SED. The concept is further complicated by the fact that when high-velocity plasma flows through the more sluggish neutral medium, it tends to heat the plasma and increase loss rates (Schunk *et al.*, 1975). The high-velocity plasma can therefore also be associated with a trough, in direct contradiction to the appearance of the SED. The details of the feature have yet to be simulated realistically in a physical model in order to be able to analyze and test the theories, and to understand the balance between the various physical processes.

12.2.9 Storm-time electrodynamics

SAPS is an example of an electrodynamic response to a geomagnetic storm. Although a distinct mechanism, it is often lumped together with the general high-latitude magnetospheric convection driven by the solar-wind dynamo, so is part of the high-latitude source, at least as far as the thermosphere–ionosphere system is concerned. This, of course, is an oversimplification, but it is sometimes difficult to separate the SAPS from the main solar-wind driven dynamo when the patterns are complex. The main part of this section, however, addresses the storm-time neutral wind dynamo and changes in the electrodynamics at mid and low latitudes, the so-called "disturbance dynamo." Volume I, Section 12.6 described the quiet time electrodynamics, at mid and low latitudes, driven by neutral wind dynamo processes. This section also briefly addresses the "leakage" of high-latitude convection electric fields to low latitudes during geomagnetic disturbances, the so-called "prompt-penetration" electric field. Understanding the balance, separation, and interaction between this prompt-penetration (PP) and the disturbance-dynamo (DD) fields still remains a challenge.

Electrodynamics at low latitudes is important because of the configuration of the magnetic field, as described in Vol. I, Section 12.6. During quiet times, the mid- and low-latitude electric fields are driven by a combination of the E- and F-region dynamo processes (Richmond and Roble, 1987; Richmond, 1995; Fesen *et al.*, 2000; Millward *et al.*, 2001; Heelis, 2004). The net result at the magnetic equator is for a zonal eastward electric field by day, westward at night, and a post-sunset pre-reversal enhancement (PRE). As described above, plasma transport from the eastward field is responsible for the formation of the most prominent quiet-time ionospheric feature, the equatorial ionization anomaly (EIA). Just as the quiet-time electrodynamics plays a major role in plasma structure, so too do changes to this field by PP or DD processes during geomagnetic storms.

For a given wind system and in the absence of magnetospheric-penetration electric fields, the ionospheric electric fields **E** and current density **J** are determined by the "dynamo equations" (Blanc and Richmond, 1980):

$$\mathbf{J} = \sigma(\mathbf{E} + \mathbf{v} \times \mathbf{B}), \tag{12.1}$$

$$\mathbf{E} = -\nabla\Phi, \tag{12.2}$$

where σ is the conductivity tensor, **v** is the neutral wind, **B** is the Earth's magnetic field, and Φ is the electrostatic potential.

To understand the storm-time response it is useful to divide the current density **J** into components driven by the wind field alone (subscript u) and by the electric field alone (subscript E), as was done by Blanc and Richmond. The horizontal components of **J** in the magnetic equatorward (θ) and eastward (ϕ) direction are then given by

$$J_{\theta u} = -\frac{\sigma_{\mathrm{P}}}{\sin(i)} u_\phi B + \sigma_{\mathrm{H}} u_\theta B, \tag{12.3}$$

$$J_{\phi u} = \sigma_{\mathrm{P}} \sin(i) u_\theta B + \sigma_{\mathrm{H}} u_\phi B, \tag{12.4}$$

$$J_{\theta E} = \frac{\sigma_{\mathrm{P}}}{\sin(i)} E_{\mathcal{E}} + \frac{\sigma_{\mathrm{H}}}{\sin(i)} E_\phi, \tag{12.5}$$

$$J_{\phi E} = -\sigma_{\mathrm{H}} E_{\mathcal{E}} + \sigma_{\mathrm{P}} E_\phi, \tag{12.6}$$

where i is the magnetic inclination below the horizontal, σ_{P} and σ_{H} are the Pedersen and Hall conductivities respectively (see Vol. I, Section 12.6), $E_{\mathcal{E}}$ is the equatorward component of the electric field perpendicular to the magnetic field, and u_θ and u_ϕ are components of the neutral wind.

During geomagnetic storms, the dynamo electric fields are altered because the normal quiet-day thermospheric neutral winds are disrupted. Blanc and Richmond (1980) describe the characteristics of the storm-time disturbance dynamo, and their results are strongly supported by observations (Scherliess and Fejer, 1997; Fejer and Emmert, 2003). The Blanc and Richmond scenario relies on the buildup of zonal winds at mid latitude under the action of the Coriolis force in response to the increased equatorward winds. The meridional winds are forced by high-latitude heating as discussed previously. The dynamo action of the zonal winds drives an equatorward Pedersen current. Positive charge builds up at the equator producing a poleward-directed electric field, which balances the wind-driven equatorward current. The poleward electric field subsequently drives an eastward Hall current, which causes positive charge to build up at the dusk terminator and negative charge to build up at dawn. The zonal electric field driven by the disturbance dynamo opposes the normal day-side eastward and night-side westward quiet-time dynamo electric field and magnetic perturbations (the normal solar quiet current is referred to as the Sq current system). The disturbance dynamo therefore acts as a reverse Sq current vortex, reducing or even reversing the eastward electric field on the day side, and reducing or reversing the normal westward electric field on the night side.

More recent three-dimensional numerical simulations support the basic Blanc and Richmond scenario, but suggest significantly more local-time structure at night (Fuller-Rowell *et al.*, 2008). In particular, rather than being a uniform reduction in the downward plasma drift on the night side, the response is much more localized in local time, and even reverses the direction of the drift to upward in the post-midnight, or pre-dawn, sector. The other significant feature in the simulations is the apparent reduction in the magnitude of the PRE. Simulations of real events (Maruyama *et al.*, 2007) show that localized post-midnight, or pre-dawn, electric field reversals from westward to eastward are fairly typical, and are in reasonable agreement with some of the observations of storm-time response seen by the Jicamarca incoherent scatter radar facility on the magnetic equator in Peru (Fejer and Scherliess, 1997). Figure 12.12 shows the vertical plasma drift observed at Jicamarca during the storm in November 2004. The observations show strong upward drift at night and significant changes in the PRE. It should be stressed that separating the impact of PP and DD during these types of events is very difficult, and is the subject of significant debate.

It is interesting to note that one of the documented storm responses is that irregularities that are normally associated with post-sunset enhancement in vertical plasma drift during quiet times often appear post-midnight, or pre-dawn, during a storm. The height of the ionosphere will be raised in the pre-dawn sector by the upward drift, whether from PP or DD, leading to conditions that are ripe for the initiation of plasma bubbles, or irregularities, from the Rayleigh–Taylor instability mechanism.

The Blanc and Richmond scenario predicts that the disturbance dynamo is slow to develop, due to the gradual build up of the zonal winds, and also slow to

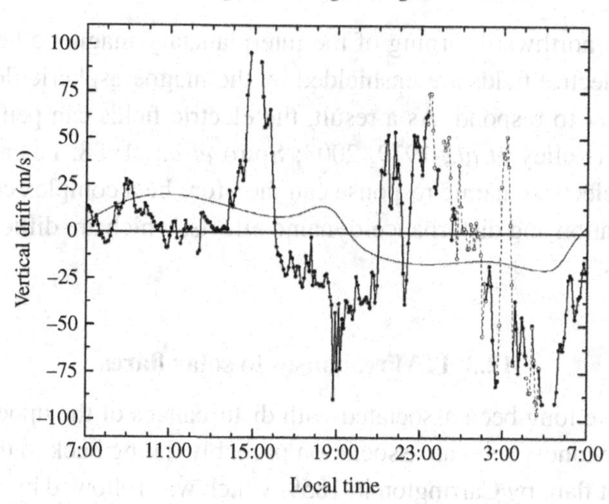

Fig. 12.12. Vertical plasma drift measured at the Jicamarca incoherent scatter radar facility in Peru on the magnetic equator for a storm in November 2004. The thin line is the quiet-day climatological drift. (From Fejer *et al.*, 2007.)

abate. An additional mechanism was mentioned by Blanc and Richmond, and was explored by Fuller-Rowell *et al.* (2002) in numerical simulations. This new mechanism appears to provide a means of generating a disturbance dynamo response about an hour or two after the onset of a geomagnetic storm, and is driven by the meridional wind surges that respond within an hour or two of the high-latitude heating. The mechanism for the rapid disturbance dynamo onset comprises a combination of two effects. The first follows the Blanc and Richmond scenario. The meridional wind surges in the geographic frame have components in both the meridional and zonal magnetic frame. The zonal component produces the same response as the Blanc and Richmond scenario, except that it does not require the slow buildup of the zonal wind via the Coriolis force. The second arises from a direct effect of the meridional wind at mid latitudes. Equations (12.3)–(12.6) show that an equatorward wind in the magnetic frame drives an eastward-directed zonal Pedersen current at mid latitudes. In both cases the electrodynamic response is to the wind surge, which drives the dynamo at mid latitudes within one to two hours of storm onset at high latitudes, and is experienced at the equator on the same time scale.

As mentioned earlier, in addition to the dynamo fields, prompt penetration electric fields are also a major source of disruption of the low-latitude ionosphere during geomagnetic storms. During quiet conditions, inner-magnetosphere plasma flow tends to shield the low latitudes from the high-latitude convection, although some leakage can always occur (Richmond *et al.*, 2003). When the high-latitude magnetospheric convection increases or decreases rapidly, usually associated with a southward or northward turning of the interplanetary magnetic field (IMF), the high-latitude electric fields are unshielded by the magnetospheric flows, since the plasma is slower to respond. As a result, the electric fields can penetrate directly to the equator (Kelley *et al.*, 1979, 2003; Spiro *et al.*, 1988; Fejer *et al.*, 1990). The observed electrodynamic response can therefore be a complex combination of prompt-penetration and disturbance-dynamo effects, which are difficult to separate in observations.

12.3 ITM responses to solar flares

Solar flares have long been associated with disturbances of the upper atmosphere, particularly the ionosphere, an association probably dating back to the observation of a white-light flare by Carrington in 1859, which was followed by a geomagnetic storm the next day (see Section 2.2). The modern concept of the solar origins of geomagnetic activity focuses instead on the coronal mass ejection and consequent disturbances in the solar wind, interplanetary magnetic field, and magnetosphere. However, flares still cause significant intensification of the Sun's photon emission spectrum, particularly in the X-ray region, and observations now show how these intensifications can extend through the ultraviolet wavelengths, and can result in small increases in the total solar irradiance. These enhancements of the photon flux received by the terrestrial atmosphere can have impulsive, though short-lived, effects on the upper atmosphere and ionosphere. The most important and largest effects are increases in the ionization rate in particular altitude ranges, leading to localized increases in ionospheric density, but detectable changes in temperature, density, and minor-species chemistry can also result from large flares.

Solar flares also often cause, or are associated with, energetic particle events that travel to the Earth much faster than solar wind disturbances, and which can have important consequences extending from the radiation belts to stratospheric chemistry, and are a potential hazard for human space flight and high-altitude aviation. This topic is not covered in this chapter other than to note that it is a motivating factor for flare research and space-based observational monitoring.

12.3.1 Flare spectra

Measurements of changes in the solar emission spectrum during flares are generally only possible from space, since the Earth's atmosphere absorbs X-ray and ultraviolet photons that are primarily enhanced during these events. Despite a long history of monitoring flare activity using X-ray sensors on orbiting satellites, these instruments have generally employed broad-band detectors, so it is only fairly recently that a quantitative understanding of solar flare spectra has begun

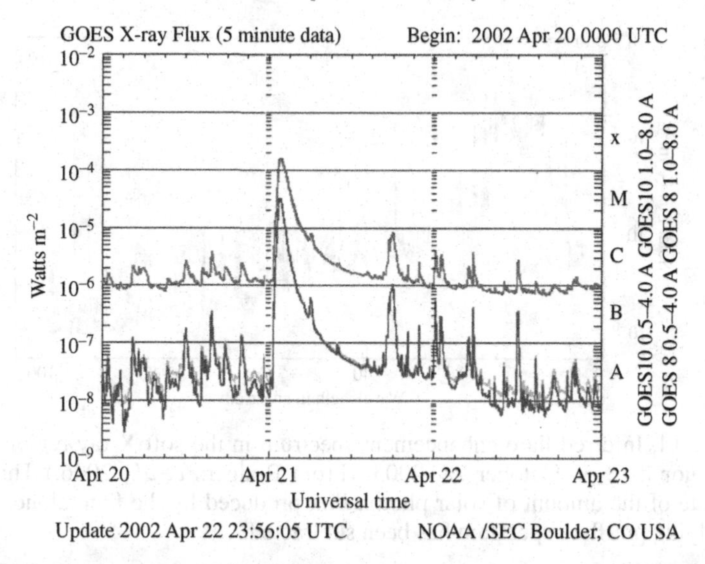

Fig. 12.13. Example of GOES XRS measurements during a large (X1.5) solar flare. See also Plate 18 in the color-plate section.

to emerge. This is due not only to measurements, but also to atomic emission calculations such as the CHIANTI model (e.g. Dere *et al.*, 2009, and references therein) that enable interpretation of broad-band measurements in a theoretical context.

Figure 12.13 shows an example of measurements during a solar flare by the XRS instrument carried on GOES satellites that are routinely provided by the NOAA Space Weather Prediction Center. The instrument has two broad-band detectors covering the approximate ranges from 0.05 nm to 4 nm and 0.1 nm to 0.8 nm. Considering the logarithmic scale, it is apparent that the main enhancements only last a few hours, even during the "gradual" phase of the flare. The initial "impulsive" phase is even shorter, generally on the scale of minutes. These X-ray enhancements enable us to track the time dependence of the flare but are actually only a small part of the total energy impacting the Earth.

For purposes of this discussion, "X-ray" will refer to the spectral region from 0.01 nm to 1.0 nm, "soft X-ray" to the range from 1.0 nm to 10 nm, "extreme ultraviolet" (EUV) to the 10 nm to 100 nm region, and "ultraviolet" (UV) to irradiance between 100 nm and 300 nm (see Fig. 4.1). According to the confluence of measurement and modeling methods, it is the soft X-ray range that is most important for large flare effects on the upper atmosphere. Figure 12.14 shows a flare enhancement spectrum inferred by fitting a CHIANTI differential emission measure model to measurements from the TIMED and SORCE satellites using

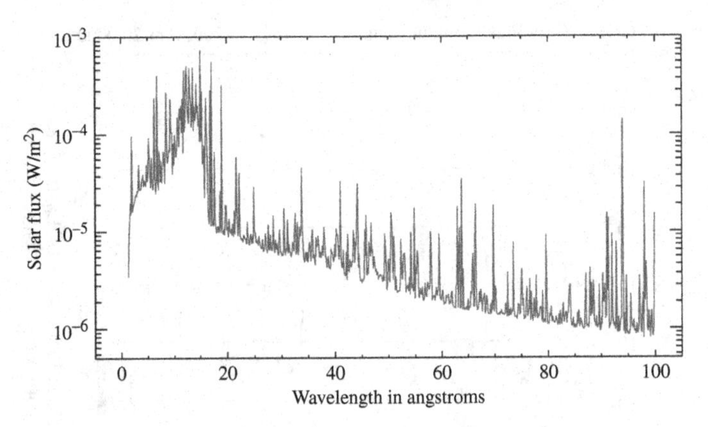

Fig. 12.14. Inferred flare enhancement spectrum in the soft X-ray region during the major flare on October 28, 2003. (From Rodgers *et al.*, 2006.) This is an estimate of the amount of solar photon flux produced by the flare alone, i.e. the underlying pre-flare spectrum has been subtracted.

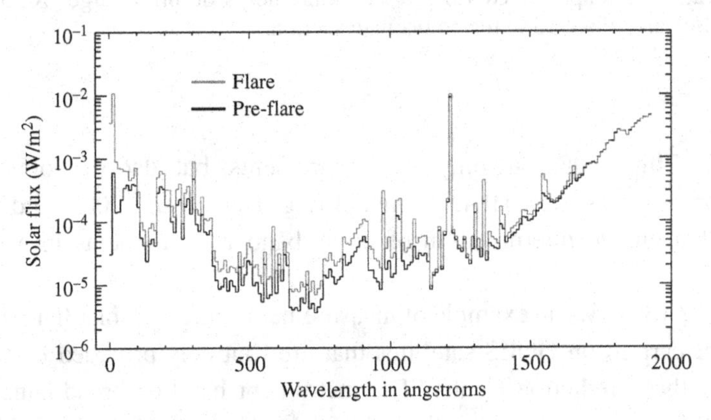

Fig. 12.15. Solar emission spectrum near the peak of the October 28, 2003 flare obtained from measurements by the TIMED/SEE instrument, compared to a spectrum obtained shortly before the event.

multiple photodiode passbands. This method is prone to model-dependent uncertainties, but the result is in agreement with ionospheric modeling and other lines of evidence. This is, however, one of the largest solar flares ever observed by any method, so the following analyses should be considered as an extreme example of flare effects.

In the EUV range, changes to the spectrum are better quantified by the Solar EUV Experiment (SEE) on the TIMED satellite (Woods *et al.*, 2005), because it performs a spectrally resolved measurement in this range using a grating spectrograph. Figure 12.15 shows the flare and pre-flare spectrum for the October 28, 2003 event using the Rodgers *et al.* (2006) method in the soft X-ray range and

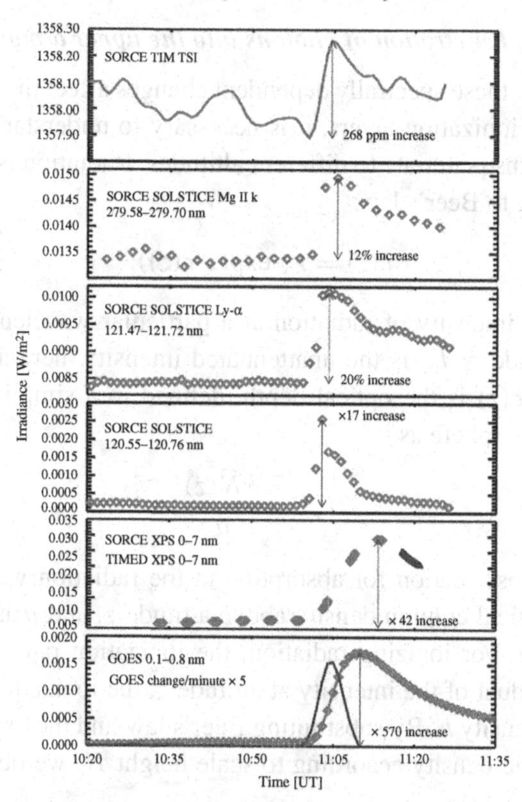

Fig. 12.16. Measurements from instruments on the SORCE satellite (Woods *et al.*, 2004, 2008) showing the time dependence of flare enhancements in various spectral regions. (From Woods *et al.*, 2008.)

the SEE spectrograph measurement in the EUV and UV. During this measurement near the peak of the flare, the irradiance is approximately twice the pre-flare spectrum through most of the EUV, and then diminishes through the UV to be almost undetectable at the longer wavelengths. Although SEE observed many significant EUV flare enhancements, this is the largest EUV enhancement ever measured. The global effects of this particularly large solar flare are shown in Fig 12.16. Time series of the initial phase of the event obtained from instruments on the SORCE satellite are shown for various spectral bands, including total solar irradiance (TSI), where the enhancement is extremely small but still detectable, and at H Lyman-α 121.6 nm, where the increase is commensurate with the TIMED/SEE observation, about 20%. Taken together, these measurements support the general idea that, like solar-cycle and solar-rotational variability, the magnitude of the change decreases with increasing wavelength. However, the notion of solar flares as merely X-ray phenomena is obsolete, because in some cases the EUV enhancements can be quite significant.

12.3.2 Penetration of photons into the upper atmosphere

To understand how these spectrally dependent changes affect the atmosphere, especially by creating ionization layers, it is necessary to understand how photons of varying wavelengths penetrate to different altitudes. Radiation is attenuated in any medium according to Beer's law:

$$I(z) = I_\infty \exp(-\tau(z)), \tag{12.7}$$

where $I(z)$ is the intensity of radiation at a particular wavelength at some location, here at altitude z, I_∞ is the unattenuated intensity, here at the "top" of the atmosphere, and $\tau(z)$ is the optical depth, defined in a simplified plane-parallel single-species atmosphere as

$$\tau(z) = \frac{\sigma N(z)}{\mu}, \tag{12.8}$$

where σ is the cross section for absorption at the radiation wavelength, $N(z)$ is the integrated vertical column density above altitude z, and μ is the cosine of the solar zenith angle. For ionizing radiation, the ionization rate q as a function of altitude is the product of the intensity at altitude z, the ionization cross section σ_{i}, and the number density n. By substituting Beer's law and the exponential distribution of atmospheric density according to scale height H, we obtain the Chapman function:

$$\begin{aligned} q(z) &= I(z)\sigma_{\mathrm{i}} n(z) \\ &= I_\infty \exp(-\tau(z))\sigma_{\mathrm{i}} n_0 \exp\left(-\frac{z-z_0}{H}\right) \\ &= I_\infty \sigma_{\mathrm{i}} n_0 \exp\left(-\frac{z-z_0}{H} - \tau(z)\right). \end{aligned} \tag{12.9}$$

This function has the property of having a peak at that altitude where $\tau = 1$, and reducing rapidly with decreasing altitude below that level (see Vol. I, Section 12.4.1 and Fig. 12.4). Thus, although ionization, even from a single wavelength of ionizing radiation, generally occurs in fairly broad regions, historical parlance often refers to ionization "layers" at characteristic altitudes. The key element in controlling these altitudes is clearly the absorption cross section(s) of the atmospheric gas(es), which is highly wavelength dependent. Thus, the actual ionization rate occurs in a superposition of many Chapman functions. Figure 12.17 illustrates this using a solar spectrum during a flare, an empirical model of major species density in the atmosphere (N_2, O_2, and O), and cross sections obtained from laboratory measurements. Note that longward of 103 nm, the energy is deposited as dissociation of O_2 rather than as ionization, with the

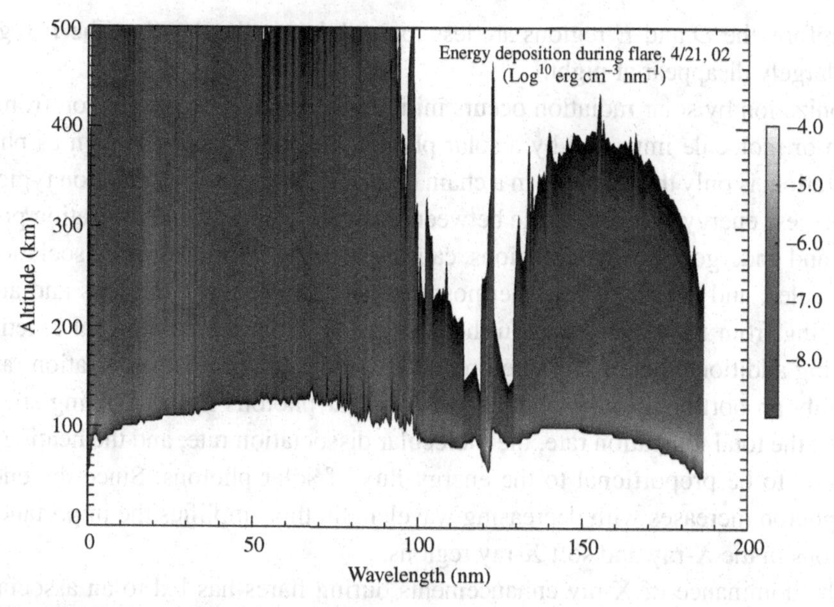

Fig. 12.17. Energy deposition in the upper atmosphere as a function of wavelength and altitude during a solar flare. See also Plate 19 in the color-plate section.

exception of H Lyman-α at 121.6 nm, which ionizes nitric oxide (NO) at low altitude.

12.3.3 Ionization in the D, E, and F regions

The ionosphere has historically been categorized, based on characteristic altitudes of ground-based radio wave reflection, into the *D*, *E*, and *F* regions (see Vol. I, Section 12.5 and Fig. 12.5). The *D* region, from about 80 to 100 km altitude, is caused by the penetration of solar H Lyman-α emission at 121.6 nm and X-rays shortward of 1 nm into the middle atmosphere, and is usually very weak. The *E* region, from about 100 to 150 km altitude, is ionized by a combination of solar H Lyman-α at 102.6 nm and other nearby solar emissions, soft X-rays from 1 to 10 nm, and, in the polar regions, ionization by auroral particles. The *F* region, above 150 km, generally has a peak around 300 km altitude, and is by far the most dominant feature of the terrestrial ionosphere. It is caused by solar ionization throughout the extreme-ultraviolet and soft X-ray region, and is sufficiently long-lived that it lasts throughout the night. This is because it is primarily composed of atomic oxygen ions, which have a long lifetime with respect to neutralization. The *D* and *E* regions are primarily composed of molecular ions, mostly the nitric oxide ion NO^+ and the molecular oxygen ion O_2^+, which are rapidly neutralized in reactions with electrons through a process known as dissociative recombination.

Therefore, the *D* and *E* regions are less ionized during the day than the *F* region, and largely disappear at night.

Ionization by solar radiation occurs initially by removal of an electron from the atom or molecule impacted by a solar photon, but this process, known as photo-ionization, is only the first step in a chain of events. The resulting electron typically has excess energy (the difference between the photon energy and ionization potential) and undergoes further collisions, causing additional ionization, dissociation of molecules, and heating of the thermosphere and ionosphere. The ions and atoms resulting from these processes further engage in exothermic chemical reactions, causing additional heating. Therefore, although the initial photo-ionization rate is roughly proportional to the number flux of solar photons in the ionizing spectral range, the total ionization rate, the molecular dissociation rate, and the heating rate all tend to be proportional to the energy flux of solar photons. Since the energy per photon increases with decreasing wavelength, this amplifies the importance of photons in the X-ray and soft X-ray regions.

The dominance of X-ray enhancements during flares has led to an association between flare effects and the altitude at which X-rays are absorbed in the atmosphere, mostly in the mesopause region near 90 km, which also corresponds to the *D* region of the ionosphere. Although ionization at this altitude is important due to its potential to disrupt radio communications, it has very little effect on atmospheric processes because the *D* region is so weakly ionized, less than one part per billion, even during flares. However, soft X-ray enhancements during flares, especially in the 1–3 nm range, have a very significant effect on the *E* region, not unlike auroral events that cause particle ionization near and above 100 km altitude, but distributed over a much wider region. This is shown by a model calculation in Fig. 12.18, using the spectra plotted in Fig. 12.15 as input. Although there are enhancements throughout the ionosphere, the most important occur in the 100–110 km range. At *F*-region altitudes, increases in ionization also occur, but they are smaller because the high-energy photons that are most enhanced during flares tend to pass through the *F* region and deposit their energy at lower altitude.

The global effects of this particularly large flare are shown in Fig. 12.19. Here, a model calculation of change in total electron content (TEC), which is the vertical column integral of electron density, is superimposed on measurements deduced from dual-frequency GPS receivers at globally distributed stations. The approximate magnitude and morphology of the enhanced TEC are reproduced, with the effect centered on the sub-solar point and declining with increasing solar zenith angle. The model only extends to about 600 km altitude, while the measurement technique integrates to several Earth radii, so the very-high-altitude component of the enhancement is not included, resulting in a slight underestimate of the enhancement. However, this provides a basic validation of the spectral enhancement

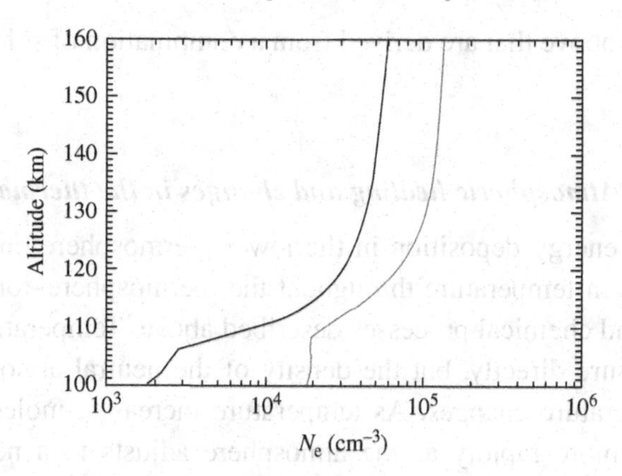

Fig. 12.18. Model calculation of electron density enhancement in the E region at high solar zenith angle (at the Sondrestrom radar site in Greenland) for the October 28, 2003 flare, using the spectrum shown in Fig. 12.15 as input to a photo-ionization/photo-electron model. Black, lower-density curve: pre-flare; grey, higher-density curve: flare. The enhancements seen are commensurate with radar observations of the flare effect.

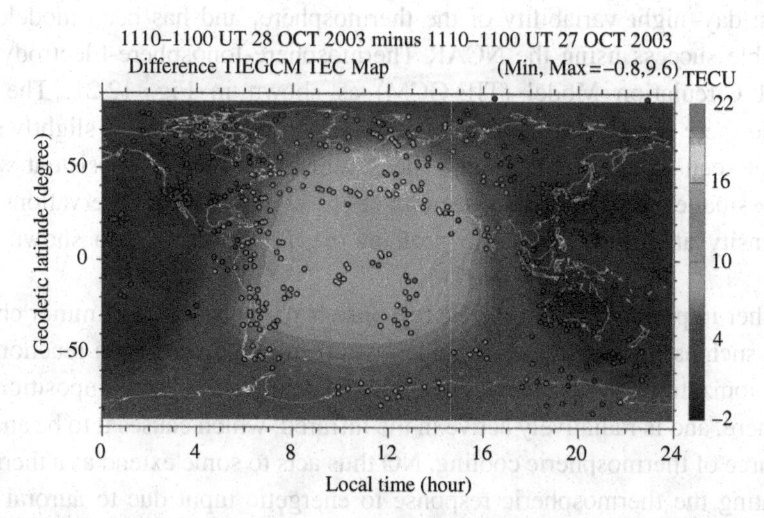

Fig. 12.19. Comparison of total electron content enhancements during the October 28, 2003 flare, observed by the global network of differential GPS stations, and modeled using the NCAR Thermosphere-Ionosphere-Electrodynamics General Circulation Model (TIE-GCM). Total electron content is the vertically integrated column electron content in units of 10^{16} m^{-2}. See also Plate 20 in the color-plate section.

estimates shown above that are derived from a combination of solar measurements and models.

12.3.4 Atmospheric heating and changes in the thermosphere

The increase in energy deposition in the lower thermosphere during large flares causes increases in temperature throughout the thermosphere–ionosphere, due to the secondary and chemical processes described above. Temperature increases are difficult to measure directly, but the density of the neutral atmosphere responds rapidly to temperature changes. As temperature increases, molecules and atoms diffuse upward more rapidly as the atmosphere adjusts to a new scale height, effectively causing the thermosphere to expand. Thus, at a constant altitude, the density is seen to increase as heating (at and below that altitude) increases. This effect has been observed by accelerometer measurements from the CHAMP and GRACE satellites for the October 28, 2003 event analyzed above. In Fig. 12.20, changes in density during the daytime segments of the orbits of these satellites are shown. The increases seen are on the order of 40% and are fairly transient, with decay lifetimes on the order of a day. This is commensurate with the day–night variability of the thermosphere, and has been modeled with reasonable success using the NCAR Thermosphere-Ionosphere-Electrodynamics General Circulation Model (TIE-GCM) as shown in Fig. 12.21. The model response (here sampled at local noon at the equator, at 400 km) is slightly smaller than that seen in the accelerometer measurements, which is consistent with the slightly smaller response also seen in TEC compared to observations. Electron density at 300 km, near the peak of the F region, is also shown in this figure.

Another important thermospheric response is the production of minor chemical species such as nitric oxide, NO, which is a result of dissociation reactions initiated by ionization. NO is chemically active in determining the composition of the ionosphere, and is radiatively active in the infrared, which causes it to be an important source of thermospheric cooling. NO thus acts to some extend as a thermostat, modulating the thermospheric response to energetic input due to auroral storms and flares. The bottom panel of Fig. 12.21 shows the modeled NO response to this event. NO is fairly long-lived in the thermosphere, but also exhibits considerable global structure, and as the atmospheric region affected by the initial phases of the flare rotates away from local noon, the NO density reduces. This recovery is not as rapid as seen in electron density, but is faster than the temperature/density reversion, as the neutral atmosphere as a whole takes longer to respond, and longer to return to its pre-flare state, than its ionospheric component, or than minor species such as NO.

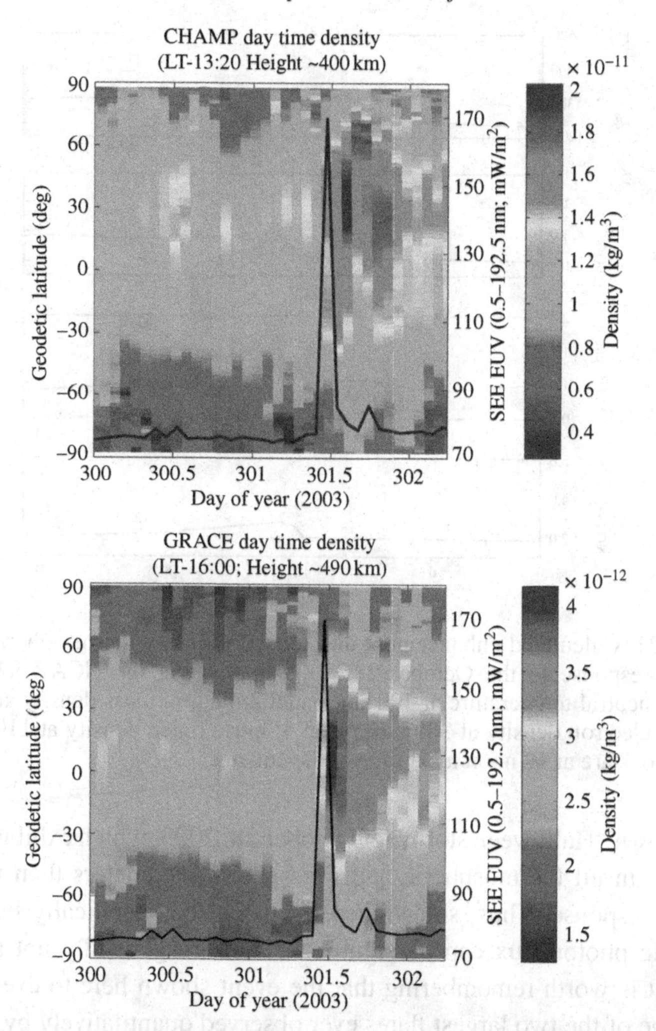

Fig. 12.20. Thermospheric density enhancements measured by accelerometers on the CHAMP satellite (altitude ~400 km) and GRACE satellite (altitude ~490 km) during the October 28, 2003 flare (Sutton *et al.*, 2006). See also Plate 21 in the color-plate section.

Compared to geomagnetic storms, flares are brief, impulsive events that have significant thermosphere–ionosphere effects on the order of hours, but detectable changes that can last on the order of a day. Since flares may occur in conjunction with coronal mass ejections, but the energetic photons travel much more rapidly than the solar wind and interplanetary magnetic field perturbations, they appear more suddenly and with less warning than geomagnetic storms. However, their effects on the thermosphere/ionosphere are much smaller and less enduring than storms. The accelerometer perturbations shown in Fig. 12.20 are those just before

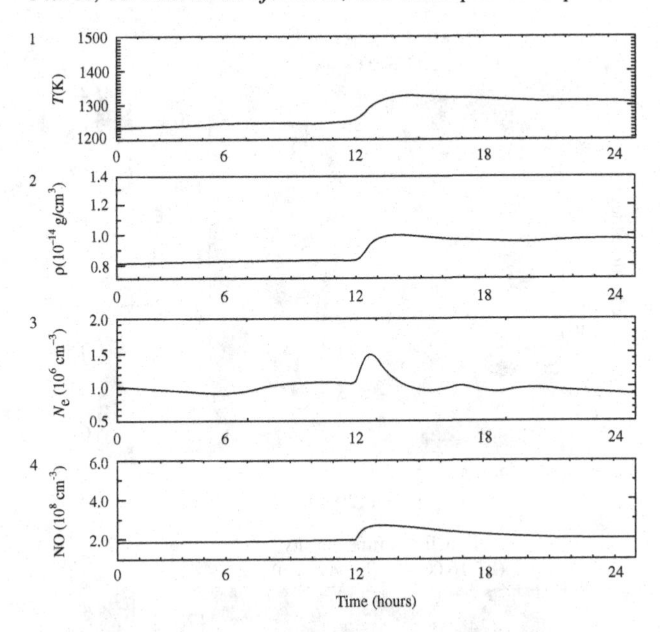

Fig. 12.21. Calculated enhancement and recovery of key thermospheric parameters in response to the October 28, 2003 flare using the NCAR TIE-GCM. Panel 1: neutral temperature at 400 km; panel 2: neutral mass density at 400 km; panel 3: electron density at 300 km; panel 4: nitric oxide density at 110 km. All calculations are at 12 noon local time at the equator.

the well-known "Halloween storm" of October 28, 2003, which exhibited far larger disturbances in all thermospheric and ionospheric parameters than those shown in the flare response. Thus, solar flare effects caused specifically by changes in the energetic photon flux can be significant, but are generally not as important as storms. It is worth remembering that the event shown here to exemplify these effects is one of the two largest flares ever observed quantitatively by space-based instruments (the other originating a week later from the same active region on the Sun).

12.4 Conclusions

Explosive events on the Sun can have a dramatic impact on planetary upper atmospheres. Our brief history of observations has likely seen only a small part of the potential spectrum of variability of our neighboring star. Hints of more substantial geomagnetic storm events are evident in our recent past and suggest bigger things to come. As we become increasingly reliant on space infrastructure and fragile technology, the impact of these potentially giant events is humbling. Our goal is to understand the heliophysical processes that we can observe, and so extrapolate the likely consequences of these unusually large events (at least for the present-day Sun, see Vol. III).

Flare response highlights and challenges our understanding of the complex photochemistry of the region. The energy involved is relatively modest compared with a full-blown geomagnetic storm. It is clear that the change in thermospheric winds and electrodynamics during a storm can be a conduit through which many of the global ionospheric and thermospheric storm-time changes flow. The dynamical and physical processes involved in the response of the thermosphere and ionosphere to a geomagnetic storm are reasonably well understood, and simulation models have illustrated how the complexity of the response is influenced by interactions among dynamics, composition, and electrodynamics. The separation, balance, and interaction between the two components of storm-time electric fields at low latitudes, and their relative importance during the different phases of a storm on the day and night sides, are fundamental unresolved questions.

Many of the physical processes in the upper atmosphere have not been treated in depth in this chapter, such as the complexities in the photochemistry, plasma outflow and magnetospheric loading, plasma instabilities, and details of the dynamics and electrodynamics in the critical interface region between space and the planet. The Earth's upper atmosphere, however, is an excellent laboratory to explore our understanding of the complex physical processes and their mutual interaction at work. We appear to have a high level of understanding, but we still need to reach a capability equivalent to the terrestrial weather community, where physically based data assimilation models can predict six hours ahead with the same accuracy as the best of the observations.

This chapter has addressed the response of the upper atmosphere to external input originating from the Sun. At times, huge increases in plasma density occur and flow through the system. We have yet to be able to trace the flow of plasma through geospace during these events, or to quantify fully the interactions between the upper atmosphere and magnetosphere in this critical interface region between the planet and space. We must still explore and comprehend the range of physical processes related to magnetosphere–ionosphere coupling, together with the range of plasma instabilities. Finally, we must still address the complexities in the connection between meteorology (or terrestrial weather), climate, and space weather.

13

Energetic particles and manned spaceflight

STEPHEN GUETERSLOH AND NEAL ZAPP

13.1 Radiation protection: introduction

Ionizing radiation is radiation that has enough energy to cause ionization in matter, and when it passes through the tissues of the body it has sufficient energy to damage DNA (Hall, 1994). Examples are α-particles (helium nuclei), β-particles (electrons or positrons), γ-rays, X-rays and neutrons. While there are many benefits to the use of X-rays, radioisotopes, and other radioactive materials in industry, research, and power generation, their use entails exposure of personnel from normal use as well as accidents. Though some small amounts of radioisotopes are used in manned space missions for instrument calibration and research, the vast majority of crew exposures are due to the environment in which they work.

Whether an activity is controlled by the Nuclear Regulatory Commission (NRC), the Department of Energy (DOE), or the Occupational Safety and Health Administration (OSHA), an operational radiation protection program is required so that doses to personnel and members of the public are monitored and documented in order that exposures may be kept at a minimum. NASA's program includes active and passive personnel dosimetry, vehicle shielding design requirements, as well as real-time active monitoring of the heliosphere to watch for changes in the environment that would be indicative of an impending solar particle event (SPE).

13.1.1 Units

When considering the amount of radiation absorbed by living tissue, the standard unit known as the gray (Gy) is employed, in which 1 Gy equals 1 J of radiation energy absorbed per kilogram of tissue (the older unit of 1 rad = 0.01 Gy). In order to factor how energetic a source of radiation is, the gray (or rad) is multiplied by a "quality factor". This factor is an experimentally determined value that defines excess damage as a proportion of γ-ray damage for the same dose (with γ-rays

Heliophysics: Space Storms and Radiation: Causes and Effects, eds. Carolus J. Schrijver and George L. Siscoe. Published by Cambridge University Press.

having a quality factor of 1.0) to produce a value indicating relative biological effectiveness of a dose of radiation. This value, dose equivalent, is measured in sieverts (Sv; 0.01 Sv = 1 rem). For example, quality factors for the South Atlantic Anomaly (SAA) vary from 1.6 to 1.9, with galactic cosmic rays having a quality factor ranging from 2.9 to 3.5 inside the space shuttle or the International Space Station (ISS), depending on orbital inclination.

13.1.2 Linear No-Threshold model

After the atomic bomb explosions in Hiroshima and Nagasaki, studies concerning life span of survivors showed a linear relationship between cancer mortality and high doses of radiation. Since there is a specific natural incidence of cancer without a specific exposure to radiation, it cannot be determined with certainty, for low level exposures, that a given case was induced by radiation or would have occurred naturally (NRC BEIR VII, 2006). In 1958, the United Nations Scientific Committee on the Effects of Atomic Radiation (UNSCEAR) therefore proposed the Linear No-Threshold (LNT) model, which was adopted by the International Commission on Radiation Protection (ICRP) one year later (ICRP, 1959). Under this model, the effects of low doses are estimated by linear extrapolation from observed high dose effects where the incidence is not likely due to statistical fluctuation. As a practical hypothesis, the usual assumption is that any amount of radiation, no matter how small, entails some risk.

13.1.3 Hormesis

As discussed above, it is the general belief, applying the LNT model, that even very low doses of ionizing radiation produce some amount of detrimental effects, proportional to the dose received. Some studies, however, suggest that very low doses of ionizing radiation are not only harmless but often have beneficial, or hormetic effects.† Early studies in animals exposed to low levels of whole-body radiation showed a longer life expectancy for the irradiated subjects than in the controls (Turner, 1995). The mechanism behind the possibility of a beneficial effect was further investigated and proponents of this hypothesis suggest that very low doses can stimulate the immune system and facilitate DNA repair by inducing the production of special proteins involved in the repair process (Ikushima *et al.*, 1996). Since some of the data are subject to criticism, and others are merely circumstantial, the hormetic effect of low-level, whole-body irradiation is generally not accepted as fact.

† A process with a favorable effect in small doses, but harmful in large doses.

13.1.4 ALARA

Regardless of whether the LNT model is correct, or whether there is some hormetic effect of very small doses of radiation, it is generally accepted that any exposure should be justified based on a benefit/burden analysis (Turner, 1995). NASA has therefore mandated that radiation exposures of astronaut crews must be maintained As Low As Reasonably Achievable (ALARA; see Table 14.2 for "damage thresholds" for hardware compared to that for human tissue). This requirement is intended to ensure that each exposure to radiation is justified and limited so that astronauts do not approach radiation limits and that such limits are not considered tolerance values. The primary functional application of the NASA ALARA program is to prevent mission-jeopardizing flight risks and minimize long-term risks to levels as low as possible based on moral and financial issues. Given the uncertain consequences of radiation exposures and risk models, cost-effective methods to ensure exposures are maintained ALARA are essential.

13.1.5 National and international regulatory structure

On the justification that peaceful uses of atomic energy require that exposure limits be specified, several national and international bodies have been established for guidance. Some of these merely make recommendations, while others have legal authority to ensure compliance with the standards that have been adopted based in whole or in part on the recommending agencies. The maximum levels of exposures permitted are deemed acceptable in view of the benefits of the activity and different permissible exposure criteria are usually applied to different groups of persons and activities. There are limits set for members of the public, occupational radiation workers, and even different limits for space exploration.

13.1.5.1 National and international organizations

ICRP The International Commission on Radiological Protection (ICRP) was established in 1928 and has close official relationships with a number of organizations that include the International Commission on Radiation Units and Measurements (ICRU), the International Atomic Energy Agency (IAEA), the National Council on Radiation Protection and Measurements (NCRP) and the World Health Organization (WHO). The ICRP, like its national US counterpart, the NCRP, has no legal authority. Its recommendations are made to provide guidance for the setting of radiation protection criteria, standards, practices, and limits by other (regulatory) agencies.

ICRU The International Commission on Radiation Units and Measurements, originally known as the International X-ray Unit Committee, was conceived at the

First International Congress of Radiology (ICR) in London in 1925 but officially came into existence at ICR-2 in Stockholm in 1928. The ICRU's principal objective is the development of internationally accepted recommendations regarding quantities and units of radiation and radioactivity as well as specifying procedures suitable for the measurement and application of these quantities in diagnostic radiology, radiation therapy, radiation biology, nuclear medicine, radiation protection, and industrial and environmental activities.

NCRP The National Council on Radiation Protection and Measurements is a non-profit corporation chartered by the US Congress in 1964. One of its most important charges is the dissemination of information and recommendations on radiation in the public interest. It is also charged with the scientific development, evaluation, and application of basic radiation concepts, measurements, and units. The NCRP maintains close working relationships with a large number of organizations, nationally and internationally, that are dedicated to various facets of radiation research, protection, and administration. Its role is primarily that of making recommendations and it has some 80 scientific committees that review, comment, and approve selected experts' recommendations before publication.

IEEE The Institute of Electrical and Electronics Engineers (IEEE) has two committees of interest to radiation protection, the Radiation Instrumentation Technical Committee (RITC) and the Radiation Effects Committee (REC). The purpose of the RITC is to promote the development and application of radiation detectors. The REC's purpose is the advancement of the study and theory of radiation effects on electronics and hardware, and to develop and promulgate standards. It is the electronic analog of the three previous bodies.

13.1.5.2 United States federal law

NRC The US Nuclear Regulatory Commission was created as an independent agency by Congress in 1974 under the Energy Reorganization Act to enable the nation to safely use radioactive materials for beneficial civilian purposes while ensuring that people and the environment are protected (42 USC 5801). As such it has statutory authority to not only set limits on exposures, but also the power to enforce its regulations. The NRC's scope of responsibility includes regulation of commercial nuclear power plants; research, test, and training reactors; nuclear fuel cycle facilities; medical, academic, and industrial uses of radioactive materials; and the transport, storage, and disposal of radioactive materials and wastes. The NRC licenses and regulates the nation's civilian use of byproduct, source, and special nuclear materials to ensure adequate protection of public health and safety, promote the common defense and security, and protect the environment.

10CFR835: DOE Title 10, Code of Federal Regulations (CFR), Part 835 regarding Occupational Radiation Protection establishes specific requirements for the development, content, revision, and approval of documented Radiation Protection Programs (RPP) for all Department of Energy (DOE) activity (10 CFR 835). As discussed in the Radiological Health and Safety Policy, the DOE has established a system of regulatory policy and guidance reflective of national and international radiation protection standards and recommendations presented by the agencies and bodies discussed above. The requirements of 10 CFR 835 are enforceable under the provisions of the Atomic Energy Act of 1954.

29CFR1910: OSHA When an activity does not fall under the purview of either the NRC or the DOE, it is neither an industrial or civilian use of byproduct, source, or special nuclear material, nor a DOE facility, it is covered by the Occupational Safety and Health Administration (29 CFR 1910). Since NASA cannot be classified as an industrial, civilian, or DOE facility, radiation protection guidelines fall under OSHA. For terrestrial workers the permissible exposure limits (PELs) are defined by OSHA. For space flight, the applicable PELs are set by NASA's chief medical officer under authority defined in 29 CFR 1960.18 (29 CFR 1960).

13.2 Sources of radiation exposure during spaceflight

In biological systems, ionizing radiation can have acute and chronic effects, depending on the magnitude of the radiation absorbed, the species of ionizing radiation, and the tissues affected. Biological responses and the various types of ionizing radiation make the task of protecting humans in space challenging.

The ionizing radiation in space comprises charged particles, uncharged particles, and high-energy photons. The particles of concern vary in size from electrons and protons to high-energy heavy nuclei. It is the combination of charge and mass that determines how quickly these particles lose energy when interacting with matter (see Section 3.4, e.g. Eq. (3.13); also Guetersloh *et al.*, 2006). For equal energies, an electron will penetrate farther into aluminum than a proton and an X-ray much farther than either one. In addition, it is the combination of mass, charge, and energy that influences where in space and how much radiation exposure a vehicle may encounter.

The radiation encountered in space may be generally attributed to three primary sources: particles making up the galactic cosmic ray (GCR) environment, particles trapped in the Earth's magnetic field, and particles from energetic solar events. Radiation levels from each of these sources vary both with solar activity and with distance from the Earth. The temporal and spatial fluctuations must be taken into account in the planning of space missions if hazardous effects of radiation exposures are to be minimized.

13.2.1 Galactic cosmic radiation

Galactic cosmic radiation originates outside our solar system but generally within our Milky Way galaxy and is treated as an isotropic radiation source. This radiation consists of atomic nuclei of hydrogen to uranium that have been ionized and accelerated to very high energies, probably by supernova remnants. The GCR population consists of about 87% protons and 12% α-particles, with the remaining 1–2% heavier nuclei with charges ranging from 3 (lithium) to about 28 (nickel; see Simpson, 1983). Ions heavier than nickel are also present, but they are rare. Electrons and positrons constitute about 1% of the overall GCR, but are a minor biological hazard as compared to the bulk of GCR since they are easily shielded.

The galactic cosmic radiation varies as a function of the level of solar activity, which follows an 11-year cycle (see Chapter 9). The number of solar particles is directly related to the number of observed solar events. When the characteristic solar wind speed increases during cycle minima, there are more particles and higher interplanetary magnetic field to interact with the influx of GCR. This interaction removes some of the lower energy GCR particles. The result is that in a time of solar maximum the GCR environment in the inner solar system has a higher energy but lower fluence than during solar minimum (see Vol. III). Doses from the GCR in interplanetary space are estimated to range from 0.3 Sv/year during solar maximum to about 1 Sv/year during solar minimum (Townsend *et al.*, 1992; Adams *et al.*, 1991).

13.2.2 Trapped radiation

The Earth's magnetic field extends thousands of kilometers into space. As charged particles interact with this magnetic field their original direction is altered according to their speed (energy), charge, and mass along the magnetic field lines of the Earth. The two main types of particles in these trapped belts are electrons and protons (see Chapter 11). Of these, the protons tend to be of greater concern since the higher energies found in these trapped environments can penetrate typical spacecraft hull materials, whereas the electrons are much more easily shielded. Other ions have been detected in these trapped regions, such as helium, carbon, and oxygen nuclei; however, they are of much less concern than protons due to their scarcity (Walt, 1994; Roederer, 1970).

Because the magnetic dipole is offset from the center of the Earth as well as tilted relative to the radiation axis, the radiation belts come closest to the surface of the Earth off the coast of Brazil in an area known as the South Atlantic Anomaly (SAA). The SAA lies roughly at 35 degrees east longitude and 35 degrees south. In the SAA the inner trapped radiation belt dips down to about 200 km (Sawyer and Vette, 1976). This is important for two reasons. First, this low altitude allows some protons to be absorbed in the atmosphere. This creates a strong anisotropy in particle flux at low-earth orbits (LEO), where over a factor of two difference exists between the proton flux from the east compared to the flux from the west. Second, the concentration of protons in this region exceeds the intensity measured at the same altitude at any other part of the globe.

Experience with Earth orbital missions to date indicates that nearly all of the accumulated radiation exposure can be attributed to passages through the SAA. In addition to altitude and orbital inclination, the integrated dose is a function of solar cycle. Increases in solar activity expand the atmosphere and increase the loss of trapped protons. Therefore, trapped radiation doses in LEO decrease during solar maximum and increase during solar minimum. Although high-inclination flights pass through the SAA maximum intensity regions, less time is spent in the SAA than is the case for low-inclination flights. Crews in high-inclination flights receive less net exposure to trapped radiation than in low-inclination flights for a given altitude.

13.2.3 Solar particle events and solar energetic particles

Solar radiation can be divided into two groupings: a steady stream of solar material called the solar wind, and solar particle events (SPE), which are associated with solar flares and coronal mass ejections (CME) (see Chapter 5). Most solar radiation is of lower energy than the GCR, but CMEs release on average 10^{12} particles so the fluence rates can be very severe (Howard *et al.*, 1985)

The solar wind is composed of approximately 95% protons, 4% α-particles and about 1% other nuclei consisting primarily of carbon, nitrogen, oxygen, neon, magnesium, silicon, and iron (Bame *et al.*, 1968). These particles contain high (~800 km/s) and low (~400 km/s) speed components. In general the low-speed winds contain a factor of three higher numbers of heavier nuclei; however, the speed and composition of the solar winds will change over the solar cycle. The solar wind may be used as an indicator of the Sun's activity, but the particles are not energetic enough to penetrate even thin habitats and are not considered a hazard.

Solar flares and CMEs are eruptions from the Sun's surface and are associated with solar active regions (Cohen, 2006). These events are much more likely to occur in the time of solar maximum than at solar minimum activity. SPEs associated with solar flares develop rapidly and can last for days. They produce intense electromagnetic radiation as well as protons, electrons, and plasmas of helium to iron, of which approximately 97.8% consists of protons and 2.1% is helium. SPEs are rather unpredictable in occurrence, intensity, and duration. This is due to the physics behind both the formation and transport of the particles. As a consequence, an intense SPE can arrive at Earth and be complete within hours, or the SPE can last for more than a week, during which there are bursts of radiation lasting a few hours.

A large number of the particles from a SPE are protons at energies (<10 MeV per nucleon) that can be relatively easily shielded by spacecraft hulls. The very high density of protons with energies greater than 10 MeV can still be a particular source of concern for external operation, while protons of more than 30 MeV can be of concern to thinly shielded habitats.

Accurate prediction of the time and intensity of individual SPEs is not currently possible. Modern data on SPEs have only been collected since 1956, which corresponds to the beginning of cycle 19 (Hathaway *et al.*, 1994). These data indicate that about 30 to 50 major SPE events occur per cycle, most during the middle five years corresponding to solar maximum (Kim *et al.*, 2006) Of particular note are the occasional very large CMEs, which have the potential for effects on crew health. One such SPE, commonly known as the August 1972 event, is among the largest recorded events. Although this event is often used in radiation protection planning of possible future SPEs, it may not be the worst case scenario. By examining nitrates in ice core samples, it has been determined that solar events of up to ten times the intensity of the 1972 event have occurred in the past 500 years (Shea and Smart, 2004).

13.3 Spaceflight operations

The Space Radiation Analysis Group (SRAG) at NASA's Johnson Space Center is responsible for ensuring that the radiation exposure received by astronauts remains below established safety limits. This responsibility includes making preflight and extravehicular activity (EVA) crew exposure projections, maintaining a comprehensive crew exposure modeling capability, providing radiological support during missions that includes active and passive dosimetry, monitoring on-board radiation instruments to characterize and quantify the radiation environment inside and outside the spacecraft, and carrying out real-time comprehensive space weather monitoring.

Factors affecting crew exposures include the structure of the spacecraft, the materials used in construction, the altitude and inclination, the status of the outer electron belts, the interplanetary proton flux, geomagnetic field conditions, solar cycle position, and EVA start time and duration.

When missions take crew members beyond LEO, impacts from radiation exposures become much larger due to the loss of protection afforded by the Earth's geomagnetic field. As such, SRAG's responsibility to Mission Control will become much more important for planning mission operation time-lines. The development of operationally robust tools derived from physics-based models to provide timely forecasts of the space environment will contribute to the overall risk mitigation architecture.

13.3.1 Shuttle and International Space Station

Low-inclination, high-altitude flights during solar minimum produce higher dose rates than high-inclination, low-altitude flights during solar maximum. At high altitudes, the area of the SAA is larger and the flux of protons is higher. Although trajectories of high-inclination flights pass through the regions of maximum intensities within the SAA, less time is spent there than during low-inclination flights, and crews on high-inclination flights typically receive less net exposure to trapped radiation for the same altitude (NRC, 2001).

During solar maximum, increases in the Sun's activity expand the atmosphere; this expansion causes losses of some of the protons in the radiation belts owing to interactions with atmospheric gases. Therefore, trapped radiation doses decrease during solar maximum and increase during solar minimum. The impact of GCR is also lower during solar maximum, because the increased speed and density of the solar wind intensifies the interplanetary magnetic field generated by the Sun, making it more difficult for GCR to penetrate the inner solar system (NRC, 2001).

The ISS poses significant challenges over shuttle missions for radiation protection of the crew due to several factors including the extended duration of the mission, the dynamic nature of the radiation environment in ISS orbit, and the necessity for many planned EVAs for station construction and maintenance. Shuttle missions typically last 7 to 10 days while ISS crew members may be on orbit for 6 months or longer.

13.3.2 Monitoring scheme

The radiation console in the Mission Control Center is staffed 4 hours per day during nominal space weather conditions, and continuously during EVAs and significant space weather activity. SRAG receives data and alerts from NOAA's Space Weather Prediction Center (SWPC) in Boulder, Colorado, and monitors its own instrumentation to make real-time assessments of the changing environment, helpful for mission operations planning.

In addition to external data and alerts, SRAG monitors a suite of its own instruments and dosimeters during missions, including active and passive dosimeters (shuttle and ISS), radiation area monitors (shuttle and ISS) as well as internal and external charged particle detectors (ISS).

13.4 The Constellation Program

Major elements of the Constellation Program are currently focused on providing the capability to transport humans and cargo first to the ISS, and then at a later date to the Moon in support of lunar exploration missions. These activities would provide the framework for future human exploration of the Moon, Mars, and other destinations in the solar system in the decades to come. Present plans call for operational missions to the ISS no later than 2015 and human missions to the Moon no later than 2020.†

The Orion crew vehicle will be capable of carrying crew and cargo to the space stations and will rendezvous with a lunar landing module, Altair, to carry crews to the Moon and beyond, and will also serve as the Earth entry vehicle for lunar and Mars returns. The Ares-I is an in-line, two-stage rocket with a primary mission of carrying Orion into orbit with a crew of four to six, it will also deliver resources and supplies to the ISS and be able to "park" payloads into orbit for later retrieval. The primary mission of Ares-V will be to transport large-scale hardware and supplies needed to extend a human presence beyond Earth orbit.

Orion and Ares-I would be used on missions to support the ISS once the space shuttle has been retired. It is anticipated that they would be used to transport crew and cargo to the ISS no later than 2015 with missions continuing throughout the life of the ISS. Orion, Ares-I, and Ares-V will all be used for lunar missions to be undertaken no later than 2020 (NASA, 2005).

The radiation factor is one of the main hazards for beyond-LEO exploration. Due to the long duration and very large distance from the Earth, mission termination to exclude radiation exposure becomes very difficult. Radiation protection requirements and architecture become much more important.

In 2005 NASA released a report entitled "NASA's Systems Exploration Architecture Study" that outlines several Design Reference Missions (DRM; NASA, 2005). One DRM was for the transportation of crew to and from the ISS, one for transportation of crew and cargo to and from anywhere on the lunar surface in support of 7-day "sortie" missions, one for the transportation of crew and cargo to and from an outpost at the lunar south pole and one DRM was also established

† http://www.nasa.gov/mission_pages/constellation/main/

for transporting crew and cargo to and from the surface of Mars for an 18-month stay.

13.4.1 ISS mission phases

The timeline for an Orion-ISS mission is proposed to be similar to current STS missions. Once locked in and all systems are cleared by controllers at both the Cape and Mission Control in Houston, the Ares-I will be launched, with staging occurring in a little over 2 minutes and orbital insertion only 4.5 minutes later. A second, circularization burn, using the onboard J-2X engine on the Ares-I second stage, places the Orion spacecraft on a proper course to the ISS.

After a two-day chasedown, the Orion spacecraft will meet up with the ISS, and then dock with the it. The six-person crew, the largest number that can fly on an Orion spacecraft, will enter the station for typically a 6-month mission. Once completed, the crew will then re-enter the Orion, which has been kept attached to the station as an emergency "lifeboat", seal off the hatches between it and the ISS, and undock from the station for the return to Earth.

While aboard the ISS, space weather and radiation exposure monitoring follows current ISS mission procedures. For these operations within Earth's geomagnetic field, little or no supplemental shielding is needed to ensure astronaut safety in a capsule or habitat. On EVA, however, it will be essential that the proper environmental monitoring instruments are in place and that research into forecasting methods continues to protect from the natural GCR environment and SPEs. For longer-duration lunar and Mars missions the currently large uncertainties in radiological risk predictions could be reduced by future research.

13.4.2 Lunar mission phases

Astronauts will be exposed to radiation hazards during all phases of a lunar mission. Some of the transit phases will be afforded protection of the Earth's geomagnetic field just as current LEO missions; some surface phases will offer shielding of the habitat. However, other mission operational phases, such as surface exploration, will pose a higher risk.

The Orion Crew Module (CM) will be the primary crew cabin for the majority of the lunar mission. It will contain the crew during launch, Earth-orbital operations, trans-lunar cruise, and in lunar orbit. For lunar orbit rendezvous missions, the crew will transfer to the Altair for the duration of surface operations, but will return to the Orion CM for additional lunar orbit operations, trans-Earth coast, and Earth entry. For direct return missions, the crew will remain in the Orion for lunar descent, surface operations, and ascent. At a minimum, the crew will spend 9 days in the

CM beyond the protection of Earth's magnetosphere in the interplanetary radiation environment (NASA, 2005)

Preliminary analyses by NASA and Lockheed Martin indicate that the Orion capsule provides adequate shielding from its structure, avionics, life support, other hardware, consumables, and waste storage such that lower-energy SPEs would not be a threat. However, for the rarer, higher energy events, the Orion capsule itself must either incorporate sufficient shielding or else have the capability to reconfigure shielding and functional hardware to provide a radiation storm shelter for the astronauts. Since the duration of the most hazardous portion of an SPE or a close series of SPEs can be hours to a few days, the Orion capsule must be capable of providing the storm-shelter capability for a somewhat extended period of time (NRC, 2008).

Low-Earth orbit assembly and checkout Altair will be capable of landing four astronauts on the Moon, providing life support and a base for week-long initial surface exploration missions, and returning the crew to the Orion spacecraft that will bring them home to Earth. Because of the spacecraft's size and weight, Altair, and its associated Earth Departure Stage, will be launched into LEO using the heavy-lift Ares-V launch vehicle, followed by a separate launch of an Orion spacecraft lifted by an Ares-I. After rendezvous and docking with Altair in LEO, the crew will then configure the Orion/Altair for the journey to the Moon. SPE and GCR exposures will be minimized by the Earth's magnetic field, but crew will still be exposed to the LEO component of the trapped radiation.

Lunar transfer orbit The Orion/Altair combination's trans-lunar coast will be about 3 days. Approaching the lunar far side, the Altair's engines will orient the vehicle in the proper direction for the lunar orbit insertion (LOI) burn to begin. Once in orbit, the crew will refine the trajectory and configure the Orion Crew and Service Module (CSM) for unmanned flight, then all crew members will transfer to the Altair, undocking from the Orion CSM, which will be left in a 95 to 110 km orbit awaiting Altair's return. Crew will be exposed briefly to the trapped radiation on transit as was encountered by the Apollo missions and will have no benefit of the geomagnetic field.

Lunar "sortie" missions The first missions will be on the order of 5 to 7 days for collecting samples and deploying experiments. As lunar missions continue, more time will be spent on the surface constructing the lunar habitation. Once complete, extended stay missions are expected to be on the order of 6 months.

The first "sortie" missions will be analogous to the Apollo surface missions and demonstrate the capability of the architecture to land humans on the Moon, operate for a limited period on the surface, and safely return them to Earth. Sortie missions also allow for exploration of high-interest science sites or scouting of future lunar outpost locations. Such a mission is assumed not to require the aid of pre-positioned lunar surface infrastructure, such as habitats or power stations, to perform the mission.

During a sortie, the crew has the capability to perform daily EVAs with all crew members egressing from the vehicle through an airlock. Performing EVAs in pairs with all four crew members on the surface every day maximizes the scientific and operational value of the mission. Surface EVAs could be as long as 8 hours, protected from the environment only by the EVA suit. Crew will be exposed to one-half of the free space environment owing to the shadowing of the Moon, but will have little shielding from SEPs so that during EVA operations contingency plans will be important during an SPE. Additionally, interaction of the GCR with the lunar surface will slightly increase the background exposures due to neutron production.

Lunar base habitation A primary objective of the lunar architecture is to establish a continuous human presence on the lunar surface to accomplish exploration and science goals. This capability will be established as quickly as possible following the return of humans to the Moon. To best accomplish science and resource utilization goals, the outpost is expected to be located at the lunar south pole. The primary purpose of the mission is to transfer up to four crew members and supplies in a single mission to the outpost site for expeditions lasting up to 6 months. Every 6 months, a new crew will arrive at the outpost, and the crew already stationed there will return to Earth. During the 6 months, multiple EVAs are planned lasting up to 8 hours.

The habitat is the most suitable module to afford radiation protection to the crew. The advantage of making the habitat serve as a radiation storm shelter is that it already contains the accommodation to sustain the crew in relative comfort over the course of a solar storm. The accommodation includes life support, food systems, hygiene and waste management systems, sleep stations, and stowage for clothing and personal articles (NRC, 2008). When the habitation is complete it will afford crews additional shielding from the GCR and serve as a storm shelter during SPEs.

13.4.3 Martian exploration mission

As with the return to the Moon, detailed trade studies will be needed to determine the optimal scenarios and timelines. Passage through the radiation belts should contribute minimally to the overall mission dose. However, since the spacecraft will be the only source of shelter on a Mars mission for the hundreds of days that the crew will be in deep space, it is critical that it provide adequate shielding to protect the crew from the steady GCR exposure and the sudden impact of one or multiple large SPEs.

The Mars exploration mission plan is to focus on long-stay missions to minimize the exposure of the crew to the deep-space radiation and zero-gravity environment, while at the same time maximizing the scientific return from the mission. This is accomplished by taking advantage of optimum alignment of Earth and Mars for both the outbound and return trajectories by varying the stay time on Mars, rather than forcing the mission through non-optimal trajectories, as in the case of the short-stay missions.

The surface exploration capability will be implemented through a split mission concept in which cargo is transported in manageable units to the surface, or Mars orbit, and checked out in advance of committing the crews to their mission. The split mission approach also allows the crew to be transported on faster, more energetic trajectories, minimizing their exposure to the deep-space environment, while the vast majority of the material sent to Mars will be sent on minimum energy trajectories. This approach allows the crew to transfer to and from Mars in about 6 months while allowing them to stay on the surface of Mars for a majority of the mission, on the order of 18 months.

During the 6-month transit most of the crew exposure is expected to come from GCR as mission timing relative to the solar cycle will likely be planned for solar minimum, reducing the risk of encountering large SPEs. Mars has no significant magnetic field to offer protection, but exposures during orbit trajectories will be reduced by up to 40% due to the planet's mass. The exploration of the Martian surface will likely take the same operational scenario as lunar explorations except that the Martian atmosphere offers an additional 16 g/cm^2 of shield, mostly carbon dioxide, that is not seen on the lunar surface. The projected dose from a typical 2.5-year Martian mission is expected to be greater than 1.0 Sv (Connoly, 2004).

13.5 Environmental characterization

A comprehensive strategy for radiation protection goes well beyond the selection and arrangement of material to be incorporated in Orion, Altair, and elements of the lunar habitat. Monitoring the variable conditions on the Sun and in the space environment that can influence the performance and reliability of space-borne and ground-based technological systems as well as endanger life or health will be crucial to a comprehensive radiation protection plan in keeping with the principles of ALARA (AMS, 2008). Currently the environment is characterized with measurements from free space and geosynchronous orbit as well as from instruments inside and outside the ISS.

13.5.1 Environmental Observations

The joint NASA–European Space Agency Solar and Heliospheric Observatory (SOHO) became operational in 1995 and became heavily relied upon as a near real-time source of solar data. It was joined in 1998 by the NASA Advanced Composition Explorer (ACE) mission, which carries a space weather beacon for continuous transmission of relevant *in-situ* space environment data. SOHO and ACE are located near the L1 Lagrangian point, 1% of the Earth–Sun distance upstream of the Earth, where it measures solar wind plasma and magnetic field approximately one hour before it reaches the Earth.

Along with SOHO and ACE, the joint NOAA–NASA Geostationary Operational Environmental Satellites (GOES) sit in geosynchronous orbits about 36 000 km above the Earth. As well as providing information about terrestrial weather, GOES contains a suite of instruments known as the Space Environment Monitor (SEM), which also provides real-time data on the Sun. The SWPC in Boulder, Colorado, receives, monitors, and interprets the GOES–SEM data and issues reports, alerts, warnings, and forecasts for special events such as solar flares and geomagnetic storms.

GOES The Geostationary Operational Environmental Satellites[†] measure *in situ* and provide real-time data of the near-Earth space environment and are used for observing the solar X-ray output. An energetic particles sensor (EPS) and high-energy proton and α detector (HEPAD) monitor the incident flux density of protons, α-particles, and electrons, while solar output is monitored by an X-ray sensor (XRS). Magnetometers are also on board and they monitor Earth's geomagnetic field strength in the vicinity of the spacecraft. X-ray spectral irradiance from GOES is critical in flare monitoring and spectral variability and is used as a potential precursor for SPE.

ACE The Advanced Composition Explorer[‡] provides data to continuously monitor the solar wind and to allow the SWPC in Boulder to produce warnings of impending major geomagnetic activity, up to one hour in advance. High time resolution data include local magnetic field vector, speed vector, density, temperature, elemental composition, bidirectional electrons and energetic particle flux (protons, electrons, ions). The data are broadcast as a continuous low-rate bit stream and with a combination of dedicated ground stations the data can be received 24 hours per day throughout the year, downloaded, processed, and dispersed within five minutes from the time they leave ACE.

[†] http://goespoes.gsfc.nasa.gov/goes/instruments/
[‡] http://www.srl.caltech.edu/ACE/

SOHO The Solar and Heliospheric Observatory[†] monitors and sends visual images of the Sun. The Large Angle and Spectrometric Coronagraph (LASCO) monitors the solar corona above the Sun's limb in a similar way as we perceive the corona during a solar eclipse and provides an understanding of the speed and angular span of CMEs. The Comprehensive Suprathermal and Energetic Particle Analyzer (COSTEP) is an array of solid-state detectors with anti-coincidence to measure energy spectra of electrons in the range 250 keV to >8.7 MeV. It is the only reliable measurement of relativistic electrons in an operations favorable position. The Extreme-ultraviolet Imaging Telescope (EIT) studies the low coronal structure and activity, returning extremely vital information about active regions, coronal holes, filaments and prominences, coronal dimmings, EIT waves and flares. The Michelson Doppler Imager (MDI) measures velocity and magnetic fields in the photosphere to give information about the magnetic field complexity of active regions, and can provide far-side imaging capabilities using photospheric dopplergrams.

Ground-based coronagraphs Coronagraphs provide images of the Sun's corona, the outermost layer of the solar atmosphere. Ground-based coronagraphs complement space-based instruments. A white light coronameter is located at the Mauna Loa Solar Observatory in Hawaii, operated by the High Altitude Observatory, National Center for Atmospheric Research. A hydrogen-α coronagraph is located at the Pic du Midi Observatory in France. Coronagraphs provide observations of CMEs by monitoring the activity, eruption, and disappearance of filaments and flares.

Ground-based radio observatories The US Air Force's Solar Electro-Optical Network (SEON) consists of both solar radio and optical telescopes. The radio telescopes provide information on the level of solar activity by monitoring radio noise. Intensities of solar radio emission during quiet and flare times are measured over the frequency range of 30 kHz to 100 GHz. Interpretation of the data is done on-site and forwarded to the forecast center in terms of Type I to Type V radio flare emission and in terms of impulsive (seconds to 10 minutes) or gradual (10 minutes to days) events.

Several new NASA scientific missions are also of interest to the operations communities because of their potential and location. The launch of the NASA–ESA Solar-Terrestrial Relations Observatory (STEREO) added a space weather data stream that covers the region between the Sun and the Earth with stereoscopic imagery. Planned for late 2009, the Solar Dynamics Observatory (SDO) will be placed in a geosynchronous orbit, at the altitude of the current GOES satellites,

[†] http://sohowww.nascom.nasa.gov/

and will stream data back to the Earth at a continuous rate of about 2 TB per day for 10 years.

13.5.2 Operational use

Prompt analyses of the Sun, solar wind, magnetosphere, geostationary orbits, and ionosphere are required so that the limited number of analysts can make informed decisions about the changing environment. Ground-based sunspot region observations, flares and solar wind monitored by ACE, the geomagnetic K-index, and electron and proton intensity observed by GOES are all indicators of change. Characterizations can be made by analyzing the interrelation between several space environmental data streams.

White light images are used for sunspot region classification. A simple scheme of sunspot classification due to McIntosh is commonly used as a starting point for flare prediction (Bornmann and Shaw, 1994). Regions called delta spots produce most large flares. Predictions are usually stated in terms of probabilities for occurrence of flares above M or X GOES class with 24 or 48 hours and are issued by SWPC. The magnetic complexity of active regions is observed by line-of-sight vector magnetograms. Extreme-ultraviolet and Hα images give further detail to better characterize the evolution of these regions and can be used at all stages of evolution to monitor the activity, eruption and disappearance of filaments and flares.

The mean velocity of particles in the solar corona's plasma is about 145 km/s, which is well below the solar escape velocity of 618 km/s. However, a few of the particles will achieve energies sufficient to reach the terminal velocity of 400 km/s, allowing them to feed the solar wind. At the same temperature, electrons, due to their much smaller mass, obtain escape velocity and build up an electric field, which tends to further accelerate protons and heavier charged ions away from the Sun (Encrenaz *et al.*, 2003). Solar wind velocities above about 600 km/s, coupled with sharp changes in the local magnetic field vector (indicating magnetic connectivity) are therefore indicators of energy coupling between the Sun and the Earth's magnetosphere (Horne, 2001). Increased electron flux may be an indicator that a large release of energy is imminent.

The sudden release of magnetic energy from the Sun is first seen as a flare. Solar flares are classified as A, B, C, M, or X according to the peak flux (in watts per square meter, W/m^2) of 100 to 800 picometer X-rays near Earth, as measured on the GOES spacecraft (cf., Section 5.2 and Table 5.1). Each class has a peak flux ten times greater than the preceding one, with X class flares having a peak flux of order 10^{-4} W/m^2. Within a class there is a linear scale from 1 to 9, so an X2 flare is twice as powerful as an X1 flare, and is four times more powerful than an M5 flare

(Machado *et al.*, 1998) SOHO–LASCO images give an immediate indicator of the speed and angular span of an associated CME.

Flares affect the ionosphere immediately, with adverse effects on communications and radio navigation (GPS and LORAN). Accompanying radio bursts from the Sun are expected to exceed cell phone system noise tolerances two or three times per solar cycle. If the flaring region is magnetically connected to the Earth and releases charged particles, the energetic particles arrive between 20 minutes and several hours later, threatening the electronics of spacecraft and unprotected astronauts, as they can potentially rise to 10 000 times the quiet background flux.

The ejected bulk plasma and its pervading magnetic field from a CME arrive in 30–72 hours (depending upon initial speed and deceleration) setting off a geomagnetic storm, causing currents to flow in the magnetosphere and particles to be energized. The currents cause atmospheric heating and increased drag for satellite operators; they also induce voltages and currents in long conductors at ground level, adversely affecting pipelines and electric power grids. The energetic particles can augment the already increased charged particle flux by as much as 50%.

13.5.3 Forecasting

At present, the capability to predict space weather events is comparable to Earth weather forecasting of about a half-century ago. Many space weather events are forecasted, but with minimal lead time because of a lack of real-time data and limited model capabilities. Investments by the USA and the global community into space weather-related research and technologies are rapidly advancing the state of knowledge and show great promise for producing improved forecasting capabilities (AMS, 2008).

13.5.3.1 Current short-term forecasts

Currently, NASA uses AP-8 and AE-8 models to estimate the trapped proton and electron environments, respectively (Sawyer and Vette, 1976). In general, AP-8 estimates of the orbital proton environment as compared to orbital dose measurements show predictions to within a factor of two. In outer regions where the magnetic field is more unstable, the differences between AP-8 and measured values can approach a factor of ten. Comparisons of AE-8 results are complicated by SPEs and geomagnetic storms and the fact that there are several electron source regions. In general, AE-8 overestimates the electron component across all regions. Accuracies at geosynchronous orbits are 10 to 50 times higher than measured values, depending on magnetic activities. In LEO, the AE-8 accuracy is generally within a factor of two, with regions at the inner belt, inner edge being of least accuracy (approaching a factor of ten Armstrong, 2000a; Armstrong and Colborn,

2000)). The SWPC issues alerts and indices based on current conditions. They also provide a running summary of key solar-geophysical indices for the current day, allowing customers to routinely appraise solar-geophysical activity. The prediction algorithm is driven by ACE real time solar wind data to predict the geomagnetic activity index, Kp (Costello, 1997). The current prediction output consists of two consecutive 1-hour averages of three solar wind parameters, velocity (V), IMF magnitude (B-total), and the IMF B_z component.

Also provided is a 3-day forecast of solar-geophysical conditions including solar flare, geomagnetic field, and satellite altitude proton activity. The NOAA Wang–Sheley model is an empirical model for forecasting solar wind speeds 3 to 4 days in advance (Arge and Pizzo, 2000), while three-dimensional magnetohydrodynamic models map velocity and magnetic field structures in the heliosphere to specific structures in the solar corona (Riley *et al.*, 2001).

13.5.3.2 Climatology

Due to the chaotic nature of the Sun, historical eruptions are all very different with respect to size, duration, and intensity. This makes a climatological study of the Sun very difficult. Currently, the only real climatology that is done with respect to the heliosphere is that of predicting the average monthly number of sunspots.

13.5.3.3 Development avenues

Methods involving artificial intelligence, Bayesian inference, and locally weighted regression have demonstrated promise in providing "nowcasting" capabilities after SEP event particles begin to arrive. These methods are capable of predicting, with reasonable accuracy, total doses and the future temporal evolution of the dose as particles arrive very early in the evolution of the event. However, these methods are at present unable to forecast SEP event fluence levels and their associated doses until after particles begin to arrive.

Active regions contain complex magnetic structures that often erupt to produce flares and CMEs. While models exist that describe the magnetic-field evolution leading up to the eruption, the timing of the eruption is not yet predictable with these models. One method for predicting the probability of eruption is to identify S-shaped magnetic-field structures, called sigmoids, which are observed in X-ray and extreme-ultraviolet (EUV) images of the corona (Canfield *et al.*, 1999). Other tools for predicting active region eruptions use photospheric magnetograms to determine the degree of non-potentiality (overall twist and shear) and the amount of free energy in coronal magnetic fields (Falconer *et al.*, 2003).

Flares are one of the primary sites for the acceleration of electrons, protons, and heavy ions but the nature of the acceleration is still under investigation. Current research offers promising results by focusing on probabilistic methods of flare forecasting using vector magnetogram data focusing on the next 24 hours (Barnes *et al.*, 2007). Without attempting to predict the happening of a flare, others use the same data to look at complexity of the active region in an attempt at predicting whether or not a flare event will be associated with an SPE (Wang and Zhang, 2007).

Groups are developing large-scale models to predict the plasma and magnetic-field environments of the global heliosphere and perfect the current models being used. The possibility for longer-term predictions may come from helioseismological models of active region formation beneath the photosphere, before their appearance on the surface (Braun and Lindsey, 2000). Helioseismology is also being used to track active regions while they are on the far side of the Sun (Braun and Lindsey, 2001).

13.6 Summary

NASA has established limits on the risk to humans that may be incurred by exposure to space radiation. These limits are specified for missions in LEO. The limiting risk for career exposure to space radiation is an increase of 3% in the probability of developing a fatal cancer. Thirty-day and annual limits are based on keeping radiation exposure below the threshold level for deterministic effects, and have incorporated the policies underlying ALARA in the designs used and the operations conducted in space (NASA, 2005).

Unlike LEO exposures, which are often dominated by solar protons and trapped radiation, interplanetary exposures may be dominated by the GCR for which there are insufficient data on biological effects. Consequently, risk prediction for interplanetary space is subject to very large uncertainties, which impact all aspects of mission design. This is especially true since ALARA requires the use of appropriate safety margins, which are directly related to the uncertainty in risk estimates (NASA, 2005).

The most commonly used means of protecting terrestrial radiation workers is through the use of shielding. In principle, shielding alone should be able to reduce exposure by attenuating the radiation and reducing the dose rates. For deep space missions, however, shielding alone cannot guarantee protection in all situations owing to the very high energies of the incident ions and the production of highly penetrating secondary particles, such as neutrons and light ions, coupled with mass constraints on the spacecraft and the large uncertainties in biological risk (NRC, 2008).

Operational protocols will also be critical to any radiation safety program, especially for any EVA beyond the Earth's magnetic field. Future concepts of operations will involve measuring the indicators of solar activity, as well as actual radiation dose as experienced by the astronauts. The second can only be accomplished by including one or more active dosimeters on their persons or on nearby equipment, such as a rover or robotic assistant. When solar activity is observed that could result in an SPE, the EVA will likely have specific contingency plans already in place for seeking shelter or for using available shielding, much as there are contingency procedures for anomalous launch or other hazardous operational situations (NRC, 2008). Though it is envisioned that a surface EVA may be as long as 8 hours, access to shielding on a short time scale, possibly less than 1 hour, is preferable in order to avoid possible excessive exposure.

Meeting these goals for the future will require an architecture that incorporates a broad range of solar, heliospheric, and energetic particle monitoring, as well as the communications support to telemeter the data in real-time. Mission planning support to lunar and future Mars missions will require forecast tools that estimate the probability of an SPE within the next few hours to days. Real-time mission operations support will substantially benefit from predictions of the expected peak flux, time to peak flux, total fluence, and duration of ongoing events within the first hour of event onset. The environmental observations are not only required for characterization, but also serve as inputs for the forecasting models and tools being developed.

The transitioning of research understanding, models, and observational capabilities from the solar and space physics communities emerges repeatedly as an issue of importance for the future of radiation protection during exploration class missions. Operational support could and should be enhanced by: ensuring the continuation of scientific spacecraft in operationally viable positions (Sun–spacecraft/crew line of sight, for example) sending real-time data both to the vehicles and to the ground; ensuring availability of active and passive personal dosimeters with well-characterized charged particle and neutron measurements; expanding the current real-time proton characterization measurements; and working closely with the scientific communities in the continual development of quiet-time operations enabling forecasts.

US government agencies such as NASA, NSF, NOAA, and the DOD support the study of space weather as a natural component of their research programs. In doing so, they have recognized that a space-age nation requires a broader view of the environment than perhaps was necessary before. In the past, both the study and applications of space weather have not generally been coordinated across activities and agencies. A change has occurred within the last few years during which efforts have been initiated to bring together the various programs and to create a broader awareness of space weather. This collective effort has come to be known as the National Space Weather Program (NAS, 1997).

14
Energetic particles and technology

ALAN TRIBBLE

14.1 Introduction

The purpose of this chapter is to provide space scientists with detailed knowledge of how the environment of space interacts with, and degrades, spacecraft systems. In particular, the goal is to highlight how these interactions are tied to the parameters that describe the environment in order to show how uncertainties in knowledge of the environment can lead to uncertainties in the prediction of the effects themselves. This in turn leads the designer to over-engineer spacecraft systems in order to ensure that the various effects are properly mitigated throughout the life of a spacecraft. As a result, improvements in models of the space environment could lead to better predictions of these space environmental effects.

The field of space environment effects is split into five separate categories depending on the nature of the environment itself. Two of these categories are directly related to the energetic particle environment: plasma and radiation. Two environments are indirectly dependent on solar conditions: neutral and micrometeoroid/orbital debris. The final environment is essentially independent: vacuum. The vacuum, neutral, and micrometeoroid/orbital debris categories will be examined briefly for completeness. The plasma and radiation effects will be examined in more detail. In particular, it will be seen how keV energy particles lead to spacecraft charging; MeV energy particles lead to total-dose radiation effects; while GeV energy particles lead to single-event effects in electronic devices.

14.2 Overview of space environment effects

The field of space environment effects is relatively new, having not been an area of concern before the first spacecraft launches some 50 years ago (Tribble, 2003; Hastings and Garrett; 1996). The field loosely defines the five areas of interest listed in Table 14.1. These areas are of interest to spacecraft designers and operators

Heliophysics: Space Storms and Radiation: Causes and Effects, eds. Carolus J. Schrijver and George L. Siscoe. Published by Cambridge University Press.

Table 14.1. *Environmental effects in space*

Environment	Associated phenomena	Section
Vacuum	Absence of a substantial atmosphere	14.2.1
Neutral	Tenuous neutral atmosphere	14.2.2
Plasma	keV energy charged particles	14.3
Radiation	MeV or GeV particles (usually charged)	14.4, 14.5
Micrometeoroid, orbital debris	Hypervelocity impact of μm-sized particles	14.2.3

owing to their ability to degrade spacecraft systems and materials, often leading to an early (and in some cases abrupt) degradation in performance or even an end to the mission. The plasma and radiation environments are primarily due to energetic particles so they will be discussed in detail. The remaining three environmental effects are discussed first.

14.2.1 Vacuum environment effects

Vacuum environment effects include solar ultraviolet (UV) degradation of materials, and molecular and particulate contamination. Solar UV degradation is directly dependent on the solar UV output, defined loosely as the energy below about 0.2 μm (2000 Å) wavelength, with photon energy given by $E = h\nu$. A photon with a wavelength of 0.2 μm has an energy of about 25 eV, which is of the same order as the binding energy of many chemical bonds. As such, photons in the UV range can break chemical bonds and degrade material properties. It is quite common for materials exposed for many months, or years, in space to be visibly darkened. This in turn can lead to increased absorption of heat from the Sun, and overheating of spacecraft subsystems. This overheating indirectly affects many system-level concerns. For example, in the design phase it may influence the decision to use passive or active thermal control. During operation, overheating may degrade the efficiency of some systems, resulting in lower performance. Finally, overheating may accelerate mission end of life. All of these factors may force a re-examination of risk and cost. Although solar conditions may lead to short-term variations in the solar UV environment, solar UV effects are typically dependent on the total hours of Sun exposure received on any particular surface, and surface material properties.

Molecular contamination is the result of outgassing by materials in the vacuum of space. These outgassed molecules may condense on spacecraft surfaces, leading to a degradation in signal strength (optical devices), power production (solar arrays), or increased heat load (thermal control surfaces). Particulate contamination is the result of micrometer-sized particles in the air depositing onto surfaces during ground processing and launch. Particulate contamination is mainly a concern for optical devices, where the particles can scatter light or obscure pixels on focal planes. As contamination is not related to the energetic particle environment, this level of detail will suffice, but a great deal of additional information is available (Tribble, 2000, and references therein).

It is important to emphasize that the vacuum environment effects are essentially independent of orbit. That is, since they are due to the absence of a substantial atmosphere, they are the same in low-Earth orbit, geosynchronous orbit, lunar orbit, Mars transfer orbit, etc., but for the scaling of the degrading effects of the solar UV as $1/r^2$, with distance r from the Sun.

14.2.2 Neutral environment effects

Neutral environment effects are those phenomena associated with the presence of a tenuous neutral atmosphere. These effects are confined to low-Earth orbit (LEO) – loosely defined as altitudes below 1000 km. Two interactions are due to the chemical properties of the LEO environment, while two are mechanical in nature.

The most predominant atmospheric constituent in LEO is atomic oxygen,[†] which is chemically very reactive. Atomic oxygen may erode surface materials and can also give rise to an optical emission near the surface of materials, known as spacecraft glow. Glow may be of concern if optical instruments are on board the spacecraft, but the more significant concern is atomic oxygen erosion. The rate at which material may be eroded by atomic oxygen is given by

$$\frac{\mathrm{d}x}{\mathrm{d}t} = \mathcal{R}_E \, n \, v, \tag{14.1}$$

where $\mathcal{R}_E$ is an experimentally determined reaction efficiency – a measure of how the material in question is eroded by atomic oxygen; n is the density of atomic oxygen; and v is the impact velocity, which is essentially the spacecraft orbital velocity of about 8 km/s. (In LEO, the thermal velocity of atomic oxygen is about 1 km/s, so the spacecraft orbital velocity dominates.) As is seen from the equation, a better understanding of the atmospheric density would lead to a better prediction of material lifetimes.

At an orbital velocity of 8 km/s, a great deal of kinetic energy is available in collisions with atmospheric particles. Two interactions that are related to the mechanical kinetic energy of impact are aerodynamic drag and sputtering. Sputtering is the physical erosion of material from a surface, and is rarely a concern for most

[†] At altitudes well above 100 km, the density of the Earth's atmosphere is low enough that the chemical composition is stratified, with increasing scale heights for decreasing molecular masses, and as solar UV radiation dissociates molecular O_2, atomic neutral O is the dominant constituent between about 200 km and 600 km.

spacecraft materials. The more significant concern is aerodynamic drag. The drag force is given by

$$F = \frac{1}{2}\rho A_n v^2 C_d, \tag{14.2}$$

where ρ is the mass density of the environment; A_n is the surface area of the spacecraft measured normal to the direction of travel; v is the spacecraft orbital velocity; and C_d is the drag coefficient – a measure of the momentum transfer to the spacecraft during collisions with atmospheric particles. For most spacecraft C_d is about 2.20. The importance of the drag force is that it imparts an acceleration to the spacecraft that causes it to slow and, eventually, to re-enter the atmosphere. To counter the drag force, the spacecraft typically carries extra fuel to burn in order to boost itself to a higher orbit and increase its lifetime. The mass of fuel required is given by

$$\delta m = \frac{A_n v^2 C_d}{2 I_{sp} g}\rho\,\delta t, \tag{14.3}$$

where I_{sp} is the specific impulse of fuel – a measure of how much force is obtained from the fuel; g is the acceleration due to gravity; and t is the orbital lifetime. As the mass of fuel is directly related to the mass density of the environment, better knowledge of the environment itself directly correlates with better knowledge of the drag makeup fuel required. As shown in Fig. 2.4, orbital decay is indirectly a function of solar cycle, as the atmospheric density n increases during solar maximum when more energy is deposited in the Earth's atmosphere, causing it to expand and increase the mass density encountered by an orbiting spacecraft.

14.2.2.1 Modeling the neutral environment

Since the advent of the space age, dozens of separate models of the Earth's neutral environment have been created (ANSI, 2003). However, it is important to emphasize that the type of model that is of interest to individuals who study the effects of the space environment is not necessarily the same type of model that is of interest to individuals who specialize in knowlege of the environment itself. Space environment effects analysis often requires a model of the particulate environment that will interact with the spacecraft. That is, the need here is for a model of how the environment is characterized, and not a model of the physics behind the environment. Engineering studies are much more concerned with "what" the environment is, and less concerned with "how" it got to be that way (Rainey, 2004).

The challenge to modeling the Earth's atmosphere is that it is a gas. If additional energy is added to a gas, the gas will expand. As the Sun undergoes its 11-year sunspot cycle, moving from solar minimum to solar maximum, the amount of total energy input to the neutral environment changes. As such, the environment expands and contracts slightly in phase with the solar cycle. Consequently, modern models of the neutral environment contain algorithms that utilize the F10.7 value (the solar flux at 10.7 cm wavelength – a parameter that is seen to follow the changes in the solar ultraviolet that influence the Earth's upper atmosphere and ionosphere) as input, and generate density models that are a function of solar conditions.

14.2.3 The micrometeoroid/orbital-debris environment

Micrometeoroid/orbital-debris (MMOD) effects are those phenomena associated with the hypervelocity impact of MMOD particles. Since a spacecraft in orbit is moving at about 8 km/s, typical impact velocities may be 10 km/s or more. As a result, even small, micrometer sized, particles are capable of inflicting severe damage. The micrometeoroid environment is the natural component of particles found in space. The orbital debris environment is the man-made component, made of the bits and pieces of spacecraft systems that remain in orbit after previous launches.

The main concern from MMOD impacts is that a single impact may damage a spacecraft system, bringing the mission to an immediate end. This is especially a significant concern on manned missions such as the space station or space shuttle. The amount of shielding required to effectively stop these particles is rarely practical, so as a rule spacecraft typically live with the risk.

A better prediction of the true MMOD environment, especially in geosynchronous Earth Orbit (GEO) where no data are available, would benefit designers by allowing them to better quantify the risk associated with these particles. This would not be an easy feat to accomplish, as it would require either post-mortem examination of defunct GEO spacecraft or sophisticated *in situ* instrumentation that would be capable of generating size versus mass distribution data from microscopic impact sites. To date, the cost of such an experiment has not been justifiable, since most GEO spacecraft seem to reach their end of life without MMOD being the limiting factor.

14.3 Effects of keV energy particles: spacecraft charging

Low-energy (keV range) charged particles make up what is known as the plasma environment. Charged particles in this energy range do not penetrate deeply into materials, less than 1 μm, hence their main effect is to electrically charge the surface. This gives rise to spacecraft charging. While this is possibly of concern to scientific instrumentation that uses the spacecraft chassis as an electrical ground (Tribble *et al.*, 1988), it usually is not a concern for the health and safety of a spacecraft. The main concern is that materials which have different electrical properties will charge to different electrical potentials. This, in turn, can lead to arc discharging, which may damage electronic equipment or degrade surface properties. Different arcing mechanisms are possible, such as electro-static discharge (ESD) or dielectric breakdown (DB), but all are dependent on an electrical potential difference between, or through, surfaces.

The key criterion in determining the potential that a surface charges to is simply the current that strikes that surface. This may be the direct current that strikes the surface as the result of the natural environment itself – the electrons and ions – or the current may be the indirect result of photo-emission from surfaces, or secondary electrons that are released by the high-energy impact of other particles. These currents will continue to charge a surface until equilibrium is reached, when the sum of the currents is zero. The potential that a surface is charged to is known as the floating potential, because this is the potential the surface will float to when exposed to the environment. Surface conductivity plays a significant role in that conductors and dielectrics will typically charge to different potentials, leading to large potential differences where different materials meet. (Charging to 10 000 volts has been seen in GEO, during periods of geomagnetic storms that can occur within a few minutes time.) These intersections are the most likely location for arc discharging.

14.3.1 Charging in low-Earth orbit

In LEO, about 1% of the neutral environment is ionized by the solar UV, so the resulting plasma is relatively low energy (0.1 eV), but also relatively high in density (10^{11} m^{-3}). The LEO environment is mainly atomic oxygen, so the ions in the resulting plasma have a thermal velocity approximated by that of the neutrals, or about 1 km/s. As this is less than the spacecraft orbital velocity of 8 km/s, ions are typically collected only in the direction of travel.

Conversely, due to their lower mass the thermal velocity of electrons is about 200 km/s, so electrons are capable of striking any spacecraft surface. The result is that spacecraft in LEO typically become slightly negatively charged ("float slightly negative"). This can be seen from simplistic current balance relations. The ion current is given by

$$I_{\mathrm{i}} = q n_{\mathrm{i}} v_{\mathrm{i}} A_{\mathrm{i}} = q n_{\mathrm{o}} v_{\mathrm{o}} A_{\mathrm{cs}}, \tag{14.4}$$

where q is the charge on the ion, n_{i} is the ion plasma density (or simply the plasma density n_{o}), v_{i} is the speed with which ions impact (the spacecraft velocity v_o), and A_{i} is the surface area that can collect ions (the cross-sectional area of the spacecraft, A_{cs}, due to the fact that the spacecraft is moving faster than the ions). Conversely, the electron current is given by

$$I_e = qn_e v_e A_e = \frac{1}{4} qn_o v_{e,th} A_{tot} \exp(-eV/kT_e), \tag{14.5}$$

where n_e is the electron density (or simply the plasma density n_o), v_e is the speed with which electrons impact (the electron thermal velocity $v_{e,th}$), and A_e is the surface area that can collect electrons (the total surface area of the spacecraft, A_{tot}, because the electron thermal velocity is greater than the orbital velocity). The exponential term describes electrostatic repulsion of low-energy electrons from the negatively biased surface. That is, because electron current will dominate, charging the spacecraft negatively, only those electrons that have sufficient kinetic energy to overcome the electrostatic repulsion to the surface will be collected. Finally, the factor of 1/4 is needed because half of the electrons are actually moving away from the spacecraft, and another factor of one-half accounts for the net velocity of those that are moving toward the spacecraft.

Setting the electron current equal to the ion current allows us to solve for the floating potential, which is given by

$$V_{fl} = -\frac{kT_e}{q} \ln\left(\frac{4n_i v_i A_i}{n_e v_{e,th} A_e}\right). \tag{14.6}$$

Solving this equation shows that V_{fl} is about 0.1 volts in LEO.

A floating potential of 0.1 V would not be expected to give rise to significant charging concerns, but this overly simplistic example only hints at the complexity that is necessary to solve the problem completely. It has not yet factored into account the interaction that would arise due to the spacecraft's electrical power system. Most spacecraft generate electrical power via the use of solar arrays. These arrays are tailored to generate a specific voltage by stringing together individual solar cells, which may generate about 1 volt each. Although each solar cell is covered with dielectric cover slides, which protect the cell from radiation damage, there is typically a small gap between cells to allow for thermal expansion and contraction. In these gaps the thin wires used to connect the cells together may collect charge from the environment. The charging situation now changes due to the fact that there is a fixed potential difference between the two ends of the solar array. That potential difference remains constant, but the entire array will now "float" relative to the plasma until the net charge (current) collected by the array is zero. Because of the dominance of the electron thermal velocity in LEO, a solar array in LEO will float so that it is mostly negative. The exact value is dependent on the spacecraft geometry and a number of other factors, but the rule of thumb is that approximately 90% of the array will float negative in a worst case scenario. (Of course at night, when the array is not generating a potential difference, the entire array again floats about 0.1 volts negative.)

In the USA, 28 volts has been a typical voltage supply because 28-volt power systems were well known to the aircraft manufacturers who began to build spacecraft in the late 1950s. Because power is the product of current and voltage, a high-power system at 28 volts would need to generate a lot of current. More recently, larger voltages have been used in an attempt to minimize the I^2R power loss. The International Space Station, for example, uses 160-volt arrays. So a 28-volt solar array may float so that the negative end of the array is at −25 volts, relative to the plasma, and the positive end is at +3 volts. For the space station, the negative end is closer to −144 volts while the positive end is at about +16 volts.

We have examined the case of a solar array, and that of the spacecraft body, separately. But we must examine them as a complete system. The spacecraft designer must make engineering decisions on how to connect the array to the spacecraft itself. These grounding options are: negative ground, positive ground, or floating ground. With a negative ground, the spacecraft is connected to the end of the array that floats negative with respect to the plasma. (This is sometimes called a "positive array" in that the array is more positive than the structures.) With a positive ground, the spacecraft is connected to the end of the array that floats positive, and with a floating ground a deliberate electrical ground is avoided. With a negative ground, the spacecraft structures are now biased negatively, and collect ions, forcing the spacecraft potential slightly more positive than −90% of the array voltage. With a positive ground, the spacecraft structures are now biased positively, and collect electrons, forcing the spacecraft potential slightly more negative than +10% of the array voltage. With a floating ground, structures would remain at about −0.1 volts, (Fig. 14.1).

The most well-known example of spacecraft charging in LEO is the International Space Station. Due to the 160-volt array the spacecraft may float −140 volts worst case. This is now a significant concern in that the space shuttle, which is

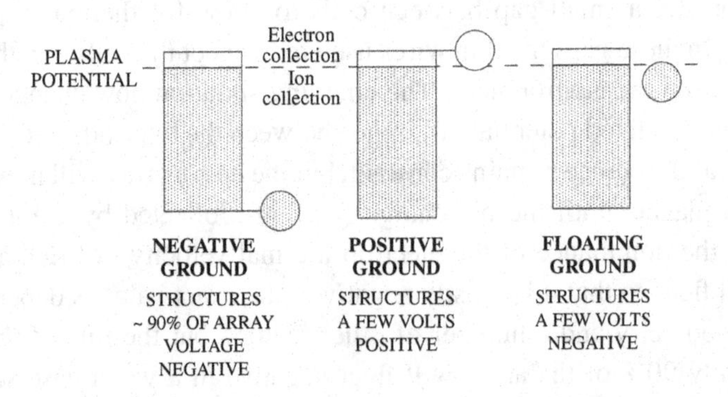

Fig. 14.1. Spacecraft floating potential vs. grounding options. (From Tribble, 2003. Reprinted by permission of Princeton University Press.)

powered by fuel cells, would tend to float about −0.1 volts. Is a 140-volt potential difference sufficient to lead to arcing or other concerns when the shuttle comes to dock? Fortunately, there seems to be no concern for shuttle-to-station arcing. However, another phenomenon is seen to arise at −140 volts and that is dielectric breakdown. The structural elements on the space station are aluminum, and the exterior surface is covered in a thin anodized coating. The coating is a dielectric and there can be a potential difference between the exterior surface, which would be charged by the environment, and the interior surface, which would be at the floating potential of the station itself. That potential difference could approach −140 volts. Dielectric breakdown is seen to occur if the electric field strength exceeds about 10 000 volts/cm (the exact answer value is dependent on the specific material in question). For a potential difference of 140 volts, the material would need to be 0.014 cm (140 μm) thick not to break down. Unfortunately, the process used to construct the anodized aluminum coating is only able to make the coating about 50 μm thick. Arcing was observed to onset on the structures at a voltage of about −60 volts; as such this was a significant issue in the design and operation of the space station. The solution in this case was to fly a "plasma contactor" – a device that generates a low-energy plasma to establish better electrical contact with the natural plasma environment. The plasma contactor alters the current balance equation and forces the station back to a potential of about −40 volts so that arcing does not occur. Note that the concern here is for dielectric breakdown on materials. Laboratory studies have confirmed that electrostatic discharge will occur on solar arrays themselves (arcing between dielectric cover slides and the electrical wires connecting solar cells) at higher voltages on the order of 200 volts or higher. For this reason the station voltage was limited to 160 volts.

14.3.2 Charging in the geosynchronous environment

In GEO, the plasma environment is much different from that found in LEO. In GEO the plasma density is much lower ($10^6\,\mathrm{m}^{-3}$) but plasma energies are much higher (130 eV). Due to the higher thermal energies, both ions and electrons can strike any spacecraft surface. Note also that at this higher altitude the plasma is now primarily electrons and protons as most heavier ions are gravitationally confined to lower altitudes.

Adding in other phenomena like photo-emission of electrons and secondary emission, one would find that a spacecraft in GEO would typically float a few volts positive. However, the GEO environment is much more dynamic than the LEO environment and charging to several thousand volts is sometimes seen.

Severe spacecraft charging in GEO is the result of geomagnetic storms, which allow more energetic plasma to reach the GEO environment. As illustrated in

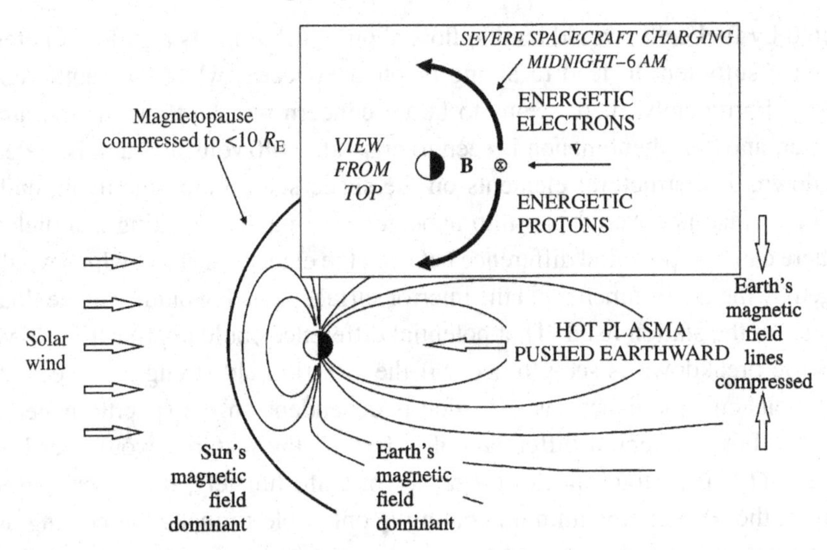

Fig. 14.2. Spacecraft charging in the geosynchronous environment at times of ICME-induced magnetic storms. (From Tribble, 2003. Reprinted by permission of Princeton University Press.)

Fig. 14.2, ICME-induced magnetic storms compress the Earth's magnetic field lines causing (some of) the more energetic plasma from higher altitudes to be injected toward the Earth. Due to the resultant $\mathbf{v} \times \mathbf{B}$ force, electrons are deflected from midnight toward sunrise, while protons are deflected from midnight toward sunset. Due to their higher mass, the protons are slower and the ambient electrons can easily counter the additional charging current. However, the high-energy electrons cannot easily be countered by the ambient protons and very high charging to several thousand volts negative has been observed. Under these conditions potential differences between spacecraft surfaces may now be hundreds, or thousands, of volts so arc discharging is much more likely.

14.3.3 Modeling the plasma environment

Improved models of the plasma environment would result in better predictions of spacecraft charging events. The most significant need is for an increased ability to predict severe spacecraft charging events in GEO. Forecasts of these events would alert spacecraft operators that conditions may be conducive to significant charging, and the resulting arc discharging. Depending on the spacecraft mission, it may be wise to power down some systems and put the spacecraft into "safe" mode until severe charging conditions subside.

The most popular spacecraft charging code is the NASA Charging Analysis Program, NASCAP (Mandell *et al.*, 2006). The latest version of the code illustrates spacecraft charging by way of four examples: charging of a geostationary satellite, self-consistent potentials for a negative probe in a low-Earth orbit spacecraft wake, potentials associated with thruster plumes, and particle-in-cell calculations of plasma effects on a very-low-frequency (about 1 to 20 kHz) antenna.

14.4 Effects of MeV energy particles: total-dose effects

Radiation environment effects are those phenomena associated with the presence of high-energy (MeV–GeV range) particles. Strictly speaking, any energetic particle (electrons, protons, heavy ions, or neutrons) or photon (γ-rays, X-rays) can be considered radiation. However, the main concern for spacecraft is from electrons or protons, the two most abundant species in the natural particle radiation environment.

Total-dose effects are the result of the cumulative interaction with the radiation environment over the life of the mission (see Chapter 3 for additional insight into the physics of the interaction of particles with matter). In contrast, single-event effects may arise as the result of the passage of a single particle through microelectronic devices. Total-dose effects are primarily due to the MeV energy particles present in the environment, which are far more numerous than the few GeV energy particles encountered (see Fig. 3.1 for the heliosheric energy distribution). Conversely, single-event effects are rarely initiated by MeV energy particles, and are primarily associated with GeV energy particles. Single-event effects will be examined in more detail in Section 14.5.

The SI unit of absorbed radiation dose is the *gray*, which is the absorption of one joule of energy by one kilogram of matter. When a charged particle passes through matter the most common interaction is with atomic electrons, which orbit the nucleus at about 10^{-10} m. Interactions with the nucleus, 10^{-15} m, are far less likely. If a charged particle and an atomic electron interact it is straightforward to understand that in the electrical interaction it would be possible for the atomic electron to gain sufficient energy to be liberated from the nucleus. For this reason, space radiation is often called ionizing radiation, and the total dose is called total ionizing dose. Interactions with the nucleus are indeed possible, and displacement of atomic ions from their lattice structure can occur. This type of interaction is called displacement damage. Although total ionizing dose and displacement damage may vary in terms of the physics of the interaction, the bottom line is that the properties of the system absorbing the dose will be degraded.

Depending on the physical characteristics of the material, it may be relatively easily degraded by radiation, or very resistant, see below. As shown in Table 14.2, the crew (see Chapter 13) will inevitably be the most susceptible to radiation effects on long-term space missions. Next in line are spacecraft electronics, which are the focus of the remainder of this discussion.

Table 14.2. *Damage thresholds for radiation in different materials*

Material	Damage threshold (gray)
Biological matter	0.1–1
Electronics	1–10^4
Lubricants, hydraulic fluid	10^3–10^5
Ceramics, glasses	10^4–10^6
Polymers	10^5–10^7
Structural metals	10^7–10^9

Total-dose effects usually lead to a slow, gradual degradation in spacecraft systems, because the radiation is often deposited over a period of months or years. An exception to this is the significant dose of radiation that often accompanies a SPE. In this case, total-dose effects may occur very rapidly, within hours or days. Degradation of solar array power production capability is a classic example of a total-dose radiation effect (cf. Fig. 2.6). The capability degrades with radiation dose, so the spacecraft designer must oversize the arrays at the beginning of mission in order to allow margin for the degradation, so that at the end of mission the arrays can still produce the minimum amount of power required to support the payload.

Individual solar cells are constructed from basic n- and p-type semiconductor materials. The np junction creates a space charge depletion region (voltage), and when charge carriers are liberated by photons from the Sun (current), they are accelerated by the internal electric field and the result is power production. A solar array is constructed by connecting individual cells together in series to create the voltage required (28 V is typical for most previous US spacecraft), and replicating the strings increases the current delivered.

An example of power loss versus radiation dose is shown in Fig. 14.3 (see also Fig. 2.6). As the figure indicates, different technologies are more or less resistant to radiation. Solar cells made of silicon are relatively soft (i.e. easily degraded), but are also relatively inexpensive. Solar cells made from gallium arsenide are more resistant, but also more costly, while cells made from indium phosphide are very resistant, but also very expensive.

A significant factor in the design of a spacecraft is the sizing of the solar array to ensure that sufficient power is produced. A key input to this design activity is the predicted knowledge of what the radiation environment (MeV particle environment) will be during the mission lifetime. As most spacecraft are designed to operate in LEO, knowledge of the trapped-radiation belts is crucial, and models of the belts are available from the NASA Community Coordinated Modeling Center

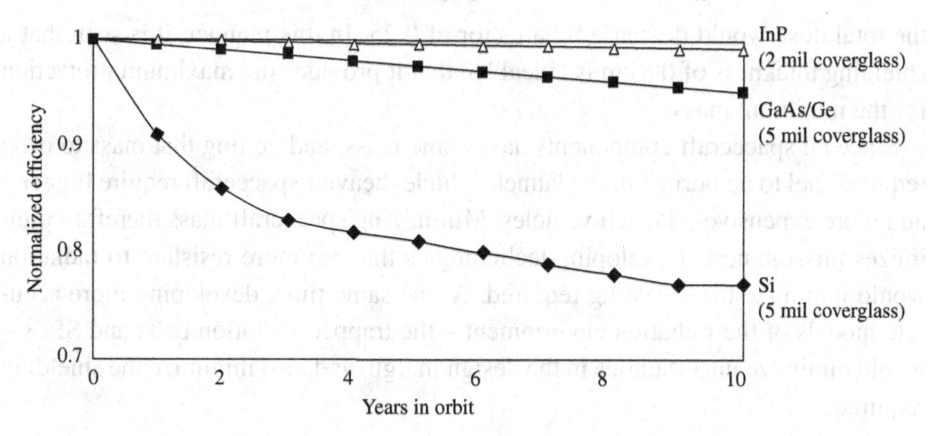

Fig. 14.3. Solar cell power loss vs. radiation (700 km altitude, 30° inclination orbit).

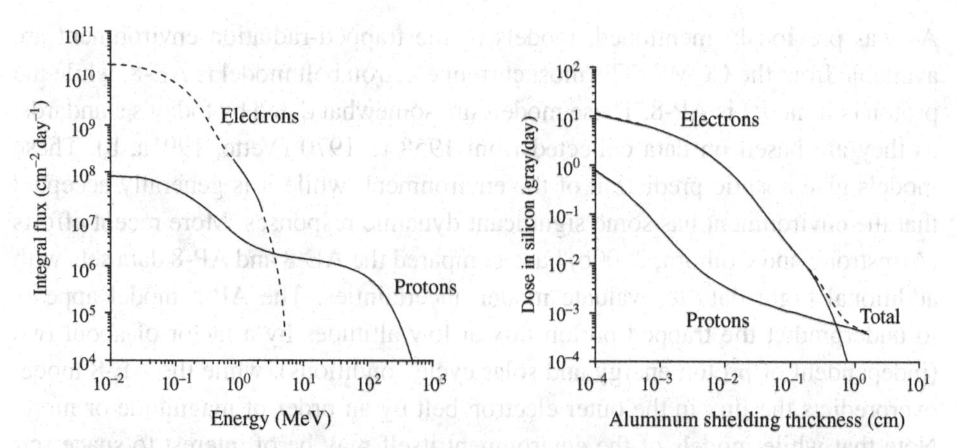

Fig. 14.4. International Space Station trapped-radiation environment *(left)* and total dose vs. shielding depth *(right)*.

(CCMC). An example of the radiation environment for the International Space Station is shown in Fig. 14.4. For engineering use, the radiation environment itself is not of primary concern as it is the effect of this environment that influences spacecraft design. As a result, many government (and a few commercial) modeling codes are available that use the radiation belt models as input, and add in the physics of how the environment interacts with materials, to generate dose versus depth curves, see Fig. 14.4.

Dose versus depth curves such as these play a key role in spacecraft design in that they often specify the ideal amount of shielding that a spacecraft must provide. As shown above, there is a "knee" at about 0.8 cm material thickness. If the effective shielding thickness were reduced by a factor of two, the total dose would increase by a factor of ten. If the shielding thickness were increased by a factor of two, the total dose would decrease by a factor of 0.25. In this manner, it is seen that a shielding thickness of 0.8 cm is "ideal" in that it provides the maximum protection for the minimum mass.

Since all spacecraft components have some mass, and getting that mass to orbit requires fuel to be burned in the launch vehicle, heavier spacecraft require bigger – and more expensive – launch vehicles. Minimizing spacecraft mass therefore minimizes mission cost. Developing technologies that are more resistant to radiation would minimize the shielding required. At the same time, developing more accurate models of the radiation environment – the trapped radiation belts and SPEs – would minimize uncertainties in the design margin and also minimize the shielding required.

14.4.1 Modeling the trapped-radiation environment

As was previously mentioned, models of the trapped-radiation environment are available from the CCMC. The most current electron belt model is AE-8, while the proton belt model is AP-8. These models are somewhat dated by today's standards, as they are based on data collected from 1958 to 1970 (Vette, 1991a, b). These models give a static prediction of the environment, while it is generally accepted that the environment has some significant dynamic responses. More recent efforts (Armstrong and Colborn, 2000b) have compared the AE-8 and AP-8 data sets with additional flight data to evaluate model uncertainties. The AP-8 model appears to underpredict the trapped proton flux at low altitudes by a factor of about two (independent of proton energy and solar cycle conditions), while the AE-8 model overpredicts the flux in the outer electron belt by an order of magnitude or more. Note that while models of the environment itself may be of interest to space scientists, spacecraft design engineers need tools that generate dose versus depth curves. Additional tools, such as SHIELDOSE, CRESSRAD, or SPACERAD, use the AE-8 and AP-8 models as input, and then calculate total radiation dose versus depth of shielding material based on the orbital parameters provided by the user.

14.5 Effects of GeV energy particles: single-event effects

Single-event effects encompass a variety of different phenomena in electronic devices, such as single-event upset, latchup, or burn out. All of these single-event effects can occur for example in a metal-oxide semicondutor field-effect transistor, MOSFET (Fig. 14.5). A MOSFET device operates as a micro-switch. When a voltage is applied to the gate, strong inversion is created in the region under the gate. Strong inversion implies that the minority carriers are attracted to the region under the gate. The minority carriers in the p-region (electrons) are also the majority

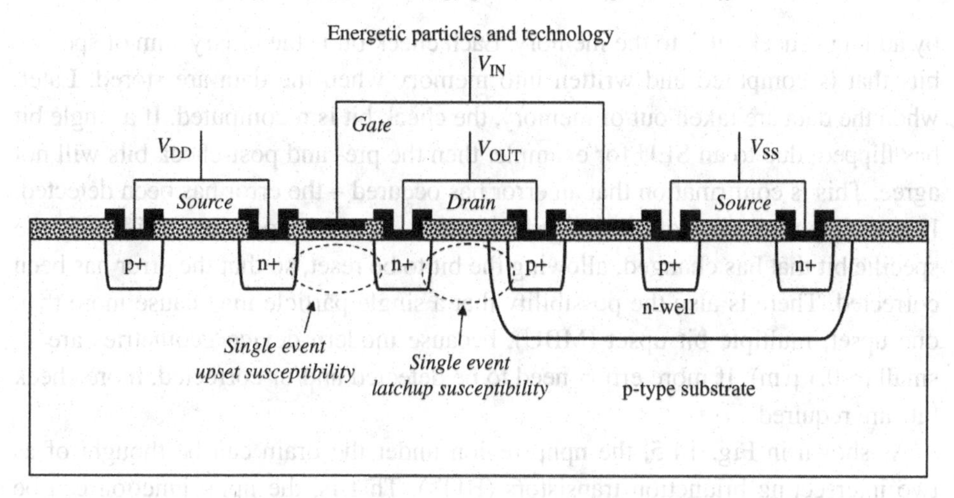

Fig. 14.5. Metal-oxide semiconductor field-effect transistor (MOSFET).

carriers in the n-regions (the source and drain). When inversion occurs the electrons under the gate allow current to flow from the source to the drain, closing the circuit. In this manner, applying an input voltage to the gate acts like a switch closing the circuit from source to drain.

As previously discussed, the most likely result of a charged particle interacting with matter would be energy transfer from the incident particle to the target material, which – if the incident particle were of sufficient energy – would result in many atoms in the target being ionized. This routinely occurs with the passage of GeV energy particles.

If a GeV energy particle were to strike the area under the gate of the MOSFET as shown, an abundance of ionized particles (i.e. free charge carriers) could be created under the gate. This could electrically connect the source and drain. The result would be equivalent to turning the gate current on, and the state of the device would change. For example, a memory bit could flip from 1 to 0. This is known as a single-event upset (SEU): A single particle passed through the device (the event) and changed the logical state (the upset).

Single-event effects are dependent on the geometry of the electronic devices and the voltages under which they operate. Smaller devices, operating at lower voltages, should be more susceptible. As the trend to smaller and smaller devices, often referred to as Moore's law, has been known for some time, it is perhaps not surprising to note that single-event effects have been predicted for almost 50 years (Wallmark and Marcus, 1962). Today, SEU are expected to occur regularly on most spacecraft, and robust error detection and correction (EDAC) techniques are incorporated to both detect these errors when they occur, and correct their effects, by resetting the upset bits to their original value. Most EDAC software codes operate by adding "check bits" to the memory. Each check bit is the binary sum of specific bits that is computed and written into memory when the data are stored. Later, when the data are taken out of memory, the check bit is recomputed. If a single bit has flipped, due to an SEU for example, then the pre- and post-check bits will not agree. This is confirmation that an error has occured – the error has been detected. If a sufficient number of check bits are added, it will be possible to identify the specific bit that has changed, allowing the bit to be reset, so that the error has been corrected. There is also the possibility that a single particle may cause more than one upset, multiple bit upset (MBU), because modern device geometries are so small ($\sim$0.1 μm). If more errors need to be detected and/or corrected, more check bits are required.

As shown in Fig. 14.5, the npnp region under the drain can be thought of as two intersecting bijunction transistors (BJTs). That is, the npnp junction can be treated as a single npn transistor colocated with a pnp transitor. The BJT can also operate as a micro-switch, when the emitter–base and base–collector junctions are properly biased. In this case, the base–collector of the first npn transistor is also the collector–base of the pnp transitor. The result can be a current loop that is the equivalent of feedback in a speaker system. The device becomes preoccupied in feeding current back into itself and is said to be "latched up" because it will remain in this feedback state until powered off. Like the SEU, the latchup may not permanently damage the device, but unlike the SEU it is of greater concern in that the device will remain non-responsive until the latchup state is detected, and corrected by the user.

While the SEU may be a transient event, and latchup may be a temporary event, some single-event effects are permanent. A single particle may deposit so much energy in an electronic device that the insulating layer under the gate is ruptured (single-event gate rupture) leading to permanent failure of the device. The single-event effects are very dependent on the design of the device, and expensive testing at radiation facilities is the best source of data on their susceptibility. However, modeling tools such as Cosmic Ray Effects on Micro-Electronics (CREME) can also be used to predict device susceptibilities, at least to first order (Tylka *et al.*, 1997). The name of this tool indicates that it is often the few particles in the galactic cosmic-ray environment that are the main source of single-event effects (Fig. 14.6), but the energetic particles seen in SPEs are also a significant concern (Fig. 14.7). In this manner, SPEs are often the most significant radiation concern for a spacecraft, being a contributing factor in both total-dose and single-event effects.

For total-dose effects, the dose versus depth curve was instrumental in estimating total-dose effects. For single-event effects, the key parameter is linear energy transfer (LET), the amount of energy deposited in the material (electronic part) per unit path length. An example of an LET plot is shown in Fig. 14.8. LET in this diagram

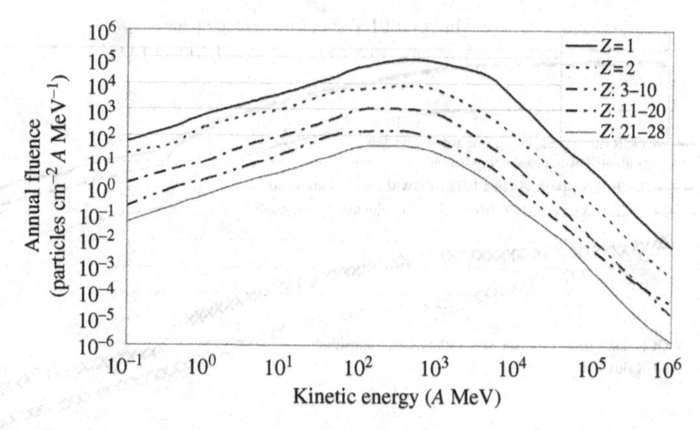

Fig. 14.6. The galactic cosmic-ray annual fluence (Z is the atomic mass number; A MeV is energy per nucleon). (From Wilson *et al.*, 1997.)

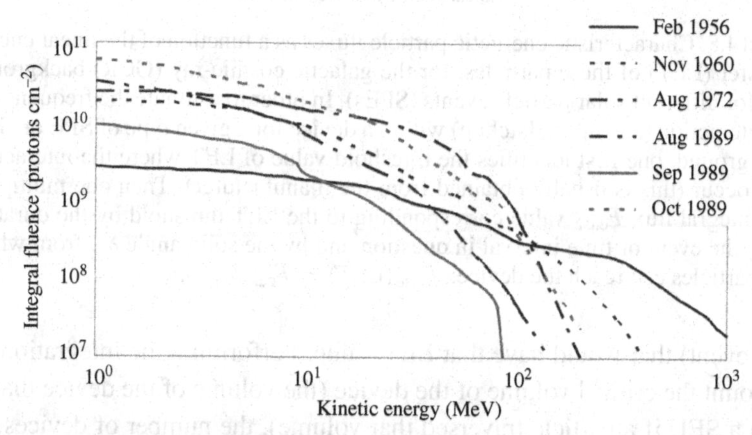

Fig. 14.7. Examples of total fluence of particles from "worst case" solar particle events. From Wilson *et al.* (1997).

is presented in units that are independent of the absorbing material; multiplying by the density of the absorbing material, (for example, silicon is 2.33 g/cm^3) gives energy per unit path length. Note first the significant differences between the various curves, depending on the nature of the solar event in question. The lowest curve, the galactic cosmic ray flux, can be thought of as the noise floor; the environment is never better than this continual, low-level flux. The higher curves occur during a solar event, but are highly dependent on the nature of the event itself. To use the LET curve to predict system effects, the designer needs to have detailed knowledge of the part itself, and some means (usually provided by the part manufacturer) of knowing the critical LET value when a particular SEU may occur. For example, if upsets occur at a specific LET value, then the LET graph can be used to predict the number of particles present in the environment (during a particular type

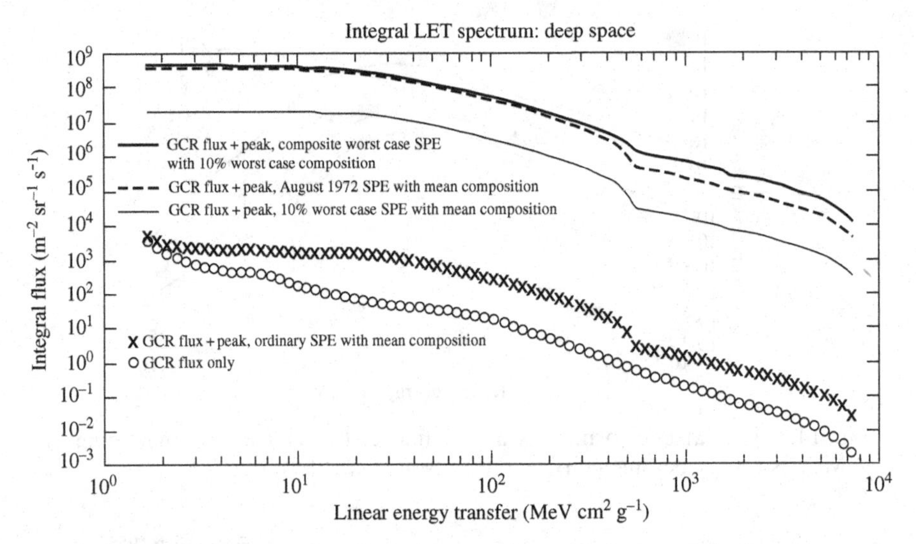

Fig. 14.8. Characteristic energetic particle fluxes as a function of the linear energy transfer (LET) of these particles, for the galactic cosmic-ray (GCR) background and for different solar particle events (SPEs). In order to estimate the frequency of an interaction (e.g. upset, latchup) within a device for a given type of SPE or GCR background, one first identifies the threshold value of LET where the interaction will occur (this is usually obtained from the manufacturer). Then one multiplies the integral flux F_{case} value corresponding to the LET threshold by the duration δt of the event or time interval in question and by the solid angle $\delta\omega$ from which the particles can reach the device: $f_{\text{case}}(\text{LET}) = F_{\text{case}}\delta t\,\delta\omega$.

of solar event) that would have that LET value. Performing the integration to take into account the critical volume of the device (the volume of the device that would result in a SEU if a particle traversed that volume), the number of devices, and so on, generates a prediction of the total number of events that would be expected. If this value is acceptably low, no further action may be required. For example, if the resulting prediction of SEU rates is low enough that the spacecraft's EDAC mechanisms can be expected to detect and correct these errors, then the design is judged sufficient for the environnment. On the other hand, if the prediction indicates that latchup, or permanent failures like gate rupture, may occur, then the design is re-examined to determine if additional shielding, redundancy, different electronic parts, etc. are required.

14.6 Modeling the GCR/SPE environment

In addition to the trapped-radiation belts, modeling the high-energy galactic cosmic-ray/solar-proton environment is also a major concern. Several models of proton fluence are available (Gussenhoven *et al.*, 1988; Feynman *et al.*, 1993;

Wilson *et al.*, 1997; Xapsos *et al.*, 1999, 2000), but the largest uncertainty is how large the events will be. Large events, which generate large numbers of very energetic particles, may be extremely damaging, while small events, which generate small numbers of less energetic particles, may have very little effect. A key decision point in the design of a spacecraft is – how large an event must the spacecraft be designed to accommodate? At the same time, depending on the nature of the mission, the spacecraft may be required simply to "survive" the event, in which case it may go off line into a "safe" mode and then return to operation later, or it may be required to "operate through" the event. Requirements to operate through solar proton events, or even hostile radiation events (i.e. nuclear weapons), will force the designer to implement a much more robust, and costly, solution.

Appendix I Authors and editors

Tim Bastian
National Radio Astronomy Observatory (NRAO)
520 Edgemont Road
Charlottesville, VA 22903

Terry Forbes
Space Science Center
University of New Hampshire
Morse Hall
8 College Road
Durham, NH 03824-2600

Timothy Fuller-Rowell
CIRES University of Colorado and NOAA Space Weather Prediction Center
325 Broadway
Boulder, CO 80303

Joe Giacalone
Theoretical Astrophysics Program
University of Arizona
Tucson, AZ 85721

George Gloeckler
Atmospheric, Oceanic & Space Sciences
University of Michigan
2237 Space Research Building
Ann Arbor, MI 48109

Janet C. Green
NOAA Space Weather Prediction Center
325 Broadway
Boulder, CO 80305

Stephen Guetersloh
Department of Nuclear Engineering
Texas A&M University
3133 TAMU
College Station, TX 77843-2133

Hugh Hudson
Space Sciences Laboratory
University of California, Berkeley
7 Gauss Way
Berkeley, CA 94720-7450

Dietmar Krauss-Varban
Space Sciences Laboratory
University of California, Berkeley
7 Gauss Way
Berkeley, CA 94720-7450

Sten Odenwald
NASA Goddard Space Flight Center
Code 612.4
Greenbelt, MD 20771

Merav Opher
Physics and Astronomy Department
George Mason University
4400 University Drive
Fairfax, VA 22030-4444

Carolus J. Schrijver (editor)
Solar and Astrophysics Laboratory
Lockheed Martin
3251 Hanover Street, Bldg. 252
Palo Alto, CA 94304-1191

George L. Siscoe (editor)
Boston University
725 Commonwealth Avenue
Boston, MA 02215

Stanley C. Solomon
NCAR High Altitude Observatory
3450 Mitchell Lane
Boulder, CO 80301

Alan Tribble
Advanced Technology Center
Rockwell Collins
400 Collins Road N.E.
Cedar Rapids, IA 52498

Vytenis Vasyliūnas
Max Planck Institute for Solar System Research
Max-Planck-Str. 2
Katlenburg-Lindau 37191
Germany

Neal Zapp
Space Radiation Analysis Group (SRAG)
Johnson Space Center
2400 NASA Road 1
Houston, TX 77058

List of illustrations

List of tables

References

10 CFR 835, *Occupational Radiation Protection*
29 CFR 1910, *Occupational Safety and Health Standards*
29 CFR 1960.18, *Supplemental Standards*
42 USC 5801: 1974, *Energy Reorganization Act of 1974*
Abbett, W. P.: 2007, ApJ 665, 1469, doi:10.1086/519788
Abbett, W. P. & Fisher, G. H.: 2003, ApJ 582, 475
Adams, W.: 1881, Nature, November 17.
Adams, Jr., J. H., Silberberg, R., & Tsao, C. H.: 1981, NASA STI/Recon Technical Report 4506 81, 34134
Adams, L., Harboe Sorensen, R., Holmes Siedle, A. G., Ward, A. K., & Bull, R.: 1991, IEEE Trans. Nucl. Sci. NS 38(6), 1686
Akabane, K.: 1956, Publ. Ast. Soc. Japan 8, 173
Akasofu, S. I.: 1964, Planet. Space Sci. 12, 273
Akasofu, S. I.: 1977, *Physics of Magnetospheric Substorms*, D. Reidel Publishing Co.
Akasofu, S.-I.: 1979, SPh 64, 333
Akasofu, S.-I.: 2001, Space Sci. Rev. 95, 613
Akasofu, S.-I. & Chapman, S.: 1972, *Solar-Terrestrial Physics: An Account of the Wave and Particle Radiations from the Quiet and the Active Sun, and of the Consequent Terrestrial Phenomena*, The International Series of Monographs on Physics, Clarendon Press
Albert, J. M.: 2003, JGR (Space Phys.) 108, 1249, doi:10.1029/2002JA009792
Alexander, R. C. & Brown, J. C.: 2002, SPh 210, 407, doi:10.1023/A:1022465615445
Alfvén, H. & Carlquist, P.: 1967, SPh 1, 220
Aly, J. J.: 1991, ApJ 375, L61
Amari, T., Luciani, J. F., & Aly, J. J.: 2005, ApJ 629, L37
Amari, T., Luciani, J. F., Aly, J. J., Mikić, Z., & Linker, J. A.: 2003, ApJ 585, 1073
Amari, T., Luciani, J. F., Mikić, Z., & Linker, J. A.: 2000, ApJ 529, L49
Amenomori, M., Ayabe, S., Caidong, Danzengluobu, *et al.*: 2000, ApJ 541, 1051, doi:10.1086/309479
AMS: 2008, Bull. Amer. Meteor. Soc. 89
Anderson, B., Decker, R., Paschalidis, N., & Sarris T.: 1997, JGR 102(A8), 17553
Andrews, G. B., Zurbuchen, T. H., Mauk, B. H., *et al.*: 2007, Space Sci. Rev. 131, 523, doi:10.1007/s11214-007-9272-5
ANSI: 2003, American National Standard Guide to Reference and Standard Atmosphere Models, ANSI/AIAA G-003-1990.

Antiochos, S. K., DeVore, C. R., & Klimchuk, J. A.: 1999, ApJ 510, 485
Antonucci, E., Dodero, M. A., Peres, G., Serio, S., & Rosner, R.: 1987, ApJ 322, 522
Archontis, V., Hood, A. W., & Brady, C.: 2007, A & A 466, 367
Arge, C. N. & Pizzo, V. J.: 2000, JGR 105, 10465, doi:10.1029/1999JA900262
Armstrong, T. W. & Colborn, B. L.: 2000a, *Trapped Radiation Model Uncertainties: Model-Data and Model-Model Comparisons*, NASA/CR-2000-210071
Armstrong, T. W. & Colborn, B. L.: 2000b, *Evaluation of Trapped Radiation Model Uncertainties for Spacecraft Design*, NASA/CR-2000-210072
Arnoldy, R. L., Kane, S. R., & Winkler, J. R.: 1968, in K. O. Kiepenheuer (ed.), *Structure and Development of Solar Active Regions, IAU Symp. 35*, 490–509, Reidel
Asai, A., Yokoyama, T., Shimojo, M., *et al.*: 2004a, ApJ 611, 557, doi:10.1086/422159
Asai, A., Yokoyama, T., Shimojo, M., & Shibata, K.: 2004b, ApJL 605, L77, doi:10.1086/420768
Aschwanden, M. J.: 1998, ApJ 502, 455, doi:10.1086/305890
Aschwanden, M. J.: 2004, *Physics of the Solar Corona. An Introduction*, Praxis
Aschwanden, M. J. & Alexander, D.: 2001, SPh 204, 91, doi:10.1023/A:1014257826116
Aschwanden, M. J., Schwartz, R. A., & Dennis, B. R.: 1998, ApJ 502, 468, doi:10.1086/305891
Aschwanden, M. J., Fletcher, L., Sakao, T., Kosugi, T., & Hudson, H.: 1999, ApJ 517, 977, doi:10.1086/307230
Aschwanden, M. J., Stern, R. A., & Güdel, M.: 2008, ApJ 672, 659, doi:10.1086/523926
Athay, R. G. & Moreton, G. E.: 1961, ApJ 133, 935
Aurass, H., Klein, K.-L., & Martens, P. C. H.: 1994, SPh 155, 203
Avrett, E. H.: 1981, in R. M. Bonnet & A. K. Dupree (eds.), *NATO ASIC Proc. 68: Solar Phenomena in Stars and Stellar Systems*, p. 173
Axford, I. W.: 1981, in *International Cosmic Ray Conference*, Vol. 12, p. 155
Axford, W. I.: 1985, SPh 100, 575
Axford, W. I., Leer, E., & Skadron, G.: 1977, in *International Cosmic Ray Conference*, Vol. 11, p. 132
Bai, T. & Ramaty, R.: 1978, ApJ 219, 705, doi:10.1086/155830
Baker, D.: 2000, IEEE Trans. Plasma Sci. 28, 6
Baker, D. N.: 2001, in I. A. Daglis (ed.), *Space Storms and Space Weather Hazards*, Kluwer, p. 285
Baker, D. N., Kanekal, S. G., Blake, J. B., & Pulkkinen, T. I.: 2001, JGR 106, 19169, doi:10.1029/2000JA003023
Bale, S. D., Mozer, F. S., & Horbury, T. S.: 2003, Phys. Rev. Lett. 91(26), 265004, doi:10.1103/PhysRevLett.91.265004
Bale, S. D., Reiner, M. J., Bougeret, J.-L., *et al.*: 1999, GRLe 26, 1573, doi:10.1029/1999GL900293
Bame, S. J., Hundhausen, A. J., Asbridge, J. R., & Strong, I. B.: 1968, Phys. Rev. Lett. 20, 393
Barghouty, A. F. & Jokipii, J. R.: 1996, ApJ 470, 858
Barnes, G., Leka, K. D., Schumer, E. A., & Della-Rose, D. J.: 2007, Space Weather 5, S09002
Bastian, T. S.: 2001, Ap&SS 277, 107, doi:10.1023/A:1012232111843
Bastian, T. S., Benz, A. O., & Gary, D. E.: 1998, A&A 36, 131, doi:10.1146/annurev.astro.36.1.131
Bastian, T. S., Pick, M., Kerdraon, A., Maia, D., & Vourlidas, A.: 2001, ApJL 558, L65, doi:10.1086/323421
Bastian, T. S., Fleishman, G. D., & Gary, D. E.: 2007, ApJ 666, 1256, doi:10.1086/520106
Basu, S., Kudeki, E., Basu, S., *et al.*: 1996, JGR 101, 26795, doi:10.1029/96JA00760

Basu, S., Groves, K. M., Yeh, H., *et al.*: 2001, *AGU Spring Meeting Abstracts* 61
Bateman, G.: 1978, *MHD Instabilities*, MIT Press.
Baumjohann, W. & Treumann, R. A.: 1996, *Basic Space Plasma Physics*, Imperial College Press
Bedingfield, K. L., Leach, R.D., & Alexander, M.B.: 1996, NASA Reference Publication 1390
Behannon, K. W., Lepping, R. P., Sittler, Jr., E. C., *et al.*: 1987, JGR 92, 15354
Bell, A. R.: 1978a, MNRAS 182, 147
Bell, A. R.: 1978b, MNRAS 182, 443
Benka, S. G. & Holman, G. D.: 1992, ApJ 391, 854, doi:10.1086/171394
Benz, A. (ed.): 2002, *Plasma Astrophysics, second edition*, Vol. 279 of Astrophysics and Space Science Library, Kluwer
Benz, A. & Murdin, P.: 2000, *Solar Flare Observations*
Berger, T. E., Shine, R. A., Slater, G. L., *et al.*: 2008, ApJ 676, L89, doi:10.1086/587171
Bieber, J. W. & Matthaeus, W. H.: 1997, ApJ 485, 655
Biesecker, D. A., Myers, D. C., Thompson, B. J., Hammer, D. M., & Vourlidas, A.: 2002, ApJ 569, 1009
Billings, D. E.: 1966, *A Guide to the Solar Corona*, Academic Press
Binder, D., Smith, E. C., & Holman, A. B.: 1975, IEEE Trans. Nucl. Sci., NS-22
Birn, J., Thomsen, M. F., Borovsky, J. E., *et al.*: 1998, JGR 103, 9235, doi:10.1029/97JA02635
Birn, J., Forbes, T. G., & Hesse, M.: 2006, ApJ 645, 732
Blanc, M. & Richmond, A. D.: 1980, JGR 85, 1669, doi:10.1029/JA085iA04p01669
Blanco-Cano, X., Omidi, N., & Russell, C. T.: 2006, JGR (Space Phys.) 111(A10), 10205, doi:10.1029/2005JA011421
Blandford, R. D. & Ostriker, J. P.: 1978, ApJL 221, L29, doi:10.1086/182658
Boggs, S. E., Coburn, W., & Kalemci, E.: 2006, ApJ 638, 1129, doi:10.1086/498930
Bolton, S. J., Gulkis, S., Klein, M. J., de Pater, I., & Thompson, T. J.: 1989, JGR 94, 121, doi:10.1029/JA094iA01p00121
Bonifazi, C. & Moreno, G.: 1981, JGR 86, 4397, doi:10.1029/JA086iA06p04397
Bornmann, P. L. & Shaw, D.: 1994, SPh 150, 127, doi:10.1007/BF00712882
Boyd, T. J. M. & Sanderson, J. J.: 2003, *The Physics of Plasmas*, Cambridge University Press
Bratenahl, A. & Baum, P. J.: 1976, SPh 47, 345
Braun, D. C. & Lindsey, C.: 2000, SPh 192, 285
Braun, D. C. & Lindsey, C.: 2001, ApJL 560, L189, doi:10.1086/324323
Brekke, P.: 2004, in I. Daglis (ed.), *Effects of Space Weather on Technology Infrastructure*, Kluwer
Brice, N.: 1964, JGR 69, 4515, doi:10.1029/JZ069i021p04515
Brosius, J. W. & White, S. M.: 2006, ApJL 641, L69, doi:10.1086/503774
Brown, J. C.: 1971, SPh 18, 489, doi:10.1007/BF00149070
Brown, J. C.: 1972, SPh 26, 441, doi:10.1007/BF00165286
Brown, J. C., Kontar, E. P., & Veronig, A. M.: 2007, in K.-L. Klein & A. L. MacKinnon (eds.), *The High Energy Solar Corona: Waves, Eruptions, Particles*, Vol. 725 of Lecture Notes in Physics, Springer Verlag, p. 65
Brown, J. C., Emslie, A. G., Holman, G. D., Johns-Krull, C. M., Kontar, E. P., Lin, R. P., Massone, A. M., & Piana, M.: 2006, ApJ 643, 523, doi:10.1086/501497
Buonsanto, M. J.: 1999, Space Sci. Rev. 88, 563, doi:10.1023/A:1005107532631
Burbank, J. E.: 1905, Terrestrial Magnetism 10
Burgess, D.: 1989, GRLe 16, 345, doi:10.1029/GL016i005p00345

Burkepile, J. T., Hundhausen, A. J., Stanger, A. L., St. Cyr, O. C., & Seiden, J. A.: 2004, JGR (Space Phys.) 109(A18), 3103, doi:10.1029/2003JA010149
Burlaga, L. F.: 1995, *Interplanetary Magnetohydrodynamics*, Oxford University Press
Burlaga, L. F., Ness, N. F., Acuña, M. H., *et al.*: 2005, Science 309, 2027, doi:10.1126/science.1117542
Burns, A. G., Killeen, T. L., & Roble, R. G.: 1991, JGR 96, 14153, doi:10.1029/91JA00678
Burrows, D. N., Hill, J. E., Nousek, J. A., *et al.*: 2005, Space Sci. Rev. 120, 165, doi:10.1007/s11214-005-5097-2
Burton, R. K., McPherron, R. L., & Russell, C. T.: 1975, JGR 80, 4204
Cai, M. J., Shang, H., Lin, H.-H., & Shu, F. H.: 2008, ApJ672, 489, doi:10.1086/523788
Cane, H. V., Stone, R. G., Fainberg, J., *et al.*: 1981, GRLe 8, 1285, doi:10.1029/GL008i012p01285
Cane, H. V., Reames, D. V., & von Rosenvinge, T. T.: 1988, JGR 93, 9555, doi:10.1029/JA093iA09p09555
Cane, H. V., von Rosenvinge, T. T., & McGuire, R. E.: 1990, JGR 95, 6575, doi:10.1029/JA095iA05p06575
Canfield, R. C., Hudson, H. S., & McKenzie, D. E.: 1999, GRLe 26, 627, doi:10.1029/1999GL900105
Cargill, P. J. & Klimchuk, J. A.: 2004, ApJ 605, 911, doi:10.1086/382526
Cargill, P. J. & Priest, E. R.: 1982, SPh 76, 357, doi:10.1007/BF00170991
Cargill, P. J., Mariska, J. T., & Antiochos, S. K.: 1995, ApJ 439, 1034
Carovillano, R. L. & Siscoe, G. L.: 1973, Rev. Geophys. Space Phys. 11, 289
Carrington, R. C.: 1859, MNRAS 20, 13
Cassak, P. A., Shay, M. A., & Drake, J. F.: 2005, Phys. Rev. Lett. 95(23)
Cassak, P. A., Drake, J. F., & Shay, M. A.: 2007, Phys. Plasmas 14, doi 054502
Cattell, C., Wygant, J. R., Goetz, K., *et al.*: 2008, GRLe 35, 1105, doi:10.1029/2007GL032009
Chapman, S.: 1969, in B. M. McCormac & A. Omholt (eds.), *Atmospheric Emissions*, Van Nostrand Reinhold Co., p. 11
Chapman, S. & Bartels, J.: 1940, *Geomagnetism*, Oxford University Press
Chen, J.: 1989, ApJ 338, 453
Chen, P. F. & Shibata, K.: 2000, ApJ 545, 524
Chen, P. F., Shibata, K., & Yokoyama, T.: 2001, Earth, Planets and Space 53, 611
Chen, Y., Reeves, G. D., & Friedel, R. H. W.: 2007, Nature Phys. 3, 614, doi:10.1038/nphys655
Chenette, D. L., Cooper, J. F., Eraker, J. H., Pyle, K. R., & Simpson, J. A.: 1980, JGR 85, 5785, doi:10.1029/JA085iA11p05785
Cheng, C. Z.: 2004, Space Sci. Rev. 113, 207
Chifor, C., Tripathi, D., Mason, H. E., & Dennis, B. R.: 2007, A&A 472, 967, doi:10.1051/0004-6361:20077771
Cho, M. & Nozaki, Y.: 2005, J. Spacecr. Rockets 42 (4), 740
Christe, S., Hannah, I. G., Krucker, S., McTiernan, J., & Lin, R. P.: 2008, ApJ 677, 1385, doi:10.1086/529011
Chupp, E. L., Forrest, D. J., Higbie, P. R., *et al.*: 1973, Nature 241, 333, doi:10.1038/241333a0
Ciaravella, A. & Raymond, J. C.: 2008, ApJ 686, 1372, doi:10.1086/590655
Ciaravella, A., Raymond, J. C., & Kahler, S. W.: 2006, ApJ 652, 774, doi:10.1086/507171
Clarke, J. T., Grodent, D., Cowley, S. W. H., *et al.*: 2004, in F. Bagenal, T. Dowling, & W. McKinnon (eds.), *Jupiter: The Planet, Satellites and Magnetosphere*, Cambridge University Press, p. 639
Cliver, E. W.: 1995, SPh 157, 285

Cliver, E. W., Kahler, S. W., Shea, M. A., & Smart, D. F.: 1982, ApJ 260, 362, doi:10.1086/160261
Cliver, E. W., Dennis, B. R., Kiplinger, A. L., *et al.*: 1986, ApJ 305, 920, doi:10.1086/164306
Cliver, E. W., Nitta, N. V., Thompson, B. J., & Zhang, J.: 2004, SPh 225, 105, doi:10.1007/s11207-004-3258-1
Cohen, M. H.: 1960, ApJ 131, 664, doi:10.1086/146878
Cohen, C.: 2006, Adv. Space Res. 38, 389, doi:10.1016/j.asr.2006.09.022
Cohen, C. M. S., Stone, E. C., & Selesnick, R. S.: 2001, JGR 106, 29871, doi:10.1029/2001JA000008
Connoly, J. F.: 2004, *Estimating the Integrated Radiation Dose for a Conjunction-Class Mars Mission Using Early MARIE Data*, ASCE, Ninth Biennial Conference of the Aerospace Division
Costello, K. A.: 1997, Ph.D. thesis, Rice University, Houston, TX
Coster, A. J., Foster, J., & Erikson, P.: 2003, Space Weather, GPS World 14(5), 42
Cowley, S. W. H.: 1977, Planet. Space Sci. 26, 539
Cowley, S. W. H. & Bunce, E. J.: 2001, Planet. Space Sci. 49, 1067
Crockett, W. R., Purcell, J. D., Schumacher, R. J., Tousey, R., & Patterson, N. P.: 1977, Appl. Opt. 16, 893
Crosby, N. B., Aschwanden, M. J., & Dennis, B. R.: 1993, SPh 143, 275
Crowley, G. & Meier, R .R.: 2008, in M. Kintner, A. J. Coster, T. J. Fuller-Rowell, *et al.* (eds.), *Midlatitude Ionospheric Dynamics and Disturbances*, AGU Geophysical Monograph 181, p. 221
Crowley, G., Schoendorf, J., Roble, R. G., & Marcos, F. A.: 1996, JGR 101, 211, doi:10.1029/95JA02584
Culhane, J. L., Harra, L. K., James, A. M., *et al.*: 2007, SPh 243, 19, doi:10.1007/s01007-007-0293-1
Davis, L. J. & Chang, D. B.: 1962, JGR 67, 2169, doi:10.1029/JZ067i006p02169
de Hoffmann, F. & Teller, E.: 1950, Phys. Rev. 80, 692
de Jager, C.: 1975, Space Sci. Rev. 17, 645, doi:10.1007/BF00727574
de Pater, I. & Goertz, C. K.: 1990, JGR 95, 39, doi:10.1029/JA095iA01p00039
de Pater, I., Schulz, M., & Brecht, S. H.: 1997, JGR 102, 22043, doi:10.1029/97JA00311
Decker, R. B., Krimigis, S. M., Roelof, E. C., *et al.*: 2005, Science 309, 2020, doi:10.1126/science.1117569
Delannée, C., Török, T., Aulanier, G., & Hochedez, J.-F.: 2008, SPh 247, 123, doi:10.1007/s11207-007-9085-4
Dellinger, J. H.: 1935, Physical Review 48, 705, doi:10.1103/PhysRev.48.705
Dennis, B. R.: 1988, SPh 118, 49, doi:10.1007/BF00148588
Dennis, B. R. & Zarro, D. M.: 1993, SPh 146, 177
Dere, K. P., Brueckner, G. E., Howard, R. A., *et al.*: 1997a, SPh 175, 601, doi:10.1023/A:1004907307376
Dere, K. P., Landi, E., Mason, H. E., Monsignori Fossi, B. C., & Young, P. R.: 1997b, A&ASS 125, 149, doi:10.1051/aas:1997368
Dere, K. P., Landi, E., Young, P. R., *et al.*: 2009, A&A 498, 915, doi:10.1051/0004-6361/200911712
Desai, M. I., Mason, G. M., Dwyer, J. R., *et al.*: 2000, JGR 105, 61, doi:10.1029/1999JA900406
Dessler, A. J.: 1980, Icarus 44, 291
Dessler, A. J. & Parker, E. N.: 1959, JGR 24, 2239
Dodson, H. W. & Hedeman, E. R.: 1968, SPh 4, 229

Dorman, L. I.: 2005, Ann. Geophys. 23, 2997
Doschek, G. A. & Warren, H. P.: 2005, ApJ 629, 1150
Drake, J. F.: 1971, SPh 16, 152, doi:10.1007/BF00154510
Drake, F. D. & Hvatum, S.: 1959, AJ 64, 329, doi:10.1086/108047
Drury, L. O.: 1983, Rep. Prog. Phys. 46, 973, doi:10.1088/0034-4885/46/8/002
Dryer, M., Wu, S. T., Steinolfson, R. S., & Wilson, R. M.: 1979, ApJ 227, 1059
Dubey, G., van der Holst, B., & Poedts, S.: 2006, A & A 459, 927
Dulk, G. A.: 1985, ARA&A 23, 169, doi:10.1146/annurev.aa.23.090185.001125
Dulk, G. A. & Marsh, K. A.: 1982, ApJ 259, 350, doi:10.1086/160171
Dulk, G. A., Leblanc, Y., Bastian, T. S., & Bougeret, J.-L.: 2000, JGR 105, 27343, doi:10.1029/2000JA000076
Duncan, R. C.: 2005, Sky & Telescope 109(1), 010000
Duncan, R. C. & Thompson, C.: 1996, in R. E. Rothschild & R. E. Lingenfelter (ed.), *High Velocity Neutron Stars*, Vol. 366 of *American Institute of Physics Conference Series*, p. 111
Dwyer, J. R., Mason, G. M., Mazur, J. E., *et al.*: 1997, ApJ 490, L115
Earl, J. A.: 1974, ApJ 193, 231
Eather, R. H.: 1980, *Majestic Lights*, American Geophysical Union
Echer, E., Gonzalez, W. D., Vieira, L. E. A., *et al.*: 2003, Braz. J. Phys. 33, 115
Egeland, A.: 1984, in *Magnetospheric Currents*, AGU Geophysical Monograph 28, p. 1
Elkington, S. R., Hudson, M. K., & Chan, A. A.: 1999, GRLe 26, 3273, doi:10.1029/1999GL003659
Elkington, S. R., Hudson, M. K., & Chan, A. A.: 2003, JGR (Space Phys.) 108, 1116, doi:10.1029/2001JA009202
Ellis, W.: 1879, J. Soc. Telegraphic Engineers and Electricians 8, 214
Ellison, D. C. & Ramaty, R.: 1985, ApJ 298, 400, doi:10.1086/163623
Emmert, J. T., Fejer, B. G., Fesen, C. G., Shepherd, G. G., & Solheim, B. H.: 2001, JGR 106, 24701, doi:10.1029/2000JA000372
Emmert, J. T., Fejer, B. G., Shepherd, G. G., & Solheim, B. H.: 2002, JGR (Space Phys.) 107, 1483, doi:10.1029/2002JA009646
Emmert, J. T., Fejer, B. G., Shepherd, G. G., & Solheim, B. H.: 2004, GRLe 31, 22807, doi:10.1029/2004GL021611
Emslie, A. G. & Brown, J. C.: 1985, ApJ 295, 648, doi:10.1086/163408
Emslie, A. G. & Sturrock, P. A.: 1982, SPh 80, 99
Emslie, A. G., Miller, J. A., & Brown, J. C.: 2004, ApJL 602, L69, doi:10.1086/382350
Emslie, A. G., Dennis, B. R., Holman, G. D., & Hudson, H. S.: 2005, JGR (Space Phys.) 110(A9), 11103, doi:10.1029/2005JA011305
Emslie, A. G., Bradsher, H. L., & McConnell, M. L.: 2008, ApJ 674, 570, doi:10.1086/524983
Encrenaz, Th., Bibring, J.-P., & Blanc, M.: 2003, *The Solar System*, Springer
Engvold, O.: 1989, in E. R. Priest (ed.), *Dynamics and Structure of Quiescent Solar Prominences*, Vol. 150 of Astrophysics and Space Science Library, Kluwer, p. 47
Eschard, G. & Manley, B. W.: 1971, Acta Electronica 14, 19
Evans, D. S., Fuller-Rowell, T. J., Maeda, S., & Foster, J.: 1988, in P. F. Wercinski (ed.), *Astrodynamics 1987*, p. 1649
Evans, R. M., Opher, M., Manchester, IV, W. B., & Gombosi, T. I.: 2008, ApJ 687, 1355, doi:10.1086/592016
Eviatar, A. & Siscoe, G. L.: 1980, GRLe 7, 1085
Falcone, A., Atkins, R., Benbow, W., *et al.*: 2003, ApJ 588, 557, doi:10.1086/373950

Falconer, D. A., Moore, R. L., & Gary, G. A.: 2003, JGR (Space Phys.) 108, 1380, doi:10.1029/2003JA010030
Falthammer, C.-G.: 1965, JGR 70, 2503
Falthammer, C.-G.: 1966, JGR 71, 1487
Fan, Y. & Gibson, S. E.: 2004, ApJ 609, 1123
Fan, Y. & Gibson, S. E.: 2007, ApJ 668, 1232
Fan, Y., Zweibel, E. G., Linton, M. G., & Fisher, G. H.: 1999, ApJ 521, 460
Fárník, F. & Savy, S. K.: 1998, SPh 183, 339
Fárník, F., Hudson, H., & Watanabe, T.: 1996, SPh 165, 169, doi:10.1007/BF00149096
Fárník, F., Hudson, H. S., Karlický, M., & Kosugi, T.: 2003, A&A 399, 1159, doi:10.1051/0004-6361:20021852
Farthing, W. H., Brown, J. P., & Bryant, W. C.: 1982, NASA Tech. Memo 83908
Fei, Y., Chan, A. A., Elkington, S. R., & Wiltberger, M. J.: 2006, JGR (Space Phys.) 111(A10), 12209, doi:10.1029/2005JA011211
Fejer, B. G. & Emmert, J. T.: 2003, JGR (Space Phys.) 108, 1454, doi:10.1029/2003JA010190
Fejer, B. G. & Kelley, M. C.: 1980, Rev. Geophys. Space Phys. 18, 401
Fejer, B. G. & Scherliess, L.: 1997, JGR 102, 24047, doi:10.1029/97JA02164
Fejer, B. G., Kelley, M. C., Senior, C., de La Beaujardiere, O., & Lepping, R.: 1990, JGR 95, 2367, doi:10.1029/JA095iA03p02367
Fejer, B. G., Emmert, J. T., & Sipler, D. P.: 2002, JGR (Space Phys.) 107, 1052, doi:10.1029/2001JA000300
Fejer, B. G., Jensen, J. W., Kikuchi, T., Abdu, M. A., & Chau, J. L.: 2007, JGR (Space Phys.) 112(A11), 10304, doi:10.1029/2007JA012376
Feldman, U., Laming, J. M., & Doschek, G. A.: 1995, ApJL 451, L79+, doi:10.1086/309695
Fennell, J. F., Koons, H. C., Chen, M. W., & Blake, J. B.: 2000, IEEE Trans. Plasma Phys. 28, 2029
Ferraro, V. C. A. & Plumpton, C.: 1966, *An Introduction to Magneto-Fluid Mechanics*, Clarendon Press
Ferris, D. L.: 2001, *Characterization of Operator-Reported Discrepancies in Unmanned On-Orbit Space Systems*, Masters Thesis, Department of Aeronautics and Astronautics, Massachusetts Institute of Technology
Fesen, C. G., Crowley, G., Roble, R. G., Richmond, A. D., & Fejer, B. G.: 2000, GRLe 27, 1851, doi:10.1029/2000GL000061
Feynman, J., Spitale, G., Wang, J., & Gabriel, S.: 1993, JGR 98, 13281, doi:10.1029/92JA02670
Field, P. R. & Rishbeth, H.: 1997, J. Atmos. Sol.-Terr. Physics 59, 163, doi:10.1016/S1364-6826(96)00085-5
Filippov, B. P.: 2001, Astron. Astrophys. Trans. 20, 445
Fishman, G. J. & Inan, U. S.: 1988, Nature 331, 418, doi:10.1038/331418a0
Fisk, L. A. & Gloeckler, G.: 2006, ApJ 640, L79
Fisk, L. A. & Gloeckler, G.: 2008, ApJ 686, 1466
Fisk, L. A. & Lee, M. A.: 1980, ApJ 237, 620
Fisk, L. A., Kozlovsky, B., & Ramaty, R.: 1974, ApJ 190, L35
Fleishman, G. D. & Melnikov, V. F.: 2003a, ApJ 587, 823, doi:10.1086/368252
Fleishman, G. D. & Melnikov, V. F.: 2003b, ApJ 584, 1071, doi:10.1086/345849
Fleishman, G. D., Bastian, T. S., & Gary, D. E.: 2008, ApJ 684, 1433, doi:10.1086/589821
Fletcher, L. & Hudson, H.: 2001, SPh 204, 69, doi:10.1023/A:1014275821318
Fletcher, L. & Hudson, H. S.: 2008, ApJ 675, 1645, doi:10.1086/527044

Fletcher, L., Hannah, I. G., Hudson, H. S., & Metcalf, T. R.: 2007, ApJ 656, 1187, doi:10.1086/510446
Forbes, T. G.: 1991, Geophys. Astrophys. Fluid Dyn. 62, 15
Forbes, T. G.: 2000, JGR 105, 23153
Forbes, T. G.: 2003, Adv. Space Res. 32, 1043
Forbes, T. G. & Acton, L. W.: 1996, ApJ 459, 330
Forbes, T. G. & Isenberg, P. A.: 1991, ApJ 373, 294
Forbes, T. G. & Priest, E. R.: 1995, ApJ 446, 377
Forbes, J. M., Lu, G., Bruinsma, S., Nerem, S., & Zhang, X.: 2005, JGR (Space Phys.) 110(A9), 12, doi:10.1029/2004JA010856
Forbes, T. G., Linker, J. A., Chen, J., *et al.*: 2006, Space Sci. Rev. 123, 251
Forman, M. A.: 1977, Astrophys. Space Sci. 49, 83
Forman, M. A. & Webb, G. M.: 1985, in *Collisionless Shocks in the Heliosphere: A Tutorial Review*, American Geophysical Union, p. 91
Forman, M. A., Jokipii, J. R., & Owens, A. J.: 1974, ApJ 192, 535
Foster, J. C.: 2008, in P. M. Kintner, A. J. Coster, T. J. Fuller-Rowell, *et al.* (eds.), *Mid-Latitude Dynamics and Disturbances*, AGU Geophysical Monograph 181
Foster, J. C. & Coster, A. J.: 2007, J. Atmos. Sol.-Terr. Phys. 69, 1241, doi:10.1016/j.jastp.2006.09.012
Foster, J. C., Erickson, P. J., Coster, A. J., Goldstein, J., & Rich, F. J.: 2002, GRL 29(13), 130000, doi:10.1029/2002GL015067
Foster, J. C., Coster, A. J., Erickson, P. J., *et al.*: 2005, AGU Geophysical Monograph 159, p. 277
Fraser, G. W. & Mathieson, E.: 1981, Nucl. Instrum. Methods 179, 591
Freeman, T. J. & Parks, G. K.: 2000, JGR 105, 15715, doi:10.1029/1999JA900501
Friedel, R. H. W., Reeves, G. D., & Obara, T.: 2002, J. Atmos. Terr. Phys. 64
Friedman, H., Lichtman, S. W., & Byram, E. T.: 1951, Physical Review 83, 1025, doi:10.1103/PhysRev.83.1025
Frisch, P. C.: 1996, Space Sci. Rev. 78, 213
Frost, K. J. & Dennis, B. R.: 1971, ApJ 165, 655
Fuller-Rowell, T. J.: 1995, in R. M. Johnson & T. L. Killeen (eds.), *The Upper Mesosphere and Lower Thermosphere: A Review of Experiment and Theory*, AGU Geophysical Monograph 87, p. 23
Fuller-Rowell, T. J. & Rees, D.: 1984, Planet. Space Sci. 32, 69, doi:10.1016/0032-0633(84)90043-6
Fuller-Rowell, T. J., Codrescu, M. V., Moffett, R. J., & Quegan, S.: 1994, JGR 99, 3893, doi:10.1029/93JA02015
Fuller-Rowell, T. J., Codrescu, M. V., Risbeth, H., Moffett, R. J., & Quegan, S.: 1996a, JGR 101, 2343, doi:10.1029/95JA01614
Fuller-Rowell, T. J., Codrescu, M. V., Roble, R. G., & Richmond, A. D.: 1997, AGU Geophysical Monograph Chapman Conference on Magnetic Storms, 98, 203
Fuller-Rowell, T. J., Millward, G. H., Richmond, A. D., & Codrescu, M. V.: 2002, J. Atmos. Solar-Terr. Phys. 64,1383
Fuller-Rowell, T. J., Rees, D., Quegan, S., *et al.*: 1996b, in R. W. Schunk (ed.), *Handbook of Ionospheric Models, STEP Report*, p. 217
Fuller-Rowell, T. J., Richmond, A. D., & Maruyama, N.: 2008, in P. M. Kintner, A. J. Coster, T. J. Fuller-Rowell, *et al.* (eds.), *Mid-Latitude Dynamics and Disturbances*, AGU Geophysical Monograph 181, p. 187

Fuselier, S. A.: 1994, in M. J. Engebretson, K. Takahashi, & M. Scholer (eds.), *Solar Wind Sources of Magnetospheric Ultra-Low-Frequency Waves*, Washington, DC: AGU p. 107
Futron Corporation: 2003, www.satelliteonthenet.co.uk/white/futron5.html
Gaizauskas, V.: 1989, SPh 121, 135
Galinsky, V. L. & Shevchenko, V. I.: 2000, Phys. Rev. Lett. 85, 90, doi:10.1103/PhysRevLett.85.90
Gallagher, P. T., Dennis, B. R., Krucker, S., Schwartz, R. A., & Tolbert, A. K.: 2002, SPh 210, 341, doi:10.1023/A:1022422019779
Gannon, J. L., Li, X., & Temerin, M.: 2005, JGR (Space Phys.) 110(A9), 12206, doi:10.1029/2004JA010679
Garrett, H. B., Levin, S. M., Bolton, S. J., Evans, R. W., & Bhattacharya, B.: 2005, GRLe 32, 4104, doi:10.1029/2004GL021986
Gary, D. E. & Keller, C. U. (eds.): 2004, *Solar and Space Weather Radiophysics: Current Status and Future Developments*, Vol. 314 of Astrophysics and Space Science Library, Kluwer
Gary, G. A.: 1989, ApJ Suppl. 69, 323, doi:10.1086/191316
Gendrin, R.: 1981, Rev. Geophys. Space Phys. 19, 171
Gerard, E.: 1970, A&A 8, 181
Gerard, E.: 1976, A&A 50, 353
Giacalone, J.: 2004, GRLe 21, 2441
Giacalone, J.: 2005, in G. Li, G. P. Zank, & C. T. Russell (eds.), *The Physics of Collisionless Shocks: 4th Annual IGPP International Astrophysics Conference*, Vol. 781 of American Institute of Physics Conference Series, p. 213
Giacalone, J., Burgess, D., Schwartz, S. J., Ellison, D. C., & Bennett, L.: 1997, JGR 102, 19789, doi:10.1029/97JA01529
Giacalone, J. & Jokipii, J. R.: 1999, ApJ 520, 204
Giacalone, J. & Jokipii, J. R.: 2004, ApJ 616, 573
Giacalone, J., Jokipii, J.R., & Kota, J.: 2002, ApJ 573, 845
Giacalone, J., Jokipii, J. R., & Mazur, J. E.: 2000, ApJ 532, L75
Gibbons, P. E. & Blamires, N. G.: 1965, J. Sci. Instrum. 42, 862
Gibson, S. E. & Fan, Y.: 2006, JGR (Space Phys.) 111(A10), 12103, doi:10.1029/2006JA011871
Gibson, S. E. & Low, B. C.: 1998, ApJ 493, 460
Gilbert, H. R., Alexander, D., & Liu, R.: 2007, SPh 245, 287
Gloeckler, G.: 1970, in *Introduction to Experimental Techniques of High-Energy Astrophysics*, NASA SP-243, p. 1
Gloeckler, G. & Geiss, J.: 1996, Nature 381, 210, doi:10.1038/381210a0
Gloeckler, G. & Hsieh, K. C.: 1979, Nucl. Instrum. Methods 165, 537
Gloeckler, G., Geiss, J., Balsiger, H., *et al.*: 1992, A&ASS 92, 267
Gloeckler, G., Balsiger, H., Bürgi, A., *et al.*: 1995, Space Sci. Rev. 71, 79, doi:10.1007/BF00751327
Gloeckler, G., Cain, J., Ipavich, F. M., *et al.*: 1998, Space Sci. Rev. 86, 497, doi:10.1023/A:1005036131689
Gloeckler, G., Fisk, L. A., Mason, G. M., & Hill, M. E.: 2008, in *American Institute of Physics Conference Series*, Vol. 1039, p. 367
Goedbloed, J. P. H. & Poedts, S.: 2004, *Principles of Magnetohydrodynamics*, Cambridge University Press
Goldstein, J., Sandel, B. R., Hairston, M. R., & Reiff, P. H.: 2003, GRLe 30(24), 240000, doi:10.1029/2003GL018390

Gonzalez, W. D.: 1990, Planet. Space Sci. 38, 627
Gonzalez, W. D., Joselyn, J. A., Kamide, Y., *et al.*: 1994, JGR 99, 5771
Gopalswamy, N., Lara, A., Kaiser, M. L., & Bougeret, J.-L.: 2001, JGR 106, 25261, doi:10.1029/2000JA004025
Gopalswamy, N., Yashiro, S., Michałek, G., *et al.*: 2002, ApJL 572, L103, doi:10.1086/341601
Gopalswamy, N., Yashiro, S., Lara, A., *et al.*: 2003, GRL 30(12), 120000, doi:10.1029/2002GL016435
Gordon, B. E., Lee, M. A., Möbius, E., & Trattner, K. J.: 1999, JGR 104, 28263, doi:10.1029/1999JA900356
Gosling, J. T.: 1993, JGR 98, 18937
Gosling, J. T. & Robson, A.E.: 1985, in B. Tsurutani & R. Stone (eds.), *Collisionless Shocks in the Heliosphere: Reviews of Current Research*, American Geophysical Union, p. 141
Gosling, J. T., Thomsen, M. F., Bame, S. J., *et al.*: 1982, GRLe 9, 1333, doi:10.1029/GL009i012p01333
Gosling, J. T., Bame, S. J., Feldman, W. C., *et al.*: 1984, JGR 89, 5409, doi:10.1029/JA089iA07p05409
Gray, P. C., Jr., Pontius, D. H., & Matthaeus, W. H.: 1996, GRL 23, 965
Green J. C., Onsager, T. G., O'Brien, T. P., & Baker, D. N.: 2004, JGR 109, A12211, doi:10.1029/2004JA010579.
Green, J. C.: 2006, in K. Takahashi, P. J. Chi, R. E. Denton, & R. L. Lysak (eds.), *Magnetospheric ULF Waves: Synthesis and New Directions*, AGU Geophysical Monograph 169, p. 225
Green, J. L. & Boardsen, S.: 2006, Adv. Space Res. 38, 130, doi:10.1016/j.asr.2005.08.054
Green, J. C. & Kivelson, M. G.: 2004, JGR (Space Phys.) 109(A18), 3213, doi:10.1029/2003JA010153
Groth, C. P. T., De Zeeuw, D. L., Gombosi, T. I., & Powell, K. G.: 2000, JGR 105, 25053, doi:10.1029/2000JA900093
Gruntman, M.: 1997, Rev. Sci. Instrum. 68, 3617
Guedel, M., Benz, A. O., Schmitt, J. H. M. M., & Skinner, S. L.: 1996, ApJ 471, 1002, doi:10.1086/178027
Guetersloh, S., Zeitlin, C., Heilbronn, L., *et al.*: 2006, NIMB 252, 319
Gulkis, S. & Gary, B.: 1971, AJ 76, 12, doi:10.1086/111075
Guo, W. P. & Wu, S. T.: 1998, ApJ 494, 419
Gurnett, D. A. & Bhattacharjee, A.: 2005, *Introduction to Plasma Physics*, Cambridge University Press
Gussenhoven, M. S., Brautigam, D. H., & Mullen, E. G.: 1988, IEEE Trans. Nucl. Sci., 35(6), 1412
Haerendel, G.: 2006, Space Sci. Rev. 124, 317, doi:10.1007/s11214-006-9092-z
Haggerty, D. K. & Roelof, E. C.: 2002, ApJ 579, 841, doi:10.1086/342870
Hale, G. E.: 1930, ApJ 71, 73
Hall, D. T., Feldman, P. D., McGrath, M. A., & Strobel, D. F.: 1998, ApJ 499, 475, doi:10.1086/305604
Hall, E. J.: 1994, *Radiobiology for the Radiologist*, J. B. Lippincott
Hannah, I. G., Christe, S., Krucker, S., *et al.*: 2008, ApJ 677, 704, doi:10.1086/529012
Hansen, R. T., Garcia, C. J., Hansen, S. F., & Yasukawa, E.: 1974, Publ. Astron. Soc 86, 500
Hara, H., Watanabe, T., Matsuzaki, K., *et al.*: 2008, Publ. Astron. Soc. Japan 60, 275
Harvey, K. L. & Recely, F.: 1984, Solar Phys. 91, 127

Harvey, K. L., Sheeley, N. R., & Harvey, J. W.: 1986, *Solar-Terrestrial Predictions*, Vol. 2, Meudon, France, 18–22 June, 1984, P. A. Simon *et al.*, (eds.), p. 198
Harwit, M.: 1981, *Cosmic Discovery: The Search, Scope, and Heritage of Astronomy*, Harvester Press, 1981
Hastings, D. & Garrett, H. (eds.): 1996, *Spacecraft–Environment Interactions*, Cambridge University Press
Hathaway, D. H., Wilson, R. M., & Reichmann, E. J.: 1994, SPh 151, 177, doi:10.1007/BF00654090
Haug, E.: 1972, SPh 25, 425, doi:10.1007/BF00192340
Haug, E.: 1975, SPh 45, 453, doi:10.1007/BF00158461
Haug, E.: 1985, Phys. Rev. D 31, 2120, doi:10.1103/PhysRevD.31.2120
Haug, E.: 1998, SPh 178, 341
Haug, E.: 2003, A&A 406, 31, doi:10.1051/0004-6361:20030782
Hawley, S. L. & Fisher, G. H.: 1994, ApJ 426, 387, doi:10.1086/174075
Hawley, S. L., Fisher, G. H., Simon, T., *et al.*: 1995, ApJ 453, 464, doi:10.1086/176408
Hayes, A. P., Vourlidas, A., & Howard, R. A.: 2001, ApJ 548, 1081
Heelis, R. A.: 2004, J. Atmos. Solar-Terr. Phys. 66, 825, doi:10.1016/j.jastp.2004.01.034
Heelis, R. A., Sojka, J. J., David, M., & Schunk, R. W.: 2009, JGR (Space Phys.) 114(A13), 3315, doi:10.1029/2008JA013690
Hellinger, P., Trávnícek, P., & Matsumoto, H.: 2002, Geophys. Res. Lett. 29(24), 240000, doi:10.1029/2002GL015915
Hénoux, J.-C.: 1986, in V. E. Stepanov and V. N. Obridko (eds.), *Solar Maximum Analysis*, VNU Science Press, p. 109
Heristchi, D.: 1986, ApJ 311, 474, doi:10.1086/164787
Hey, J. S., Parsons, S. J., & Phillips, J. W.: 1948, MNRAS 108, 354
Heyvaerts, J.: 1974, SP 38, 419
Heyvaerts, J., Priest, E. R., & Rust, D. M.: 1977, ApJ 216, 123
Hiei, E., Hundhausen, A. J., & Sime, D. G.: 1993, Geophys. Res. Lett. 20, 2785
Hill, T. W.: 1979, in B. Battrick and J. Mort (eds.), *Magnetospheric Boundary Layers*, ESA SP-148, Noordwijk p. 325
Hill, T. W.: 2001, JGR 106, 8101
Hill, T. W., Dessler, A. J., & Rasbach, M. E.: 1983, Planet. Space. Sci. 31, 1187
Hill, T. W., Thomsen, M. F., Henderson, M. G., *et al.*: 2008, JGR 113(A01214), doi:10.1029/2007JA012626
Hines, C. O.: 1963, Planet. Space Sci. 10, 239
Hirayama, T.: 1974, SPh 34, 323
Holman, G. D., Sui, L., Schwartz, R. A., & Emslie, A. G.: 2003, ApJL 595, L97, doi:10.1086/378488
Hones, E. W., Jr.: 1977, JGR 82, 5633
Hori, K., Yokoyama, T., Kosugi, T., & Shibata, K.: 1997, ApJ 489, 426
Hori, K., Yokoyama, T., Kosugi, T., & Shibata, K.: 1998, ApJ 500, 492
Horne, R. B.: 2001, *Space Weather Parameters Required by Users*, ESA Space Weather Program Study, WP1300 and WP1400
Horne, R. B., Thorne, R. M., Shprits, Y. Y., *et al.*: 2005, Nature 437, 227, doi:10.1038/nature03939
Horne, R. B., Thorne, R. M., Glauert, S. A., *et al.*: 2008, Nature Phys. 4, 301
Howard, R. A., Sheeley, Jr., N. R., Michels, D. J., & Koomen, M. J.: 1985, JGR 90, 8173, doi:10.1029/JA090iA09p08173
Hoyos, S. E., Evans, H. D. R., & Daly, E.: 2004, IEEE Trans. Nucl. Sci. 51, 2927
Hu, Y. Q., Li, G. Q., & Xing, X. Y.: 2003, JGR 108, 1072

Huang, T. S. & Hill, T. W.: 1989, JGR 94, 3761
Hudson, H. S.: 1972, SPh 24, 414, doi:10.1007/BF00153384
Hudson, H. S.: 1978, ApJ 224, 235, doi:10.1086/156370
Hudson, H. S.: 1987, SPh 113, 1
Hudson, H. S.: 1988, A&A 26, 473, doi:10.1146/annurev.aa.26.090188.002353
Hudson, H. S.: 1991, SPh 133, 357
Hudson, H. S.: 2000, ApJL 531, L75, doi:10.1086/312516
Hudson, H. S.: 2007a, in P. Heinzel, I. Dorotovič, and R. J. Rutten (eds.), *The Physics of Chromospheric Plasmas*, Vol. 368 of Astronomical Society of the Pacific Conference Series, p. 365
Hudson, H. S.: 2007b, ApJL 663, L45, doi:10.1086/519923
Hudson, H. S. & Cliver, E. W.: 2001, JGR 106, 25199, doi:10.1029/2000JA004026
Hudson, H. S. & Micela, G.: 2006, in *14th Cambridge Workshop on Cool Stars, Stellar Systems and the Sun* (poster)
Hudson, H. S. & Ohki, K.: 1972, SPh 23, 155, doi:10.1007/BF00153899
Hudson, H., Haisch, B., & Strong, K. T.: 1995a, JGR 100, 3473
Hudson, H. S., Acton, L. W., Alexander, D., *et al.*: 1995b, in *Solar Wind Eight*, American Institute of Physics, p. 58
Hudson, H. S., Labonte, B. J., Sterling, A. C., & Watanabe, T.: 1998a, in *Observational Plasma Astrophysics : Five Years of YOHKOH and Beyond*
Hudson, M. K., Kotelnikov, A. D., Li, X., *et al.*: 1995c, GRL 22, 291, doi:10.1029/95GL00009
Hudson, M. K., Elkington, S. R., Lyon, J. G., *et al.*: 1997, JGR 102, 14087, doi:10.1029/97JA03995
Hudson, M. K., Marchenko, V. A., Roth, I., et al.: 1998b, Adv. Space Res. 21(4), 597
Hudson, H. S., Kosugi, T., Nitta, N. V., & Shimojo, M.: 2001, ApJL 561, L211, doi:10.1086/324760
Hudson, H. S., Khan, J. I., Lemen, J. R., Nitta, N. V., & Uchida, Y.: 2003, SPh 212, 121, doi:10.1023/A:1022904125479
Hudson, M. K., Kress, B. T., Mazur, J. E., Perry, K. L., & Slocum, P. L.: 2004, J. Atmos. Solar-Terr. Phys. 66/15-16, 1389, doi:doi:10.1016/j.jastp.2004.03.024
Hudson, H. S., Wolfson, C. J., & Metcalf, T. R.: 2006, SPh 234, 79, doi:10.1007/s11207-006-0056-y
Hudson, H. S., Hannah, I. G., DeLuca, E. E., & Weber, M.: 2008, in *First Results from Hinode*, ASP Conference Series 397, 130
Hundhausen, A. J.: 1988, in V. Pizzo, T. Holzer, and D. G. Simes (eds.), *Proceedings of the Sixth International Solar Wind Conference*, 181–241, NCAR/TN 306, Boulder
Hundhausen, A.: 1999, in K. T. Strong, J. L. R. Saba, B. M. Haisch, and J. T. Schmelz (eds.), *The Many Faces of the Sun: A Summary of the Results from NASA's Solar Maximum Mission*, Springer, p. 143
Hundhausen, A. J., Burkepile, J. T., & St. Cyr, O. C.: 1994, JGR 99, 6543
Hunsucker, R. D.: 1982, Rev. Geophys. 20, 293, doi:10.1029/RG020i002p00293
Hurford, G. J., Schwartz, R. A., Krucker, S., *et al.*: 2003, ApJL 595, L77, doi:10.1086/378179
Hurford, G. J., Krucker, S., Lin, R. P., *et al.*: 2006, ApJL 644, L93, doi:10.1086/505329
Hurley, K., Boggs, S. E., Smith, D. M., *et al.*: 2005, Nature 434, 1098, doi:10.1038/nature03519
Hurricane, O. A., Fong, B. H., Cowley, S. C., *et al.*: 1998, in S. Kokubun and Y. Kamide (Eds.), *Substorms-4*, Terra Scientific Publishing Company and Kluwer Academic Publishers, p. 373

Hysell, D. L., Kudeki, E., & Chau, J. L.: 2005, Ann. Geophys. 23, 2647
ICRP: 1959, *Recommendations of the International Commission on Radiological Protection, (adopted 1958), ICRP Publication 1*, Pergamon Press
Ikushima, T., Aritomi, H., & Morisita, J.: 1996, Mutation Res. 358, 193
Inan, U. S., Lehtinen, N. G., Lev-Tov, S. J., Johnson, M. P., Bell, T. F., & Hurley, K.: 1999, Geophys. Res. Lett. 26, 3357, doi:10.1029/1999GL010690
Ingraham, J. C., Cayton, T. E., Belian, R. D., *et al.*: 2001, JGR 106, 25759, doi:10.1029/2000JA000458
Inoue, S. & Kusano, K.: 2006, ApJ 645, 742
Ipavich, F. M., Lundgren, R. A., Lambird, B. A., & Gloeckler, G.: 1978, Nucl. Instrum. Methods 154, 291
Ipavich, F. M., Galvin, A. B., Gloeckler, G., Scholer, M., & Hovestadt, D.: 1981, JGR 86, 4337, doi:10.1029/JA086iA06p04337
Ipavich, F. M., Ma Sung, L. S., & Gloeckler, G.: 1982, *Technical Report TR-82-172*, University of Maryland
Isbell, J., Dessler, A. J., & Waite, Jr., J. H.: 1984, JGR 89, 10716
Isenberg, P. A.: 1997, JGR 102, 4719
Isenberg, P. A. & Forbes, T. G.: 2007, ApJ 670, 1453
Isenberg, P. A. & Jokipii, J. R.: 1979, ApJ 234, 746
Isenberg, P. A., Forbes, T. G., & Démoulin, P.: 1993, ApJ 417, 368
Isola, C., Favata, F., Micela, G., & Hudson, H. S.: 2007, A&A 472, 261, doi:10.1051/0004-6361:20077643
Jackman, C. M., Russell, C. T., Southwood, D. J., *et al.*: 2007, GRLe 34(L11203), doi:10.1029/2007GL029764
Jacobs, C., Poedts, S., & van der Holst, B.: 2006, A&A 450, 793
Jacobs, C., Poedts, S., van der Holst, B., Dubey, G., & Keppens, R.: 2007a, in C. Dumitrache, V. Mioc, & N. A. Popescu (eds.), *Flows, Boundaries, and Interactions*, AIP Conference Proceedings 934, p. 101
Jacobs, C., van der Holst, B., & Poedts, S.: 2007b, A&A 470, 359
Jokipii, J. R.: 1966, ApJ 146, 480
Jokipii, J. R.: 1986, JGR 91, 2929
Jokipii, J. R. & Coleman, P. J.: 1968, JGR 73, 5495
Jokipii, J. R. & Giacalone, J.: 2004, ApJ 605, L145
Jokipii, J. R. & Parker, E. N.: 1969, ApJ 155, 777
Jokipii, J. R. & Thomas, B. T.: 1981, ApJ 243, 1115
Jokipii, J. R., Levy, E. H., & Hubbard, W. B.: 1977, ApJ 213, 861
Jokipii, J. R., Kota, J., & Giacalone, J.: 1993, GRL 20, 1759
Jokipii, J. R., Giacalone, J., & Kota, J.: 2004, ApJ 611, L141
Jones, F. C. & Ellison, D. C.: 1991, Space Sci. Rev. 58, 259, doi:10.1007/BF01206003
Jones, F. C., Jokipii, J. R., & Baring, M. G.: 1998, ApJ 509, 238
Kahler, S. W.: 1982, JGR 87, 3439
Kahler, S. W.: 1992, A&A 30, 113, doi:10.1146/annurev.aa.30.090192.000553
Kahler, S. W. & Hundhausen, A. J.: 1992, JGR 97, 1619, doi:10.1029/91JA02402
Kahler, S.: 1994, ApJ 428, 837, doi:10.1086/174292
Kahler, S. W., Sheeley, Jr., N. R., Howard, R. A., *et al.*: 1984, JGR 89, 9683, doi:10.1029/JA089iA11p09683
Kai, K., Melrose, D. B., & Suzuki, S.: 1985, in D. J. McLean & N. R. Labrun (eds.), *Solar Radiophysics*, Cambridge University Press, p. 415
Kaiser, M. L., Desch, M. D., & Farrell, W. M.: 1993, Planet. Space Sci. 41, 1073
Kallenrode, M.-B.: 1998, *Space Physics*, Springer-Verlag Berlin Heidelberg New York

Kallenrode, M.-B.: 2004, *Space Physics: An Introduction to Plasmas and Particles in the Heliosphere and Magnetospheres*, Springer
Kamide, Y., Baumjohann, W., Daglis, I. A., *et al.*: 1998, JGR 103, 17,705
Kan, J. R., Akasofu, S.-I., & Lee, L. C: 1983, SPh 84, 153
Kane, S. R. & Anderson, K. A.: 1970, ApJ 162, 1003
Kane, S. R. & Donnelly, R. F.: 1971, ApJ 164, 151
Kappenman, J.: 2004, in I. A. Daglis (ed.), *Effects of Space Weather on Technology Infrastructure*, Kluwer Academic Publishers, p. 257
Kappenman, J., Albertson, V. D., & Mohan, N.: 1981, IEEE PAS Trans. PAS-100, 1078
Karimabadi, H. & Krauss-Varban, D.: 1992, in G. P. Zank & T. K. Gaisser (eds.), *Particle Acceleration in Cosmic Plasmas*, AIP Conference Proceedings 264, p. 118
Karimabadi, H., Krauss-Varban, D., & Terasawa, T.: 1992, JGR 97, 13853, doi:10.1029/92JA00997
Karimabadi, H., Krauss-Varban, D., & Omidi, N.: 1995, JGR 100, 11957, doi:10.1029/94JA03035
Kaufmann, P., Raulin, J.-P., de Castro, C. G. G., *et al.*: 2004, ApJL 603, L121, doi:10.1086/383186
Kelley, M. C.: 1989, *The Earth's Ionosphere: Plasma Physics and Electrodynamics*, Elsevier
Kelley, M. C., Fejer, B. G., & Gonzales, C. A.: 1979, GRLe 6, 301, doi:10.1029/GL006i004p00301
Kelley, M. C., Makela, J. J., Chau, J. L., & Nicolls, M. J.: 2003, GRLe 30(4), 040000, doi:10.1029/2002GL016321
Kellogg, P. J.:1959, Van Allen radiation of solar origin, Nature,183, 1295
Kennel, C. F.: 1995, *Convection and Substorms*, Oxford University Press
Kennel, C. F., Edmiston, J. P., & Hada, T.: 1985, AGU Geophysical Monograph 34, p. 1
Kennel, C. F., Coroniti, F. V., Scarf, F. L., *et al.*: 1986, JGR 91, 11917, doi:10.1029/JA091iA11p11917
Khan, J. I. & Aurass, H.: 2002, A&A 383, 1018, doi:10.1051/0004-6361:20011707
Killeen, T. L., Hays, P. B., Carignan, G. R., *et al.*: 1984, JGR 89, 7495, doi:10.1029/JA089iA09p07495
Killeen, T. L., Ponthieu, J.-J., Craven, J. D., Frank, L. A., & Spencer, N. W.: 1988, JGR 93, 2675, doi:10.1029/JA093iA04p02675
Kim, H.-J. & Chan, A. A.: 1997, JGR 102, 22107
Kim, H.-J., Chan, A. A., Wolf, R. A., & Birn, J.: 2000, JGR 105, 7721, doi:10.1029/1999JA900465
Kim H.-J., Rostoker, G., & Kamide, Y.: 2002, JGR 107(A11), 1378, doi:10.1029/2001JA007513
Kim, M.-H.Y., Cucinotta, F.A., & Wilson, J.W.: 2006, Radiation Meas. 41(9-10), 1115
Kiplinger, A. L.: 1995, ApJ 453, 973, doi:10.1086/176457
Klecker, B.: 1995, Space Sci. Rev. 72, 419
Klein, K.-L.: 1987, A&A 183, 341
Klein, M. J., Gulkis, S., & Stelzried, C. T.: 1972, ApJL 176, L85, doi:10.1086/181026
Kleinknecht, K.: 1998, *Detectors for Particle Radiation*, 2nd edn., Cambridge University Press
Kletzing, C. A.: 1994, JGR 99, 11095, doi:10.1029/94JA00345
Kliem, B. & Török, T.: 2006, Phys. Rev. Lett. 96(25), 255002
Kliem, B., Titov, V. S., & Török, T.: 2004, A&A 413, L23
Knoll, G. F.: 2000, *Radiation Detection and Measurement*, 3rd edn., Wiley

Koch, H. W. & Motz, J. W.: 1959, Rev. Mod. Phys. 31, 920, doi:10.1103/RevModPhys.31.920
Kolmogorov, A.: 1941, Dokl. Akad. Nauk SSSR 30, 301
Kontar, E. P., Emslie, A. G., Massone, A. M., *et al.*: 2007, ApJ 670, 857, doi:10.1086/521977
Kontar, E. P., MacKinnon, A. L., Schwartz, R. A., & Brown, J. C.: 2006, A&A 446, 1157, doi:10.1051/0004-6361:20053672
Koons, H. C., Mazur, J. E., Selesnick, R. S., *et al.*: 2000, in *Aerospace Report No. TR-99(1670)-1*, The Aerospace Corporation, El Segundo, CA.
Koskinen, H. E. J. & Tanskanen, E. I.: 2002, JGR 107(A11), 1415, doi:10.1029/2002JA009283
Kosovichev, A. G. & Zharkova, V. V.: 1998, Nature 393, 317
Kosovichev, A. G. & Zharkova, V. V.: 1999, SPh 190, 459, doi:10.1023/A:1005226802279
Kota, J.: 2000, JGR 105, 2403
Kota, J. & Jokipii, J. R.: 1983, ApJ 265, 573
Kota, J. & Jokipii, J. R.: 2000, ApJ 531, 1067
Kóta, J., Manchester, W. B., Jokipii, J. R., de Zeeuw, D. L., & Gombosi, T. I.: 2005, in G. Li, G. P. Zank, & C. T. Russell (eds.), *The Physics of Collisionless Shocks: 4th Annual IGPP International Astrophysics Conference*, Vol. 781 of American Institute of Physics Conference Series, p. 201
Krakowski, H.: 2008, *Evolving Aviation Technologies and the Need for Space Weather Prediction and Specification*, Space Weather Enterprise Forum, May 21, 2008, National Press Club, Washington, DC.
Krall, J., Chen, J., & Santoro, R.: 2000, ApJ 539, 964
Krauss-Varban, D. & Karimabadi, H.: 2006, in G. P. Zank & N. V. Pogorelov (eds.), *Numerical Modeling of Space Plasma Flows*, Vol. 359 of Astronomical Society of the Pacific Conference Series, p. 264
Krauss-Varban, D. & Omidi, N.: 1991, JGR 96, 17715, doi:10.1029/91JA01545
Krauss-Varban, D. & Welsch, B. T.: 2007, Highlights Astron. 14, 89, doi:10.1017/S1743921307009921
Krauss-Varban, D. & Wu, C. S.: 1989, JGR 94, 15367, doi:10.1029/JA094iA11p15367
Krauss-Varban, D., Burgess, D., & Wu, C. S.: 1989, JGR 94, 15089, doi:10.1029/JA094iA11p15089
Krauss-Varban, D., Omidi, N., & McKean, M. E.: 1994, in J. D. Brown, M. T. Chu, D. C. Ellison, & R. J. Plemmons (eds.), *Proceedings of the Cornelius Lanczos International Centenary Conference*, SIAM, Philadelphia, PA, p. 464
Krauss-Varban, D., Karimabadi, H., & Omidi, N.: 1995, JGR 100, 11981, doi:10.1029/94JA03034
Krauss-Varban, D., Li, Y., & Luhmann, J. G.: 2008, in G. Li, Q. Hu, O. Verkhoglyadova, (eds.), *Particle Acceleration and Transport in the Heliosphere and Beyond*, Vol. 1039 of American Institute of Physics Conference Series, p. 307
Kress, B. T., Hudson, M. K., & Slocum, P. L.: 2005, GRLe 32, 6108, doi:10.1029/2005GL022373
Kress, B. T., Hudson, M. K., Looper, M. D., *et al.*: 2007, JGR (Space Phys.) 112(A11), 9215, doi:10.1029/2006JA012218
Krimigis, S. M., Carbary, J. F., Keath, E. P., *et al.*: 1983, JGR 88, 8871, doi:10.1029/JA088iA11p08871
Krimigis, S. M., Mitchell, D. G., Hamilton, D. C., *et al.*: 2004, Space Sci. Rev. 114, 233, doi:10.1007/s11214-004-1410-8

Krimigis, S. M., Sergis, N., Mitchell, D. G., Hamilton, D. C., & Krupp, N.: 2007, Nature 450, 1050, doi:10.1038/nature06425
Krucker, S., Kontar, E. P., Christe, S., & Lin, R. P.: 2007, ApJL 663, L109, doi:10.1086/519373
Krucker, S., Battaglia, M., Cargill, P. J., *et al.*: 2008, Astron. Astrophys. Rev. 16, 155
Krupp, N., Vasyliūnas, V. M., Woch, J., *et al.*: 2004, in F. Bagenal, T. Dowling, & W. McKinnon (eds.), *Jupiter: The Planet, Satellites and Magnetosphere*, Cambridge University Press, p. 617
Krymskii, G. F.: 1977, Akad. Nauk SSSR Dokl. 234, 1306
Kucharek, H. & Scholer, M.: 1992, in G. P. Zank & T. K. Gaisser (eds.), *Particle Acceleration in Cosmic Plasmas*, Vol. 264 of American Institute of Physics Conference Series, p. 364
Kulsrud, R. M.: 2005, *Plasma Physics for Astrophysics*, Princeton University Press
Kundu, M. R.: 1965, *Solar Radio Astronomy*, Interscience
Kundu, M. R., Raulin, J. P., Nitta, N., *et al.*: 1995, ApJL 447, L135, doi:10.1086/309567
Kurz, E. A.: 1979, Am. Lab. 11, 67
Kwak, Y.-S. & Richmond, A. D.: 2007, JGR (Space Phys.) 112(A11), 1306, doi:10.1029/2006JA011910
Lampton, M.: 1981, Sci. Am. 245, 62
Lampton, M. & Carlson, C. W.: 1979, Rev. Sci. Instrum. 50, 1093
Lampton, M., Siegmund, O., & Raffanti, R.: 1987, Rev. Sci. Instrum. 58, 2298
Landi, E., Del Zanna, G., Young, P. R., *et al.*: 2006, ApJSS 162, 261, doi:10.1086/498148
Lario, D., Hu, Q., Ho, G. C., *et al.*: 2005, in B. Fleck, T. H. Zurbuchen, & H. Lacoste (eds.), *Solar Wind 11/SOHO 16, Connecting Sun and Heliosphere*, Vol. 592 of ESA Special Publication, p. 81
Larosa, T. N., Moore, R. L., & Shore, S. N.: 1994, ApJ 425, 856, doi:10.1086/174031
Lee, M. A.: 1982, JGR 87, 5063, doi:10.1029/JA087iA07p05063
Lee, M. A.: 1983, JGR 88, 6109, doi:10.1029/JA088iA08p06109
Lee, M. A.: 1997, in N. Crooker, J. Joselyn, & J. Feynman (eds.), *Coronal Mass Ejections*, AGU Geophysical Monograph 99
Lee, J. & Gary, D. E.: 2000, ApJ 543, 457
Lee, M. A.: 2005, ApJ Suppl. 158, 38, doi:10.1086/428753
Lee, J. & Gary, D. E.: 2000, ApJ 543, 457, doi:10.1086/317080
Lei, J., Wang, W., Burns, A. G., *et al.*: 2008, JGR (Space Phys.) 113(A12), 1314, doi:10.1029/2007JA012807
Lengyel-Frey, D., Thejappa, G., MacDowall, R. J., Stone, R. G., & Phillips, J. L.: 1997, JGR 102, 2611, doi:10.1029/96JA02871
leRoux, J. A. & Potgieter, M. S.: 1995, JGR 442, 847
Leroy, M. M. & Mangeney, A.: 1984, Ann. Geophys. 2, 449
Lever, E. L., Quest, K. B., & Shapiro, V. D.: 2001, GRLe 28, 1367, doi:10.1029/2000GL012516
Levin, S. M., Bolton, S. J., Gulkis, S. L., *et al.*: 2001, GRLe 28, 903, doi:10.1029/2000GL012087
Li, G., Zank, G. P., & Rice, W. K. M.: 2003, JGR (Space Phys.) 108, 1082, doi:10.1029/2002JA009666
Li, X., Baker, D. N., Temerin, M., Reeves, G. D., & Belian, R. D.: 1998, GRLe 25, 3763, doi:10.1029/1998GL900001
Li, X., Roth, I., Temerin, M., *et al.*: 1993, GRLe 20, 2423, doi:10.1029/93GL02701
Li, X., Baker, D. N., Kanekal, S. G., Looper, M., & Temerin, M.: 2001, GRLe 28, 3827, doi:10.1029/2001GL013586

Lin, H., Kuhn, J. R., & Coulter, R.: 2004, ApJL 613, L177, doi:10.1086/425217
Lin, H., Penn, M. J., & Tomczyk, S.: 2000, ApJL 541, L83, doi:10.1086/312900
Lin, J. & Forbes, T. G.: 2000, JGR 105, 2375
Lin, J., Forbes, T. G., Isenberg, P. A., & Démoulin, P.: 1998, ApJ 504, 1006
Lin, J., Ko, Y.-K., Sui, L., *et al.*: 2005, ApJ 622, 1251
Lin, R. P.: 1974, Space Sci. Rev. 16, 189, doi:10.1007/BF00240886
Lin, R. P. & Hudson, H. S.: 1971, SPh 17, 412, doi:10.1007/BF00150045
Lin, R. P. & Hudson, H. S.: 1976, SPh 50, 153
Lin, Y.: 2003, JGR (Space Phys.) 108, 1390, doi:10.1029/2003JA009991
Lindsey, C., Kopp, G., Clark, T. A., & Watt, G.: 1995, ApJ 453, 511, doi:10.1086/176412
Linker, J. A., Mikić, Z., Lionello, R., & Riley, P.: 2003, Plasma Phys. 10, 1971
Liu, C., Deng, N., Liu, Y., *et al.*: 2005, ApJ 622, 722, doi:10.1086/427868
Liu, H. & Lühr, H.: 2005, JGR (Space Phys.) 110(A9), 9, doi:10.1029/2004JA010908
Liu, Y. C.-M., Opher, M., Cohen, O., Liewer, P. C., & Gombosi, T. I.: 2008, ApJ 680, 757, doi:10.1086/587867
Longcope, D. W.: 1996, SPh 169, 91
Longcope, D. W.: 2001, Phys. Plasmas 8, 5277
Longcope, D. W., Beveridge, C., Qiu, J., *et al.*: 2007, SPh 244, 45
Lorentzen, K. R., Mazur, J. E., Looper, M. D., Fennell, J. F., & Blake, J. B.: 2002, JGR 107(A9), 1231, doi:10.1029/2001JA000276
Lovering, J.: 1857, in *The American Almanac and Repository of Useful Knowledge*, Crosby, Nichols, and Co., p. 67
Low, B. C.: 1996, SPh 167, 217
Low, B. C.: 2001, JGR 106, 25141
Low, B. C. & Smith, D. F.: 1993, ApJ 410, 412
Lu, G., Goncharenko, L. P., Richmond, A. D., Roble, R. G., & Aponte, N.: 2008, JGR (Space Phys.) 113(A12), 8304, doi:10.1029/2007JA012895
Lugaz, N., Manchester, W. B. IV, & Gombosi, T. I.: 2005a, ApJ 627, 1019
Lugaz, N., Manchester, W. B. IV, & Gombosi, T. I.: 2005b, ApJ 634, 651, doi:10.1086/491782
Luhmann, J. G.: 1976, JGR 81, 2089
Lui, A. T. Y.: 1996, JGR 101, 13067
Lui, A. T. Y.: 2004, Space Sci. Rev. 113, 127
Lutz, G.: 1999, *Semiconductor Radiation Detectors, Device Physics*, Springer-Verlag
Lynch, B. J.: 2006, Ph.D. thesis, University of Michigan
Lyons, L. R. & Thorne, R. M.: 1973, JGR 78, 2142
Machado, M. E. & Linsky, J. L.: 1975, SPh 42, 395
Machado, M. E., Moore, R. L., Hernandez, A. M., *et al.*: 1988, ApJ 326, 425, doi:10.1086/166106
Mackay, D. H. & van Ballegooijen, A. A.: 2005, ApJ 621, L77
MacNeice, P., Antiochos, S. K., Phillips, A., *et al.*: 2004, ApJ 614, 1028
Magara, T.: 2001, ApJ 549, 608
Maia, D. J. F., Gama, R., Mercier, C., *et al.*: 2007, ApJ 660, 874, doi:10.1086/508011
Manchester, W. B.: 2003, JGR (Space Phys.) 108, 1162, doi:10.1029/2002JA009252
Manchester, W. B. IV: 2007, ApJ 666, 532
Manchester, W. B., Gombosi, T. I., Roussev, I., *et al.*: 2004a, JGR (Space Phys.) 109(A18), 1102, doi:10.1029/2002JA009672
Manchester, W. B., Gombosi, T. I., Roussev, I., *et al.*: 2004b, JGR (Space Phys.) 109(A18), 2107, doi:10.1029/2003JA010150

Manchester, W. B., Gombosi, T. I., De Zeeuw, D. L., *et al.*: 2005, ApJ 622, 1225, doi:10.1086/427768

Mandea, M. & Balasis, G.: 2006, Geophys. J. Int. 167, 586, doi:10.1111/j.1365-246X.2006.03125.x

Mandell, M. J., Davis, V. A., Cooke, D. L., Wheelock, A. T., & Roth, C. J.: 2006, IEEE Trans. Plasma Sci. 34, 2084

Mandrini, C. H., Demoulin, P., Henoux, J. C., & Machado, M. E.: 1991, A&A 250, 541

Mannucci, A. J., Tsurutani, B. T., Iijima, B. A., *et al.*: 2005, GRLe 32, 12, doi:10.1029/2004GL021467

Mariska, J. T.: 1987, ApJ 319, 465

Marsh, D. R., Solomon, S. C., & Reynolds, A. E.: 2004, JGR (Space Phys.) 109(A18), 7301, doi:10.1029/2003JA010199

Martens, P. C. H. & Zwaan, C.: 2001, ApJ 558, 872

Martin, C., Jelinsky, P., Lampton, M., Malina, R. F., & Anger, H. O.: 1981, Rev. Sci. Instrum. 52, 1067

Martin, S., Panasenco, O., Engvold, O., & Lin, Y.: 2008, Ann. Geophys. 26, 3061

Maruyama, N., Sazykin, S., Spiro, R. W., *et al.*: 2007, J. Atmos. Solar-Terr. Phys. 69, 1182, doi:10.1016/j.jastp.2006.08.020

Mason, G. M.: 2000, AIP Conf. Proc. 528, 234

Mason, G. M., Gloeckler, G., & Hovestadt, D.: 1984, ApJ 280, 902, doi:10.1086/162066

Mason, G. M., Gold, R. E., Krimigis, S. M., *et al.*: 1998, Space Sci. Rev. 86, 409, doi:10.1023/A:1005079930780

Mason, G. M., Leske, R. A., Desai, M. I., *et al.*: 2008, ApJ 678, 1458, doi:10.1086/533524

Masuda, S., Kosugi, T., Hara, H., Tsuneta, S., & Ogawara, Y.: 1994, Nature 371, 495, doi:10.1038/371495a0

Mathie, R. A. & Mann, I. R.: 2000, GRLe 27, 3261, doi:10.1029/2000GL003822

Matthaeus, W. H., Gray, P. C., Jr., D. H. Pontius, & Bieber, J. W.: 1995, Phys. Res. Lett. 75, 2136

Matthaeus, W. H., Qin, G., Bieber, J. W., & Zank, G. P.: 2003, ApJ 590, L53

Mattsson, C.-G. & Holmén, G.: 1971, Phys. Scr. 3, 101

Mauk, B. H., Krimigis, S. M., Keath, E. P., *et al.*: 1987, JGR 92, 15283

Mayaud, P. N.: 1980, *Derivation, Meaning, and Use of Geomagnetic Indices*, AGU Geophysical Monograph 22

Mazur, J. E., Mason, G. M., Dwyer, J. R., *et al.*: 2000, ApJ 35, L79

McComas, D. J. & Schwadron, N. A.: 2006, GRLe 33, 4102, doi:10.1029/2005GL025437

McCracken, K. G., Dreschhoff, G. A. M., Zeller, E. J., Smart, D. F., & Shea, M. A.: 2001, JGR 106, 21585, doi:10.1029/2000JA000237

McDonald, F. B., Schardt, A. W., & Trainor, J. H.: 1980, JGR 85, 5813, doi:10.1029/JA085iA11p05813

McKenzie, D. E.: 2000, SPh 195, 381

McKenzie, D. E. & Hudson, H. S.: 1999, ApJL 519, L93, doi:10.1086/312110

McKibben, R. B. & Simpson, J. A.: 1980, JGR 85, 5773, doi:10.1029/JA085iA11p05773

McLean, D. J. & Labrum, N. R.: 1985, *Solar Radiophysics: Studies of Emission from the Sun at Metre Wavelengths*, Cambridge University Press

McNutt, Jr., R. L., Selesnick, R. S., & Richardson, J. D.: 1987, JGR 92, 4399

Mead, G. D. & Hess, W. N.: 1973, JGR 78, 2793, doi:10.1029/JA078i016p02793

Meier, R. R., Crowley, G., Strickland, D. J., *et al.*: 2005, JGR (Space Phys.) 110(A9), 9, doi:10.1029/2004JA010990

Melnikov, V. F., Shibasaki, K., & Reznikova, V. E.: 2002, ApJL 580, L185, doi:10.1086/345587

Melrose, D. B.: 1980, *Plasma Astrophysics: Nonthermal Processes in Diffuse Magnetized Plasmas. Volume 2, Astrophysical Applications*, Gordon and Breach Science Publishers
Melrose, D. B.: 1983, SPh 89, 149
Melrose, D. B.: 1985, in D. J. McLean & N. R. Labrum (eds.), *Solar Radiophysics: Studies of Emission from the Sun at Metre Wavelengths*, Cambridge University Press, p. 177
Melrose, D. B.: 1992, ApJ 387, 403, doi:10.1086/171092
Melrose, D. B.: 1995, ApJ 451, 392
Melrose, D. B. & Brown, J. C.: 1976, MNRAS 176, 15
Melrose, D. B. & McClymont, A. N.: 1987, SPh 113, 241
Mendillo, M., Papagiannis, M. D., & Klobuchar, J. A.: 1970, Radio Sci. 5, 895, doi:10.1029/RS005i006p00895
Meredith, N. P., Horne, R. B., & Anderson, R. R.: 2001, JGR 106, 13165, doi:10.1029/2000JA900156
Meredith, N. P., Horne, R. B., & Iles, R. H. A.: 2002, JGR 107, 1144
Metcalf, T. R., Alexander, D., Hudson, H. S., & Longcope, D. W.: 2003, ApJ 595, 483, doi:10.1086/377217
Mewaldt, R. A., Cohen, C. M. S., Mason, G. M., *et al.*: 2003, in *International Cosmic Ray Conference*, Vol. 6, p. 3229
Mewaldt, R. A., Cohen, C. M. S., Haggerty, D. K., *et al.*: 2007, in *Turbulence and Nonlinear Processes in Astrophysical Plasmas*, Vol. 932 of American Institute of Physics Conference Series, p. 277
Meyer, A. & Murray, R. B.: 1962, Phys. Rev. 128, 98
Meziane, K., Hull, A. J., Hamza, A. M., & Lin, R. P.: 2002, JGR (Space Phys.) 107, 1243, doi:10.1029/2001JA005012
Millan, R.M. and Thorne, R. M.: 2007, J. Atmos. Solar-Terr. Phys. 69, 362
Millward, G. H., Moffett, R. J., Quegan, S., & Fuller-Rowell, T. J.: 1996, in R. W. Schunk (ed.), *Handbook of Ionospheric Models, STEP Report*, p. 239
Millward, G. H., Müller-Wodarg, I. C. F., Aylward, A. D., *et al.*: 2001, JGR 106, 24733, doi:10.1029/2000JA000342
Moses, D., Clette, F., Delaboudiniere, J.-P., *et al.*: 1997, SPh 175, 571
Murphy, R. J., Dermer, C. D., & Ramaty, R.: 1987, ApJSS 63, 721, doi:10.1086/191180
Murphy, R. J., Share, G. H., Letaw, J. R., & Forrest, D. J.: 1990, ApJ 358, 298, doi:10.1086/168987
Murphy, R. J., Share, G. H., Skibo, J. G., & Kozlovsky, B.: 2005, ApJSS 161, 495, doi:10.1086/452634
Murray, R. B. & Meyer, A.: 1961, Phys. Rev. 122, 815
Nagashima, K. & Yokoyama, T.: 2007, in K. Shibata, S. Nagata, and T. Sakurai (eds.), *New Solar Physics with Solar-B Mission*, Vol. 369, Astronomical Society of the Pacific, p. 429
Nakajima, H. & Yokoyama, T.: 2002, ApJL 570, L41, doi:10.1086/340832
Narukage, N. & Shibata, K.: 2006, ApJ 637, 1122
NAS: 1997, *A Space Weather Research Perspective*, National Academies of Science
NASA: 2005, *NASA's Exploration Systems Architecture Study*, NASA-TM-2005-224062
Nelson, G. J. & Melrose, D. B.: 1985, in D. J. McLean & N. R. Labrum (eds.), *Solar Radiophysics: Studies of Emission from the Sun at Metre Wavelengths*, Cambridge University Press, p. 333
Neugebauer, M. & Snyder, C. W.: 1962, Science 138, 1095, doi:10.1126/science.138.3545.1095-a
Neupert, W. M.: 1968, ApJ 153, L59

Ng, C. K. & Reames, D. V.: 1994, ApJ 424, 1032, doi:10.1086/173954
Ng, C. K., Reames, D. V., & Tylka, A. J.: 1999, GRL 26, 2145, doi:10.1029/1999GL900459
Ng, C. K., Reames, D. V., & Tylka, A. J.: 2003a, ApJ 591, 461, doi:10.1086/375293
Ng, C. K., Reames, D. V., & Tylka, A. J.: 2003b, in *International Cosmic Ray Conference*, Vol. 6, p. 3339
Ning, Z.: 2008, Astrophys. Space Sci. 314, 137
Nitta, N. V. & Hudson, H. S.: 2001, GRLe 28, 3801, doi:10.1029/2001GL013261
Northrop, T. G.: 1963, Rev. Geophys. 1, 283
Northrop, T. G. & Teller, E.: 1960, Phys. Rev. 117, 215, doi:10.1103/PhysRev.117.215
NRC: 2001, *Space Radiation Hazard and the Vision for Space Exploration*, National Research Council, The National Academies Press
NRC: 2006, *Health Risks to Exposure to Low-Levels of Ionizing Radiation, BEIR VII*, National Research Council, The National Academies Press
NRC: 2008, *Managing Space Radiation Risk in the New Era of Space Exploration*, National Research Council, The National Academies Press
Obayashi, T.: 1975, SPh 40, 217
O'Brien, T. P., McPherron, R. L., Sornette, D., *et al.*: 2001, JGR 106, 15533, doi:10.1029/2001JA000052
O'Brien, T. P., Lorentzen, K. R., Mann, I. R., *et al.*: 2003, JGR 108, 1329
Odenwald, S. F.: 2000, *The 23rd Cycle: Learning to Live with a Stormy Star*, Columbia University Press
Odenwald, S. F.: 2007, Space Weather 38, s11005, doi:10.1029/2007SW000344
Odenwald, S. F. & Green, J. L.: 2007, Space Weather 5, S06002, doi:10.1029/2006SW000262
Odenwald, S. F., Green, J. L., & Taylor, W.: 2005, Adv. Space Res. 38, 280, doi:10.1016/j.asr.2005.10.046
Odstrcil, D., Linker, J. A., Lionello, R., *et al.*: 2002, JGR (Space Phys.) 107, 1493, doi:10.1029/2002JA009334
Ogilvie, K. W., Scudder, J. D., Vasyliūnas, V. M., Hartle, R. E., & Siscoe, G. L.: 1977, JGR 82, 1807
Ogilvie, K. W., Chornay, D. J., Fritzenreiter, R. J., *et al.*: 1995, Space Sci. Rev. 71, 55, doi:10.1007/BF00751326
Olmstead, D.: 1837, Am. J. Sci. VII, 127
Onsager T. G., Rostoker, G., Kim, H.-J. *et al.*: 2002, JGR 107(A11), 1382, doi:10.1029/2001JA000187
Ontiveros, V. & Vourlidas, A.: 2009, ApJ 693, 267, doi:10.1088/0004-637X/693/1/267
Opher, M.: 2009, *Lectures 11, 12 13 of Space Physics*, http://physics.gmu.edu/~mopher/Lecture##.pdf
Opher, M., Liewer, P. C., Gombosi, T. I., *et al.*: 2003, ApJL 591, L61, doi:10.1086/376960
Opher, M., Liewer, P. C., Velli, M., *et al.*: 2004, ApJ 611, 575, doi:10.1086/422165
Opher, M., Stone, E. C., & Liewer, P. C.: 2006, ApJL 640, L71, doi:10.1086/503251
Opher, M., Stone, E. C., & Gombosi, T. I.: 2007, Science 316, 875, doi:10.1126/science.1139480
Opher, M., Richardson, J. D., Toth, G., & Gombosi, T. I.: 2009, Space Sci. Rev. 143, 43, doi:10.1007/s11214-008-9453-x
Osten, R. A., Drake, S., Tueller, J., *et al.*: 2007, ApJ 654, 1052, doi:10.1086/509252
Pallavicini, R., Serio, S., & Vaiana, G. S.: 1977, ApJ 216, 108
Palmer, I. D. & Smerd, S. F.: 1972, SPh 26, 460, doi:10.1007/BF00165287

Palmer, D. M., Barthelmy, S., Gehrels, N., *et al.*: 2005, Nature 434, 1107, doi:10.1038/nature03525
Parker, E. N.: 1960, JGR 65, 3117, doi:10.1029/JZ065i010p03117
Parker, E. N.: 1965, Planet Space Sci. 13, 9
Parker, E. N.: 1974, ApJ 191, 245
Parker, E. N.: 1979, *Cosmical Magnetic Fields: Their Origin and their Activity*, Oxford University Press
Parker, E. N.: 1988, ApJ 330, 474, doi:10.1086/166485
Parker, E. N.: 1994, *Spontaneous current sheets in magnetic fields: with applications to stellar x-rays*, Oxford University Press
Parker, E. N.: 1996, JGR 101, 10587
Parker, E. N.: 2007, *Conversations on Electric and Magnetic Fields in the Cosmos*, Princeton University Press
Paschmann, G., Sckopke, N., Papamastorakis, I., *et al.*: 1981, JGR 86, 4355, doi:10.1029/JA086iA06p04355
Paschmann, G., Sckopke, N., Bame, S. J., & Gosling, J. T.: 1982, Geophys. Res. Le. 9, 881, doi:10.1029/GL009i008p00881
Paschmann, G., Haaland, S., & Treumann, R. (eds.): 2003, *Auroral Plasma Physics*, Kluwer Academic Publishers
Paularena, K. I., Wang, C., von Steiger, R., & Heber, B.: 2001, GRLe 28, 2755, doi:10.1029/2001GL013122
Paxton, L. J., Christensen, A. B., Humm, D. C., *et al.*: 1999, in A. M. Larar (ed.), *Society of Photo-Optical Instrumentation Engineers (SPIE) Conference Series*, Vol. 3756, p. 265
Perkins, F.: 1973, JGR 78, 218, doi:10.1029/JA078i001p00218
Perrault, P. & Akasofu, S.-I.: 1978, Geophys. J. R. Astron. Soc. 54, 547
Perry, K. L., Hudson, M. K., & Elkington, S. R.: 2005, JGR (Space Phys.) 110(A9), 3215, doi:10.1029/2004JA010760
Pesses, M. E., Eichle, D., & Jokipii, J. R.: 1981, ApJ 246, L85
Peterson, L. E. & Winckler, J. R.: 1959, JGR 64, 697, doi:10.1029/JZ064i007p00697
Petrosian, V.: 1981, ApJ 251, 727, doi:10.1086/159517
Petrosian, V. & Liu, S.: 2004, ApJ 610, 550, doi:10.1086/421486
Petschek, H. E.: 1964, NASA Special Publication 50, 425
Piana, M., Massone, A. M., Kontar, E. P., *et al.*: 2003, ApJL 595, L127, doi:10.1086/378171
Pirjola, R., Viljanen, A., Pulkkinen, A., Kilpua, S., & Amm, O.: 2004, in I. A. Daglis (ed.), *Effects of Space Weather on Technology Infrastructure*, Kluwer Academic Publishers, p. 235
Pizzo, V.: 1985, in *Collisionless Shocks in the Heliosphere*, AGU Geophysical Mongraph 35, p. 51
Poletto, G. & Kopp, R. A.: 1986, in *The Lower Atmosphere of Solar Flares: Proceedings of the Solar Maximum Mission Symposium*, p. 453
Pontius, Jr., D. H.: 1995, JGR 100, 19531
Potgieter, M. S.: 1998, Space Sci. Rev 83, 147
Potgieter, M. S. & Moraal, H.: 1985, ApJ 294, 425
Prescott, G. B.: 1860, *History, Theory and Practice of the Electric Telegraph, Chapter XV*, Tiknor
Priest, E. R.: 1982, *Solar Magnetohydrodynamics*, Reidel, Hingham, M
Priest, E. R., van Ballegooijen, A. A., & Mackay, D. H.: 1996, ApJ 460, 530
Prölss, G. W.: 1997, in B. T. Tsurutani, W. D. Gonzalez, Y. Kamide, & J. K. Arballo (eds.), *Magnetic Storms*, AGU Geophysical Monograph 98, p. 227

Qiu, J.: 2007, J. Atmos. Solar-Terr. Phys. 69, 129
Qiu, J., Lee, J., & Gary, D. E.: 2004, ApJ 603, 335, doi:10.1086/381353
Rainey, L.: 2004, *Space Modeling and Simulation*, Kluwer Academic Publishers
Ramaty, R.: 1969, ApJ 158, 753, doi:10.1086/150235
Ramaty, R.: 1986, in P. A. Sturrock (ed.), *Physics of the Sun*, Reidel, p. 291
Ramaty, R. & Mandzhavidze, N.: 1994, in J. Ryan & W. T. Vestrand (eds.), *High-Energy Solar Phenomena: A New Era of Spacecraft Measurements*, Vol. 294 of American Institute of Physics Conference Series, p. 24
Ramaty, R. & Petrosian, V.: 1972, ApJ 178, 241, doi:10.1086/151783
Ramaty, R., Mandzhavidze, N., Kozlovsky, B., & Murphy, R. J.: 1995, ApJLe 455, L193, doi:10.1086/309841
Reames, D. V.: 1999, Space Sci. Rev. 90, 413, doi:10.1023/A:1005105831781
Reames, D. V.: 2002, ApJL 571, L63, doi:10.1086/341149
Reames, D. V. & Ng, C. K.: 1998, ApJ 504, 1002, doi:10.1086/306124
Reeves, K. K. & Forbes, T. G.: 2005, ApJ 610, 1133
Reeves, K. K. & Warren, H. P.: 2002, ApJ 578, 590
Reeves, G. D., McAdams, K. L., Friedel, R. H. W., & O'Brien, T. P.: 2003, GRLe 30(10), 100000, doi:10.1029/2002GL016513
Reeves, K. K., Warren, H. P., & Forbes, T. G.: 2007, ApJ 668, 1210
Reid, E. J.: 1963, in R. M. Foster (ed.), *Satellite Communications Physics*, Bell Telephone Laboratories, 78
Reiner, M. J. & Kaiser, M. L.: 1999, JGR 104, 16979, doi:10.1029/1999JA900143
Reiner, M. J., Kaiser, M. L., Fainberg, J., & Stone, R. G.: 1998, JGR 103, 29651, doi:10.1029/98JA02614
Reiner, M. J., Kaiser, M. L., Fainberg, J., & Stone, R. G.: 1999, in S. R. Habbal, R. Esser, J. V. Hollweg, & P. A. Isenberg (eds.), Vol. 471 of American Institute of Physics Conference Series, p. 653
Ricchiazzi, P. J. & Canfield, R. C.: 1983, ApJ 272, 739, doi:10.1086/161336
Rice, W. K. M., Zank, G. P., & Li, G.: 2003, JGR (Space Phys.) 108, 1369, doi:10.1029/2002JA009756
Richardson, J. D., Paularena, K. I., Wang, C., & Burlaga, L. F.: 2002, JGR (Space Phys.) 107, 1041, doi:10.1029/2001JA000175
Richardson, J. D., Liu, Y., Wang, C., & Burlaga, L. F.: 2006, Adv. Space Res. 38, 528, doi:10.1016/j.asr.2005.06.049
Richardson, J. D., Kasper, J. C., Wang, C., Belcher, J. W., & Lazarus, A. J.: 2008, Nature 454, 63, doi:10.1038/nature07024
Richmond, A. D.: 1995, in H. Volland (ed.), *Handbook of Atmospheric Electrodynamics, Vol. II*, p. 249
Richmond, A. D. & Matsushita, S.: 1975, JGR 80, 2839, doi:10.1029/JA080i019p02839
Richmond, A. D. & Roble, R. G.: 1987, JGR 92, 12365, doi:10.1029/JA092iA11p12365
Richmond, A. D., Peymirat, C., & Roble, R. G.: 2003, JGR (Space Phys.) 108, 1118, doi:10.1029/2002JA009758
Riley, P., Linker, J. A., & Mikić, Z.: 2001, JGR 106, 15889, doi:10.1029/2000JA000121
Riley, P., Linker, J. A., Mikić, Z., *et al.*: 2002, ApJ 578, 972, doi:10.1086/342608
Riley, P., Lionello, R., Mikić, Z., *et al.*: 2007, ApJ 655, 591
Rishbeth, H., Fuller-Rowell, T. J., & Rees, D.: 1987, Planet. Space Sci. 35, 1157, doi:10.1016/0032-0633(87)90022-5
Robinson, P. A.: 1985, ApJ 298, 161, doi:10.1086/163595
Robinson, P. A. & Cairns, I. H.: 1998, SPh 181, 395

Roble, R. G.: 1977, *The Upper Atmosphere and Magnetosphere*, National Academy of Science
Roble, R. G., Ridley, E. C., & Dickinson, R. E.: 1987, JGR 92, 8745, doi:10.1029/JA092iA08p08745
Rodger, A. S., Wrenn, G. L., & Rishbeth, H.: 1989, J. Atmos. Terr. Phys. 51, 851
Rodgers, D. J., Murphy, L. M., & Dyer, C. S.: 2000, Benefits of a European Space Weather Programme. DERA Report No. DERA/KIS/SPACE/ TR000349. ESWPS-DER-TN-0001. Issue 2.1 December 19, 2000. ESA Space Weather Program Study (ESWPS). http://www.wdc.rl.ac.uk/SWstudy/public/tr110v2_1b.pdf
Rodgers, E. M., Bailey, S. M., Warren, H. P., Woods, T. N., & Eparvier, F. G.: 2006, JGR (Space Phys.) 111(A10), 10, doi:10.1029/2005JA011505
Roederer, J. G.: 1970, *Dynamics of Geomagnetically Trapped Radiation*, Physics and Chemistry in Space, Springer
Roelof, E. C.: 1967, in H. Ogelman & J. R. Wayland (eds.), *Lectures in High-Energy Astrophysics*, NASA p. 111
Rossi, B.: 1961, *High Energy Particles*, Prentice Hall
Rossi, B. & Olbert, S.: 1970, *Introduction to the Physics of Space*, McGraw-Hill
Rostoker, G., Akasofu, S. I., Foster, J., *et al.*: 1980, JGR 85, 1663
Rostoker, G., Akasofu, S. I., Baumjohann, W., Kamide, Y., & McPherron, R. L.: 1987, Space Sci. Rev. 46, 93
Rostoker, G., Skone, S., & Baker, D. N.: 1998, GRLe 25, 3701, doi:10.1029/98GL02801
Roussev, I. I., Forbes, T. G., Gombosi, T., *et al.*: 2003, ApJL 588, L45
Roussev, I. I., Sokolov, I. V., Forbes, T. G., *et al.*: 2004, ApJ 605, L73
Roussev, I. I., Lugaz, N., & Sokolov, I. V.: 2007, ApJ 668, L87
Ruffalo, D.: 1995, ApJ 442, 861
Russell, C. T.: 1985, in *Collisionless Shocks in the Heliosphere: Reviews of Current Research*, American Geophysical Union, p. 109
Russell, C. T., Mewaldt, R. A., & von Rosenvinge, T. T.: 1998, Space Sci. Rev. 86
Rust, D. M.: 1979, in *Physics of Solar Prominences*, IAU Colloq. 44, p. 252
Rust, D. M., Sakurai, T., Gaizauskas, V., *et al.*: 1994, SPh 153, 1
Ryabov, B.: 2004, in *Solar and Space Weather Radiophysics: Current Status and Future Developments*, D. E. Gary & C. U. Keller (eds.), ASSL Series Vol. 314, Springer, Chapter 7
Rybicki, G. B. & Lightman, A. P.: 1979, *Radiative Processes in Astrophysics*, Wiley-Interscience
Saba, J. L. R., Gaeng, T., & Tarbell, T. D.: 2006, ApJ 641, 1197
Sabine, E.: 1852, Philos. Trans. R. Soc. Lond. 142, 103
Sagdeev, R. Z.: 1966, Rev. Plasma Phys. 4
Samson, J. C.: 1998, in S. Kokubun & Y. Kamide (eds.), *Substorms-4*, Terra Scientific Publishing Company and Kluwer Academic Publishers, p. 505
Sanderson, T. R., Reinhard, R., van Nes, P., & Wenzel, K.-P.: 1985, JGR 90, 19, doi:10.1029/JA090iA01p00019
Santos-Costa, D., Blanc, M., Maurice, S., & Bolton, S. J.: 2003, GRLe 30(20), 200000, doi:10.1029/2003GL017972
Sarris, E. T. & van Allen, J. A.: 1974, JGR 79, 4157, doi:10.1029/JA079i028p04157
Sawyer, D. M. & Vette, J. I.: 1976, NASA STI/Recon Technical Report N 77, 18983
Sazykin, S., Spiro, R. W., Wolf, R. A., *et al.*: 2005, in T. I. Pulkkinen, N. A. Tsyganenko, & R. H. W. Friedel (eds.), *The Inner Magnetosphere: Physics and Modeling*, AGU *Geophysical Monograph* 155, p. 263
Schaefer, B. E., King, J. R., & Deliyannis, C. P.: 2000, ApJ 529, 1026, doi:10.1086/308325

Scherliess, L. & Fejer, B. G.: 1997, JGR 102, 24037, doi:10.1029/97JA02165
Schindler, K.: 1974, JGR 79, 2803
Schindler, K.: 1976, SPh 47, 91
Schindler, K.: 2007, *Physics of Space Plasma Activity*, Cambridge University Press
Schnoor, P. W., Welch, D. L., Fishman, G. J., & Price, A.: 2003, *Gamma-Ray Collaboration Network Bulletin, GCN-2176*, http://gcn.gsfc.nasa.gov/gcn3/2176.gcn3
Scholer, M., Hovestadt, D., Ipavich, F. M., & Gloeckler, G.: 1983, JGR 88, 1977, doi:10.1029/JA088iA03p01977
Scholer, M., Kucharek, H., & Giacalone, J.: 2000, JGR 105, 18,285
Schrijver, C. J.: 2007, ApJL 655, L117, doi:10.1086/511857
Schrijver, C. J.: 2009, Adv. Space Res. 43, 739, doi:10.1016/j.asr.2008.11.004
Schrijver, C. J., DeRosa, M. L., Metcalf, T., *et al.*: 2008, ApJ 675, 1637, doi:10.1086/527413
Schrijver, C. J., Hudson, H. S., Murphy, R. J., Share, G. H., & Tarbell, T. D.: 2006, ApJ 650, 1184, doi:10.1086/506583
Schulz, M. & Lanzerotti, L. J.: 1974, *Particle Diffusion in the Radiation Belts*, Physics and Chemistry in Space, Springer
Schunk, R. W., Banks, P. M., & Raitt, W. J.: 1975, JGR 80, 3121, doi:10.1029/JA080i022p03121
Schwartz, S. J., Burgess, D., Wilkinson, W. P., *et al.*: 1992, JGR 97, 4209, doi:10.1029/91JA02581
Schwenn, R. W., Raymond, J. C., Alexander, D., *et al.*: 2006, Space Sci. Rev. 123, 127
Sckopke, N.: 1966, JGR 71, 3125
Scudder, J. D., Aggson, T. L., Mangeney, A., Lacombe, C., & Harvey, C. C.: 1986a, JGR 91, 11019, doi:10.1029/JA091iA10p11019
Scudder, J. D., Aggson, T. L., Mangeney, A., Lacombe, C., & Harvey, C. C.: 1986b, JGR 91, 11053, doi:10.1029/JA091iA10p11053
Selesnick, R. S. & Blake, J. B.: 2000, JGR 105, 2607, doi:10.1029/1999JA900445
Sen, H. K. & White, M. L.: 1972, SPh 23, 146
Shafranov, V. D.: 1966, Rev. Plasma Phys. 2, 103
Share, G. H., Murphy, R. J., Smith, D. M., Schwartz, R. A., & Lin, R. P.: 2004, ApJL 615, L169
Shea, M. A. & Smart, D. F.: 2004, SPh 224, 483, doi:10.1007/s11207-005-4138-z
Sheeley, N. R., Walters, J. H., Wang, Y.-M., & Howard, R. A.: 1999, JGR 104, 24739, doi:10.1029/1999JA900308
Sheeley, Jr., N. R., Warren, H. P., & Wang, Y.-M.: 2004, ApJ 616, 1224, doi:10.1086/425126
Shibata, K. & Yokoyama, T.: 1999, ApJL 526, L49, doi:10.1086/312354
Shibata, K., Ishido, Y., Acton, L. W., *et al.*: 1992, Publ. Astron. Soc. Japan 44, L173
Shibata, K., Eto, S., Narukage, N., *et al.*: 2002, in P. C. H. Martens & D. Cauffman (eds.), *Multi-Wavelength Observations of Coronal Structure and Dynamics*, Elsevier Science, p. 279
Shibata, K., Nakamura, T., Matsumoto, T., *et al.*: 2007, Science 318, 1591, doi:10.1126/science.1146708
Shimojo, M., Hashimoto, S., Shibata, K., *et al.*: 1996, Publ. Ast. Soc. Japan 48, 123
Shimojo, M., Shibata, K., & Harvey, K. L.: 1998, SPh 178, 379
Shiokawa, K., Otsuka, Y., Ogawa, T., *et al.*: 2002, JGR (Space Phys.) 107, 1088, doi:10.1029/2001JA000245
Shprits, Y. Y., Thorne, R. M., Friedel, R., *et al.*: 2006, JGR (Space Phys.) 111(A10), 11214, doi:10.1029/2006JA011657

Shu, F. H., Shang, H., Glassgold, A. E., & Lee, T.: 1997, Science 277, 1475
Sicard, A. & Bourdarie, S.: 2004, JGR (Space Phys.) 109(A18), 2216, doi:10.1029/2003JA010203
Siegmund, O. H. W. *et al.*: 1994, in *Proc. SPIE 2280*, p. 89
Silva, A. V. R., Share, G. H., Murphy, R. J., *et al.*: 2007, SPh 245, 311, doi:10.1007/s11207-007-9044-0
Silva, A. V. R., White, S. M., Lin, R. P. *et al.*: 1996, ApJ 458, L49
Simnett, G. M., Roelof, E. C., & Haggerty, D. K.: 2002, ApJ 579, 854, doi:10.1086/342871
Simpson, J. A.: 1983, Ann. Rev. Nucl. Part. Sci. 33(1), 323
Simpson, J. A., Bastian, T. S., Chenette, D. L., McKibben, R. B., & Pyle, K. R.: 1980, JGR 85, 5731, doi:10.1029/JA085iA11p05731
Siscoe, G. L.: 1966, Planet. Space Sci. 14, 947
Siscoe, G. L.: 1980, JGR 85, 1643
Siscoe, G. L.: 1983, in R. L. Carovillano and J. M. Forbes (eds.), *Solar-Terrestrial Physics*, D. Reidel Publishing Co., p. 11
Siscoe, G. L. & Cummings, W. D.: 1969, Planet. Space Sci. 17, 1795
Siscoe, G. & Odstrcil, D.: 2008, JGR 113, 2008JA013142
Siscoe, G. L. & Siebert, K. D.: 2006, J. Atmos. Solar-Terr. Phys. 68, 911, doi:10.1016/j.jastp.2005.11.012
Siscoe, G. L., Ness, N. F., & Yeates, C. M.: 1975, JGR 80, 4359
Sittler, E. & Guhathakurta, M.: 1999, in S. T. Suess, G. A. Gary, & S. F. Nerney (eds.), *Empirical model of the corona-solar wind with multiple current sheets*, Vol. 471 of American Institute of Physics Conference Series, p. 401
Skoblin, M. G. & Förster, M.: 1993, Ann. Geophys. 11, 1026
Skoug, R. M., Winglee, R. M., McCarthy, M. P., *et al.*: 1996, GRL 23, 1223, doi:10.1029/96GL00301
Smart, D. F., Shea, M. A., & McCracken, K. G.: 2006, Adv. Space Res. 38, 215, doi:10.1016/j.asr.2005.04.116
Smith, A. J., Horne, R. B., & Meredith, N. P.: 2004, JGR (Space Phys.) 109(A18), 2205, doi:10.1029/2003JA010204
Sokolov, I. V., Roussev, I. I., Fisk, L. A., *et al.*: 2006, ApJL 642, L81, doi:10.1086/504406
Song, P., Singer, H. J., & Siscoe, G. L. (eds.): 2001, *Space Weather*, AGU Geophysical Monograph 125
Sonnerup, B. U. Ö.: 1969, JGR 74, 1301, doi:10.1029/JA074i005p01301
Sparks, L., Komjathy, A., Mannucci, A.J., Altshuler, E., Walter, T., Blanch, W., Bakry El-Arini, M. and Lejeune, R.: 2005, "Extreme ionospheric storms and their impact on WAAS," Ionospheric Effects Symposium 2005, Alexandria, VA, May 3–5, 2005, http://trs-new.jpl.nasa.gov/dspace/bitstream/2014/37807/1/05-0934.pdf.
Speiser, T. W.: 1965, JGR 70, 4219
Spieler, H.: 2005, *Semiconductor Detector Systems*, Oxford University Press
Spiro, R. W., Wolf, R. A., & Fejer, B. G.: 1988, Ann. Geophys. 6, 39
Stassinopoulos, E. G., Brucker, G. J., Adolphsen, J. N., & Barth, J.: 1996, J. Spacecr. Rockets 33, 877
Steiner, O.: 2007, ArXiv e-prints 709
Stenquist, D.: 1914, Ph.D. thesis, University of Stockholm
Stern, D. P.: 1991, in *Magnetospheric Substorms*, AGU Geophysical Monograph 64, p. 11
Stone, E. C., et al.: 1990, AIP Conf. Proc. 203, 48
Stone, E. C., Cohen, C. M. S., Cook, W. R., *et al.*: 1998, Space Sci. Rev. 86, 357, doi:10.1023/A:1005027929871

Stone, E. C., Cummings, A. C., McDonald, F. B., *et al.*: 2005, Science 309, 2017, doi:10.1126/science.1117684

Stone, E. C., Cummings, A. C., McDonald, F. B., *et al.*: 2008, Nature 454, 71

Strickland, D. J., Daniell, R. E., & Craven, J. D.: 2001, JGR 106, 21049, doi:10.1029/2000JA000209

Strong, K. T., Harvey, K., Hirayama, T., *et al.*: 1992, Publ. Ast. Soc. Japan 44, L161

Sturrock, P. A.: 1991, ApJ 380, 655

Sturrock, P. A., Kaufmann, P., Moore, R. L., & Smith, D. F.: 1984, SPh 94, 341

Sturrock, P. A., Dixon, W. W., Klimchuk, J. A., & Antiochos, S. K.: 1990, ApJL 356, L31, doi:10.1086/185743

Suarez-Garcia, E., Hajdas, W., Wigger, C., *et al.*: 2006, SPh 239, 149, doi:10.1007/s11207-006-0268-1

Sudol, J. J. & Harvey, J. W.: 2005, ApJ 635, 647

Sui, L. & Holman, G. D.: 2003, ApJL 596, L251, doi:10.1086/379343

Sui, L., Holman, G. D., & Dennis, B. R.: 2004, ApJ 612, 546, doi:10.1086/422515

Summers, D. & Thorne, R. M.: 2003, JGR 108(A4), 1143, doi:10.1029/2002JA009489

Summers, D. & Omura, Y.: 2007, GRLe 34, 24205, doi:10.1029/2007GL032226

Sutton, E. K., Forbes, J. M., & Nerem, R. S.: 2005, JGR (Space Phys.) 110(A9), 9, doi:10.1029/2004JA010985

Sutton, E. K., Forbes, J. M., Nerem, R. S., & Woods, T. N.: 2006, GRL 33, 22101, doi:10.1029/2006GL027737

Švestka, Z.: 1976, *Solar Flares*, D. Reidel Dordrecht

Švestka, Z. & Cliver, E. W.: 1992, in Z. Švestka, B. V. Jackson, & M. E. Machado (eds.), *Eruptive Solar Flares*, Springer-Verlag, p. 1

Švestka, Z. & Simon, P.: 1969, SPh 10, 3

Švestka, Z. F., Fontenla, J. M., Machado, M. E., Martin, S. F., & Neidig, D. F.: 1987, SPh 108, 237

Syrjäsuo, M. & Donovan, E. (eds.): 2007, in *Proceedings of the Eighth International Conference on Substorms (ICS-8)*, University of Calgary, Alberta, Canada

Takakura, T., Ohki, K., Shibuya, N., *et al.*: 1971, SPh 16, 454, doi:10.1007/BF00162487

Tandberg-Hanssen, E.: 1995, *The Nature of Solar Prominences*, Kluwer Academic Publishers

Tandberg-Hanssen, E. & Emslie, A. G.: 1988, *The Physics of Solar Flares*, Cambridge University Press

Terasawa, T.: 1999, in *International Cosmic Ray Conference*, Vol. 6, p. 528

Thayer, J. P., Vickrey, J. F., Heelis, R. A., & Gary, J. B.: 1995, JGR 100, 19715, doi:10.1029/95JA01159

Thomas, V. A. & Brecht, S. H.: 1988, JGR 93, 11341

Thomas, N., Bagenal, F., Hill, T. W., & Wilson, J. K.: 2004, in F. Bagenal, T. Dowling, & W. McKinnon (eds.), *Jupiter: The Planet, Satellites and Magnetosphere*, Cambridge University Press, p. 561

Thompson, C. & Duncan, R. C.: 1995, MNRAS 275, 255

Thompson, B. J., Gurman, J. B., Neupert, W. M., *et al.*: 1999, ApJ 517, L151, doi:10.1086/312030

Thomsen, M. F.: 1985, in R. G. Stone & B. Tsurutani (eds.), *Collisionless Shocks in the Heliosphere*, AGU Geophysical Monograph 34, p. 253

Thomson, N. R., Rodger, C. J., & Clilverd, M. A.: 2005, JGR (Space Phys.) 110(A9), 6306, doi:10.1029/2005JA011008

Titov, V. S. & Démoulin, P.: 1999, A&A 351, 707

Tomblin, F. F.: 1972, ApJ 171, 377, doi:10.1086/151288

Tomczyk, S., Card, G. L., Darnell, T., *et al.*: 2008, SPh 247, 411, doi:10.1007/s11207-007-9103-6
Török, T., Kliem, B., & Titov, V. S.: 2004, A & A 413, L27 doi: 10.1051/0004-6361:20031691
Townsend, W., Cucinotta, F. A., & Wilson, J. W.: 1992, Rad. Res. 129, 48
Townsend, L. W., Stephens, D. L., Hoff, J. L., *et al.*: 2006, Adv. Space Res. 38, 226, doi:10.1016/j.asr.2005.01.111
Tribble, A. C.: 2000, *Fundamentals of Contamination Control*, SPIE Press
Tribble, A. C.: 2003, *The Space Environment: Implications for Spacecraft Design*, 2nd Edn. Princeton University Press
Tribble, A. C., D'Angelo, N., Murphy, G., Pickett, J., & Steinberg, J. T.: 1988, J. Spacecr. 25, 64
Trulsen, J. & Fejer, J. A.: 1970, J. Plasma Phys. 4, 825
Tsuneta, S. & Naito, T.: 1998, ApJL 495, L67+, doi:10.1086/311207
Tsurutani, B. T. & Lin, R. P.: 1985, JGR 90, 1, doi:10.1029/JA090iA01p00001
Tsurutani, B. T. & Rodriguez, P.: 1981, JGR 86, 4317, doi:10.1029/JA086iA06p04317
Tsurutani, B. T., Gonzalez, W. D., Kamide, Y., & Arballo, J. K. (eds.): 1997, *Magnetic Storms*, AGU Geophysical Monograph 98
Tsurutani, B. T., Gonzalez, W. D., Gonzalez, A. L. C., *et al.*: 2006, JGR 111(A07S01), doi:10.1029/2005JA011273
Turner, J. E.: 1995, *Atoms, Radiation and Radiation Protection*, 2nd Edn., John Wiley and Sons
Tylka, A. J.: 2001, JGR 106, 25333, doi:10.1029/2000JA004028
Tylka, A. J., Adams Jr., J. H., Boberg, P. R., *et al.*: 1997, IEEE Trans. Nucl. Sci. 44, 2150
Tylka, A. J., Reames, D. V., & Ng, C. K.: 1999, GRLe 26, 2141, doi:10.1029/1999GL900458
Tylka, A. J., Cohen, C. M. S., Dietrich, W. F., *et al.*: 2003, in *28th International Cosmic Ray Conference*
Tylka, A. J., Cohen, C. M. S., Dietrich, W. F., *et al.*: 2005, ApJ 625, 474, doi:10.1086/429384
Uchida, Y.: 1970, Pub. Astron. Soc. Japan 22, 341
Uchida, Y.: 1974, SPh 39, 431
Ukhorskiy, A. Y., Anderson, B. J., Brandt, P. C., & Tsyganenko, N. A.: 2006, Journal of Geophysical Research (Space Physics) 111, 11, doi:10.1029/2006JA011690
Usmanov, A. V. & Dryer, M.: 1995, SPh 159, 347, doi:10.1007/BF00686537
Vampola, A. L.: 1987, J. Electrostatics 20, 21
Vampola, A. L., Osborne, J. V., & Johnson, B. M.: 1992, J. Spacecr. Rockets 29, 592
Van Allen, J. A., Ludwig, G.H., Ray, E. C., & McIlwain, C. E.: 1958, Jet Propul. 28, 588
Van Allen, J. A., Thomsen, M. F., Randall, B. A., Rairden, R. L., & Grosskreutz, C. L.: 1980, Science 207, 415, doi:10.1126/science.207.4429.415
van Ballegooijen, A. A. & Mackay, D. H.: 2007, ApJ 659, 1713
van Ballegooijen, A. A. & Martens, P. C. H.: 1989, ApJ 343, 971
van der Linden, R. A. M., Hood, A. W., & Goedbloed, J. P.: 1994, SPh 154, 69
van Nes, P., Reinhard, R., Sanderson, T. R., Wenzel, K.-P., & Zwickl, R. D.: 1984, JGR 89, 2122, doi:10.1029/JA089iA04p02122
van Tend, W. & Kuperus, M.: 1978, SPh 59, 115
Vasyliūnas, V. M.: 1976, in B. M. McCormack (ed.), *Magnetospheric Particles and Fields*, D. Reidel Publishing Co., p. 99
Vasyliūnas, V. M.: 1983, in A. J. Dessler (ed.), *Physics of the Jovian Magnetosphere*, Cambridge University Press, p. 395

Vasyliūnas, V. M.: 1987, in A. T. Y. Lui (ed.), *Magnetotail Physics*, The Johns Hopkins University Press, p. 411
Vasyliūnas, V. M.: 1998, in S. Kokubun & Y. Kamide (eds.), *Substorms-4*, Terra Scientific Publishing Company and Kluwer Academic Publishers, p. 9
Vasyliūnas, V. M.: 2005, Ann. Geophys. 23, 2589
Vasyliūnas, V. M.: 2006, Ann. Geophys. 24, 1085
Vasyliūnas, V. M.: 2007, Ann. Geophys. 25, 255
Vasyliūnas, V. M. & Song, P.: 2005, JGR 110(A02301), doi:10.1029/2004JA010615
Veltri, P., Mangeney, A., & Scudder, J. D.: 1992, Nuovo Cimento C 15(5), 607
Vernazza, J. E., Avrett, E. H., & Loeser, R.: 1981, ApJ Suppl. 45, 635, doi:10.1086/190731
Veronig, A. M. & Brown, J. C.: 2004, ApJL 603, L117, doi:10.1086/383199
Veronig, A. M., Karlický, M., Vršnak, B., *et al.*: 2006a, A&A 446, 675, doi:10.1051/0004-6361:20053112
Veronig, A. M., Temmer, M., Vršnak, B., & Thalmann, J. K.: 2006b, ApJ 647, 1466, doi:10.1086/505456
Vette, J. I.: 1991a, *The AE-8 Trapped Electron Model Environment*, NSSDC/WDC-A-R&S 91-24
Vette, J. I.: 1991b, *The NASA/National Space Science Data Center Trapped Radiation Environment Model Program (1964-1991)*, NSSDC/WDC-A-R&S 91-29
Vinas, A. F., Goldstein, M. L., & Acuna, M. H.: 1984, JGR 89, 3762, doi:10.1029/JA089iA06p03762
von Steiger, R. & Richardson, J. D.: 2006, Space Sci. Rev. 123, 111, doi:10.1007/s11214-006-9015-z
Vorpahl, J. A.: 1973, SPh 28, 115, doi:10.1007/BF00152916
Vourlidas, A.: 2006, in H. Lacoste & L. Ouwehand (eds.), *SOHO-17: 10 Years of SOHO and Beyond*, ESA SP-617, 23.1, European Space Agency, Nordwiijk
Vourlidas, A., Subramanian, P., Dere, K. P., & Howard, R. A.: 2000, ApJ 534, 456
Vršnak, B. & Cliver, E. W.: 2008, SPh 253, 215, doi:10.1007/s11207-008-9241-5
Vršnak, B. & Lulić, S.: 2000, SPh 196, 157
Wahlund, J.-E., Wedlin, L. J., Carrozi, T., *et al.*: 1999, in *WP 130 Technical Note (SPEE-WP130-TN)*
Wallmark, J.T. & Marcus, S. M.: 1962, *Minimum Size and Maximum Packing Density of Non-Redundant Semi-Conductor Devices*, Proc. IRE, 286.
Walt, M.: 1994, 2005, *Introduction to Geomagnetically Trapped Radiation*, Cambridge University Press
Wang, H.: 2006, ApJ 649, 490
Wang, X. & Bhattacharjee, A.: 1993, JGR 98, 19419
Wang, Y. & Zhang, J.: 2007, ApJ 665, 1428, doi:10.1086/519765
Wang, H., Spirock, T. J., Qiu, J., *et al.*: 2002, ApJ 576, 497, doi:10.1086/341735
Wang, L., Lin, R. P., Larson, D. E., & Luhmann, J. G.: 2008, Nature 454, 81, doi:10.1038/nature07068
Wang, Y., Zheng, H., Wang, S., & Ye, P.: 2005, A&A 434, 309
Warren, H. P.: 2006, ApJ 637, 522, doi:10.1086/497904
Warren, H. P. & Doschek, G. A.: 2005, ApJ 618, L157
Webb, D. F.: 1985, SPh 97, 321
Webb, G. M., Zank, G. P., Ko, C. M., & Donohue, D. J.: 1995, ApJ 453, 178, doi:10.1086/176379
Webb, D. F., Cliver, E. W., Gopalswamy, N., Hudson, H. S., & St. Cyr, O. C.: 1998, GRLe 25, 2469, doi:10.1029/98GL00493

Webb, D. F., Burkepile, J., Forbes, T. G., & Riley, P.: 2003, JGR 108(A12), 1440 doi:10.1029/2003JA009923
Weimer, D. R.: 2005, JGR (Space Phys.) 110(A9), 5306, doi:10.1029/2004JA010884
Weiss, L. A., Reiff, P. H., Moss, J. J., Heelis, R. A., & Moore, B. D.: 1992, in *Substorms 1*, ESA SP-335, Noordwijk, p. 309
Wenzel, K.-P., Reinhard, R., Sanderson, T. R., & Sarris, E. T.: 1985, JGR 90, 12, doi:10.1029/JA090iA01p00012
Wesson, J. A.: 1987, *Tokamaks*, Oxford University Press
Wheatland, M. S., Sturrock, P. A., & Roumeliotis, G.: 2000, ApJ 540, 1150, doi:10.1086/309355
White, S. M.: 2005, in D. E. Innes, A. Lagg, and S. A. Solanki (eds.), *Chromospheric and Coronal Magnetic Fields*, Vol. 596 of ESA Special Publication
White, S. M., Kundu, M. R., Bastian, T. S., *et al.*: 1992, ApJ 384, 656, doi:10.1086/170907
Wild, J. P. & Smerd, S. F.: 1972, ARA&A 10, 159, doi:10.1146/annurev.aa.10.090172.001111
Wild, J. P., Smerd, S. F., & Weiss, A. A.: 1963, ARA&A 1, 291, doi:10.1146/annurev.aa.01.090163.001451
Wilkinson, D.: 1994, J. Spacecr. Rockets 31, 160
Williams, D. R., Török, T., Démoulin, P., van Driel-Gesztelyi, L., & Kliem, B.: 2005, ApJ 628, L163
Wilson, J. W., Cucinotta, F. A., Simonsen, L. C., *et al.*: 1997, *Galactic and Cosmic Ray Shielding in Deep Space*, NASA TP 3682
Wisa, J. L.: 1979, Nucl. Instrum. Methods 162, 587
Witte, M., Rosenbauer, H., Keppler, E., *et al.*: 1992, A&ASS 92, 333
Wolfson, R. & Low, B. C.: 1992, ApJ 391, 353
Woods, T. N., Eparvier, F. G., Fontenla, J., *et al.*: 2004, GRL 31, 10802, doi:10.1029/2004GL019571
Woods, T. N., Eparvier, F. G., Bailey, S. M., *et al.*: 2005, JGR (Space Phys.) 110(A9), 1312, doi:10.1029/2004JA010765
Woods, T. N., Kopp, G., & Chamberlin, P. C.: 2006, JGR (Space Phys.) 111(A10), 10, doi:10.1029/2005JA011507
Woods, T. N., Chamberlin, P. C., Peterson, W. K., *et al.*: 2008, SPh 250, 235, doi:10.1007/s11207-008-9196-6
Wrenn, G. L., Rodgers, D. J., & Ryden, K. A.: 2002, Ann. Geophys. 20, 953
Wu, C. S.: 1984, JGR 89, 8857, doi:10.1029/JA089iA10p08857
Wu, C. C.: 1990, JGR 95, 8149
Wu, S. T., Guo, W. P., Michels, D. J., & Burlaga, L. F.: 1999, JGR 104, 14789, doi:10.1029/1999JA900099
Wu, C. C., Fry, C. D., Wu, S. T., Dryer, M., & Liou, K.: 2007, JGR 112(A9 CiteID A09104), doi 10.1029/2006JA012211
Wurz, P.: 2000, in *The Outer Heliosphere: Beyond the Planets*, Copernicus Gesellschaft
Xapsos, M. A., Summers, G. P., Barth, J. L., Stassinopoulos, E. G., & Burke, E. A.: 1999, IEEE Trans. Nucl. Sci. 46, 1481
Xapsos, M. A., Summers, G. P., Barth, J. L., Stassinopoulos, E. G., & Burke, E. A.: 2000, IEEE Trans. Nucl. Sci. 47, 486
Xu, Y., Cao, W., Liu, C., *et al.*: 2004, ApJL 607, L131, doi:10.1086/422099
Yashiro, S., Gopalswamy, N., Akiyama, S., Michalek, G., & Howard, R. A.: 2005, JGR (Space Phys.) 110(A9), 12, doi:10.1029/2005JA011151
Yashiro, S., Akiyama, S., Gopalswamy, N., & Howard, R. A.: 2006, ApJ 650, L143
Yashiro, S., Michalek, G., & Gopalswamy, N.: 2008, Ann. Geophys. 26, 3103

Ye, Z.-Y., Wei, F.-S., Wang, C., Feng, X.-S., & Zhong, D.-K.: 2007, Chinese Astron. Astrophys. 31, 128
Yokoyama, T., Akita, K., Morimoto, T., Inoue, K., & Newmark, J.: 2001, ApJL 546, L69, doi:10.1086/318053
Young, S. L., Denton, R. E., Anderson, B. J., & Hudson, M. K.: 2008, JGR 113, A03210, doi:10.1029/2006JA012133
Zank, G. P. & Müller, H.-R.: 2003, JGR (Space Phys.) 108, 1240, doi:10.1029/2002JA009689
Zank, G. P., Rice, W. K. M., & Wu, C. C.: 2000, JGR 105, 25079, doi:10.1029/1999JA000455
Zank, G. P., Li, G., & Verkhoglyadova, O.: 2007, in R. von Steiger, G. Gloeckler, & G. M. Mason (eds.), *The Composition of Matter*, Springer Science and Business Media, p. 255
Zarro, D. M., Sterling, A. C., Thompson, B. J., Hudson, H. S., & Nitta, N.: 1999, ApJL 520, L139, doi:10.1086/312150
Zhang, Y.-Z. & Wang, J.-X.: 2007, ApJ 663, 592
Zhang, J., Dere, K. P., Howard, R. A., & Vourlidas, A.: 2004, ApJ 604, 420, doi:10.1086/381725
Zhou, Q. & Mathews, J. D.: 2006, JGR (Space Phys.) 111(A10), 12309, doi:10.1029/2006JA011696
Zirin, H.: 1988, *Astrophysics of the Sun*, Cambridge University Press
Zirin, H. & Lackner, D. R.: 1969, SPh 6, 86
Zirin, H. & Liggett, M. A.: 1987, SPh 113, 267
Zirker, J. B., Engvold, O., & Martin, S. F.: 1998, Nature 396, 440
Zurbuchen, T. H., Raines, J. M., Gloeckler, G., *et al.*: 2008, Science 321, 90

Index